Temperate Tree Fruits and Nuts

Temperate Tree Fruits and Nuts

Edited by

George Manganaris

ACADEMIC PRESS
An imprint of Elsevier

Academic Press is an imprint of Elsevier
125 London Wall, London EC2Y 5AS, United Kingdom
50 Hampshire Street, 5th Floor, Cambridge, MA 02139, United States

ISBN: 978-0-443-29137-1

For information on all Academic Press publications visit our website at https://www.elsevier.com/books-and-journals

Publisher: Peter Linsley
Acquisitions Editor: Rakhshan Rizwan
Editorial Project Manager: Aleksandra Packowska
Production Project Manager: Prasanna Kalyanaraman
Cover Designer: Mark Rogers

Typeset by TNQ Tech

Contents

Preface

The cultivation and management of temperate fruit and nut crops represent a vital facet of global horticulture across numerous regions worldwide. This comprehensive textbook provides an in-depth exploration of key temperate tree fruits and nuts, meticulously structured to serve as an authoritative resource, primarily for University students, as well as for researchers, orchard managers and practitioners interested in the diverse and dynamic world of fruits and nuts of temperate origin.

The book covers 14 major crops: Apple, pear, peach, sweet cherry, sour cherry, apricot, plum/prune, kiwifruit, fig, persimmon, pomegranate, almond, walnut, and pistachio. Each chapter has been authored by a number of experts in the species under study who shared their expertise and extensive knowledge from a range of perspectives.

Structured to provide in-depth insights, each chapter adheres to a uniform format, ensuring clarity and consistency, which facilitates comparative analysis across different species. The chapters comprehensively address critical aspects such as:

- **Origin, domestication and spread**: Evolutionary journey, historical and geographical distribution insights.
- **Economic importance and relevance**: Global production trends, main producing countries, contribution to local and global markets.
- **Botany and relevant species**: Botanical aspects, taxonomic details and morphological features.
- **Breeding**: Breeding advancements on rootstocks and cultivars, selection criteria, opportunities to investigate complex traits and improve breeding efficiency through the employment of molecular biology and biotechnology.
- **Propagation**: Techniques on improving sexual and asexual propagation efficiency, rootstock characteristics and scion compatibility.
- **Orchard management**: Procedures for orchard establishment, training systems and planting densities.
- **Cultivation techniques towards sustainable production**: Fertilization, irrigation, pruning, thinning and crop load management.
- **Fruit maturation and ripening:** Harvest criteria and guidelines for optimal harvest timing.
- **Postharvest physiology and technology**: Precooling methods, supply chain management protocols, handling of physiological disorders and means to alleviate chilling injury symptoms
- **Main diseases and pests and their management**: Identification, management strategies to combat main diseases and pests with special reference to integrated control methods.
- **Current challenges, future outlook and directions**: Insights into advancements on production protocols and emerging threats.

This structured approach allows for a holistic understanding of the cultivation and utilization of temperate fruits and nuts, rendering it an essential resource for advancing sustainable fruit production.

The authors' meticulous attention to detail ensures the integration of current research and technological advancements, positioning this textbook as a vital reference dedicated to the development and dissemination of knowledge in temperate horticulture.

Prof. George Manganaris

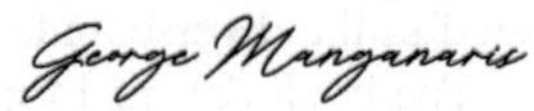

Editor

The book is being dedicated to the esteemed colleagues Dr. Takuya Testumura and Dr. Moshe Flaishman, contributors of this endeavor, who unexpectedly passed away during the period of its preparation.

CHAPTER

1

Apple

Fabrizio Costa[1], Susan Brown[2], Massimo Tagliavini[3], Alessandro Botton[4], Lee Kalcsits[5], Andreas Peil[6], Christopher B. Watkins[7]

[1]*Center Agriculture Food Environment (C3A), University of Trento, Italy*
[2]*Horticulture Section, School of Integrative Plant Science, College of Agriculture and Life Sciences, Cornell University, Geneva, NY, United States*
[3]*Faculty of Agricultural, Environmental and Food Sciences, Free University of Bozen-Bolzano, Italy*
[4]*DAFNAE - Department of Agronomy, Food, Natural resources, Animals and Environment University of Padova, Agripolis, V.le dell'Università, 16 | 35020, Legnaro (PD) – Italy*
[5]*Tree Fruit Research and Extension Center, Department of Horticulture, Washington State University, Wenatchee, WA, United States*
[6]*Julius Kühn Institute (JKI) – Federal Research Centre for Cultivated Plants, Institute for Breeding Research on Fruit Crops, Pillnitzer Platz 3a, 01326 Dresden, Germany*
[7]*Horticulture Section, School of Integrative Plant Science, College of Agriculture and Life Sciences, Cornell University, Ithaca, NY, United States*

Chapter outline

Temperate Tree Fruits and Nuts. https://doi.org/10.1016/B978-0-443-29137-1.00001-3

1. Origin and spread

Apple is one of the most popular and familiar fruits worldwide. Apple trees are grown around the globe, especially in temperate areas, and its historical evolution is tightly connected with humans. Wild ancestors were found in the Tian Shan Mountains in Central Asia and the regions of China, Kazakhstan, and Kyrgyzstan. From this center of origin for apple, several archaeobotanical studies revealed the evolution and domestication of the modern apple by humans' spreading it along the Silk Road.

It has been hypothesized that the phenomenon of domestication, which made the plant more adapted for human consumption, paralleled the initiation of agriculture about 12,000 years ago. The domestication syndrome is marked with discontinuities between wild relatives and related domesticated species or a divergent evolutionary lineage resulting from the pressure of artificial selection. Domestication is both an initial unconscious selection by farmers of the best performing and productive plants, followed by an intentional breeding phase, with parental accessions selected to pyramid or introgressive favorable and important traits (Gross and Olsen, 2009; Olsen and Wendel, 2013).

In apple, domestication led to fruit enlargement associated with adaptation to megafaunal mammals and seed dispersal. An important contribution to the expanded growing and selection of apple was attributed to the domestication of horses that also occurred in the region of Kazakhstan about 7000 years ago. Domesticated horses, traveling with humans along the Silk Road, contributed to the dissemination of apple pips over a large area. The role of humans in apple domestication started only after the last Ice Age, when, due to commercial trading, fruit and seeds were transported across Eurasia along the Silk Road (Figure 1.1).

This type of fruit was discovered by Ivan Sievers in 1793, and in 1830, Carl Friedrich von Ledebour named this apple *Malus sieversii* in his honor. However, the same apple was described by Philip Miller in the UK in 1768 as *Malus pumila*, which has been largely used as the botanical nomenclature of apple. Humans/animals traveling westward facilitated the hybridization of apple. *Malus sieversii* hybridized with *Malus baccata* in Siberia, *Malus orientalis* in the Caucasus region, and *Malus sylvestris* in Europe, the closest relatives to *Malus domestica* (Harris et al., 2002), which shaped chloroplast and nuclear genomes (Cornille et al., 2012). The genome resequencing of a large apple collection composed of 117 diverse accessions (Duan et al., 2017) validated that the origin of *Malus domestica* was largely derived from *Malus sieversii*, followed by an important introgression with *Malus sylvestris* (Cornille et al., 2012). Genome-wide molecular inference proposed that major alleles of *Malus domestica* derived for 46% by *Malus sieversii* and

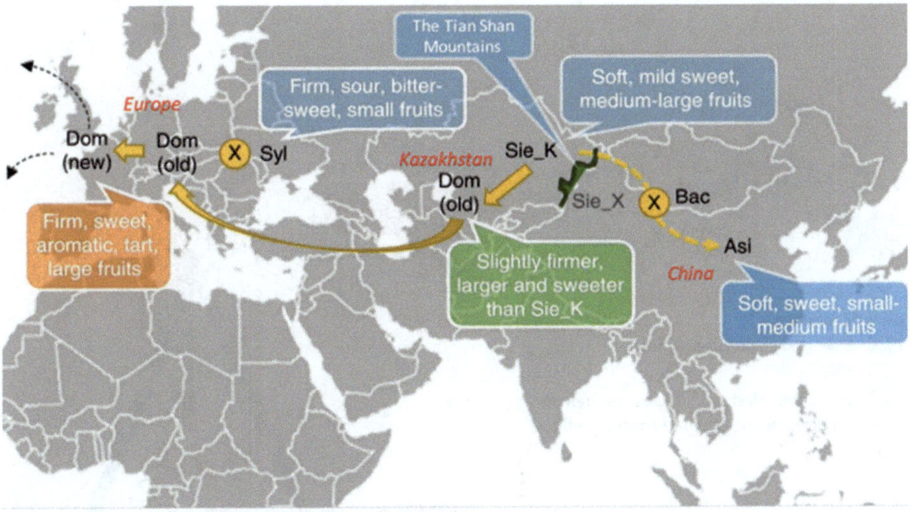

FIGURE 1.1

Evolutionary map of apple along the Silk Road. *Asi, Malus asiatica*; *Bac, Malus baccata*; *Sie, Malus Sieversii*; *Syl, Malus sylvestris*; *Dom, Malus domestica.*

(Figure from Duan et al., 2017.)

21% by *Malus sylvestris*, with the remaining genomic contribution of uncertain region (Harris et al., 2002; Velasco et al., 2010) (Figure 1.2).

Hybridization was facilitated by the high fertility among *Malus domestica* and the aforementioned species (Cornille et al., 2013). Molecular analysis with microsatellite markers revealed a relatively weak genetic population structure underlining high gene flow admixture (Richards et al., 2009; Cornille et al., 2013). Interspecific introgressions were facilitated by a self-incompatibility (SI) system and interspecific fertility, promoting outbreeding and hybridization (Castric et al., 2008). Analysis of selection sweep patterns identified important loci associated with quality traits, such as flavor, texture, and size, for which a two-step evolution process was proposed. The domestication of apple was initiated by an already large fruit; therefore, the domestication process of apple had relatively lower evolutionary pressure with regards to other species. Analysis of the nucleotide diversity indicated that the domestication process in apple did not affect initial genetic diversity, showing a very weak domestication bottleneck, confirmed by the rapid linkage disequilibrium (LD) decay observed in domesticated accessions (Cornille et al., 2012; Duan et al., 2017). The first characteristics that greatly defined the domestication syndrome (Doebley et al., 2006) are represented by the improvement of apple size, with fruit size ranging from small in *M. sylvestris*, medium-large in *M. sieversii,* and large in *M. domestica.* Quantitative genetic analysis identified two quantitative trait loci (QTLs) associated with fruit weight, mapped to LG 8 and 15, respectively (Devoghalaere et al., 2012), which

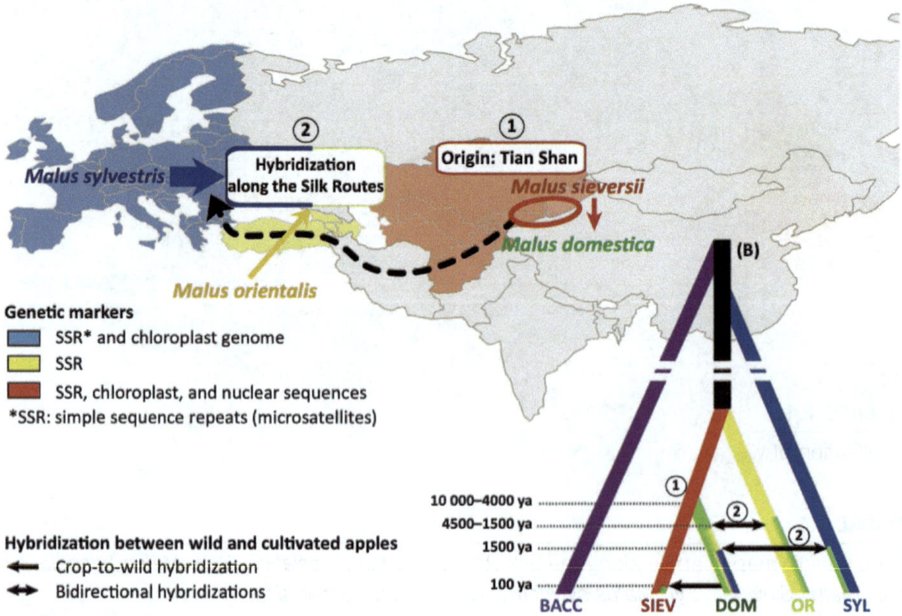

FIGURE 1.2

Evolution of apple and genealogical relationship among the main apple wild relatives. *Bacc,* Malus baccata; *Dom,* Malus domestica; *Or,* Malus orientalis; *Siev,* Malus Sieversii; *Syl,* Malus sylvestris.

(Picture from Cornille et al., 2014.)

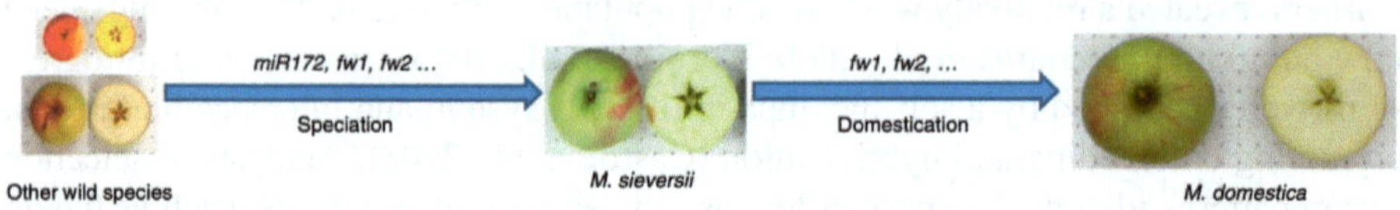

FIGURE 1.3

Fruit size two-step domestication process in apple.

(Picture from Duan et al., 2017.)

colocalize with the selective sweep regions. These data fully supported the two-step model for the fruit size enlargement (Figure 1.3).

The technique of grafting, described in the Hippocratic treatise (424 BCE) and by Theophrastus (371–287 BCE), played a role in the evolution and selection of superior accessions (Harris et al., 2002; Gardiner et al., 2007). The already enlarged apple types were further hybridized and grafted. In the Tian Shan area, large sweet apples were selected for eating by animals, which released the still-viable seeds into their waste (Figure 1.4).

A domestication model suggested that to complete the domestication of species from wild to cultivated, a cycle of 2 to 3000 generations is required. Apple has a generation of about 20 years, and apple has been cultivated ~3000 years, so apple did

FIGURE 1.4

Collection of wild *(left)* and domesticated *(right)* apple accessions.

(Photo courtesy Nicola Busatto.)

not have time to accomplish the domestication relative to other species (McKey et al., 2010). The reduction in gene flow observed between progenitors and cultivated apple promoted an initial divergent selection, followed by a biased selection on domesticated accessions outside the center of origin and vegetative propagation, all contributing to the evolution of cultivated apples (Cornille et al., 2012; Gross et al., 2012).

2. Economic importance

The world production of apples has progressively increased from about 60 million tons in 2000 to 96 million tons in 2022. Production from China makes up approximately half of global apple production, where acreage markedly increased starting in the 1990s. The United States, Turkey, Poland, and India comprise the top five apple-producing countries. Regionally, Eastern Asia was the largest producer with 51.7 million tonnes in 2022. Then, the European Union, Western Asia, North America, Africa, South America, and Oceania produced 12.6 million, 6.1 million, 5.8 million, 3.8 million, 3.2 million, and 0.9 million tons of apples in 2022, respectively. Total production area spanned more than 5 million hectares (ha) worldwide and is among the most widely distributed temperate fruit crops in the world (FAO.org).

In the EU, Poland, Italy, and France are important apple producers. Also in Eastern Europe, there are countries with significant apple production potential, like Russia, Ukraine, Serbia, and Moldova. In the last decade, many countries in the Southern Hemisphere have increased their production: South Africa is the largest producer, followed by Brazil, Chile, New Zealand, Argentina, and Australia. A small number of countries are responsible for a large portion of the world's exports of fresh apples, totaling about 8.5 million tons per year. They are China (that exports a very small percentage of its production), Italy, Poland, France, New Zealand, United States, South Africa, and Chile.

Apples are fruits largely consumed fresh (dessert apples) thanks to their well-known health properties, but they are also suitable for processing, such as juice, nectars, purée, mustard, vinegar, cider, dehydrated products, frozen for pastry, and concentrate for the preparation of many beverages. Apples remain the most deciduous fruit consumed worldwide, although trends show a decrease: for example, in Europe from 2013 to 2023, per capita annual apple consumption decreased from 16.4 kg to 13.7 kg (approx. 69 apples). Similarly, consumption in the United States was approximately 11.4 kg. Approximately half of this amount was fresh consumption, and the other half were processed products.

Although there are about 3000 apple varieties, at a global level, the apple is subject to strong varietal concentration, with less than 10 polyclonal cultivars covering approximately 80% of the trade. Besides traditional varieties that have been grown for many decades, more than 60 club cultivars with their own brand and distinctive identity are seeking space in international markets, creating added value. These cultivars are constantly changing and being introduced from private and public breeding programs around the world.

Regarding varietal assortment, the five most widespread cultivars in Europe are Golden Delicious (19%), Gala (13%), Red Delicious (5%), Idared (5%), and Shampion (5%), followed by Granny Smith, Fuji, and Elstar (8%), all accounting for more than 100,000 tons (WAPA, 2023). About 70% of the apples produced in China are Fuji. In the United States, five cultivars make up more than 50% of production: Gala (18.1%), Red Delicious (12.5%), Honeycrisp (11.1%), Fuji (10.1%), and Granny Smith (9.9%).

3. Botany and relevant species

The cultivated apple *(Malus domestica)* belongs to the genus *Malus* Mill., which is a member of the Rosaceae family. The genus *Malus* is divided into six sections (*Chloromeles*, *Docyniopsis*, *Eriolobus*, *Gymnomeles*, *Malus,* and *Sorbomalus*). Nevertheless, the taxonomy of the genus *Malus* is still not fully clarified. The number of species in the genus Malus ranges from approximately 25 to 47 according to Robinson et al. (2001), whereas Forsline et al. (2003) report 26 species. The composition of the genus *Malus* is dynamic. Qian et al. (2006) proposed transferring species from the section *Sorbomalus* to a new section, *Yunnanensis*, and new species such as *M. shizongensis* are still being discovered. Li et al. (2022) published a revision of the genus *Malus* based on morphological data, comprising 26 wild species, with cultivated *M. domestica* excluded due to its complicated domestication history. The genome of *Malus* has $x = 17$ chromosomes, which according to Velasco et al. (2010) probably originated from an autopolyploidization of an ancestor around 50 million years ago (Mya). Most *Malus* species are diploid, but some are tri- or tetraploid, or have different ploidy levels within the species (Höfer and Meister, 2010).

M. domestica is cultivated for fresh consumption, juice, cider, wine, and processing. *M. domestica* is an economically important species with a production of 96 million tons and ranks third in the world fruit production in 2022 (FAO, 2023). This domesticated species originates from *M. sieversii*, which is native in Kazakhstan, with severe introgressions from *M. sylvestris* (Duan et al., 2017) and contributions from other species (Cornille et al., 2012). *M. domestica* is distributed and cultivated in the temperate regions of the world. *Malus baccata* L. (Borkh.; crabapple) are used as ornamentals. Genotypes of wild apple species are often used in apple breeding as donors for certain traits, e.g., *M. floribunda*, the Japanese flowering crab, is a donor of the scab-resistance gene *Rvi6* or *M.* × *robusta* as a donor for resistance to fire blight (Peil et al., 2007).

The medium-sized trees are deciduous with a single trunk and a spreading canopy. Selected cultivars are propagated vegetatively by budding or grafting onto rootstocks. The size and shape of the tree depends mainly on the type of rootstock and the planting and pruning system. The green leaves are alternate, stalked, and undivided, with distinct stipules that fall off early. Bud break and flowering behavior of apple trees in spring strongly depends on the preceding phase of winter dormancy. During endo-dormancy, a genotype-specific amount of cold needs to be perceived before eco-dormancy is initiated. Chilling requirements range from 600-800 chilling hours for most cultivars, but some cultivars suitable for production in low-chill environments can have chilling requirements as low as 200 chilling hours (Pertille et al., 2021). For the phase transition of eco-dormancy to bud break the heat-requirement of a genotype has to be fulfilled (Lempe et al., 2022). Showy, short-stalked flowers vary in color from white to red and are arranged in corymbose racemes with five to seven flowers that emerge at the same time as vegetative growth. Flowers usually have 5 sepals and 5 petals, around 20 stamens, and 5 carpels. The fruits are pomes ranging from 3 to 12 cm in diameter, with a red, yellow, or green skin and typically white fruit flesh but can also be red, pink, or yellow. The fleshy hypanthium develops from the base of the flower and surrounds the core, which contains brownish seeds. *Malus* species are cross-pollinated; interspecific hybridization is possible and occur naturally. Self-fertilization is prevented by gametophytic SI, but a few cultivars are parthenocarpic, e.g., 'Spencer Seedless' and 'Wellington Bloomless'; apomixis has been observed in several *Malus* species (Liu et al., 2023).

4. Breeding and propagation

New apple varieties represent cultivars bred to improve key characteristics. There are many public and private breeding programs worldwide. Molecular biology and biotechnology provide apple breeders with opportunities to investigate complex traits, to conduct research across disciplines, to benefit from collaborative efforts, and to improve apple quality, tree performance, and adaptation to new challenges.

Apple breeding programs may have certain objectives to fit regional needs, but all programs stress improving fruit quality and incorporating resistance to disease and disorders as key traits to improve.

Breeding and the choice of cultivars to be used as parents in crossing is tightly dependent on the availability of germplasm, as a pool of potential favorable alleles. The gene bank accessions available at the USDA Apple Germplasm Collection in the U.S. and in clonal collections in other countries, including diverse *Malus* species and material from the center of origin, needs additional collections, characterization, and use (Volk et al., 2015). While the use of wild species in breeding is occurring, it needs to be expanded.

4.1 Molecular apple breeding

Classical breeding is an expensive and time-consuming activity. Selection can be supported by the employment of specific molecular markers that if associated with a specific trait can promote a more efficient identification of the desired genotypes. Molecular markers associated with specific traits have been a valuable resource, yet every trait reveals its complexity the more we study it, so that additional markers are often needed. The reviews of Moriya (2024) and Choi et al. (2024) on allele mining practices, tools, and techniques provide valuable updates.

4.1.1 Apple breeding goal: disease resistance

Breeding activities are oriented to improve the plant/fruit to meet different needs, from quality, to provide consumers with a novel and more appreciated fruit, to disease resistance, and to release novel cultivars prone to a more sustainable horticulture. Climate change, spread of pathogens, loss of certain pesticides, and our ability to detect pathogens has changed breeders' focus from the "big 3"—scab, fire blight, and powdery mildew—to diseases becoming much more problematic in a short period of time. The following are the most relevant diseases for which breeding research is focusing.

Fire blight *(Erwinia amylovora).* Rezzonico et al. (2024) studied host pathogen interactions and the evolution of *Erwinia amylovora*. They emphasized that the number of different regulators controlling virulence regulation, their environmental cues, and their locations are key to furthering our understanding. Bühlmann-Schütz et al. (2023) compared artificial fire blight shoot and flower inoculations in apple. Eleven of the 18 tested genotypes showed no significant difference between the results of the two methods. Flower shedding was identified as an important trigger mechanism for flower resistance to fire blight. Acimovic et al. (2023) developed molecular detection methods of strains to study the role of rootstock infections for apple tree survival, the presence of latent fire blight infections, and the value and utility of molecular detection methods to assist in effective apple tree removal after epidemics. They stressed the use of resistant rootstocks to lessen epidemics. Breeding for resistance to fire blight was reviewed by Peil et al. (2021).

Scab *(Venturia inaequalis).* Resistance to scab was first denoted by the V from *Venturia* and a subscript denoting the source of resistance. Later, this was changed to *Rvi* for resistance to *Venturia inaequalis* (Bus et al., 2011), with 17 *Rvi* genes and *Rvi*6 *(V_f)* the most intensively used in breeding, but this resistance has eroded in several locations. Parisi et al. (1993) reported one of the first substantiated reports of erosion in France. Pyramiding genes for scab resistance is strongly suggested (Patocchi et al., 2021). *Rvi*4 and *Rvi*15 were found to be the same gene, and the use of *Rvi*4 as its designation was suggested.

Powdery mildew *(Podosphaera leucotricha).* Strickland et al. (2021) review highlights the biology of the pathogen and lists many references as to source of resistance and characterization of commercial cultivars and as to their susceptibility to this pathogen.

Alternaria. There are different opinions on *Alternaria* as a primary versus secondary pathogen. Prechsl et al. (2023) reported on nonfungicidal control, indicating that *Alternaria* is a secondary colonizer. Yet, Cao et al. (2024) found that *Alternaria alternata* (ALT) secretes AM-toxin and secretory proteins to promote fungal invasion and overcome host defense systems.

***Colletotrichum Glomerella* complex.** *Colletotrichum acutatum* species complex (CASC) and *Colletotrichum Gloeosporioides* species complex (CGSC) have changed how we view bitter rot (Martin et al., 2021), now a pathogen to both fruit and leaves. Khodadadi et al. (2023) reported that more than 20 *Colletotrichum* species can cause apple bitter rot. In their study, Honeycrisp was the most susceptible and *Malus sylvestris* PI 369855 the most resistant. McHenry and Aćimović (2024) developed species-specific PCR assays for *Colletotrichum* species causing apple bitter rot, which will benefit further research.

***Botryosphaeria obtusa* (black rot) and *Botryosphaeria dothidea* (white rot).** The first report of *Botryosphaeria dothidea* causing stem canker and dieback of apple trees in Ontario, Canada, was by Ilyukhin et al. (2022). Since then, the aggressiveness and extent of infections by these pathogens has increased worldwide.

Rapid or sudden apple decline (RAD/SAD). Once thought to be isolated in occurrence, the issue of decline is of importance globally. Singh et al. (2019) looked at the potential role of weather, soil, and plant microbial communities in the rapid decline of apple trees and did not find fungal or bacterial pathogens to play a role. They suggested cold damage and stress have important roles. However, RAD-associated fungal pathogens were surveyed on 91 declining apple trees from orchards across Korea. The predominant infections were *Botryosphaeria* canker (63%), *Phomopsis* canker (20%) caused by *Diaporthe* sp., *Cytospora* canker (9%) caused by *Cytospora mali*, white root rot (6%) caused by *Rosellinia necatrix*, and violet root rot (4%) caused by *Helicobasidium mompa*. Moreover, 30% of samples with *Phytophthora* root rot infection suspected tested positive for the disease (Lee et al., 2023). Martino et al. (2024) surveyed dying trees in Northern Italy, and 11 species were associated with dieback. All species were pathogenic when inoculated on healthy apple plants. *B. dothidea* had the most aggressive infections.

Marssonina blotch, caused by the fungal pathogen *Diplocarpin coronariae*, can cause severe defoliation and affect leaves, fruits and twigs. Control in organic planting can be especially challenging. Wöhner & Emeriewen (2019) suggested investigating apothecial development and ascospore germination on *Malus* hosts to better understand of the role of the teleomorphic stage as primary inoculum of apple blotch. More insight on time points and environmental conditions favorable for ascospore flight were required to develop a forecast model and plant protection recommendations for growers. Apple blotch resistance in wild apple germplasm (*Malus* spp.) accessions was studied, and 15 accessions had resistance, including four accessions of *M. baccata*, one of *M. prunifolia*, one of *M. platacarpa*, two of *M. sargentii*, one of *M. sieboldii*, two of *M. sublobota*, one hybrid of *M. transitora* x *M. floribunda*; 'Evereste' and 'Golden Delicious' were the control (Wöhner and Emeriewen, 2019).

Bull's eye rot (BER). *Neofabraea vagabunda* (syn. *N. alba*; *Phlyctema vagabunda*) is the agent of bull's eye rot. A transcriptome study of the susceptible variety 'Rohos,' versus the resistant 'Ariane' apple, revealed that some volatiles, such as ethanol, methanol, furan, and formaldehyde, might be potential markers for *N. vagabunda* infection. Other volatiles, such as hexenal and methyl acetate, were putatively involved in regulating apple-fungi interaction (Baldi et al., 2024).

Anthracnose ***(Neofabraea malicorticis).*** This canker is challenging to control via excision of cankers or spraying before autumn rains. Climate change may extend its spread.

European Canker ***(Neonectria ditissima).*** Infection risk is high during bloom. Harteveld et al. (2023) developed a TaqMan PCR assay that detected *N. ditissima* in 51% of asymptomatic trees that had been inoculated, compared to 11% detection with culturing and detection in branches and best pruning strategy. An 'Aroma' (resistant) x 'Discovery' (susceptible) mapping population was used to detect five QTLs regions for less canker development on linkage groups 1, 8, 15, and 16 (Skytte af Sätra et al., 2023).

Viruses, viroids, phytoplasmas. New technologies such as high throughput sequencing (HTS) to scan for viruses in apple enabled the discovery of 100s of apple viruses, like the 17 viruses discovered in research by Wright et al. (2020). However, many viruses, while identified, have not always been found to be detrimental, creating phytosanitary certification uncertainty.

4.1.2 Apple breeding goal: fruit quality

The economic success of an apple cultivar is enhanced by its quality properties. Quality is mainly defined by appearance, flavor, texture, and nutritional/nutraceutical properties.

Fruit appearance: Fruit appearance, particularly color, is the first trait evaluated by consumers and has been of great interest in breeding. Many cultivars are dominated by red skin, while a much lower proportion are characterized by yellow skin, such as Golden Delicious, or a green type, such as Granny Smith. Red color confers a high value of marketability to apples. Molecular biology investigations have elucidated the regulatory mechanism coordinating the accumulation of anthocyanins, constituted by a complex of transcription factors such as *MYB*, *bHLH*, *WD40,* and *WRKY* (Espley et al., 2007). Genetic analysis and QTL mapping approach identified a genomic region on chromosome 9 associated with the red color (Chagné et al., 2016). The markers included in this interval can be used for this high-valued trait in the early stages of selection.

Fruit flavor: Volatiles are key for breeding for flavor. Orcheski et al. (2023) profiled apple volatile organic compounds (VOCs) in a New Zealand germplasm collection as a resource for breeding desirable flavors. Esters were the main drivers of volatile differentiation, with apples often producing either ethyl or acetate esters, suggesting a different genetic control for these ester types. Esters generally had larger broad-sense heritabilities, enhancing their selection for improving apple flavor (Shimizu et al., 2024). They further identified aroma compounds responsible for apple flavor via a palatability score prediction using tree-based regression models. The release of VOC has been also investigated for cider production. Medina et al. (2020) evaluated volatile fingerprints from apple fruits and ciders to find putative biomarkers for each variety. Esters, alcohols, and aldehydes were higher in fruits and juices, yet esters, alcohols, and acids were dominant in ciders. There were unique VOCs for each variety, highlighting the importance of apple variety selection to obtain good sensory and quality ciders.

Fruit texture: Texture is a fundamental attribute of fruits for two essential aspects. The first is its capacity to greatly influence consumer preference. The second is that relying on the chemical properties of the structural polysaccharidic architecture of the cell also determines fruit storability and shelf-life. Due to its importance, this multitrait quality aspect (composed of both mechanical and acoustic parameters) still represents one of the major goals in many breeding programs worldwide (Teh et al., 2021). Consumer demands were oriented toward the improvement of crispness, one of the preferred apple quality traits. Despite its relevance and role in the determination of quality, this trait presents a complex heritability, and only recently fruit crispness was analytically determined in a reliable and efficient manner (Costa et al., 2011; Teh et al., 2021). In the last decades, several QTLs have been identified, and specific candidate markers have been designed and fine-tuned for their possible application as a DNA-informed toolkit enabling more efficient selection of texture components (Longhi et al., 2012).

Fruit functional aspects: The functional properties of apple are related to their accumulation of bioactive compounds. These molecules, mainly represented by vitamin C and phenolic compounds (such as flavanols, hydroxycinnamic acid, and dihydrochalcones), through their antioxidant activity can scavenge an excess of free radicals, promoting human health (Angeli et al., 2024). Research has elucidated the diversity and accumulation of these molecules across apple germplasm (Gutierrez et al., 2018). Wild species, especially for their reduced size, are richer in these molecules, which accumulate more in the skin than in the flesh (Farneti et al.,

2015). Research has identified several QTLs associated with specific phenolic compounds.

4.1.3 Basic biological knowledge employed in traditional breeding

Successful breeding takes advantage of comprehensive knowledge of the biological systems of the plant, especially related to the mechanisms governing autoincompatibility system and flower biology.

S alleles: The understanding of SI alleles in apples took guesswork out of making apple crosses and revealed some of the traits linked to that locus. Liu et al. (2024) identified S-RNase genotypes and used them to analyze origin and evolutionary patterns in *Malus*.

Pollination biology: Studies on the duration of stigmatic receptivity ranged from 6.3 to 8.1 days depending on the year and cultivar. Pollen tubes required 5.5 to 7.0 days to reach the ovules. Meanwhile, ovule longevity was from 8.2 to 11.3 days, resulting in an effective pollination period (EPP) of 3.0 to 5.6 days. The importance of the maternal parent was stressed (Roeder et al, 2021). EPP will become more of an issue with changing climates and will continue to be studied.

4.1.4 Novel genomic approaches

Improvement in DNA sequenced technology and affordability allowed the possibility to sequence, assemble, and interpretate the information included in the genomes, also for nonmodel species. Since 2010, the release of the first genome for apple by Velasco et al. (2010) was followed by Golden Delicious doubled-haploid (GDDH13) (Daccord et al., 2017). Other species and variety genomes were sequenced, such as *Malus baccata* (Chen et al., 2019), Galaxy Gala (Broggini et al., 2020), Gala, *M. sieversii,* and *M. sylvestris* (Sun et al., 2020), Royal Gala (Tian et al., 2022), Honeycrisp, Hanfu (Qin et al., 2023), *Malus angustifolia* (Mansfeld et al., 2023), WA 38 (Cosmic Crisp) (Zhang et al., 2024), and a scab-resistant plant introduction. In addition, apple accessions used for pan-genome assembly include four wild apple species: *M. sylvestris* (MSM), *M. sieversii* (MSR), *M. orientalis* (MO), and *M. asiatica* Zisai Pearl (MA), and nine apple cultivars *(M. domestica)* (Wang et al, 2023). The availability of high-resolution genomes paved the way to several biotechnological strategies also for apple. Initial genetic engineering transgenic approach, followed by more recent new breeding technologies, such as gene editing with CRISPR/Cas9, allowed the specific modification of traits of interest in a more specific manner and without the need to pass through sexual reproduction as occurs in classical breeding. The nonbrowning 'Arctic' apple series remain the only transgenic apples in the market. A transgenic approach has been also used to accelerate the time of flowering. T1190, a transgenic apple expressing the *BpMADS4* gene of silver birch that causes early flowering (Flachowsky et al, 2011), is used in the rapid cycle breeding of apple to shorten the juvenile period in crossing and offspring (Patocchi et al., 2021). This breeding technology was used by Luo et al. (2020) to introgress blue mold resistance into elite apple germplasm. The CRISPR/Cas9 (Clustered regularly interspaced short palindromic repeat) and CRISPR-associated protein 9 (Cas9) system, is a tool for the introduction of site-specific double-stranded DNA breaks. However, genome editing using CRISPR/Cas9 may be subject to random/off-target effects in the genome, mutations, and chimerism.

4.2 Rootstocks

Although breeding has been dedicated to the improvement of the scion, the fruit-bearing part of a grafted plant, equally important activities also led to the improvement of rootstocks. The history, physiology, management, and breeding of apple rootstocks was reviewed by Marino and Fazio (2018). Three QTLs, Dw1 Dw2, and Dw3, are responsible for the genetic control of rootstock-induced dwarfing, with Dw1 having a more significant role in dwarfing. Other rootstock-scion relationships may also contribute to dwarfing in apple rootstocks, including modifications in flowering genes, the anatomy and morphology, carbohydrate assimilation and partitioning, nutrient/water uptake and distribution, and hormonal regulation in dwarfing rootstocks compared to vigorous stocks. These cumulatively influence canopy size. There is also a need for more research on the rootstock/scion interface.

5. Orchard management

Training and cropping systems. In the past, apple orchards were grafted on vigorous rootstocks, planted with spacing ranging from 12 × 10 m to 8 × 6 m, depending on the variety, with densities of 80 to 200 plants/ha. Starting in the 1970s through the introduction of the M9 rootstock, it was possible to increase the density of orchards to 1800 - 2400 trees or even more. The following describes some traditional planting systems and those of more recent introduction.

The **slender spindle**, a form ultimately resembling a cone type of shape, has become established in many areas as the most interesting. The central axis, starting from a height of 50 to 70 cm above the ground, bears lateral branches that become progressively shorter toward the top of the plant. In fruit-growing practice, the single-row system is adopted for easy management and good balance between the quantity of fruit produced and their quality. Instead of the 1800 to 2400 plants/ha that were initially used, in some areas, investments are considered ranging from 2800 to 4500 plants/ha, considering the variety, soil characteristics, and planting elevation (Figure 1.5A).

Solaxe System. This planting system aims to reduce investment costs and labor by appropriately reducing the number of plants per hectare. With a planting density of approximately 2000 plants/ha, the plants are grown to a height of 2.5 to 3.0 m. The top is not pruned but bent downward under the weight of its fruits. Branches are also bent downward, and the resulting vegetation is allowed to grow (long pruning). Through appropriate techniques, vegetation developing from the basal part of the

FIGURE 1.5

(A) Apple trees trained according to the slender spindle system. (B) Apple trees trained according to the multiaxis system. (C) Apple trees trained according to the Y shape (Bibaum) system.

branches, near the insertion with the trunk, is removed to create a "channel" that allows light penetration into the canopy.

Multi-axis systems. These cultivation forms resemble viticulture more than classical fruit growing. Nurseries have also carefully analyzed the situation and already offer the first patented solutions of pre-formed plants (e.g., DaVorBaum, GuyotTree), which promise a faster entry into production and the possibility of performing fewer cultural operations (Figure 1.5B).

Y shape (Bibaum). Already in the nursery, young plants are prepared to assume this shape in the field. At transplanting, trees are planted in single rows, so that the two axes of the same tree are oriented along the row and cultivated to form two cordons. Consequently, the resulting wall will be thin and tall. Dividing the plants into two cordons reduces their vigor and decreases the time required for pruning. This cultivation form was already in use in the past when the stone fruit trees were forked after planting (Figure 1.5C).

Other cultivation forms are represented by V and Y systems. In the case of the V system, it is possible to invest from 3500 to 4000 plants/ha (planting distances: 3.5 × 0.7-0.8 m) and increase the leaf surface area by splitting the vegetation into two parts. The transverse Y system, similar to the previous one, allows for splitting the productive wall by exploiting two branches that bifurcate from the trunk, transversely to the row direction.

Fertilization and irrigation. The demand for mineral elements by the apple tree is not constant throughout the season. During the active growth phase of the shoots, in the spring and early summer, for example, most nitrogen absorption occurs. Apples contain significant amounts of potassium, so the highest demands for this element occur during the peak fruit growth phases and are higher in highly productive orchards. As an indication, it can reasonably be assumed that in fully productive orchards (e.g., with 60 t of fruit/ha), trees remove approximately 70 kg of nitrogen (N), 15 kg of phosphorus (P) and magnesium (Mg), 105 kg of potassium (K), and 70 kg of calcium (Ca) from the soil annually. However, net removals, i.e., the elements that leave the system and should be replenished, are lower because they must consider the nutrients in the fallen leaves returning to the soil and any other nutrients introduced into the system (e.g., through irrigation water or amendments). When the soil is well endowed with Ca, soil Ca supply is rare, but canopy sprays are frequently adopted to prevent the onset of physiological disorders like bitter pit (Figure 1.6).

To fertilize the apple tree sustainably from an environmental point of view, it is first necessary to try to synchronize the availability of nutrients in the soil with the tree's needs. Fertilization management and, in general, all practices that modify the "soil" environment (tillage, weed control, mulching, irrigation, etc.) should aim to minimize nutrient losses to the environment through leaching and volatilization. Nutrient losses through leaching significantly increase during the autumn-winter season when higher rainfall is accompanied by reduced root activity of the tree in its vegetative rest phase. Ground cover significantly reduces the risk of nitrogen leaching.

FIGURE 1.6

Bitter pit on apple fruit.

Fertilization objectives vary depending on the apple orchard's development stages: in the first years after transplantation into the field, when productivity is low or nonexistent, the tree must form permanent structures and therefore must have adequate nitrogen quantities in the soil. During the productive phase, the fertilization objective is to support fruit production, enhance its quality without excessively stimulating vegetative growth, and limit nutrient losses from the orchard through leaching, volatilization, and runoff.

In fine-tuning the application of preplanting fertilizers, soil analysis is necessary to obtain information about its chemical-physical characteristics including texture, total and active lime content, cation exchange capacity, organic matter content, total nitrogen, assimilable phosphorus, and exchangeable potassium. In coarse texture soils, preplanting fertilization is less critical because fertility elements move better along the profile than in clayey soil. Elements that may need to be applied, depending on the analysis performed, are mainly potassium, phosphorus, magnesium, and calcium (the latter in acidic soils), while nitrogen in mineral form should not be applied. Especially in sandy soils with low humus content, it is advisable to enrich the soil with organic matter through compost or manure addition.

Annual fertilization should be based on expected yields, basal soil fertility, and tree vigor (Carranca et al., 2018). Most N is applied to the soil, split into several applications to limit losses. Additional foliar N is sometimes sprayed to the canopy to make it rapidly available to the leaves. In late summer-early autumn, such intervention also favors the buildup of N storage in winter and allows for faster leaf decomposition in winter, reducing overwintering forms of apple scab, with probable benefits on the orchard's health the following year.

To prevent physiological disorders like bitter pit, calcium is sprayed through canopy treatments as calcium chloride. Apple varieties have different susceptibilities to Ca deficiencies in fruit: for example, Golden Delicious has a lower incidence of bitter pit compared to Fuji and Braeburn.

Bitter pit is affected greatly by the cultivar and growing region. For example, Gala rarely develops the disorder in more moderate climactic regions such as New Zealand but can be more susceptible in countries such as Chile. Susceptible cultivars are routinely sprayed with $CaCl_2$ or with calcium formulations that claim greater calcium ion uptake into the fruit. The number of sprays can range from 2 or 3 to as many as 12. Calcium movement into the fruit can be affected by climactic factors. Overall, it is generally better to apply calcium to the fruit at regular intervals rather than early or late. Postharvest calcium dips is routinely used by some apple industries, as these treatments could apply a season's amount of calcium directly to the fruit. However, food safety concerns have led to phasing out of postharvest applications.

Although irrigation primarily functions to moisten the soil to increase water availability, it is also used to modify the microclimate, as in the case of antifreeze irrigation. Typically, overhead sprinkler interventions are used for this purpose (Figure 1.7), with systems delivering 1 to 2 mm/hour, activated based on temperatures and phenological stage. For example, if the tree is in full bloom, with a critical temperature for damage onset around −2°C, irrigation should be activated when the temperature is around 0°C to prevent damage from further temperature drops due to evaporation when the first water droplets reach the flower.

High-quality fruit production with optimal size achievable in modern orchards on dwarfing rootstocks is typically only possible through irrigation, which helps prevent water deficiencies during the summer period (Naor and Girona, 2012). Water stresses that reduce leaf water potential below −1.5 MPa decrease photosynthesis and fruit cell size. Evapotranspiration in apple orchards reaches levels of 5 to 6 mm/day on hot, sunny summer days in several areas. If apples are grown under

FIGURE 1.7

Overhead irrigation system in action to protect against late spring frost.

hail netting, atmospheric evapotranspiration demands are significantly reduced, by 15% to 20%. Water consumption during the growing season depends heavily on leaf quantity, radiation, and vapor pressure deficit (VPD) in the cultivation environment. A significant reduction in soil water availability also reduces transpiration.

Irrigation water is frequently applied through overhead sprinkler systems or under-canopy microirrigation, including drip or microjet systems. Microirrigation reduces water losses due to evaporation and avoids leaf wetting, which can cause pesticide leaching and often creates favorable microclimatic conditions for fungal pathogen development. Irrigation schedules and volumes should be calculated based on soil hydrological characteristics.

Soil surface management. In commercial apple orchards, ground cover vegetation in the alleyway is almost always present (Figure 1.8). Typically, this consists of spontaneous species, but in some cases, specific grasses may be planted. Regardless, regular mowing of this vegetation is necessary. Ground cover vegetation offers several benefits, with improved soil structure being perhaps the most significant. Over time, ground cover increases soil organic matter, a process linked to its ability to sequester atmospheric CO_2. Additionally, leguminous species in the vegetation help increase soil nitrogen content by fixing atmospheric nitrogen through symbiotic relationships with root microbes. Overall, ground cover vegetation reduces nitrogen losses through leaching by absorbing residual nitrogen at the end of the season.

Within the row corresponding to the tree row, approximately 1 to 1.5 meters wide, weed control measures must be more meticulous in the early years of orchard life. Subsequently, as tree roots deepen, making the tree less susceptible to competition, and as weed competition decreases due to shading from the canopy, weed control becomes less critical.

Pruning. It is primarily performed during winter, although interventions during the growing season may also be necessary. The intensity and type of pruning depend on the fruiting habit of the variety and the intensity of vegetative growth in the previous season. In trees subject to alternate bearing, after a season of low fruit

FIGURE 1.8

Orchard floor management with the presence of grassed alleys.

production, more pruning wood may be removed to leave well-exposed branches at the end of the operation. In general, excess branches are thinned, overly vigorous "wood" branches are removed, and those with a too-narrow insertion angle are pruned. To promote adventitious and less vigorous shoots with a wider insertion angle, the vigorous branches and twigs are pruned, leaving a stub. Winter pruning also involves removing branches that have exhausted their productive potential and making deviation cuts to reduce their length. Rejuvenation cuts are sometimes made in the lower part of the canopy in varieties prone to producing weak branches on exhausted shoots (e.g., Gala and Braeburn). These cuts involve shortening the branch above a wood bud to stimulate vegetative activity at that point. Apple trees can also be mechanically pruned (Figure 1.9), which reduces costs and speeds up operations, compensating for the lack of specialized personnel for pruning. Mechanical pruning can be carried out both in winter and during the growing season, either laterally to limit space toward the alleyway or horizontally to lower the height. After mechanical pruning, especially if performed every year, manual pruning is necessary to refine the work and remove branches from the inside of the canopy.

In many varieties producing bicolored apples, summer pruning is common. This allows for a better light penetration inside the canopy before harvest (about 2 weeks before) and improves fruit coloration. These interventions involve removing water sprouts and shortening shoots. This is sometimes followed, especially in club varieties whose fruits must have a certain percentage of color, by the deposition of reflective sheets (white or silver) on the soil surface or the use of a pneumatic defoliators in preharvest, which further improves fruit color by better exposure to light.

Thinning. It is an indispensable practice for apple growers, aimed not only at obtaining high-quality production with a marketable fruit size, but also to avoid alternate bearing. The optimal fruit load depends on several factors, among which the cultivar, tree size, and its vigor are the most important. The number of fruits per hectare may range between 330,000, such as in Gala, and 400,000, such as in

FIGURE 1.9

Apple trees mechanically pruned *(in front)* and unpruned trees *(at the back).*

Braeburn, corresponding to approximately 5 to 7 fruits/cm^2 of trunk section area. The ideal number of fruitlets per corymb spans between one (the central fruit, also called the "king") and two (the central plus a lateral of similar size), depending on the size of the branch and its capacity to support fruit growth. Early thinning operations are the most effective in promoting floral induction in buds and, thus, a suitable return bloom.

Chemical thinning performs very well on apple and is broadly diffused thanks to the wide availability of thinning agents that are applied sequentially according to specific strategies, followed by a final fine-tuning of fruit load performed manually. While blossom thinning is usually aimed at preventing fruit set through the desiccation of flower organs, most fruitlet thinners rely on a hormone-like activity and can be applied within a range of king fruit cross-diameter of 6 to 20 millimeters, according to the thinner and cultivar sensitivity (Table 1.1).

Mechanical thinning is widely adopted in organic apple cultivation as an alternative to chemical thinners. This technique is applied through specific machines, such as the Darwin, consisting of a rotating device carried by a tractor along the row and able to damage a certain amount of flowers. The main disadvantage of this technique, however, is the low selectivity. Organic thinning can also be carried out with preparations of calcium polysulfide, vinasse (with high nitrogen content), potassium soap, or mineral oils, which are allowed in organic cultivations.

Table 1.1 Conventional chemical thinners that can be used on apple
The time of application is described as either phenological stage or the king fruit cross-diameter (Costa et al., 2018)

Type of thinner	Thinning agent	Time of application/ reference fruit size	Mechanism of action
Blossom thinner	ATS (Ammonium Thiosulfate)	Blossom stage/full bloom	Flower organ desiccation
Post-blossom thinner	NAD (Naphtaleneacetamide)	5–6 mm	Unknown
Fruitlet thinner	NAA (Naphthaleneacetic acid)	6–12 mm	Photosynthesis inhibition and nutritional stress
	6-BA (6-benzyladenine)	10–14 mm	Stimulation of vegetative activity and nutritional stress
	Metamitron	10–20 mm	Photosynthesis inhibition and nutritional stress

6. Fruit maturation and harvest criteria

Maturation of apples has a large influence on fruit quality after storage. Quality is often defined by preharvest pomologists as size/yield and, in the case of red apples, the coverage and intensity of the anthocyanin pigments. However, timing of harvest and therefore fruit maturation is a compromise between quality factors and storage potential of the fruit because as positive quality attributes are increasing, the storage potential of the fruit is decreasing (Figure 1.10). Fruits that are harvested too early may store for a long time but can lack flavor, while those harvested late can be fully flavored but have limited storage potential. Depending on the cultivar, maturity can also affect the susceptibility of fruit to physiological storage disorders; those such as bitter pit and superficial scald being associated with early harvest and senescent breakdown and soft scald being associated with late harvest (Watkins and Mattheis, 2019).

Physiological maturity is defined as "the stage of development when a plant or plant part will continue ontogeny even if detached." However, this definition is not as straightforward for apples as it is for a fruit such as tomato, where, for example, the fruit will not ripen (turn red) if picked prior to the mature green stage. An apple fruit does not show the same distinct changes of color and softening during ripening and may look visually appealing despite poor eating quality. The term "mature" is a subjective term being generally recognized as "that stage at which a commodity has reached a sufficient stage of development that after harvesting and

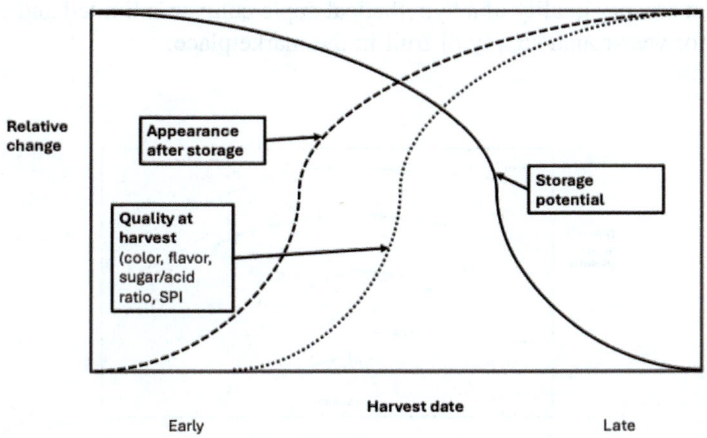

FIGURE 1.10

The relationship between the storage potential of an apple fruit and the quality characteristics desired by the consumer. As a fruit matures and ripens on the tree, the quality attributes such as color and flavor increase, but the storage potential declines. Appearance of the fruit after storage, however, is higher at earlier harvests than the desirable quality characteristics.

postharvest handling, its quality will be at least the minimum acceptable to the ultimate consumer." The minimum acceptable is subjective and affected by several variables. Internal quality is judged primarily by flavor, texture, and aroma; however, texture is the key attribute followed by flavor characteristics such as soluble solids concentration (SSC) and titratable acidity (TA), desired by consumers (Endrizzi et al., 2015). While aroma specific to the cultivar is expected, higher ester production can also be associated with overripening of the fruit. There are no formal studies available, but it is likely that the time of year when fruit are consumed affect consumer expectations; those for flavor being higher in fruit marketed closer to the harvest season than after long-term storage where the key attribute is texture rather than flavor. Postharvest technologies such as controlled atmosphere (CA) storage and 1-methylcyclopropene (1-MCP), discussed in the next section, are known to decrease volatile production but treated fruit can still be preferred by consumers (Marin et al., 2009).

In some apple industries, "commercial maturity" can supersede "physiological maturity" as an unfortunate response to market demand where an apple shortage in the market, especially early season, can lead to premature fruit harvest. This problem can be exacerbated by high color strains as they can be harvested when the required red coloration is present even though flavor is poor. The ethylene-releasing agent ethephon is also applied in some growing regions to advance red color development, but these fruit have compromised storage potential (Stover et al., 2003).

Another way of thinking about the relationship between maturation and storage potential is shown in Figure 1.11, which illustrates the effect of these factors on sensory quality as well as senescence. The harvest period and storage length that results in excellent sensory quality of a hypothetical apple cultivar is limited and cannot be achieved for year-round supply of fruit in the marketplace.

FIGURE 1.11

Relationship between quality characteristics of a hypothetical apple cultivar with increasing time in storage as affected by harvest date. The figure shows that the harvest period and the time in storage that excellent fruit quality can be obtained is limited.

(Modified from Little and Holmes, 2000.)

To further understand compromise between maturity and quality, the distinction between maturation and ripening as well as the physiology of apple fruit ripening should be considered. Ethylene production from the fruit occurs during the maturation/ripening stages, but the timing of its initiation and rate is affected greatly by cultivar. As a climacteric fruit, ethylene production can have a major influence on the ripening process, being associated with physiological changes such as loss of acidity, starch degradation, color change, flesh softening, and volatile production. However, dependency on ethylene and sensitivity to ethylene are different for individual ripening processes. Loss of acidity and starch degradation are ethylene independent, but highly sensitive to ethylene, while flesh softening is strongly dependent on ethylene, but sensitivity of the fruit to ethylene is low (Johnston et al., 2009).

The timing and rates of maturation and ripening can be affected by several preharvest factors. These include climate and orchard management practices such as crop load and nutrition disorders (Serra et al., 2016). In addition, preharvest treatments with plant growth regulators (PGRs) are available to many apple industries depending on registration and therefore fruit-growing region (Watkins, 2017). Naphthalene acidic acid (NAA) has been used extensively to delay preharvest apple fruit drop and allow red color development of cultivars such as McIntosh but can stimulate ethylene production and consequently reduce storage potential (Stover et al., 2003; Yuan and Carbaugh, 2007). More recently, aminoethoxyvinylglycine (AVG; commercial formulation ReTain) and 1-methylcyclopropene (1-MCP; commercial formulation Harvista) have been developed. AVG inhibits the activity of 1-aminocyclopropane-1-carboxylate (ACC) synthase by binding to pyridoxal 5'-phosphate. ACC catalyzes the conversion of methionine to ACC, the precursor to ethylene in the ethylene biosynthetic pathway (Boller et al., 1979), although its effects on nonethylene pathways cannot be excluded. 1-MCP inhibits ethylene perception (Sisler, 2006).

Preharvest AVG and 1-MCP delay fruit maturation as indicated by slower starch hydrolysis (higher Starch Pattern Index [SPI]), softening, loss of chlorophyll, and development of watercore (do Amarante et al., 2022). Because fruit maturity directly affects susceptibility of fruit to postharvest physiological storage disorders, those associated with less mature fruit, such as bitter pit, leather blotch, and carbon dioxide injury, can be enhanced, while those associated with more mature fruit, such as soft scald and senescent breakdown, can be decreased (DeEll and Ehsani-Moghaddam, 2010). The situation for superficial scald is more complex as the disorder is decreased by inhibition of ethylene production (McArtney et al., 2009) but can be higher in fruit with delayed maturity and reduced anthocyanin concentrations (Al Shoffe et al., 2024).

The impact of new cultivars on the relationships between quality and harvest has also become a confounding factor in fruit management. Traditional cultivars such as Red Delicious, Jonagold, Golden Delicious, Empire, and McIntosh have well-defined climacterics with high ethylene production that are associated with softening, loss of chlorophyll, and development of aroma volatiles. In contrast, cultivars such as Gala, Fuji, and several of the newer cultivars including Honeycrisp have lower ethylene production rates and maintenance of texture after storage (Chang and Tong, 2020; Harb et al., 2012) as well as complicated postharvest management procedures such as conditioning of fruit before storage (Al Shoffe et al., 2020). These new cultivars also affect the impact of postharvest technologies as will be discussed in the next section.

Common harvest indices that are used to determine harvest periods include days after full bloom, external color (red and green), ethylene production (often assessed by measurement of the internal ethylene concentration [IEC]), flesh firmness, SSC, starch content (usually by starch pattern index [SPI] rating), TA, and seed color. The Streif test, which utilizes the SPI, SSC, and flesh firmness (Streif, 1996), is used commonly in Europe.

The term *harvest indices* is preferred to maturity indices because factors such as flesh firmness and SSC are affected by orchard factors that include nutrition and fruit exposure and do not represent physiological measures of fruit maturity. Even widely used harvest indices such as the SPI can be flawed as the visual image of iodine-stained starch assumes that hydrolysis is initiated when a full complement of starch exists (Doerflinger et al., 2015).

In general, the range of methods used depends on the specific growing region and can vary from purely subjective assessments, i.e., no specific objective ones being used, to comprehensive analyses that include investment in expensive equipment such as gas chromatograph for ethylene measurements. The degree of sophistication is usually greater in large-scale storage operations and particularly when fruit from many orchard blocks is being stored, as insurance against claims if pack-out of the fruit is low, e.g., because of poor quality or presence of disorders. Other quality measurements of fruit that are assessed on incoming loads of fruit to a central storage facility include size, skin condition, and bruising.

Additional approaches to harvest decision-making have been investigated or adopted for industry use. These include a genetic test based on gene expression of ripening-associated genes that was marketed by NSure, and instrument-based measurement of chlorophyll and dry matter as harvest and quality indices. The genetic test developed by NSure provided earlier prediction of the optimum harvest dates than the Streif text for three cultivars, and the accuracy between the two approaches was similar (Neuwald and Streif, 2012).

Loss of chlorophyll is an important factor during fruit ripening (Zude-Sasse et al., 2002). Background color, a visual assessment of chlorophyll content, using color cards has often been used for bicolored apples such as McIntosh and Gala (Brookfield et al., 1993). The pressure on the industry to use high color sports has resulted in the development of strains with total red coverage, making the use of background color cards difficult. However, the development of the Difference of Absorbance (DA) meter has overcome this limitation. The DA meter produces an index of absorbance difference between 670 and 720 nm, known as the I_{AD} value. This value is an indirect measure of chlorophyll concentration in the fruit epidermal and subepidermal tissues, higher values indicating higher chlorophyll concentrations. Strong associations between I_{AD} values and chlorophyll have been found

(DeLong et al., 2020; Toivonen and Hampson, 2014), and I_{AD} values have been used as an additional harvest index. Lower I_{AD} values are associated with softer fruit, higher SPI and SSC (Moran et al., 2020; Doerflinger et al., 2024), and with ethylene production of fruit at ambient temperatures with and without postharvest 1-MCP (Cocetta et al., 2017). I_{AD} values can also provide a mechanism to investigate fruit ripening more precisely (Cai et al., 2023) and predict susceptibility of fruit to physiological storage disorders such as superficial scald (Farneti et al., 2015), soft scald (Moran et al., 2020), and flesh browning (Doerflinger et al., 2024).

The dry matter concentration (DMC) of fruit, which is basically comprised of all components except water, including both soluble and insoluble carbohydrates, has potential to predict fruit quality attributes. It has been shown that fruit DMC of several apple cultivars at harvest can be used to predict fruit SSC after cool storage (McGlone et al., 2003). DMC at harvest also has been suggested as a key metric to predict fruit SSC at harvest and after cool storage; however, the correlation between DMC and flesh firmness was weaker and more dependent on cultivar (Palmer et al., 2010).

7. Postharvest physiology and technology

Apple fruit continues to carry out anabolic and catabolic processes associated with ripening and senescence after harvest. Metabolism is more rapid for harvested fruit than for those on the tree at ambient temperatures, and stored carbohydrates and other substrates are the only energy source available to the fruit. The ripening processes include the respiratory climacteric and associated utilization of respiratory substrates and use of oxygen to produce carbon dioxide, water and energy (heat), softening, loss of chlorophyll, acidity, starch hydrolysis, and water through transpiration, as well as changes in the cuticle properties. A key factor associated with these changes is autocatalytic ethylene production, which is generally recognized as the plant hormone that is most responsible for coordination of the metabolic pathways that are involved in ripening (Johnston et al., 2009).

It is sometimes forgotten that the biological changes that occur during ripening are desirable for seed dispersal from an evolutionary perspective, rather than the anthropomorphic goal of maintaining the fruit for consumption, with or without storage. The wide range of commercial fruit phenotypes with rapid and slow ripening are a result of selection of chance crosses or breeding programs. These phenotypes have usually been selected based on appearance and sensory characteristics rather than for storability per se, although cultivars such as Idared and Fuji were originally commercially successful because of their good storage potential in common storage, i.e., without refrigeration. While consumers primarily expect visually appealing fruit, they also expect good texture, as well as the sugar/acid balance associated with each specific cultivar (Harker et al., 2008; Corollaro et al., 2014).

The development of physiological disorders in stored apple fruit is a major problem for many apple cultivars. While fruit that may be softer than desired can still be marketed, those with skin or flesh browning disorders are unacceptable in the marketplace. Major disorders include bitter pit, senescent breakdown, superficial scald, external and internal carbon dioxide injuries, low temperature breakdown, and variants such as soft scald and soggy breakdown. The development of these disorders can be associated with senescence, mineral imbalances, low oxygen and/or high carbon dioxide concentrations, and storage temperature (Watkins and Mattheis, 2019). The susceptibility of fruit to specific disorders is affected by cultivar, growing region and associated climate, orchard management, harvest maturity, postharvest treatments, storage type (air/CA) and length, and storage temperature (Little and Holmes, 2000).

Postharvest management of apple fruit for human consumption is focused on slowing down the ripening by minimizing the rate of respiration as well as the production and action of ethylene. There are four major goals of fruit management after harvest: (1) reduce metabolic rates of processes that result in undesirable changes in color, composition, texture, flavor, and nutritional status; (2) reduce water loss that can result in loss of marketable weight and shriveling; (3) minimize bruising, friction damage, and other mechanical injuries; and (4) prevent development of physiological and pathological disorders.

Harvest represents the first time that a high-quality fruit can be damaged, but all postharvest operations can subject apples to risk of injury, especially bruising (Hussein et al., 2020).

Depending on the type and size of the operation, sampling of fruit at a central storage facility can occur, especially where fruit from many farms is stored in a central storage facility. Assessment of bruising and other damage can provide a feedback system to each picking crew to ensure high performance. At the same time, average fruit weight/size, red color coverage, maturity, skin condition, and other defects such as injury, insect damage, and bitter pit are measured.

Reducing respiration rates of fruit quickly after harvest is desirable and represents the first step in the cold chain, a concept that focuses on managing cold temperatures from harvest to the consumer. The major steps in this process are field management, cooling, cold storage, refrigerated transport, and marketing in retail outlets. Each step has the potential to result in loss of fruit quality, including when fruits are exposed to nonoptimal temperatures during grading and packing. The impact of breaks in the cold chain will vary by cultivar and storage period. Loss of quality in an early season cultivar with higher rates of metabolism are more serious than with later season ones.

Optimum field management minimizes the time between harvest and getting the bins of apples into the refrigerated storage room. Formal information about temperatures of fruit exposed to direct sunlight in the field was not found for apples, but surface temperature of the fruit, especially red cultivars, are usually much higher than the ambient temperature.

Apples may be marketed immediately, but the majority of fruit are cooled and stored at low temperatures to maintain quality. Decreasing fruit temperatures and thereby their metabolic rates should be a primary focus of any storage facility.

The rapid softening cultivar McIntosh softens as much as 20 times faster at 21°C than at 0°C. However, softening rates at warmer temperatures are affected greatly by cultivar, e.g., rapid for Royal Gala and Cox's Orange Pippin but slow for Granny Smith and Pacific Rose (Johnston et al., 2001). Ethylene production in cultivars such as Golden Delicious, Granny Smith, and Pacific Rose is stimulated by exposure to low storage temperatures (Johnston et al., 2001; Knee et al., 1983) in a way that is analogous to European pear fruit. Some new cultivars such as Honeycrisp and its progeny have slow softening rates at warm temperatures (Harb et al., 2012). This characteristic of Honeycrisp allows use of conditioning in which fruit are kept at 10°C for 7 days to reduce susceptibility of fruit to chilling injury (Al Shoffe et al., 2020; Watkins et al., 2004).

Cooling apples in the storage facility is usually carried out by "room cooling," a passive method in which the fruit bins are stacked in a refrigerated room to remove the field heat (Duan et al., 2020). This handling step is highly energy intensive as the refrigeration capacity required to cool fruit immediately after harvest is much greater than for the maintenance of low temperatures during storage. Other cooling methods such as forced-air cooling or hydrocooling can be used but are typically not cost effective and not necessary for a relatively slow ripening fruit such as apples. Effective passive cooling requires appropriate distribution of bins relative to the airflow from the refrigeration as air will not move through stacks of bins without adequate airflow velocity. Although cooling of apples in refrigerated storage has been carried out for decades, research continues on improving uniform airflow and even temperature distribution as well as managing operation time of fans, which represent a source of heat in apple storages (Praeger et al., 2020).

Relative humidity (RH) in the storage is also important, the optimum RH range being from 90% to 95%. The RH in the storage naturally increases in rooms that are filled with fruit. Depending on cultivar, lower or higher RHs can increase or decrease disorders such as flesh breakdown and skin cracking (Weber et al., 2012).

Drenching or dipping of fruit bins with the antioxidant diphenylamine (DPA) (as well as fungicides) for control of superficial scald was common around the world (Lurie and Watkins, 2012). DPA application is no longer allowable in the European Union, and control of superficial scald is largely by use of low oxygen concentrations in the storage atmosphere (Zanella and Stürz, 2013). DPA remains in use in North America, and the label has been extended to approve its use to control carbon dioxide injuries. DPA is rarely applied by drenching because of food safety and pollution concerns.

Cold storage is the fundamental technology to maintain quality of apple fruit after harvest. While "common storage" where kilns, caves, and trenches are used to maintain temperatures between 0°C and 6°C are still used in parts of Asia, refrigerated storage is the basic technology used by most apple industries. Depending on the cultivar, refrigerated air storage is utilized for storage periods of up to 3 to 4 months, while cold storage with controlled atmospheres (CAs) in which oxygen concentrations in the storage atmosphere are decreased while those of carbon dioxide are increased is typically used for longer periods. The recommended storage periods in air or CA are greatly influenced by cultivar and market expectations. The recommended temperature in air or CA storage, typically 0°C to 3°C, is also a function of each cultivar and its susceptibility to chilling injuries and other physiological disorders (Watkins and Mattheis, 2019).

Conventional CA storage in which safe oxygen concentrations (2%–5%) and carbon dioxide concentrations (2%–5%) have been derived empirically for each apple cultivar were used widely around the world. The atmospheres (and storage temperatures) varied widely based on cultivar and growing region (Kupferman, 2003). These differences reflected the physiological responses of each specific cultivar and even strain but sometimes also the sophistication of the industry in a particular growing region.

Two important supplementary technologies, advanced CA technologies and 1-methylcyclopropene (1-MCP), have been commercialized over the last 20 or so years. First, significant advances in technology have revolutionized CA storage of apples. Historically, measurement of gas concentration was often by manual Orsat gas analyzers and manual adjustment of atmospheres that resulted in high fluctuations of atmospheres and subsequently recommended oxygen concentrations well above what was commonly regarded as safe. However, improvements in storage facility design, and development of equipment for oxygen and carbon dioxide removal, and of computerized monitoring and gas control systems made low oxygen storage less risky (Hoehn et al., 2009). Modern storage techniques minimized fluctuations of gas composition in CA storage rooms, thereby reducing risk of low oxygen injuries to fruit, and have led to the development use of several low oxygen storage strategies.

These strategies include ultralow oxygen (ULO) where oxygen concentrations less than 2% but well above the anaerobic compensation point (ACP) or low oxygen limit (LOL) below which anaerobic respiration can result in injury to the fruit

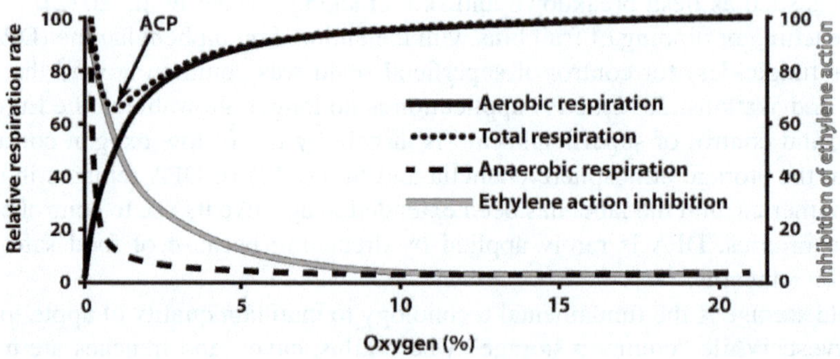

FIGURE 1.12

Aerobic, anaerobic, and overall respiration rates and relative inhibition of ethylene action of a generalized plant product in response to O_2. The anaerobic compensation point *(ACP)* approximates the lowest O_2 concentration achievable without injurious fermentation and represents the concentration that results in the maximum storage potential.

(Figure 1.12). Recommendations of oxygen concentrations in ULO remained empirically determined. The use of initial low oxygen stress (ILOS) in conjunction with ULO storage was also developed (Zanella, 2003). ILOS involves reducing oxygen concentrations to less than 0.5% for periods of up to 2 weeks, resulting in short-term ethanol accumulations that are thought to inhibit development of superficial scald. ILOS was further modified to involve repeated LOS (RLOS) during storage in ULO (Zanella and Stürz, 2013).

A major breakthrough in CA technology occurred with advances in the ability to sense the fruit responses to low oxygen in the storage environment in real time. It is possible to adjust the oxygen concentrations to those that are close to the ACP or LOL based on responses of the fruit of each cultivar to low oxygen, thereby reducing or eliminating quality loss associated with ripening without losses caused by anaerobic respiration (Figure 1.12). The tolerance of each cultivar to low oxygen is a function of factors such as skin resistance to gas diffusion, intercellular air space volume and the permeance to gas diffusion of inner tissues, and different thresholds for anaerobic metabolism (Watkins, 2020).

To date, the dynamic controlled atmosphere (DCA) technologies available to storage operators are chlorophyll fluorescence (DCA-CF) (Prange et al., 2013), measurement of the respiratory quotient (DCA-RQ) (Bessemans et al., 2016), and determination of ethanol production by the fruit (DCA-Eth) (Veltman et al., 2003). In addition, a new and simplified approach based on carbon dioxide production (DCA-CD) has been developed (Thewes et al., 2020), and it has been integrated with storage temperature adjustment that will maximize the effects of DCA. Regardless of the method, the oxygen concentration in the storage is lowered until a response signal (fluorescence, RQ, ethanol accumulation, or increased carbon dioxide production) indicates that the fruit is "stressed." Then, oxygen is typically increased to about 0.2% above the LOL/ACP to avoid low oxygen damage to the fruit. Moreover, continued measurement of the responses of fruit to low oxygen can be used to identify changes in fruit physiology during prolonged storage and the optimized storage atmospheres.

A driving factor for the implementation of DCA technology, especially in Europe, was control of superficial scald (Zanella and Stürz, 2013). Subsequently, DCA has been shown to reduce the incidence of other disorders such as flesh browning, core browning, core flush, and watercore depending on the cultivar (Nock et al., 2019; Weber et al., 2019). However, DCA storage may increase incidence of carbon dioxide-related disorders, especially if fruit has been treated with 1-MCP (as discussed later).

The second supplementary technology that has had a large impact on apple storage is postharvest use of 1-MCP. By inhibiting ethylene perception, 1-MCP prevents ethylene-induced ripening, especially softening. Although 1-MCP can be used on several horticultural products, its commercial success can be attributed in large part to results for apples. A high-quality apple is one that is close to that at the time of harvest and issues of irreversible inhibition of ripening are of less concern. This contrasts with fruits such as avocado, banana, European pear and tomato, which require ripening, e.g., softening and color change, to be acceptable by the consumer. Apples are also stored for long periods of time in large scale facilities, providing rooms that can be easily sealed for the 1-MCP treatment process. Early concerns about negative effects of 1-MCP on consumer acceptance because of reduced volatile production have not been realized (Marin et al., 2009; Stanich et al., 2014), perhaps in part because of the main effects on texture and because volatiles in untreated fruit can be associated with overripening.

The complexity of the apple industry is such that 1-MCP has affected postharvest chemical treatments, storage temperature regimes, storage atmospheres, and planned storage duration based on fruit maturity at harvest (Mattheis, 2008). In addition to the effect of 1-MCP on maintaining quality, a driving factor in its adoption as with DCA was control of superficial scald and therefore as a replacement for DPA treatment of fruit (Watkins, 2017). Depending on cultivar, however, 1-MCP can increase external and internal carbon dioxide injuries of susceptible cultivars (Fawbush et al., 2008), as well as increase incidence of flesh browning disorders of late-harvested Empire and Gala apples (Doerflinger et al., 2015; Lee et al., 2016). Management strategies have been developed to reduce risk of carbon dioxide injuries, including delayed CA, maintenance of very low carbon dioxide concentrations in storage, or ironically, increased use of DPA, which inhibits carbon dioxide injury, where allowed (Fawbush et al., 2008).

The extent of use of DCA and/or 1-MCP varies depending on factors such as growing region and cultivar. The specific effects of each technology on fruit quality and each technology has advantages and disadvantages (Watkins, 2017). These include the quality of the storage facilities available given that DCA requires storage rooms that are very airtight to maintain very low oxygen concentrations, the ease of 1-MCP application in a variety of room types or directly on bins as sachets, the unavailability of 1-MCP for organic fruit, and specific cases where 1-MCP increases susceptibility of cultivars to physiological disorders.

An important difference is that 1-MCP maintains fruit condition after removal from storage and during the marketing chain, whereas the residual effects of DCA on conditions can be lost over time. Both DCA and 1-MCP can be useful in combination during storage depending on the cultivar. 1-MCP treatment after DCA storage might also be useful to maintain fruit quality after DCA storage (DeEll and Lum, 2017).

Different decision strategies for postharvest storage technologies have been developed by various industries, but the most sophisticated decision strategies have been developed at the Laimburg Research Centre for use of CA, ULO, RLOS, DCA, or ILOS management using DCA (DCA-light) and 1-MCP treatment (Figure 1.13; Zanella and Stürz, 2013). Decisions for the technology or treatment used are based on orchard factors such as tree age, crop load, and maturity status of the fruit during the harvest window. For example, the longest-term storage periods are based on DCA at the start of the harvest window, and 1-MCP and then DCA + 1-MCP as the harvest window progresses.

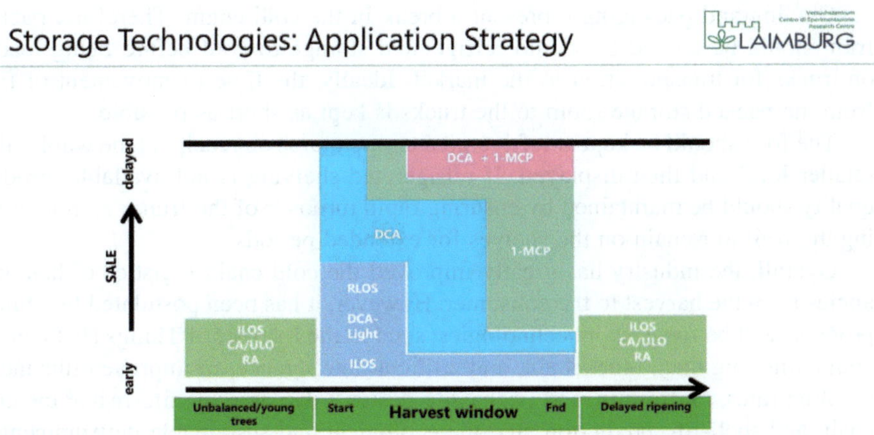

FIGURE 1.13

A stylized outline of storage technology strategies developed at the Laimburg Research Centre for apples.

(From Angelo Zanella.)

Regardless of storage or postharvest treatment method, apples are removed from storage to be graded and packed (or presized and returned to storage). As with all handling processes described in this chapter, variations in management procedures vary widely internationally with contrasts from small operations where all handling, sorting, and packing steps are done by hand to large-scale highly automated systems. In most modern packing lines, hydro handling, where apple bins are immersed in water fumes so the fruit float out, is the most common method used to empty the bins. Tipping fruit from bins directly onto conveyor belts is less common because of greater risk of injuries to the fruit. The fruit then floats through water flumes (sometimes chlorinated) to a conveyor belt and then to a sorting line where fruit that are undersized, defective, damaged, or decayed are removed.

The fruit is then graded by hand or by computer-controlled automatic optical technology into size and color (% coverage and intensity of red coloration) categories that meet specified apple grade standards. For example, in the U.S., requirements to meet U.S. Extra Fancy, U.S. Fancy, U.S. No. 1 standards by color and other criteria such as tolerances for skin blemishes and disorders are defined by regulation (USDA, 2019). Use of internal quality analyzers to detect internal disorders of the fruit is increasing. Rejected fruit are directed to processing as pies, sauce, juice, or ciders depending on the fruit quality.

The fruit is then washed with sanitizers, rinsed, and dried. Fruit may be sprayed with an edible skin coating, food grade wax, depending on the country and market. In the U.S., for example, waxing of apples is common for conventionally, but not organically, grown fruit to improve appearance, reduce water loss and respiration rates, and extend the life of the fruit (Bai et al., 2003; Ul Hasan et al., 2024).

Grading and packaging represent a break in the cold chain. Therefore, packed fruit should be moved back into refrigerated storage to cool before being loaded on trucks for transportation to the market. Ideally, the time of movement of fruit from the packed storage room to the trucks is kept as short as possible.

The fruit should be kept at refrigerated temperatures on receipt at the wholesaler/ retailer level and then displayed. If refrigerated shelving is not available, product quality should be maintained by ensuring rapid turnover of the fruit, i.e., not allowing the fruit to remain on the shelves for extended periods.

Overall, the industry has greatly improved the cold chain logistics of handling apples from the harvest to the consumer. However, it has been postulated that future progress will be focused on technologies, such as the Internet of Things (IoT), cloud computing, big data, blockchain, and artificial intelligence, to improve efficiencies (cooling rates, uniformity, and packaging design), dynamic monitoring of the cold chain and shelf-life prediction, and the ecological and sustainable development of the cold chain through reduced energy consumption, cost, and environmental impact (Han et al., 2021). The extent to which such a vision is realized will ultimately be derived from cost/benefit analysis within a very diverse apple industry, both in terms of scale and pricing.

8. Main diseases and pests

Apple production is challenged by various diseases and pests that can severely affect yields and fruit quality. Apples are attacked by fungal, bacterial including phytoplasma, and viral and viroid diseases, as well as insects and mites. Effective management of these threats is crucial to ensure yield and fruit quality. A compendium of apple and pear diseases and pests including management was presented by Sutton et al. (2014).

8.1 Fungal diseases

Apple scab caused by *Venturia inaequalis* is the most important disease of apple worldwide (Bowen et al., 2011). It occurs everywhere in the world where apples are grown and results in more losses than any other apple disease (Gauthier, 2018). The dominating primary inoculum in spring is ascospores originating from bodies in leaf litter (Philion et al., 2023). Symptoms are mainly visible on leaves, severely infected fruit dropped from the tree, and fruit (Figure 1.14).

Powdery mildew caused by *Podosphaera leucotricha* may overwinter in buds, and mycelium growth starts after bud break and spreads across the developing shoot and leaf tissue (Strickland et al., 2021). Foliar infections contribute most to disease spread and are the focus of most management actions. Leaves and twigs are often covered with a white, powdery fungal growth. Furthermore, infestation can lead to shoot malformation, stunting leaves, and fruit may develop a russeted appearance.

FIGURE 1.14

Apple scab on leaves of *Malus floribunda* 821, the donor of the scab resistance gene *Rvi6*.

This disease is more prevalent in dry conditions, unlike many other fungal diseases that thrive in wet environments like apple scab.

Other important fungal disease are apple cankers caused by *Neofabraea, Neonectria*, and other fungi (Saville and Olivieri, 2019), Alternaria leaf blotch and fruit spot caused by *Alternaria* ssp. (Prechsl et al., 2023), Marssonina blotch caused by *Diplocarpon coronariae* (Wöhner and Emeriewen, 2019) (Figure 1.15), black rot caused by *Diplodia seriata* (Trapman et al., 2008), brown rot diseases caused by *Monilinia* ssp. (Holb, 2008), and sooty blotch and flyspeck caused by a multispecies complex (Weber et al., 2016).

Sanitation measures such as removing and destroying fallen leaves and pruned branches can reduce sources of infection. Furthermore, proper pruning to ensure good air circulation can reduce humidity levels favorable to fungi. Nevertheless, management of fungal diseases relies primarily on multiple synthetic or organic fungicide applications per season (Strickland et al., 2023). Decision support systems (DSS), like NEWA (Carroll et al., 2017) or RIMpro (Wallhead and Zhu, 2017), to forecast infection periods, are a common component of integrated pest management, to reduce pesticide applications (Strickland et al., 2023).

FIGURE 1.15

Symptoms of *Diplocarpon coronariae* on apple leaves.

The growing of resistant cultivars is another possibility to reduce infestation of fungal diseases. Breeding programs start to focus on pyramided resistances for more durable resistance to diseases (Baumgartner et al., 2015). The most important targets are resistance to scab and mildew, for which several resistance genes are known (Hanke et al., 2020).

8.2 Bacterial diseases including phytoplasma

Fire blight, caused by *Erwinia amylovora*, is the most threatening bacterial disease in apple production. In 2000, a fire blight epidemic that occurred in Michigan was estimated to have caused the death of 350,000 to 450,000 trees, resulting in a total economic loss of $42 million (Norelli, 2017). *E. amylovora* can infect blossoms, stems, shoots, roots, and unripe fruits. Bacteria are distributed by insect, wind, rain, or mechanically. Characteristic symptoms are necrotic wilted shoots, forming a shepherd's crook-like appearance and the production of bacterial ooze (Peil et al., 2009). Effective management of fire blight requires the integration of several practices aimed at reducing the amount of initial inoculum, imposing barriers to successful establishment of the pathogen, and reducing host susceptibility to infection. Wild *Malus* species accessions are used in apple research and breeding to identify and introduce fire blight resistance into *M. domestica* (Peil et al., 2021). Challenges in disease management and host resistance breeding are summarized by Rezzonico et al. (2024).

Other bacterial pathogens are *Pseudomonas syringae* ssp. causing bacterial blossom blast and blister spot and *Agrobacterium rhizogenes* causing crown galls (Sutton et al., 2014).

Apple proliferation is caused by the phloem-dwelling cell wall-less bacterium *Candidatus Phytoplasma mali* (Görg et al., 2021). The proliferation of axillary buds leads to the formation of the typical symptom known as "the witches' broom." Transmission of the disease to healthy apple trees was proven for the summer apple psyllid *Cacopsylla picta*. Distribution of the disease can be prevented by controlling the vector *C. picta*. A promising approach is the use of resistant rootstocks (Seemüller et al., 2018).

8.3 Viruses and viroids

Some viruses and viroids have been described to cause diseases in apple. The three latent viruses Apple chlorotic leaf spot virus (ACLSV), Apple stem grooving virus (ASGV) and Apple stem pitting virus (ASPV) are widespread in apple-growing regions around the world. Although no symptoms are usually visible, the combinations of latent viruses in a tree can reduce the yield by 12% to 30%. The Apple scar skin viroid (ASSVd) can cause scars on fruits, the typical symptom. The use of scion and rootstock material certified free of viruses and viroids is the most efficient way to control these pathogens (Sutton et al., 2014).

8.4 Arthropod pests

Arthropods like moths, fruit flies and midges, saw flies and chalcids, leafrollers and fruitworms, bugs, aphids and psyllids, leafminers, weevils, beetles, and scales can damage leaves, fruits, stem, and trunk of apple trees (Sutton et al., 2014). A few main arthropods affecting apple are presented as examples.

Codling moth, *Cydia pomonella*, is a serious insect pest in apple production. Larvae of *C. pomonella* bore through the fruit to the core and feed on the seeds, leaving holes filled with feces (Sutton et al., 2014), making damaged apple unmarketable. Codling moth develops one to four generations per year depending on the cultivation area and the climatic conditions (Kadoić Balaško et al., 2020). Codling moth can be controlled with chemicals, granulosis virus, the use of pheromones and kairomones, sterile males to disrupt mating, and natural enemies (Kadoić Balaško et al., 2020).

Apple maggot flies, *Rhagoletis pomonella*, oviposit their eggs in the skin of developing apples. The emerging larvae tunnel throughout the apple, leaving small brownish threadlike trails. Apple maggot damage is in general not visible on the surface of apples. Fruits must be cut to see the characteristic irregular brown larval feeding trails. Insecticides and attract-and-kill strategies are used to manage the pest (Sutton et al., 2014).

The rosy apple aphid, *Dysaphis plantaginea*, causes leaf roll, which appear reddish, and fruit deformation. The aphids feed and reproduce inside the roll and are thus protected from insecticides. Management of the pest is possible by the application of synthetic pesticides or biological control with natural enemies (Sutton et al., 2014; Dib et al., 2016). The rosy leaf curling aphid, *Dysaphis devecta*, causes similar symptoms.

The San Jose scale, *Quadraspidiotus perniciosus*, a quarantine pest in several countries, feeds from tree sap. An abundant number of scales can thus weaken the plant and reduce fruit and shoot growth. Feeding on apples causes reddish spots under the scale, reduces the marketability, and prevents the export of apples in countries with a quarantine status for San Jose scale. A regular pesticide program cannot effectively control San Jose scale, while spraying mineral oil, alone or in combination with insecticides, or the use of pheromones for mating disruption are generally more effective (Maas et al., 2023).

8.5 General management practices

Production of apples with good quality and high yield relies on different orchard management systems, i.e., conventional, integrated, and ecological pest and disease management. Control of pests and diseases by conventional management is done mainly by chemical pesticides. Integrated pest management combines chemical, biological, and cultural methods to control pests and diseases sustainably using thresholds for the apple. The use of chemical pesticides is forbidden in ecological management of apple orchards' biological preparations, but also copper, natural predators, and parasitoids.

Mathis et al. (2022) reported that apple in Switzerland is among the crops with the highest Plant Protection Products (PPP) treatment frequency and the highest PPP amounts applied per hectare, also in organic systems. The reduction of PPPs is challenging because the crop cannot be rotated annually, which encourages an increase in infestation from one year to the next (Simon et al., 2010; Mathis et al., 2022).

Nevertheless, apple growers face a situation where the reduction of PPPs is required, e.g., one aim of the European Green Deal is the 50% reduction of chemical pesticides by 2030, so the availability of PPPs is decreasing and pathogens can develop resistance to chemical PPPs.

9. Current and future challenges

Apple production in Europe changed in the last century. In the 1940s, apple cultivation was about a hundred local varieties, well-adapted to different environments, but less prone to a large-scale industrialized cultivation. Breeding programs specifically oriented to select novel accessions with superior fruit quality and storability changed apple production today to a focus on about 10 cultivars. Internationally, Golden Delicious, Fuji, Gala, Cripps Pink, and Braeburn are the most current cultivated apple varieties. These well-known cultivars free from royalties changed the apple panorama represented by the recent release on the market of more than 60 club varieties generated by almost 100 ongoing breeding program worldwide.

According to the recent estimate of the World Apple and Pear Association (WAPA), 75% of the apple production in Europe is represented by five countries (in order of contribution) Poland, Italy, France, Germany, and Spain. Apple cultivation and production are interested in important innovations to face important challenges, such as the introgression of resistance to diseases, in particular apple scab. The identification and selection of resistant cultivars as parents in controlled crosses is a major focus in apple improvement, since resistant cultivars are certainly more adapted to a sustainable cultivation requiring much less chemical treatment to prevent apple scab. In addition to quality and disease resistance, attention is also given to promote plant resilience. Climate change will severely affect the apple cultivation and fruit quality. Reduced availability of water, increasing temperature, and more frequent incidence of freezing events in spring will force a redesign of a novel set of varieties to maintain plant performance and fruit quality characteristics.

Genetic research has not only focused on the improvement of scion cultivars, but also on rootstocks. In the European countries, M.9 represents the most grafted rootstock to date. However, due to the increased incidence of fire blight (bacterial pathogen *Erwinia amylovora*) and the spread of the soil fatigue disease (also known as replant disease), several resistant or less-susceptible genotypes have been tested in experimental trials, representing a possible valuable alternative rootstock option.

Apple innovation is also oriented to select novel accessions distinguished by a more singular and intense coloration for promoting the quality of apples grown at lower altitude, affected by insufficient fruit coloration, especially with the mountain environment. To this end, another important challenge is the selection of novel varieties with a reduced need of chilling units to promote and extend the apple cultivation area in Brazil and Israel, characterized by mild winter temperatures.

One of the major variety innovations is the release of red-flesh apples. The mild-to-intense red coloration of the apple flesh can be naturally retrieved from the wild apple gene bank, and this feature, lost during domestication and early selection, is now presented as an important innovation. The red coloration is determined by a tissue-specific accumulation of polyphenolic compounds, in particular anthocyanin. This class of secondary metabolites is known to play an important antioxidant role, contributing to scavenging excessive accumulation of reactive oxygen species (ROS). Because of this mechanism, the increased concentration of these molecules may represent an important biofortification of apple and its functional quality promoting human health.

References

Aćimović, S. G., Santander, R. D., Meredith, C. L., & Pavlović, Ž. M. (2023). Fire blight rootstock infections causing apple tree death: A case study in high-density apple orchards with *Erwinia amylovora* strain characterization. *Postharvest Biology and Technology, 2*, 1082204.

Al Shoffe, Y., Nock, J. F., Baugher, T. A., Marini, R. P., & Watkins, C. B. (2020). Bitter pit and soft scald development during storage of unconditioned and conditioned 'Honeycrisp' apples in relation to mineral contents and harvest indices. *Postharvest Biology and Technology, 160*, 9.

Al Shoffe, Y., Nock, J. F., Gunes, N., Maloney, K., Brown, S. K., & Watkins, C. B. (2024). Antioxidants of blushed and unblushed sides in relation to superficial scald of 'NY2' (RubyFrost®) apples treated preharvest with 1-methylcyclopropene and aminoethoxyvinylglycine. *Scientia Horticulturae, 327*, 11.

Angeli, L., Populin, F., Morozova, K., et al. (2024). Comparative analysis of antioxidant activity and capacity in apple varieties: Insights from stopped flow DPPH• kinetics, mass spectrometry and electrochemistry. *Food Bioscience*, 103729.

Büchele, F., Hivare, K., Khera, K., Thewes, F. R., Argenta, L. C., Hoffmann, T. G., et al. (2024). Novel energy-saving strategies in apple storage: A review. *Sustainability, 16*(3), 11. https://doi.org/10.3390/su16031052.

Bühlmann-Schütz, S., Hodel, M., Dorfmann, E., Vonmetz, L., Lussi, L., & Patocchi, A. (2023). Comparison between artificial fire blight shoot and flower inoculations in apple. *Journal of Plant Pathology*, 1−10.

Bai, J. H., Hagenmaier, R. D., & Baldwin, E. A. (2003). Coating selection for 'Delicious' and other apples. *Postharvest Biology and Technology, 28*(3), 381−390.

Baldi, P., Buti, M., Gualandri, V., Khomenko, I., Farneti, B., Biasioli, F., et al. (2024). Transcriptomic and volatilomic profiles reveal *Neofabraea vagabunda* infection-induced changes in susceptible and resistant apples during storage. *Postharvest Biology and Technology, 212*, 112889.

Baumgartner, I. O., Patocchi, A., Frey, J. E., Peil, A., & Kellerhals, M. (2015). Breeding elite lines of apple carrying pyramided homozygous resistance genes against apple scab and resistance against powdery mildew and fire blight. *Plant Molecular Biology Reporter, 33*(5), 1573−1583.

Bessemans, N., Verboven, P., Verlinden, B. E., & Nicolaï, B. M. (2016). A novel type of dynamic controlled atmosphere storage based on the respiratory quotient (RQ-DCA). *Postharvest Biology and Technology, 115*, 91−102.

Boller, T., Herner, R. C., & Kende, H. (1979). Assay for and enzymatic formation of an ethylene precursor, 1-aminocyclopropane-1-carboxylic acid. *Planta, 145*(3), 293−303.

Bowen, J. K., Mesarich, C. H., Bus, V. G., Beresford, R. M., Plummer, K. M., & Templeton, M. D. (2011). *Venturia inaequalis*: The causal agent of apple scab. *Molecular Plant Pathology, 12*(2), 105−122.

Brookfield, P. L., Hughes, J. G., & Watkins, C. B. (1993). Fruit maturity and quality of selectively picked Royal Gala apples. *New Zealand Journal of Crop and Horticultural Science, 21*(2), 183−187.

Broggini, G. A. L., Schlathölter, I., Russo, G., Copetti, D., Yates, S. A., Studer, B., & Patocchi, A. (2020). Chromosome-scale de novo diploid assembly of the apple cultivar 'Gala Galaxy'. *bioRxiv*, 2020-04. https://doi.org/10.1101/2020.04.25.058891

Bus, V. G., Rikkerink, E. H., Caffier, V., Durel, C. E., & Plummer, K. M. (2011). Revision of the nomenclature of the differential host-pathogen interactions of *Venturia inaequalis* and *Malus*. *Annual Review of Phytopathology, 49*, 391−413.

Cai, W. L. H., Al Shoffe, Y., Park, D., & Watkins, C. B. (2023). Harvest maturity and preharvest aminoethoxyvinylglycine treatment effects on cold-induced ethylene production of 'Gala' aples. *Hortscience, 58*, 532−538.

Cao, C., Gong, S., Li, Y., Tang, J., Li, T., & Zhang, Q. (2024). Pathogenic factors and mechanisms of the Alternaria leaf spot pathogen in apple. *Horticulturae, 10*(3), 212.

Carranca, C., Brunetto, G., & Tagliavini, M. (2018). Nitrogen Nutrition of Fruit Trees to Reconcile Productivity and Environmental Concerns. *Plants, 7*, 4. https://doi.org/10.3390/plants7010004.

Carroll, J., Weigle, T., Agnello, A., Reissig, H., Cox, K., Breth, D., et al. (2017). NEWA (Network for Environment and Weather Applications) provides fruit IPM and production tools from 400 weather stations. *New York Fruit Quarterly, 25*, 19–24.

Castric, V., et al. (2008). Repeated adaptive introgression at a gene under multiallelic balancing selection. *PLoS Genet, 4*, e1000168.

Chagné, D., Kirk, C., How, N., Whitworth, C., Fontic, C., Reig, G., et al. (2016). A functional genetic marker for apple red skin coloration across different environments. *Tree Genetics & Genomes, 12*, 1–9.

Chang, H. Y., & Tong, C. B. S. (2020). Identification of candidate genes involved in fruit ripening and crispness retention through transcriptome analyses of a 'Honeycrisp' population. *Plants (Basel), 9*(10), 1335. https://doi.org/10.3390/plants9101335

Chen, X., Li, S., Zhang, D., Han, M., Jin, X., Zhao, C., et al. (2019). Sequencing of a wild apple (*Malus baccata*) genome unravels the differences between cultivated and wild apple species regarding disease resistance and cold tolerance. *G3: Genes, Genomes, Genetics, 9*, 2051–2060.

Choi, C., Dougherty, L., & Ban, S. (2024). Tools and techniques for allele mining in apple. In *Allele mining for genomic designing of fruit crops* (pp. 80–102). CRC Press.

Cocetta, G., Beghi, R., Mignani, I., & Spinardi, A. (2017). Nondestructive apple ripening stage determination using the Delta Absorbance meter at harvest and after storage. *Hort Technology, 27*(1), 54–64.

Cornille, A., et al. (2013). Crop-to-wild gene flow and spatial genetic structure in the closest wild relatives of the cultivated apple. *Evolutionary Applications, 6*, 737–748.

Cornille, A., et al. (2012). New insight into the history of domesticated apple: Secondary contribution of the European wild apple to the genome of cultivated varieties. *PLoS Genet, 8*, e1002703.

Cornille, A., Giraud, T., Smulders, M. J., Roldán-Ruiz, I., & Gladieux, P. (2014). The domestication and evolutionary ecology of apples. *Trends in Genetics, 30*(2), 57–65.

Corollaro, M. L., Aprea, E., Endrizzi, I., Betta, E., Demattè, M. L., Charles, M., et al. (2014). A combined sensory-instrumental tool for apple quality evaluation. *Postharvest Biology and Technology, 96*, 135–144.

Costa, F., Cappellin, L., Longhi, S., Guerra, W., Magnago, P., Porro, D., et al. (2011). Assessment of apple (*Malus*× *domestica* Borkh.) fruit texture by a combined acoustic-mechanical profiling strategy. *Postharvest Biology and Technology, 61*, 21–28.

Costa, G., Botton, A., & Vizzotto, G. (2018). Fruit thinning: Advances and trends. In I. Warrington (Ed.), *Horticultural reviews*. https://doi.org/10.1002/9781119521082.ch4.

Daccord, N., Celton, J. M., Linsmith, G., Becker, C., Choisne, N., Schijlen, E., et al. (2017). High-quality de novo assembly of the apple genome and methylome dynamics of early fruit development. *Nature Genetics, 49*, 1099–1106.

Dan, C., Sestras, A. F., Bozdog, C., & Sestras, R. E. (2015). Investigation of wild species potential to increase genetic diversity useful for apple breeding. *Genetika, 47*, 993–1011.

DeEll, J. R., & Ehsani-Moghaddam, B. (2010). Preharvest 1-methylcyclopropene treatment reduces soft scald in 'Honeycrisp' apples during storage. *HortScience, 45*(3), 414–417.

DeEll, J. R., & Lum, G. B. (2017). Effects of low oxygen and 1-methylcyclopropene on storage disorders of 'Empire' apples. *HortScience, 52*, 1265–1270.

DeLong, J., Harrison, P., & Harkness, L. (2020). An optimal harvest maturity model for 'Minneiska' apple fruit based on the delta-absorbance meter. *The Journal of Horticultural Science and Biotechnology, 95*, 637–644.

Devoghalaere, F., et al. (2012). A genomics approach to understanding the role of auxin in apple (*Malus x domestica*) fruit size control. *BMC Plant Biology, 12*, 1–15.

Dib, H., Sauphanor, B., & Capowiez, Y. (2016). Effect of management strategies on arthropod communities in the colonies of rosy apple aphid, Dysaphis plantaginea Passerini (Hemiptera: Aphididae) in South-Eastern France. *Agriculture, Ecosystems & Environment, 216*, 203–206.

do Amarante, C. V. T., Argenta, L. C., de Freitas, S. T., & Steffens, C. A. (2022). Efficiency of pre-harvest application of 1-MCP (HarvistaTM 1.3 SC) to delay maturation of 'Cripps Pink' apple fruit. *Scientia Horticulturae, 293*, 5.

Doebley, J. F., Gaut, B. S., & Smith, B. D. (2006). The molecular genetics of crop domestication. *Cell, 127*, 1309–1321.

Doerflinger, F. C., Al Shoffe, Y., Sutanto, G., Nock, J. F., & Watkins, C. B. (2024). Preharvest 1-methylcyclopropene (1-MCP) treatment effects on quality of spot and strip picked 'Gala' apples at harvest and after storage as affected by postharvest 1-MCP and temperature conditioning treatments. *Scientia Horticulturae, 325*, 11.

Doerflinger, F. C., Miller, W. B., Nock, J. F., & Watkins, C. B. (2015). Relationships between starch pattern indices and starch concentrations in four apple cultivars. *Postharvest Biology and Technology, 110*, 86–95.

Doerflinger, F. C., Rickard, B. J., Nock, J. F., & Watkins, C. B. (2015). An economic analysis of harvest timing to manage the physiological storage disorder firm flesh browning in 'Empire' apples. *Postharvest Biology and Technology, 107*, 1–8.

Duan, N., Bai, Y., Sun, H., et al. (2017). Genome re-sequencing reveals the history of apple and supports a two-stage model for fruit enlargement. *Nature Communications, 8*, 249.

Duan, Y., Wang, G. B., Fawole, O. A., Verboven, P., Zhang, X. R., Wu, D., et al. (2020). Postharvest precooling of fruit and vegetables: A review. *Trends in Food Science and Technology, 100*, 278–291.

Endrizzi, I., Torri, L., Corollaro, M. L., Demattè, M. L., Aprea, E., Charles, M., et al. (2015). A conjoint study on apple acceptability: Sensory characteristics and nutritional information. *Food Quality and Preference, 40*, 39–48.

Espley, R. V., Hellens, R. P., Putterill, J., Stevenson, D. E., Kutty-Amma, S., & Allan, A. C. (2007). Red colouration in apple fruit is due to the activity of the MYB transcription factor, MdMYB10. *The Plant Journal, 49*, 414–427.

FAO. (2023). Production: Crops and livestock products. In *Rome* [Cited December 2023] https://www.fao.org/faostat/en/#data/QCL.

Farneti, B., Gutierrez, M. S., Novak, B., Busatto, N., Ravaglia, D., Spinelli, F., et al. (2015). Use of the index of absorbance difference (IAD) as a tool for tailoring post-harvest 1-MCP application to control apple superficial scald. *Scientia Horticulturae, 190*, 110–116.

Farneti, B., Masuero, D., Costa, F., Magnago, P., Malnoy, M., Costa, G., et al. (2015). Is there room for improving the nutraceutical composition of apple? *Journal of Agricultural and Food Chemistry, 63*, 2750–2759.

Fawbush, F., Nock, J. F., & Watkins, C. B. (2008). External carbon dioxide injury and 1-methylcyclopropene. *Postharvest Biology and Technology, 48*, 92–98.

Flachowsky, H., Le Roux, P. M., Peil, A., Patocchi, A., Richter, K., & Hanke, M. V. (2011). Application of a high-speed breeding technology to apple (*Malus× domestica*) based on transgenic early flowering plants and marker-assisted selection. *New Phytologist, 192*, 364–377.

Forsline, P. L., Aldwinckle, H. S., Dickson, E. E., Luby, J. J., & Hokanson, S. C. (2003). *Collection, maintenance, characterization, and utilization of wild apples of Central Asia. Horticultural Reviews* vol. 29. Westport then New York: John Wiley & Sons. https://doi.org/10.1002/9780470650868.

Görg, L. M., Gallinger, J., & Gross, J. (2021). The phytopathogen '*Candidatus Phytoplasma Mali*' alters apple tree phloem composition and affects oviposition behavior of its vector *Cacopsylla picta. Chemoecology, 31*, 31–45.

Gardiner, S. E., et al. (2007). Apple. In C. Kole (Ed.), *Genome mapping and molecular breeding in plants, fruits and nuts* (pp. 1–62). Springer.

Gauthier, N. (2018). Symptoms and signs. *The Plant Health Instructor, 18.*

Gross, B. L., et al. (2012). Identification of interspecific hybrids among domesticated apple and its wild relatives. *Tree Genetics & Genomes, 8*, 1223–1235.

Gross, B. L., & Olsen, K. M. (2009). Genetic perspectives on crop domestication. *Trends in Plant Science, 15*, 529–537.

Gutierrez, B. L., Arro, J., Zhong, G. Y., & Brown, S. K. (2018). Linkage and association analysis of dihydrochalcones phloridzin, sieboldin, and trilobatin in *Malus. Tree Genetics & Genomes, 14*, 1–11.

Harker, F. R., Kupferman, E. M., Marin, A. B., Gunson, F. A., & Triggs, C. M. (2008). Eating quality standards for apples based on consumer preferences. *Postharvest Biology and Technology, 50*(1), 70–78. https://doi.org/10.1016/j.postharvbio.2008.03.020

Höfer, M., & Meister, A. (2010). Genome size variation in *Malus* species. *Journal of Botany, 2010*, 480873.

Han, J. W., Zuo, M., Zhu, W. Y., Zuo, J. H., Lu, E. L., & Yang, X. T. (2021). A comprehensive review of cold chain logistics for fresh agricultural products: Current status, challenges, and future trends. *Trends in Food Science and Technology, 109*, 536–551.

Hanke, M.-V., Flachowsky, H., Peil, A., & Emeriewen, O. F. (2020). *Malus × domestica* apple. In R. Litz, F. Pliego-Alfaro, & J. I. Hormaza (Eds.), *Biotechnology of fruit and nut crops. Biotechnology in agricultural series,* 2nd ed., vol. 29 (pp. 440–473). Wallingford: CAB International.

Harb, J., Gapper, N. E., Giovannoni, J. J., & Watkins, C. B. (2012). Molecular analysis of softening and ethylene synthesis and signaling pathways in a non-softening apple cultivar, 'Honeycrisp' and a rapidly softening cultivar, 'McIntosh'. *Postharvest Biology and Technology, 64*(1), 94–103.

Harris, S. A., Robinson, J. P., & Juniper, B. E. (2002). Genetic clues to the origin of the apple. *Trends Genet, 18*, 426–430.

Harteveld, D. O. C., Goedhart, P. W., Houwers, I., Köhl, J., de Jong, P. F., & Wenneker, M. (2023). Detecting the asymptomatic colonization of apple branches by Neonectria ditissima, causing European canker of apple. *European Journal of Plant Pathology, 166*, 291–301.

Hoehn, E., Prange, R. K., & Vigneault, C. (2009). Storage technology and applications. In E. M. Yahia (Ed.), *Modified and controlled atmospheres for the storage, transportation, and packaging of horticultural commodities* (pp. 17–72). Baca Raton: CRC Press.

Holb, I. (2008). Brown rot blossom blight of pome and stone fruits: Symptom, disease cycle, host resistance, and biological control. *International Journal of Horticultural Science, 14*, 15–21.

Hussein, Z., Fawole, O. A., & Opara, U. L. (2020). Harvest and postharvest factors affectingbruise damage of fresh fruits. *Horticultural Plant Journal, 6*, 1–13.

Ilyukhin, E., Schneider, K., & Ellouze, W. (2022). First report of *Botryosphaeria dothidea* causing stem canker and dieback of apple trees in Ontario, Canada. *Plant Disease, 106*(11), 2.

Johnston, J. W., Gunaseelan, K., Pidakala, P., Wang, M., & Schaffer, R. J. (2009). Co-ordination of early and late ripening events in apples is regulated through differential sensitivities to ethylene. *Journal of Experimental Botany, 60*, 2689–2699.

Johnston, J. W., Hewett, E. W., Hertog, M., & Harker, F. R. (2001). Temperature induces differential softening responses in apple cultivars. *Postharvest Biology and Technology, 23*(3), 185–196.

Kadoić Balaško, M., Bažok, R., Mikac, K. M., Lemic, D., & Pajač Živković, I. (2020). Pest management challenges and control practices in codling moth: A review. *Insects, 11*(1), 38.

Khodadadi, F., Santander, R. D., McHenry, D. J., Jurick, W. M., & Aćimović, S. G. (2023). A bitter, complex problem: Causal *Colletotrichum* species in Virginia orchards and apple fruit susceptibility. *Plant Disease, 107*, 3164–3175.

Knee, M., Looney, N. E., Hatfield, S. G. S., & Smith, S. M. (1983). Initiation of rapid ethylene synthesis by apple and pear fruits in relation to storage-temperature. *Journal of Experimental Botany, 34*, 1207–1212.

Kupferman, E. (2003). Controlled atmosphere storage of apples and pears. *Acta Horticulturae, 600*, 729–735.

Lee, J., Mattheis, J. P., & Rudell, D. R. (2019). High storage humidity affects fruit quality attributes and incidence of fruit cracking in cold-stored 'Royal Gala' apples. *HortScience, 54*, 149–154.

Lee, J., Mattheis, J. P., & Rudell, D. R. (2016). Storage temperature and 1-methylcyclopropene treatment affect storage disorders and physiological attributes of 'Royal Gala' apples. *Hortscience, 51*, 84–93.

Lee, S. Y., Peter, K. A., Das, K., Diane, A. R., & Jung, H. Y. (2023). The rapid apple decline phenomenon: Current status and expected associated factors in Korea. *The Plant Pathology Journal, 39*, 538.

Lempe, J., Peil, A., & Flachowsky, H. (2022). Time-resolved analysis of candidate gene expression and ambient temperature during bud dormancy in apple. *Frontiers in Plant Science, 2995*.

Li, J-c, Liu, J-q, & Gao, X-f (2022). A revision of the genus *Malus* Mill. (Rosaceae). *European Journal of Taxonomy, 853*, 1–127.

Little, C. R., & Holmes, R. (2000). *Storage technology for apples and pears: A guide to production, postharvest treatment and storage of pome fruit in Australia*. In J. Faragher (Ed.). Victoria: Dept. of Natural Resources and Environment. https://nla.gov.au/nla.cat-vn556873.

Liu, D.-D., Wang, D.-R., Yang, X.-Y., Zhao, C.-H., Li, S.-H., Sha, G.-L., et al. (2023). Apomictic *Malus* plants exhibit abnormal pollen development. *Frontiers in Plant Science, 14*, 1065032.

Liu, Z., Gao, Y., Wang, K., Feng, J., Sun, S., Lu, X., et al. (2024). Identification of S-RNase genotype and analysis of its origin and evolutionary patterns in Malus plants. *Journal of Integrative Agriculture, 23*, 1205–1221.

Longhi, S., Moretto, M., Viola, R., Velasco, R., & Costa, F. (2012). Comprehensive QTL mapping survey dissects the complex fruit texture physiology in apple (*Malus x domestica* Borkh.). *Journal of Experimental Botany, 63*, 1107–1121.

Luo, F., Norelli, J. L., Howard, N. P., Wisniewski, M., Flachowsky, H., Hanke, M. V., et al. (2020). Introgressing blue mold resistance into elite apple germplasm by rapid cycle breeding and foreground and background DNA-informed selection. *Tree Genetics & Genomes, 16*, 1–15.

Lurie, S., & Watkins, C. B. (2012). Superficial scald, its etiology and control. *Postharvest Biology and Technology, 65*, 44–60.

Maas, J., Huang, J., Leach, H., & Wilson, J. K. (2023). Evaluating pheromone dispenser density and longevity for San Jose scale mating disruption in apple orchards. *Crop Protection, 172*, 106337.

Mansfeld, B. N., Yocca, A., Ou, S., Harkess, A., Burchard, E., Gutierrez, B., et al. (2023). A haplotype resolved chromosome-scale assembly of North American wild apple *Malus fusca* and comparative genomics of the fire blight *Mfu10* locus. *The Plant Journal, 116*(4), 989–1002.

Marin, A. B., Colonna, A. E., Kudo, K., Kupferman, E. M., & Mattheis, J. R. (2009). Measuring consumer response to 'Gala' apples treated with 1-methylcyclopropene (1-MCP). *Postharvest Biology and Technology, 51*(1), 73–79.

Marini, R. P., & Fazio, G. (2018). Apple rootstocks: History, physiology, management, and breeding. *Horticultural Reviews, 45*, 197–312.

Martin, P. L., Krawczyk, T., Khodadadi, F., Aćimović, S. G., & Peter, K. A. (2021). Bitter rot of apple in the Mid-Atlantic United States: Causal species and evaluation of the impacts of regional weather patterns and cultivar susceptibility. *Phytopathology, 111*, 966–981.

Martino, I., Agustí-Brisach, C., Nari, L., Gullino, M. L., & Guarnaccia, V. (2024). Characterization and pathogenicity of fungal species associated with dieback of apple trees in Northern Italy. *Plant Disease, 108*, 311–331.

Mathis, M., Blom, J. F., Nemecek, T., Bravin, E., Jeanneret, P., Daniel, O., et al. (2022). Comparison of exemplary crop protection strategies in Swiss apple production: Multi-criteria assessment of pesticide use, ecotoxicological risks, environmental and economic impacts. *Sustainable Production and Consumption, 31*, 512–528.

Mattheis, J. P. (2008). How 1-methylcyclopropene has altered the Washington state apple industry. *HortScience, 43*(1), 99–101.

McArtney, S. J., Obermiller, J. D., Hoyt, T., & Parker, M. L. (2009). 'Law Rome' and 'Golden Delicious' apples differ in their response to preharvest and postharvest 1-methylcyclopropene treatment combinations. *HortScience, 44*, 1632–1636.

McGlone, V. A., Jordan, R. B., Seelye, R., & Clark, C. J. (2003). Dry-matter - a better predictor of the post-storage soluble solids in apples? *Postharvest Biology and Technology, 28*, 431–435.

McHenry, D. J., & Aćimović, S. G. (2024). New species-specific real-time PCR assays for Colletotrichum species causing bitter rot of apple. *Microorganisms, 12*(5), 878.

McKey, D., Elias, M., Pujol, B., & Duputié, A. (2010). The evolutionary ecology of clonally propagated domesticated plants. *New Phytologist, 186*, 318–332.

Medina, S., Perestrelo, R., Pereira, R., & Câmara, J. S. (2020). Evaluation of volatilomic fingerprint from apple fruits to ciders: A useful tool to find putative biomarkers for each apple variety. *Foods, 9*(12), 1830.

Moran, R., DeEll, J., & Tong, C. B. S. (2020). Regional variation in the index of absorbance difference as an indicator of maturity and predictor of storage disorders in apples. *HortScience, 55*(9), 1500–1508.

Moriya, S. (2024). Practices of allele mining in apple. In Allele mining for genomic designing of fruit crops (pp. 20–79). CRC Press.

Naor, A., & Girona, J. (2012). Apple. In FAO (Ed.), *Crop yield response to water* (pp. 328–342).

Neuwald, D. A., & Streif, J. (2012). Apple ripening determined by conventional methods and the 'NSure' genetic assay. *Acta Horticulturae, 934*, 871–875.

Nock, J. F., Doerflinger, F. C., Sutanto, G., Al Shoffe, Y., Gunes, N., Zhang, Y., et al. (2019). Managing stem-end flesh browning, a physiological disorder of 'Gala' apples. *Acta Horticulturae, 1256*, 163–168.

Norelli, J. (2017). Bacterial diseases affecting apples. In E. Kate (Ed.), *Achieving sustainable cultivation of apple*, vol. 18, pp. 403–424). Burleigh Dodds Science Publishing. https://doi.org/10.19103/AS.2016.0017.19.

Olsen, K. M., & Wendel, J. F. (2013). A bountiful harvest: Genomic insights into crop domestication phenotypes. *Annual Review of Plant Biology, 64*, 47–70.

Orcheski, B., Hedderley, D., Hunt, M., Rowan, D., & Volz, R. (2023). Profiling apple volatile organic compounds in a New Zealand collection of germplasm as a resource for breeding cultivars with desirable flavors. *Euphytica, 219*, 116.

Palmer, J. W., Harker, F. R., Tustin, D. S., & Johnston, J. (2010). Fruit dry matter concentration: A new quality metric for apples. *Journal of the Science of Food and Agriculture, 90*, 2586–2594.

Parisi, L., Lespinasse, Y., Guillaumes, J., & Krüger, J. (1993). A new race of *Venturia inaequalis* virulent to apples with resistance due to the *Vf* gene. *Phytopathology, 83*, 533–537.

Patocchi, A., Keilwagen, J., Berner, T., Wenzel, S., Broggini, G. A., Altschmied, L., et al. (2021). No evidence of unexpected transgenic insertions in T1190–A transgenic apple used in rapid cycle breeding–following whole genome sequencing. *Frontiers in Plant Science, 12*, 715737.

Peil, A., Bus, V. G. M., Geider, K., Richter, K., Flachowsky, H., & Hanke, M. V. (2009). Improvement of fire blight resistance in apple and pear. *International Journal of Plant Breeding, 3*, 1–27.

Peil, A., Emeriewen, O. F., Khan, A., Kostick, S., & Malnoy, M. (2021). Status of fire blight resistance breeding in *Malus*. *Journal of Plant Pathology, 103*(1), 3–12.

Peil, A., Garcia-Libreros, T., Richter, K., Trognitz, F. C., Trognitz, B., Hanke, M. V., et al. (2007). Strong evidence for a fire blight resistance gene of *Malus* ×*robusta* located on linkage group 3. *Plant Breeding, 126*(5), 470–475.

Peil, A., Emeriewen, O. F., Khan, A., Kostick, S., & Malnoy, M. (2021). Status of fire blight resistance breeding in *Malus*. *Journal of Plant Pathology, 103*, 3–12.

Pertille, R. H., Citadin, I., Patto, L. S., Oldoni, T. L. C., Scariotto, S., Grigolo, C. R., et al. (2021). High-chilling requirement apple cultivar has more accentuated acrotony than low-chilling one in mild winter region. *Trees, 35*, 1135–1150.

Philion, V., Joubert, V., Trapman, M., & Stensvand, A. (2023). Physical modes of action of fungicides against apple scab: Timing is everything, but dose matters. *Plant Disease, 107*(10), 2949–2961.

Praeger, U., Jedermann, R., Sellwig, M., Neuwald, D. A., Hartgenbusch, N., Borysov, M., et al. (2020). Airflow distribution in an apple storage room. *Journal of Food Engineering, 269*, 11.

Prange, R. K., Wright, A. H., DeLong, J. M., & Zanella, A. (2013). History, current situation and future prospects for dynamic controlled atmosphere (DCA) storage of fruits and vegetables, using chlorophyll fluorescence. *Acta Horticulturae, 1012*, 905–915.

Prechsl, U. E., Rizzoli, W., Marschall, K., & Jasper Wubs, E. R. (2023). Fungicide-free management of Alternaria leaf blotch and fruit spot on apple indicates *Alternaria* spp. as secondary colonizer. *Scientific Reports, 13*(1), 8431.

Qian, G.-Z., Liu, L.-F., & Tang, G.-G. (2006). A new section in *Malus* (Rosaceae) from China. In *Annales botanici fennici* (pp. 68–73). JSTOR.

Qin, S., Xu, G., He, J., Li, L., Ma, H., & Lyu, D. (2023). A chromosome-scale genome assembly of *Malus domestica*, a multi-stress resistant apple variety. *Genomics, 115*(3), 110627.

Rezzonico, F., Emeriewen, O., Zeng, Q., et al. (2024). Burning questions for fire blight research: I. Genomics and evolution of *Erwinia amylovora* and analyses of host-pathogen interactions. *Journal of Plant Pathology, 106*, 797–810.

Richards, C., et al. (2009). Genetic diversity and population structure in *Malus sieversii*, a wild progenitor species of domesticated apple. *Tree Genetics & Genomes, 5*, 339–347.

Robinson, J., Harris, S., & Juniper, B. (2001). Taxonomy of the genus *Malus* Mill.(Rosaceae) with emphasis on the cultivated apple, *Malus domestica* borkh. *Plant Systematics and Evolution, 226*(1–2), 35–58.

Roeder, S., Serra, S., & Musacchi, S. (2021). Novel metrics to characterize in vitro pollen tube growth performance of apple cultivars. *Plants, 10*(7), 1460.

Saville, R., & Olivieri, L. (2019). Fungal diseases of fruit: Apple cankers in Europe. In *Integrated management of diseases and insect pests of tree fruit* (pp. 59–84). Burleigh Dodds Science Publishing.

Seemüller, E., Gallinger, J., Jelkmann, W., & Jarausch, W. (2018). Inheritance of apple proliferation resistance by parental lines of apomictic *Malus sieboldii* as donor of resistance in rootstock breeding. *European Journal of Plant Pathology, 151*, 767–779.

Serra, S., Leisso, R., Giordani, L., Kalcsits, L., & Musacchi, S. (2016). Crop load influences fruit quality, nutritional balance, and return bloom in 'Honeycrisp' apple. *HortScience, 51*, 236–244.

Shimizu, T., Hattori, Y., Ogami, S., Imayoshi, Y., Moriya, S., Okada, K., et al. (2024). Identification of aroma compounds responsible for apple flavor via palatability score prediction in tree-based regression models. *LWT, 192*, 115737.

Simon, S., Bouvier, J.-C., Debras, J.-F., & Sauphanor, B. (2010). Biodiversity and pest management in orchard systems. A Review. *Agronomy for Sustainable Development, 30*(1), 139–152.

Singh, J., Silva, K. J. P., Fuchs, M., & Khan, A. (2019). Potential role of weather, soil and plant microbial communities in rapid decline of apple trees. *PLoS One, 14*(3), e0213293.

Sisler, E. C. (2006). The discovery and development of compounds counteracting ethylene at the receptor level. *Biotechnology Advances, 24*(4), 357–367.

Skytte af Sätra, J., Odilbekov, F., Ingvarsson, P. K., et al. (2023). Parametric mapping of QTL for resistance to European canker in apple in 'Aroma' × 'Discovery. *Tree Genetics & Genomes, 19*, 12.

Stanich, K., Cliff, M. A., Hampson, C. R., & Toivonen, P. M. A. (2014). Shelf-life and sensory assessments reveal the effects of storage treatments with 1-methylcyclopropene on new and established apples. *Canadian Journal of Plant Science, 94*(8), 1427–1439.

Stover, E., Fargione, M. J., Watkins, C. B., & Iungerman, K. A. (2003). Harvest management of Marshall 'Mcintosh' apples: Effects of AVG, NAA, ethephon, and summer pruning on preharvest drop and fruit quality. *HortScience, 38*, 1093–1099.

Streif, J. (1996). Optimum harvest date for different apple cultivars in the 'Bodensee' area. In A. de Jager DJ, & E. Hohn (Eds.), *Determination and prediction of optimum harvest date of apples and pears. COST 94* (pp. 15–20). Luxembourg: The postharvest treatment of fruit and vegetables. Office for the Offic. Publ. of the European Communities.

Strickland, D. A., Ayer, K. M., Olmstead, D. L., & Cox, K. D. (2023). Refining management of apple powdery mildew in New York State with weather-based fungicide application timing programs. *Plant Disease, 107*, 1425–1432.

Strickland, D. A., Hodge, K. T., & Cox, K. D. (2021). An examination of apple powdery mildew and the biology of Podosphaera leucotricha from past to present. *Plant Health Progress, 22*, 421–432.

Sun, X., Jiao, C., Schwaninger, H., Chao, C. T., Ma, Y., Duan, N., et al. (2020). Phased diploid genome assemblies and pan-genomes provide insights into the genetic history of apple domestication. *Nature Genetics, 52*, 1423–1432.

Sutton, T. B., Aldwinckle, H. S., Agnello, A. M., & Walgenbach, J. F. (2014). *Compendium of Apple and Pear Diseases and Pests*. The American Phytopathological Society. https://doi.org/10.1094/9780890544334.

Teh, S. L., Kostick, S., Brutcher, L., Schonberg, B., Barritt, B., & Evans, K. (2021). Trends in fruit quality improvement from 15 years of selection in the apple breeding program of Washington State University. *Frontiers in Plant Science, 12*, 714325.

Thewes, F. R., Brackmann, A., Both, V., Anese, R. D., Schultz, E. E., Ludwig, V., et al. (2020). Dynamic controlled atmosphere based on carbon dioxide production (DCA - CD): Lower oxygen limit establishment, metabolism and overall quality of apples after long-term storage. *Postharvest Biology and Technology, 168*, 13.

Tian, Y., Thrimawithana, A., Ding, T., Guo, J., Gleave, A., Chagné, D., et al. (2022). Transposon insertions regulate genome-wide allele-specific expression and underpin flower colour variations in apple (*Malus spp.*). *Plant Biotechnology Journal.*

Toivonen, P. M. A., & Hampson, C. R. (2014). Relationship of I_{AD} index to internal quality attributes of apples treated with 1-methylcyclopropene and stored in air or controlled atmospheres. *Postharvest Biology and Technology, 91*, 90–95.

Trapman, M., Maxin, P., & Weber, R. W. (2008). *Diplodia seriata*, cause of black fruit rot in organically grown apples in Holland, Belgium and Northern Germany. In *Ecofruit-13th international conference on cultivation technique and phytopathological problems in organic fruit-growing: Proceedings to the conference from 18thFebruary to 20th February 2008 at Weinsberg/Germany* (pp. 177–181).

Ul Hasan, M., Singh, Z., Shah, H. M. S., Kaur, J., & Woodward, A. (2024). Water loss: A postharvest quality marker in apple storage. *Food Bioprocess Technol, 26*.

USDA. (2019). *U.S. standards for grades of apples*. https://www.federalregister.gov/documents/2019/10/01/2019-20570/us-standards-for-grades-of-apple.

Velasco, R., Zharkikh, A., Affourtit, J., Dhingra, A., Cestaro, A., et al. (2010). The genome of the domesticated apple (*Malus* ×*domestica* Borkh.). *Nature Genetics, 42*(10), 833–839.

Veltman, R. H., Verschoor, J. A., & van Dugteren, J. H. R. (2003). Dynamic control system (DCS) for apples (*Malus domestica* borkh. Cv 'Elstar'): Optimal quality through storage based on product response. *Postharvest Biology and Technology, 27*(1), 79–86.

Volk, G. M., Chao, C. T., Norelli, J., Brown, S. K., Fazio, G., Peace, C., et al. (2015). The vulnerability of US apple (*Malus*) genetic resources. *Genetic Resources and Crop Evolution, 62*, 765–794.

Wöhner, T., & Emeriewen, O. F. (2019). Apple blotch disease (*Marssonina coronaria* (Ellis & Davis) Davis)—review and research prospects. *European Journal of Plant Pathology, 153*(3), 657—669.

Wallhead, M., & Zhu, H. (2017). Decision support systems for plant disease and insect management in commercial nurseries in the midwest: A perspective review. *Journal of Environmental Horticulture, 35*(2), 84—92.

WAPA. (2023). *PROGNOSFRUIT 2023: First apple and pear crop forecast released.* https://www.wapa-association.org/asp/article_2.asp?doc_id=662.

Wang, T., Duan, S., Xu, C., Wang, Y., Zhang, X., Xu, X., et al. (2023). Pan-genome analysis of 13 *Malus* accessions reveals structural and sequence variations associated with fruit traits. *Nature Communications, 14*(1), 7377.

Watkins, C. B. (2017). Advances in postharvest handling and storage of apples. In K. Evans (Ed.), *Achieving sustainable cultivation of apples,* vol 18. *Burleigh Dodds series in agricultural science* (pp. 337—367). Sawston: Burleigh Dodds Science Publishing Ltd. https://doi.org/10.19103/as.2016.0017.16.

Watkins, C. B. (2020). Cultivar effects on CA/MA requirements of fruits and vegetables. In *Controlled and modified atmospheres for fresh and fresh-cut produce*. London: Academic Press Ltd-Elsevier Science Ltd.

Watkins, C. B., & Mattheis, J. P. (2019). Apple. In *Postharvest physiological disorders in fruits and vegetables* vol. 8, pp. 165—206). CRC Press.

Watkins, C. B., Nock, J. F., Weis, S. A., Jayanty, S., & Beaudry, R. M. (2004). Storage temperature, diphenylamine, and pre-storage delay effects on soft scald, soggy breakdown and bitter pit of 'Honeycrisp' apples. *Postharvest Biology and Technology, 32*(2), 213—221.

Weber, A., Brackmann, A., Thewes, F. R., Both, V., Anese, R. D., & Schorr, M. R. W. (2012). Relative humidity and its interaction with the storage temperature of 'Gala' apples and mutants. *Ciencia Rural, 42*(12), 2159—2165.

Weber, A., Thewes, F. R., Sellwig, M., Brackmann, A., Wünsche, J. N., Kittemann, D., et al. (2019). Dynamic controlled atmosphere: Impact of elevated storage temperature on anaerobic metabolism and quality of 'Nicoter' apples. *Food Chemistry, 298*, 10.

Weber, R. W., Späth, S., Buchleither, S., & Mayr, U. (2016). A review of sooty blotch and flyspeck disease in German organic apple production. *Erwerbs-Obstbau, 58*, 63—79.

Wright, A. A., Cross, A. R., & Harper, S. J. (2020). A bushel of viruses: Identification of seventeen novel putative viruses by RNA-seq in six apple trees. *PLoS One, 15*(1), e0227669.

Yuan, R. C., & Carbaugh, D. H. (2007). Effects of NAA, AVG, and 1-MCP on ethylene biosynthesis, preharvest fruit drop, fruit maturity, and quality of 'Golden Supreme' and 'Golden Delicious' apples. *HortScience, 42*(1), 101—105.

Zanella, A. (2003). Control of apple superficial scald and ripening - a comparison between 1-methylcyclopropene and diphenylamine postharvest treatments, initial low oxygen stress and ultra low oxygen storage. *Postharvest Biology and Technology, 27*, 69—78.

Zanella, A., & Stürz, S. (2013). Replacing DPA postharvest treatment by strategical application of novel storage technologies controls scald in 1/10th of EU's apples producing area. *Acta Horticulturae, 1012*, 419—426.

Zhang, H., Eaker, A., Ko, I., Haney, S., Khuu, N., Ryan, K., et al. (2024). A phased, chromosome-scale genome for Malus domestica 'WA 38'. *bioRxiv, 2024—01*.

Zude-Sasse, M., Truppel, I., & Herold, B. (2002). An approach to non-destructive apple fruit chlorophyll determination. *Postharvest Biology and Technology, 25*(2), 123—133.

CHAPTER

Pear

2

Akihiro Itai[1], Takuya Morimoto[1], Dragan Radivojević[2], Elizabeth J. Mitcham[3], Marcel Wenneker[4], Luca Corelli Grappadelli[5]

[1]*Kyoto Prefectural University, Japan;* [2]*University of Belgrade, Serbia;* [3]*University of California, Davis, United States;* [4]*Wageningen University and Research, The Netherlands;* [5]*University of Bologna, Italy*

Chapter outline

Temperate Tree Fruits and Nuts. https://doi.org/10.1016/B978-0-443-29137-1.00002-5

1. Basic botany of pear

1.1 Origin and relevant species

Pear species belong to the genus *Pyrus*, the subtribe *Pyrinae* (*Malinea*), the subfamily Maloideae (Pomoideae) in the family Rosaceae. There are at least 26 widely recognized primary species and 10 naturally occurring interspecific hybrid taxa (Table 2.1), which are distributed in Europe, temperate Asia, and mountainous areas

Table 2.1 *Pyrus* species, interspecific hybrids, and geographic distribution[a]

Geographic group and species	Centers of diversity
European:	
P. communis L.[b]	Western and Southeast Europe, Turkey
P. cordata Desv.	France, Spain, Turkey
P. × *nivalis* Jacq.	South Central Europe, Ukraine, France
P. × canescens Sprach	—
P. × complexa Rubtzov	Caucasus
P. × salvifolia DC	Europe (Crimea)
Circum-Mediterranean:	
P. elaeagrifolia Pall.	Southeast Europe, Ukraine, Turkey
P. syriaca Boiss.	Tunisia, Libya, Middle East, Armenia, Iran
P. cossonii Rehder	Algeria
P. gharbiana Trab.	Morocco, Algeria
P. mamorensis Trab.	Morocco
P. spinosa Forssk. (syn. *P. amygdaliformis* Vill.)	Southeast Europe, Turkey

Table 2.1 *Pyrus* species, interspecific hybrids, and geographic distribution[a]—*cont'd*

Geographic group and species	Centers of diversity
Mid-Asian:	
P. glabra Boiss.	Iran
P. korshinskyi Litv.	Afghanistan, Kyrgyzstan, Tajikistan, Uzbekistan
P. longipes Coss. and Dur.	Algeria
P. pashia D. Don.	Pakistan, India, Nepal, Bhutan, Afghanistan, China, Vietnam
P. regelii Rehd.	Afghanistan, Central Asia
P. salicifolia Pall.	Northern Iran, Turkey, Caucasus
East Asian:	
P. armeniacifolia Yu	Northwestern China, Kazakhstan
P. betulifolia Bunge	Central and Northern China, Laos, Southern Manchuria
P. × *bretschneideri* Rehd.	China, Korea
P. calleryana Decne.	Central and Southern China, Japan, Korea, Taiwan, Vietnam
P. dimorphophylla Makino	Japan
P. fauriei Schneid.	Korea
P. hondoensis Nak. and Kik.	Japan
P. × *hopeiensis* T. T. Yu	China (Hebei, Shandong)
P. koehnei C. K. Schneid.	South China, Taiwan
P. × *phaeocarpa* Rehd.	Northern China
P. pseudopashia T. T. Yu	China (Guizhou, Yunnan, Gansu)
P. pyrifolia (Burm.) Nak.	China, Japan, Korea, Taiwan
P. × *serrulata* Rehd.	Central China
P. × *sinkiangensis* T. T. Yu	Northwestern China
P. taiwanensis Iket. and Ohashi	Taiwan
P. ussuriensis Maxim.	Siberia, Manchuria, Northern China, Korea, Japan
P. × *uyematsuana* Makino	Japan
P. xerophila T. T. Yu	Northwestern China

[a] *Source: Bell and Itai, 2011*
[b] *Includes subspecies and several taxa listed as valid species names for which species status is uncertain.*

of Northern Africa (Bell and Itai, 2011). Most of the known species are native to Asia. All species of *Pyrus* are interfertile, and there are no major incompatibility barriers to interspecific hybridization, despite the wide geographic distribution of this genus. Dispersal is thought to have followed the mountain chains both east and west. Speciation probably involved geographic isolation and adaptation to colder and drier environments.

Kikuchi (1946) classified *Pyrus* species into three groups: small-fruited species with two carpels, large-fruited species with five carpels, and their hybrids with three or four carpels. Small-fruited species, sometimes known as Asian pea pears, *P. calleryana* Decne., *P. fauriei* Schneid., *P. betulifolia* Bunge, and others are used for ornamental purposes or as rootstocks. Among the large-fruited species with five carpels, there are four major species, which are grown commercially in temperate regions: *P. communis* L. (pear or European pear), *P.* × *bretschneideri* Rehd., *P. ussuriensis* Maxim. (Chinese pear), and *P. pyrifolia* (Burm.) Nakai (Japanese pear: Nashi). *P. communis* is native to Europe and is the main commercial species in Europe, North America, South America, Africa, and Australia. *P.* × *bretschneideri* is the main species in Northern and Central China. *P. pyrifolia* is the most important species in Japan, Southern and Central China, Taiwan, and Korea. *P. nivalis* Jacq., the snow pear, is cultivated in Europe for the production of perry, an alcoholic pear cider. *P. pashia* P. Don. is grown in Northern India, Nepal, and Southern China. The genome size of *Pyrus* was estimated by using the *de novo* sequencing analysis (Li et al., 2022). In total, about 45,000 protein-encoding genes were predicted in the genome of *P. pyrifolia* cultivar Nijisseiki (Shirasawa et al., 2021). Asian pears are thought to have been domesticated in prehistoric times and cultivated in China for at least 3000 years (Lombard and Westwood, 1987). European pears are thought to have been in Europe since as early as 1000 BCE. Homer referred to a large orchard of pears in the Odyssey, written between 900 and 800 BCE. The earliest written records of Japanese pears date back to the ancient manuscript of the Emperor Jito in 693 AD.

1.2 Evolution of *Pyrus*

Maloideae (Pyrinae or Malinea) include the genera *Malus* (apple), *Eriobotrya* (loquat), *Cydonia* (quince), and *Chaenomeles* (Chinese quince). The basic chromosome number of the Maloideae ($x = 17$) is high compared to other Rosaceae subfamilies ($x = 7$ to 9), indicating a polyploid origin. It has been suggested that the Maloideae arose as an amphidiploid of two primitive forms of Rosaceae, crossing a basic chromosome number of 8 and 9 (Zielinski and Thompson, 1967). These may have been possibly primitive members of the Prunoideae ($x = 8$) and Spiraeoideae ($x = 9$). A molecular study of the chloroplast gene *rbc*L suggests that Spiraeoideae is the maternal ancestor of Maloideae (Morgan et al., 1994). However, phylogenetic analyses of the granule-bound starch synthase *(GBSS)* genes, also known as waxy *(Wx)* genes, which were used as a tool to study the molecular taxonomy of Rosaceae, did not support the widespread hybridization between basic chromosome numbers of 8 and 9 (Evans and Campbell, 2002). Instead, *GBSS* gene sequences resolved Gillenia ($x = 9$) into the clade containing taxa with $x =$ 15 or 17. The monophyletic origin of *Pyrinae* and *Gillenia* was also confirmed by a molecular phylogeny of a broader set of genes (Velasco et al., 2010). They suggested that a relatively recent (more than 50 million years ago) genome-wide duplication (GWD) resulted in the transition from 9 ancestral chromosomes (similar to extant *Gillenia*) to 17 chromosomes in the *Pyrinae*.

The majority of cultivated pears are functional diploids ($2n = 34$). A few polyploid (triploid and tetraploid) cultivars of *P. communis*, *P. pyrifolia*, and *P.* × *bretschneideri* exist. Speciation in *Pyrus* has occurred without a change in chromosome number (Zielinski and Thompson, 1967). The genus is thought to have originated in the mountainous area of Western and Southwestern China and spread to the east and west. Recent genome data suggest that the divergence time of *Pyrus* and *Malus* was approximately 8.2 to 16.8 million years ago (Li et al., 2022). The divergence time of Asian pears from European pears appeared to be between 3.3 and 6.6 million years ago.

Three subcenters (Near East, Central Asia, and Eastern China) of diversity for the genus have been identified: the Chinese center, where forms of *P. pyrifolia* and *P. ussuriensis* occur; the Central Asian center, consisting of Tajikistan, Uzbekistan, India, Afghanistan, and the western Tian Shan mountains, where forms intermediate between *P. communis*, and the putative species, *P. heterophylla*, *P. korshinski*, and *P. boissieriana* occur; and the Near East center, consisting of the Caucasus Mountains and Asia Minor, where domesticated forms of *P. communis* occur.

1.3 Economic importance of pear

Pear is the third most important temperate fruit crop after apple and peach with a global world production of 26.3 million tons (MT), up 12% from 23.5 MT in 2012. Asia produces the most (20.3 MT), followed by Europe (2.8 MT), Africa (0.87 MT), and South America (0.82 MT). The European pear (*P. communis*) production is concentrated in five production areas: Europe, North America, South America, South Africa, and Oceania, while Asian pear (*P.* × *bretschneideri, P. pashia and P. pyrifolia*) production is concentrated exclusively in Asia. China produces 19.3 MT, followed by the United States (584,230 T), Argentina (566,227 T), Turkey (551,086 T), South Africa (560,000 T), Italy (519,190 T), Belgium (393,590 T), the Netherlands (351,000 T), and India (309,000 T).

1.4 Morphology

Pear tree is a temperate deciduous tree that can reach 15 m or more. The leaves are alternately arranged, simple, 2 to 12 cm long, and vary in shape from broad oval to narrow lanceolate. The flowers of *Pyrus* usually have five sepals, five petals, many stamens, and two to five pistils. Asian pea pears (*P. calleryana, P. betulaefolia, P. dimorphophylla,* etc.) have two pistils, while the major edible species (*P. communis, P.* × *bretschneideri,* and *P. pyrifolia*) have five pistils. The number of pistils is equal to the number of carpels. *Pyrus* has mixed flower buds with vegetative and flower initials, one ovary, and two to five carpels. The carpels are united with each other at the receptacle, and each locule has 2 ovules giving rise to a maximum of 10 seeds. Fruit shape varies from oblong or ovoid to pyriform, 7 to 14 cm in diameter in cultivars. The flesh is dense, white, and juicy, with a rough texture due to the presence of lignin-rich cells called stone cells.

2. Breeding system

2.1 Crossbreeding (hybridization)

Breeding a new pear cultivar often requires 20 years or more from the initial cross to the selection and release of a commercial cultivar. Prior to crossing, the flowers of the seed parent must be emasculated, and the pollen from the desired parent is then transferred to these emasculated flowers. Subsequently, the pollinated flowers are isolated using bags to prevent contamination with undesirable pollen.

Self-incompatibility: Self-incompatibility (SI) mechanisms are genetically controlled processes in plants that enable the recognition and rejection of self-pollen and pollen from closely related individuals. Typically, self/nonself recognition is governed by a single multiallelic *S* locus, which contains genes encoding pistil *S* determinants and pollen *S* determinants. In many rosaceous species, including pears, gametophytic SI is observed, where the growth of a pollen tube carrying either of the two *S* alleles present in the pistil is arrested in the style. This trait is crucial in breeding programs, as genotypes sharing one *S* allele are partially compatible, while those sharing both *S* alleles are fully incompatible. In Japanese pear, a naturally occurring self-compatible (SC) mutant, Osa-Nijisseiki, was identified as a bud sport mutation of the pistil *S* determinant in the original SI cultivar, Nijisseiki (Sassa et al., 1997). Osa-Nijisseiki has been employed in the development of new SC cultivars in Japanese pear breeding programs. SC mutants have also been identified in European and Chinese pear.

Distant hybridization: In the *Pyrus* genus, all species are cross-compatible through interspecific hybridization, with several taxa naturally exhibiting interspecific hybridization. Interspecific hybrids have been instrumental in genetic studies and agricultural advancements. For example, a new cultivar, PremP009, branded as PIQA BOO, has been released by combining genomic characteristics and traits from *P. communis, P. pyrifolia*, and *P.* × *bretschneideri*. Intergeneric hybridizations have also garnered considerable attention in pear breeding programs (Fischer et al., 2014).

2.2 Mutation breeding

Pear breeding programs have capitalized on mutants that occur naturally or are induced. However, since the frequencies of naturally occurring spontaneous mutations are very low, artificial mutagenic approaches have been employed to induce mutations. In fruit tree species, irradiation (including X-rays and gamma rays) is the most commonly used method to increase mutation frequencies. This approach aims to modify one or limited traits in well-adapted cultivars for agricultural improvements.

Self-compatible mutation: As previously mentioned, there are several SC mutants in pear. Osa-Nijisseiki is a naturally occurring SC mutant Japanese pear cultivar. The artificial induction of a SC mutant has been achieved through the pollination of irradiated Japanese pear pollen (Sawamura et al., 2013).

Disease resistance: As major Japanese pear cultivars like Nijisseiki are susceptible to black spot, caused by *Alternaria alternata* Japanese pear pathotype, the breeding objective has been to develop resistant cultivars while maintaining good fruit quality, equivalent to Nijisseiki (Saito et al., 2016). Since the 1980s, gamma-irradiated mutants showing varying degrees of resistance to black spot disease have been selected, leading to the release of several cultivars, including Gold Nijisseiki (Saito, 2016).

Fruit-related traits: Giant-fruit mutants have been selected in Chinese pear (*Pyrus ussuriensis*) (Zhang et al., 2011) and European pear (Isuzugawa et al., 2014). In the latter case, the fruit weight of this mutant, named G. LaF, is more than twice that of the original cultivar La France. Red skin color is a significant breeding target in pears due to its appeal in enhancing consumer preference in the market. While red skin mutants like Max Red Bartlett are known to revert to green skin color over time, other cultivars have been bred with solid red skin color, such as Falstaff and Lucy Red. Red skin mutants have also been observed in Asian pears, as seen in Zaosu Red, a hybrid of Asian and European pear (Ou et al., 2020).

2.3 Transgenic approach

Conventional breeding methods, such as crossbreeding, can be inefficient in pears due to their long juvenile phase, resulting in a prolonged timeline for releasing a commercial cultivar. To address this challenge, researchers have turned to genetic transformation as a promising approach to accelerate the breeding process. Agrobacterium-mediated pear transformation was first demonstrated by Mourgues et al. (1996). The frequencies of regeneration and genetic transformation in most *Pyrus* species are relatively low, emphasizing the critical need for efficient regeneration and genetic transformation systems.

Disease resistance: Fire blight, caused by *Erwinia amylovora*, stands as the most significant bacterial disease affecting pears. In a pioneering study by Reynoird et al. (1999), constitutive expression of the *attacin E* gene under the control of a CaMV35S promoter provided fire blight resistance in Passe Crassane. Scab, caused by *Venturia* species, poses a severe threat to pear cultivation in temperate regions. Transgenic pear clones constitutively expressing the *Rvi6* gene from apple exhibited strong defense reaction against *V. pyrina*, a causal pathogen of European pear, marking the first successful intergeneric transfer of a resistance gene through a transgenic approach (Perchepied et al., 2021).

Early flowering: The extended juvenile period in pear breeding programs poses challenges, as it delays the evaluation of important reproductive traits such as fruit quality. Moreover, this prolonged phase complicates the introduction of genes of interest from wild relatives, as the desired traits must be separated from unwanted characteristics through multiple rounds of backcrossing and selection efforts. One efficient strategy involves the use of transgenic approaches to induce early flowering, facilitating both breeding programs and the functional characterization of genes of interest (Matsuda et al., 2009). More recently, the CRISPR/Cas system was employed to target the *PcTFL1* gene, resulting in early flowering phenotypes observed between 4 and 12 months after micropropagation of the transgenic lines (Charrier et al., 2019).

2.4 Genomics-assisted breeding

Over the years, genomic tools have evolved and been applied to enhance the efficiency of pear breeding. DNA markers derived from genetic mapping and quantitative trait loci (QTLs) analysis have facilitated marker-assisted selection (MAS) in crossbreeding. The development of whole-genome reference sequences has significantly advanced the utilization of new genotyping methods, such as genotyping-by-sequencing (GBS) and single nucleotide polymorphism (SNP) arrays, enabling the creation of high-density molecular markers for genetic map construction and association mapping such as genome-wide association studies (GWAS). Presently, whole-genome sequences are available for several pear species, including European pear, Japanese pear, Chinese pears, and *P. betulifolia* (Li et al., 2022). Most of the pear genome sequences and annotation-related datasets are stored in the Genome Database for Rosaceae (GDR; https://www.rosaceae.org). The availability of these genome resources significantly enhances our understanding of genetics, genomics, and breeding in pear. In Japanese pears, several molecular markers have been identified and deployed in MAS within practical breeding programs at the National Agriculture and Food Research Organization (NARO) Institute of Fruit Tree Science, Japan (Yamamoto and Terakami, 2016).

2.5 Breeding target

Fruit quality and production: Achieving high-quality fruit is the primary objective of all breeding programs, encompassing various attributes such as taste, texture, appearance, shelf-life, and the incidence of physiological disorders. Fruit quality is a multifaceted trait influenced by both external and internal characteristics, including sweetness, sourness, aroma, texture, skin color, physiological disorders, and postharvest characteristics (Brewer and Volz, 2019). In addition to fruit quality, production-related traits also warrant attention as they significantly impact cultivation practices and contribute to the production of high-quality fruit. These traits encompass factors such as harvest time, yield, fruit size, shape, tree vigor, disease resistance, and adaptability to environmental conditions. Balancing the improvement of both fruit quality and production-related traits is essential for the development of superior pear cultivars that meet the needs of growers and consumers alike.

Adaptation to climate change: Unstable weather conditions pose significant concerns regarding phenological phases. Satisfying the chilling requirement of pear cultivars, typically ranging from 750 to 1000 chilling hours, is essential for breaking winter dormancy and ensuring proper flowering and fruit development. However, in regions with warmer winters or controlled environments such as greenhouses, pear trees may experience erratic flowering patterns, including delayed blooming, uneven bud break, and damaged flower buds (Tominaga et al., 2022). The occurrence of warm temperatures before the blooming period can induce early bud break in the spring, increasing the risk of exposure to subsequent cold temperatures and frost damage. Low-chill varieties are particularly important in regions experiencing milder winters, as they require fewer chilling hours to break dormancy and initiate flowering. Additionally, cultivars with later flowering times are sought after, as they can potentially avoid late frosts that may occur due to erratic weather patterns.

2.6 Rootstock breeding

The choice of rootstocks can significantly impact various aspects in pear cultivation by influencing factors such as productivity, tree size control, juvenile period, and tolerance to both biotic and abiotic stresses (da Silva, 2019). By selecting appropriate rootstocks, growers can optimize fruit production, enhance tree performance, and mitigate the impact of environmental challenges.

Genetic resources: Rootstock options for pear cultivation encompass a range of species, including those within the *Pyrus* genus as well as alternatives from other genera. These *Pyrus* species offer various traits such as vigor control, adaptability to different soils and climates, and tolerance to biotic and abiotic stresses. Quince rootstocks are commonly used for European pears due to their strong vigor control, precocity of the pear scion, and ease of propagation. *Amelanchier* spp. rootstocks are cold hardy, dwarfing, and precocious, as well as higher fruit yield (Einhorn, 2021). Rootstock breeding programs aim to develop elite clonal selections that offer fruiting precocity, reduced vigor, and adaptability to different environmental conditions. Examples include the Old Home × Farmingdale (OHxF) series in the United States, the Fox series in Italy, the Pyriam series and Brossier in France, the Pyrodwarf and Pyro 2-33 in Germany, and the 708-36 in England (da Silva, 2019). In Japanese pear cultivation, rootstocks are primarily propagated using seedlings, but several clones of *Pyrus betulifolia* and *P. calleryana* have been identified for their ability to induce dwarfing scion growth and adaptability to various environmental conditions (Tamura, 2012).

Dwarfing: Controlling tree vigor is essential for establishing high-density plantings, and certain genetic factors play a significant role in determining the growth habit of pear trees. European pear cultivar Nain Vert exhibits a dwarf growth habit controlled by a single dominant gene *PcDw* located on LG16 (Wang et al., 2011). This genetic locus has been characterized, and DNA markers closely linked to the dwarf growth habit have been developed, offering potential for MAS in breeding programs (Wang et al., 2016). While the use of quince rootstocks has been effective in reducing tree vigor in European pear production, issues such as graft incompatibility can arise in certain rootstock-scion combinations (Lombard and Westwood, 1987). Understanding the genetic mechanisms underlying tree vigor and graft compatibility is crucial for the development of improved rootstock-scion combinations and the establishment of efficient pear orchards.

Adaptation to environmental conditions: Among *Pyrus* rootstocks, *P. betulifolia* stands out for its high tolerance to both abiotic and biotic stresses, including drought, salt, cold, pear decline, and bacterial canker (Lombard and Westwood, 1987; Tamura, 2012). Although some selections of *P. betulifolia* may exhibit high vigor, certain cultivars have shown promising traits such as good rooting and size control of scion cultivars. *P. syriaca* is utilized as a rootstock specifically for adapting to low pH soils and drought conditions. Additionally, *P. calleryana* serves as a versatile rootstock due to its high tolerance to various diseases and pests, including fire blight, *Pyricola* (woolly pear aphid), and *Pratylenchus vulnus* (root lesion nematode) (Lombard and Westwood, 1987). Overall, *Pyrus* rootstocks provide growers with a range of options to address specific challenges in pear cultivation, offering resilience to various environmental stresses and contributing to the establishment of healthy and productive orchards.

Propagation: Vegetatively propagated rootstocks play a crucial role in pear cultivation, and their successful propagation is essential for orchard establishment and productivity. While hardwood cuttings are commonly used for propagation, some *Pyrus* species, such as *Pyrus betulifolia*, have shown better success rates with softwood cuttings. *In vitro* propagation techniques offer an efficient means to multiply rootstocks quickly.

3. Fruit maturation and postharvest physiology and technology

3.1 Harvest maturity

European pears are harvested mature but unripe, meaning they can be ripened to good eating quality after harvest. The best indicator of European pear maturity is flesh firmness. In some areas, soluble solids content is also used as a harvest indicator, but soluble solids content is primarily an indicator of potential sensory quality of the harvested fruit. The DA-meter nondestructively measures subtle changes in the chlorophyll content of pear fruit skin and flesh by measuring the absorption difference between 670 and 720 nm (IAD index). The IAD index has shown some promise as a nondestructive indicator of pear fruit maturity, but further work is needed (Torres et al., 2019). Starch degradation, as measured with the starch-iodine test, is important for determination of harvest maturity in some varieties and growing regions (Saquet, 2019). Pears harvested later are larger and ripen faster but also have a shorter storage life. European pear fruit require exposure to low temperatures to develop the capacity to produce ethylene and ripen. This low temperature exposure can occur before or after harvest. Harvest maturity can be delayed slightly through field applications of 1-methylcyclopropene (1-MCP), an ethylene action inhibitor, or aminoethoxyvinylglycine (AVG), an ethylene biosynthesis inhibitor. When applied as a preharvest spray, 1-MCP delayed pear fruit softening on the tree slightly but significantly reduced fruit softening during cold storage. AVG has a more significant impact on maturity development on the tree, maintaining fruit firmness and green skin color.

Asian pears are consumed when they are crisp and juicy and do not require softening before consumption; however, harvest maturity has a big effect on the storage life and poststorage quality (Kaur, et al. 2023). Some Asian pears (Nijisseiki, Kosui, Niitaka) produce very little ethylene and have a nonclimacteric respiratory pattern, while other varieties (Tsu Li, Ya Li, Chojuro, Shinsui) are climacteric and produce significant amounts of ethylene (Crisosto, 2016). Fruit firmness is not a good maturity indicator in Asian pears as in European pears. Fruit should be harvested when the skin color just starts to change to yellowish-green (Nijisseiki, 20th Century, Shinseiki, Tsu Li, Ya Li) or golden brown (Hosui, Kosui, Niitaka, Shinko) (Crisosto, 2016). Soluble solids content higher than 12.5% and a firmness between 3.6 and 5.0 kg-force indicate good quality. Immature fruit usually have poor flavor quality, while overmature fruit are more susceptible to surface marking (brown discoloration due to skin abrasions) and storage disorders (Kaur et al., 2023). Many growers determine harvest time by fruit taste and skin color.

3.2 Storage

The storage life of European pear fruit varies by variety, harvest maturity, disease pressure, field nutrition, speed of cooling, storage temperature, and storage atmosphere. Summer pears, such as Bartlett, Rocha, Blanquilla, and Conference, are harvested earlier than other varieties and require minimum exposure to low temperatures to induce ripening. Summer pears generally have a shorter storage life, at least partly due to their high ethylene production. A multivariate statistical approach utilizing a principal components analysis provides a global view of the storage behavior of six Italian pear varieties, highlighting the differences in potential storage life which ranged from 4 to 16 weeks (Pasquariello et al., 2013). Tree and fruit nutrition during production influence potential storage life. High nitrogen and low calcium nutrition are generally correlated with shorter storage life.

The temperature of the fruit during storage has a significant influence on storage life and poststorage quality of pears (Figure 2.1). Storage life can be 35% to 40% longer when fruit are stored at −1°C compared to 0°C. Relative humidity above 90% is also important to reduce water loss during extended storage, particularly in fruit stored in open bins. Most pear fruit are stored between 0°C and −1°C and will note freeze until the flesh temperature reaches −2°C or lower (Meheriuk et al., 1994). Susceptibility to freezing is influenced by the soluble solids content of the pears. Bartlett pears with 8% soluble solids were reported to freeze at −1.2°C. Soluble solids content is often lowest in the center of the pear and increases with longer time on the tree (higher maturity) before harvest.

FIGURE 2.1

Effect of storage temperature on skin color of Bartlett pears following 3 months of storage in air.

FIGURE 2.2

Skin color and incidence of senescent scald on Bartlett pears stored for 3 months in air at −1°C or 2°C in the presence of near-zero levels of ethylene (control) or 1 ppm ethylene.

The concentration of ethylene gas accumulating in the atmosphere during European pear fruit storage can impact the quality of fruit and its storage life. The effect of ethylene is reduced at low temperatures and under controlled atmosphere (CA) conditions. During air storage at 2°C, ethylene had a substantial impact on the quality of Bartlett pear fruit following 3 months of storage (Figure 2.2; Bower et al.,

2003). Even under laboratory conditions, it was very difficult to scrub ethylene to a low enough concentration to reduce its negative effects. However, pear fruit stored in air at 0°C were not affected by ethylene levels as high as 5 ppm when the fruit were stored for 3 months. The recommendation is to store fruit at as low a temperature as tolerated rather than trying to scrub ethylene from the storage room air. Similar results were found with Conference and Packham's Triumph pears. Asian pears are also stored at temperatures near 0°C, with a freezing point near −1.5°C (Crisosto, 2016). Storage life in air ranges from 1 to 5 months, depending on the variety and is usually limited by development of maturity-related disorders.

3.3 Controlled atmosphere storage

Most pear fruit stored more than 1 or 2 months are stored under CA conditions. Optimal CA storage extends storage life, reduces the development of disorders, and slows skin yellowing, decay development, and senescence (Drake and Elving, 2004). Storage under CA generally extends the storage life by approximately 50% compared to air storage at the same temperature (Saquet, 2019). The optimal atmosphere varies with variety, growing location, and fruit maturity at harvest. In traditional CA storage, oxygen levels range from 1 to 3 kPa and CO_2 from 0 to 5 kPa (Kupferman, 2003). The lower the oxygen level, the slower the respiration rate as well as the development of superficial scald. While elevated carbon dioxide also slows the respiration rate, many pear fruit are susceptible to high CO_2 injury, which manifests itself as flesh browning and sometimes flesh cavitation.

As new CA technologies, such as dynamic controlled atmosphere (DCA), initial low oxygen stress, and delayed CA, are incorporated into commercial practice, atmosphere recommendations are changing (Saquet, 2019). DCA systems control the pO_2 based on the response of the fruit (chlorophyll fluorescence, ethanol production, or respiratory quotient) to achieve the lowest safe oxygen concentration for the specific fruit in the storage room. Storage of organically grown European pears under DCA maintained firmness and reduced browning disorders and fungal decay significantly better than traditional CA. The benefits of DCA storage for Asian pears is more modest than for European pears, potentially extending storage life by 25% (Crisosto, 2016). Some varieties benefit from pO_2 levels of 1 to 3 kPa, while others do better with 3 to 5 kPa O_2. Many Asian pears are damaged by pCO_2 levels >2 kPa. Damage is manifested by core or flesh browning.

Storage of European pears under ultralow pO_2 (0.3 to 0.7 kPa) for 10 to 14 days before traditional CA storage improved European pear fruit quality after extended storage life in some varieties. Other studies have shown the benefits of storing European pears in regular air storage for 21 days before reducing pO_2, and this has become commercial practice for Conference pears (Saquet, 2019).

Use of ethylene inhibitors: The ethylene-action inhibitor, 1-MCP, has been applied to European pears either before or after harvest to delay ripening and extend storage life in air or CA storage (Villalobos-Acuña et al., 2011; Yu and Wang, 2017). Preharvest sprays of California Bartlett pears with 1-MCP had a very slight impact on fruit softening on the tree but significantly reduced fruit softening and development of internal breakdown during 3.5 months of storage, and the fruit were able to ripen normally after storage. Postharvest application of 1-MCP (0.2 to 0.6 μl L^{-1}) gas also greatly inhibits fruit softening during storage but also frequently inhibits fruit ripening after storage (Villalobos-Acuña et al., 2011; Yu and Wang, 2017). The effects of postharvest 1-MCP treatments are reduced and can be completely eliminated when similar or higher concentrations of ethylene are present during 1-MCP treatment (Yu and Wang, 2017). 1-MCP has a very strong effect on Anjou pears, and it can be very difficult to ripen the fruit after treatment. Treatment of European pears before harvest with AVG, an ethylene biosynthesis inhibitor, also extends the storage life of fruit by inhibiting ethylene production and fruit softening. Although AVG treatment slowed fruit ripening immediately after harvest, ripening of stored pears was not affected.

Studies on the benefits of 1-MCP on Asian pears are limited but show promise in extending storage life. The concentration of 1-MCP utilized is similar to that used for apples (0.5 to 1.0 μl L^{-1}), maximizing the potential benefits of 1-MCP in a fruit that does not need to soften prior to consumption. In addition, 1-MCP significantly slowed the development of core browning, which was accompanied by higher maintenance of ascorbic acid concentrations and reduced production of hydrogen peroxide.

3.4 Physiological disorders of European pears

Any fruit stored for a long period of time can develop a number of physiological disorders, and pears are no exception. Many of these disorders can be managed by effective use of storage atmospheres, but some are the result of intolerance to storage conditions.

Pink end (premature ripening) develops before harvest when the fruit, primarily Bartlett pears, are exposed to night temperatures of about 7°C or lower during the last 4 weeks of fruit growth. This low temperature exposure is sufficient to stimulate ethylene production and the onset of fruit ripening on the tree. Fruit should be cooled quickly and sold as soon as possible to avoid losses.

Superficial scald is manifested by browning of the peel tissues after extended cold storage in air or CA and results from oxidative damage to the cells that leads to cell death. The speed with which superficial scald develops is reduced by storage under low oxygen atmospheres. Low oxygen and high carbon dioxide treatments before storage are also effective. Antioxidant treatments, such as Ethoxyquin, also reduce superficial scald development but are falling out of favor as they are banned from the EU. Treatment of pears with 1-MCP before cold storage reduces the development of superficial scald, but effective concentrations often have a negative impact on the ability to ripen the fruit after storage.

Skin speckling, also called black speck, has been observed on Anjou pears after more than 4 months of storage in low pO_2, generally less than 1 kPa O_2.

Carbon dioxide injury: Pears are less tolerant to CO_2 than apples, and CO_2 injury manifests as internal flesh browning. The associated disorders brown core and pithy brown core in Anjou and Bosc pears, respectively, includes cavities along with flesh browning. Susceptibility to CO_2 injury increases with advanced maturity, slow cooling, high storage temperatures, prolonged storage, and low oxygen atmospheres.

Friction marking (skin abrasion) caused by handling may result in browning of the skin and the underlying tissues of pears. Pears are very susceptible and can be damaged by just the movement of adjacent fruit against each other in a package. The browning results from mixing of natural phenolic compounds in the fruit, such as chlorogenic acid, with polyphenol oxidase as a result of cell damage. The antioxidant Ethoxyquin, used to control scald in pears, also reduces skin browning from friction. Careful handling also reduces this disorder.

Core breakdown results from senescence and manifests as brown, watery core tissues in fruit that have been stored too long (Meheriuk et al., 1994). Yellowing of pear skin in storage often indicates high incidence of core breakdown, and further storage often leads to senescent scald. The incidence of core breakdown is influenced by harvest date, storage temperature, and delayed cooling.

Senescent breakdown manifests as very soft, brown flesh tissues inside the pear, as well as a dark brown discoloration of the peel tissues. The disorder occurs when fruit have reached the end of their storage life and is often accompanied by off-odors and off-flavors.

3.5 Physiological disorders of Asian pears

Watercore is similar to the similarly named disorder in apple fruit. Fruit develop glassy, water-soaked areas in the flesh, generally while on the tree. In storage, this tissue can turn brown and develop off-flavors, and therefore, the fruit should be marketed quickly.

Superficial scald in Asian pear is similar to that found in European pears and apples. It develops after cold storage for more than 4 months and is enhanced by ethylene. In Asian pears, the incidence is related to packaging of pears in pocket packs and develops especially on the portions of the fruit that are tightly confined in the package (Crisosto, 2016).

Flesh spot decay/internal browning is a physiological disorder and is not caused by a microorganism despite the name. Fruit develop brown areas in the flesh and sometimes cavities. The cause is unknown, but it is most associated with fruit harvested overmature and cold stored.

Core browning is related to harvest of overmature fruit and appears to be related to senescence of the fruit. Fruit develop brown and senescent tissues in the core while the rest of the flesh remains healthy.

Low O_2 injury is manifested as discolored depressions on the surface of the fruit. It is observed in some varieties stored at 1 kPa O_2 but can also develop at 2 or 3 kPa O_2 with extended storage time. **High CO_2 injury** appears as core or medial flesh browning, with cavities developing in severe disorder. The disorder is frequently seen when pCO_2 is >5 kPa but can occur at lower concentrations in extended storage.

3.6 Pathological disorders

For pome fruits, wounds are the primary entry point for postharvest decays and therefore careful handling, effective water sanitation, use of postharvest fungicides, and low temperature CA storage are key to reducing the incidence of postharvest decay in pears. Postharvest diseases limit the storage life of European pear fruit. *Penicillium expansum* is the main postharvest pathogen affecting European pears, followed by *Botrytis cinerea* (Figure 2.3). Prestorage drenching of pear fruit to apply fungicides or scald inhibitors have been associated with increased incidence of decay caused by *P. expansum* and *Mucor piriformis* (Sugar and Basile, 2011). Asian pears are susceptible to several wound-invading pathogens, particularly Botrytis, Alternaria, and Phomopsis species, which slowly develop during extended cold storage. Careful handling to reduce wounding can reduce postharvest decay.

3.7 Ripening

European pears are climacteric fruit, and their ripening behavior varies greatly by variety. Most pears are resistant to ripening at the time of harvest due to an inability to synthesize ethylene. Fruit become competent to ripen after exposure to low temperatures (≤10°C) before or after harvest induces expression of genes encoding enzymes involved in ethylene biosynthesis.

Pears grown in climates with colder temperatures before harvest and pears harvested more mature are less resistant to ripening after harvest. Within the same orchard and variety, pears harvested earlier in the season require more exposure to cold storage temperatures than pears harvested later.

Studies with Bartlett fruits were the first to investigate how pear fruit develop the capacity to ripen on the tree (Nham et al., 2015). Four maturity stages were identified: S1 – fruit are unable to soften after harvest, S2 – fruit are able to soften partially with or without ethylene treatment, S3 – fruit are able to soften with

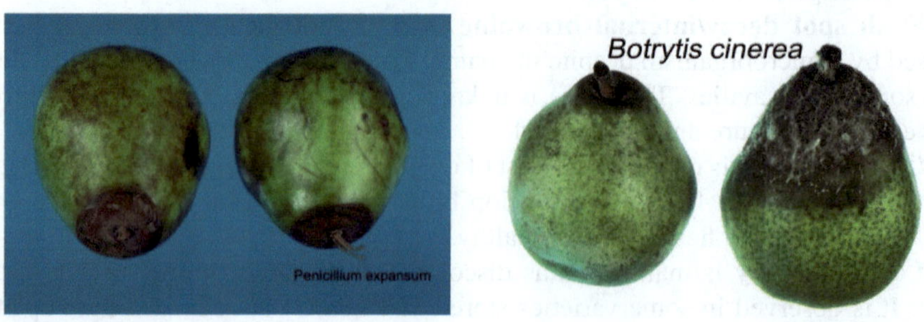

FIGURE 2.3

Penicillium expansum on Anjou pears *(left)* and *Botrytis cinerea* on pear fruit *(right)*.

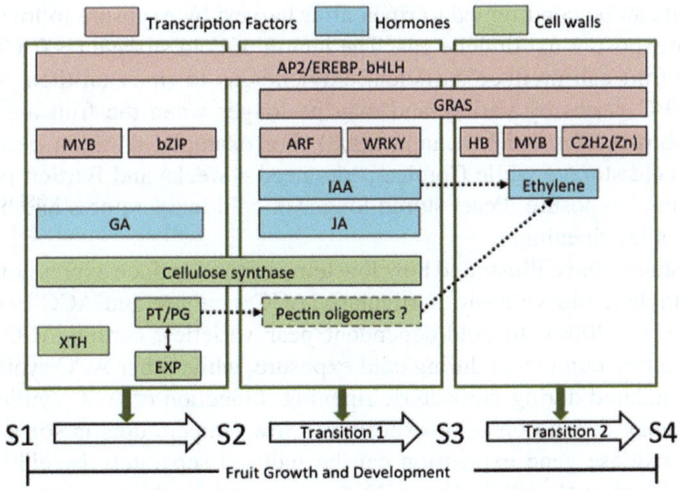

FIGURE 2.4

Proposed mechanisms regulating ripening capacity development during the final stages of pear fruit growth. Transition 1: Fruit develop ripening capacity responsive to ethylene treatment; Transition 2: Fruit develop ripening capacity without the need of ethylene treatment. *AP2/EREBP,* APETALA 2/ethylene response element binding protein; *bHLH,* basic helix-loop-helix; *bZIP,* basic region/leucine zipper; *ARF,* Auxin response factors; *HB,* homeobox; *C2H2(Zn),* Cys_2-HIS_2 Zinc finger; *GA,* Gibberellin; *JA,* Jasmonic acid; *IAA,* Auxin/Indole-3-acetic acid; *XTH,* Xyloglucan endotransglucosylase/hydrolase; *PT/PG,* pectin lyase/pectate lyase/polygalacturonase; *Exp,* Expansin. (From Nham et al., 2015).

ethylene treatment, and S4 – fruit are able to soften without ethylene treatment (Figure 2.4). The transcription profiles of these four maturity stages were analyzed, focusing on cell wall metabolism, hormone signaling, and transcriptional regulation. The analysis suggests that pectin degradation might result in fragments that act as early signal molecules to stimulate ethylene biosynthesis. In addition, auxin along with transcription factors ARF and WRKY appear to be essential in regulating the transition of pear fruit from being ethylene-unresponsive to ethylene-responsive (S2 to S3). Lindo-García et al. (2019) studied Blanquilla pears ripening on and off the tree. Ripening off the tree was initially mediated by oxidative stress, and then by ethylene, and limited by 1-aminocyclopropane-1-carboxylic acid (ACC) synthase activity. The slow ripening on the tree was associated with delayed production of ethylene and limited by ACC oxidase activity. The beginning of ethylene production on the tree was associated with increases in sucrose and decreases in hexanal levels, which the authors speculated may modulate on-tree pear ripening. In a follow-on study, the inability of Conference pears to produce ethylene at harvest was associated with the coordinated action of gibberellins, salicylic acid, and jasmonic acid, which remained at high levels in the latest stages of fruit growth (Lindo-García et al., 2020).

Pear fruit can be conditioned to ripen after harvest by exposure to low temperatures and/or exposure to ethylene gas. The length of cold storage (~0°C) required before pear fruit can produce sufficient ethylene gas to ripen on their own when placed at 20°C varies by variety and may be longer when the fruit are stored in CA (Villalobos-Acuña and Mitcham, 2008). For example, d'Anjou pears require 8 weeks of cold storage while Comice pears need 4 weeks and Bartlett pears need 2 weeks of cold exposure. Pears stored longer in cold temperatures had better sensory quality after ripening.

Several studies have illustrated how low temperatures induce expression of genes related to ethylene biosynthesis, particularly ACC synthase and ACC oxidase (El-Sharkawy et al., 2004). In cold-dependent pear varieties, certain ACC synthase gene transcripts accumulated during cold exposure, while other ACC synthase transcripts accumulated during climacteric ripening. Induction of ACC synthase genes by ethylene may depend on prior exposure to low temperatures in some varieties, while ACC oxidase gene expression can be induced separately by either cold or ethylene (Lelièvre et al., 1997). Using RNA-seq to analyze the transcriptome of Bartlett pears exposed to low temperatures, Nham et al. (2017) revealed that genes associated with jasmonic acid biosynthesis and signaling, as well as transcription factors TCP9a, TCP9b, CBF1, CBF4, AGL24, MYB1R1 and HsfB2b could be involved in low-temperature mediated enhancement of ripening, either independently or upstream of ethylene (Figure 2.5).

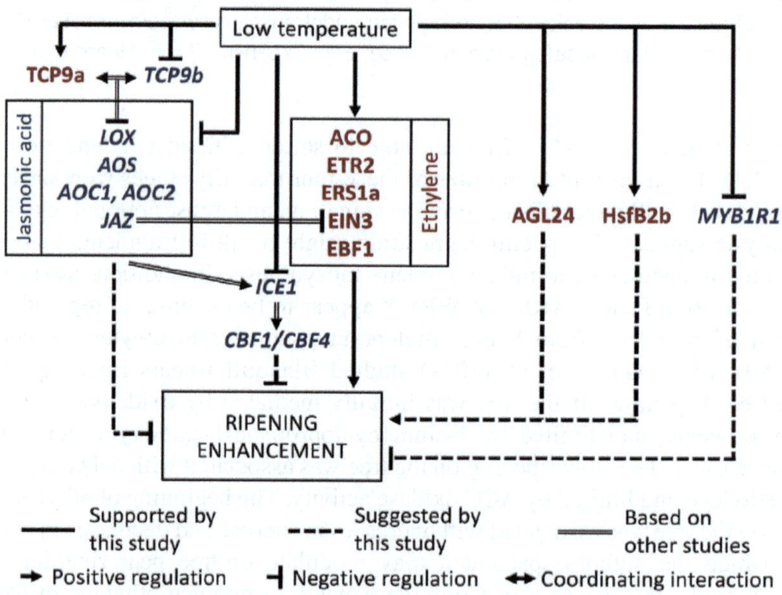

FIGURE 2.5

Proposed molecular model in the regulation of low-temperature-mediated enhancement of pear ripening. Genes in *blue italic* font were downregulated and genes in *red* font were upregulated by low temperatures. (From Nham et al., 2017).

Exposure to intermediate temperatures (5°C to 10°C) stimulates ripening capacity more quickly than low storage temperatures (−1°C to 0°C) or room temperature (20°C). Only 7 days of cold storage at 5°C or 10°C induced ethylene biosynthesis and ripening in Bartlett pears after warming; however, when pears were stored at −1°C, 14 to 21 days were required to stimulate ripening (Villalobos-Acuña and Mitcham, 2008). Similar results were found in Bosc, Anjou, and Comice pears (Sugar and Basile, 2013).

Exposure to ethylene gas can also induce production of enzymes required for ethylene biosynthesis (autocatalytic ethylene production) in pear fruit. The length of ethylene exposure required to induce sufficient ethylene production depends on the resistance of the fruit to ripening. The temperature of the fruit during ethylene exposure greatly influences the length of ethylene treatment required to induce ethylene biosynthesis in the fruit. In Bartlett pear, the required exposure to ethylene gas was twice as long when the fruit temperature was 10°C and three times as long when fruit temperature was 5°C as compared to ethylene treatment at 20°C. A summary of protocols demonstrated to induce ripening in different pear varieties can be found in Villalobos-Acuña and Mitcham (2008). These protocols have been implemented by the pear industry in some regions to ensure ripe fruit are available for sale in the market. When ripening pear fruit, it is important to understand any field or prestorage treatments that have been applied that could inhibit the ripening process. The ethylene action inhibitor, 1-MCP, has been used to extend the storage life of pear fruit, and residual effects of this treatment frequently slow the ripening of pear fruit at harvest and also after storage. In some cases, ripening is severely inhibited (Villalobos-Acuña et al., 2011), interfering with successful marketing of the fruit.

Because fruit temperature has a great effect on its response to ethylene and its rate of ripening, it is important to reduce variation in temperatures among fruit in a batch to be ripened. Commercial packaging of pears can interfere with the ability to warm and cool the fruit uniformly during ripening (Table 2.2), including boxes, box liners, bags, and tissue wraps.

Table 2.2 Days required to reach the one-half cooling temperature ($T_{1/2}$)^z for palletized Anjou pears that were place-packed or tray-packed and stored at 0°C with an air flow of 12 m/min around the pallet surface

	One-half Cooling Times (days)			
	Place-packed^y		*Tray-packed*^{x}	
Pallet level	*Outside fruit*	*Inside fruit*	*Outside fruit*	*Inside fruit*
1 bottom	2.1	3.3	1.8	2.1
2	4.9	8.7	3.4	5.5
3	4.9	12.4	4.3	7.2
4	7.0	15.0	4.6	7.2
5	6.4	15.7	4.2	7.3
6	7.0	13.0	4.1	7.4
7 top	2.0	4.9	1.7	4.2

^{x} *Start temperature = 19.6°C, Room temperature = 0.1°C,* $T_{1/2} = 9.9°C$
^{y} *Start temperature = 21.6°C, Room temperature = 0.3°C,* $T_{1/2} = 10.9°C$
^{z} $T_{1/2}$ *= (fruit starting temperature − storage room temperature) x 0.5 + storage room temperature*
Source: Faubion and Kader (1997).

3.8 Commercial pear handling

Pear fruit are hand harvested into field bins or small totes and transported to storage and packing facilities. Some varieties of European pears, such as Comice and Red Bartlett, and Asian pears must be particularly carefully handled to reduce skin scuffing and other damage that results in brown or black markings on the fruit surface. Workers may wear gloves and place fruit into small, smooth-sided, padded buckets or totes to reduce damage. Stem punctures should be avoided; in some cases, stems are clipped close to the fruit.

Pear fruit might be transferred directly into cold storage or CA rooms for storage for up to 8 months, depending on the variety and facility. In most locations, room cooling is used to cool the fruit; however, in warm climates like California, forced-air cooling is commonly used to cool pears at the start of cold storage. Some pears are sorted and packed immediately after harvest and stored in packed and palletized boxes. The disadvantage of this approach occurs when quality issues related to disorders or decay develop during storage, requiring repacking of the fruit. It is generally inadvisable to sort the fruit and re-bin for storage given the extra trip across the packing line and the ease of scuffing the skin. Pears are most frequently place-packed into cartons, either as loose, tissue-wrapped pears (Figure 2.6); on a paper or foam tray; or in a Panta Pak (Figure 2.7). Asian pears may be individually placed in foam socks. A box liner is often used to reduce water loss in fruit destined for long-term storage. Some pears are loosely filled into a specialized box called tight fill that is settled by vibration to immobilize the fruit during transit to market. Other fruit are bagged prior to placement in the box. Asian pears packed in a two-layer tray carton are often immobilized during transit by extra padding and strapping of the carton. Alternatively, individual fruit may be placed in foam punnets on a tray within the carton.

FIGURE 2.6

Hand wrapping and place-packing of sized pear fruit.

FIGURE 2.7

Pear fruit packed in a Panta Pak, showing foam padding used between each layer and Excelsior paper pad for top of box.

Successful marketing of tasty pear fruit to consumers requires attention to harvest maturity and storage conditions to avoid extending shelf-life beyond the period of good eating quality. In addition, effective ripening of the fruit, including developing typical texture and flavor profiles, is critical to attract and retain more consumers that enjoy eating pears.

4. Orchard systems

Establishing a new pear orchard requires the investment of large financial resources. The amount of investment depends primarily on the environmental conditions of the site chosen for the orchard and the orchard system. Unsuitable sites and soils that lead to crop or tree losses should not be used for planting orchards, as they rapidly reduce the profitability of pear production. The orchard system must be adapted to the agro-ecological characteristics of the chosen site. The following horticultural factors are integrated into the orchard system: (1) cultivar and cultivar arrangement, (2) rootstock, (3) tree density and tree arrangement, (4) nursery tree quality, (5) support system, and (6) tree size and tree form (Wertheim et al., 2000). Chosen orchard system has a huge implication on precocity, productivity, fruit quality, cost of production, and finally on net profit.

4.1 Cultivar

The choice of variety is primarily determined by market factors such as current market demand and product price. However, the variety chosen for planting, regardless of its economic value, must be adapted to the environmental conditions. All pear varieties are self-sterile and require pollenizers for cross-pollination by insects to develop fertilized fruits, which is challenging (and often results in low yields), since pear flowers are not very attractive to bees and other foraging insects. Usually, pear orchards contain a main variety and one or two pollenizers arranged in separate, adjacent rows. Certain pear varieties (Abate Fetel, Conference, Rocha, etc.) can offset self-sterility through natural or promoted parthenocarpy through the application of a plant growth regulator (PGR), making it possible to plant solid blocks of a single genotype. However, the lack of seeds can result in misshapen fruit of smaller size, which are in general not favored by markets.

4.2 Rootstock

The principal pear production regions around the world depend on rootstock clones from *Pyrus sp.* (pear) or *Cydonia oblonga* (quince) (Einhorn, 2021). In general, the most reliable rootstock for controlling pear growth and rapid entry into production of the orchard are the quince rootstocks. Significant disadvantages of quince rootstocks can be the high susceptibility to iron chlorosis (Figure 2.8), graft-incompatibility with many pear varieties, poor cold hardiness (Einhorn, 2021), and sensitivity to high spring and summer temperatures (Sansavini et al., 2008).

Quinces C and Adams are the most dwarfing rootstock recommended for very high-density plantings (VHDP) in Northern European countries (Sansavini et al., 2008). They are not suitable for hot regions as their roots are shallow and sensitive to high temperatures and drought. The quince rootstocks Sydo and Adams are

FIGURE 2.8

The quince rootstock MA is susceptible to iron chlorosis on calcareous soil.

moderately vigorous, whereby the rootstock BA29 is somewhat more vigorous and easily overcomes graft incompatibility and shows greater tolerance to high soil pH values.

Pyrus rootstocks are generally too vigorous and nonprecocious for many pear cultivars (Einhorn, 2021). Rootstocks from seeds of *P. communis* have a good affinity to all varieties, are suitable for almost all soil types, tolerate active lime and are drought-resistant and can be tolerant to pear decline and fire blight. However, they are very vigorous, have inhomogeneous early growth, late start to production, and fruit are generally smaller than on quinces. Vegetative (including micropropagated) rootstocks of *P. communis* derived from the cross of two fire blight and pear decline-resistant clones, Old Home × Farmingdale (OH×F), are also relatively widespread in Europe (Sansavini et al., 2008) and the U.S. (Einhorn, 2021). OH×F 87, which confers precocity and yield efficiency, is the most desirable pear rootstock in the U.S. (Einhorn, 2021). Other notable OH×F rootstocks currently in use are OH×F 40, 97, and 69 (Musacchi et al., 2021; Einhorn, 2021). Among other species, *P. calleryana* has been used as a seedling rootstock in the Southern Hemisphere because of its fire blight, pear decline and woolly aphid resistance, drought tolerance, and control of vigor, although the species is very susceptible to winter injury (McClymont et al., 2021). Selections of *Amelanchier sp.* have been developed relatively recently through intra- and interspecific hybridization. These rootstocks are cold hardy, dwarfing, and precocious with high yield efficiency and produce large fruit (Einhorn, 2021).

4.3 Planting density

The planting density is determined by the arrangement of the trees and the distance between them. There is a wide range of pear orchards, ranging from low to high density. Orchards with a very low-density planting (VLDP) (<1000 trees ha^{-1}) have low initial costs. However, this type of planting density cannot be recommended for the establishment of pear orchards due to low precocity and extremely slow entry into full production.

The low-density planting (LDP) (1000–2000 trees ha^{-1}) is still used for long-life pear orchards with freestanding central leader or vertical axe tree forms. These types of orchards are found in regions featuring specialized growing, where holdings are medium-to-large and seasonal workers are available. This planting density also has some disadvantages, such as the slow entry into full production and the higher labor costs for pruning and harvesting.

Medium-density planting (MDP) (2000–3000 trees ha^{-1}) and high-density planting (HDP) (3000–4000 trees ha^{-1}) are the most widespread plantings worldwide with the spindle training system (Musacchi et al., 2021), the V-training systems, or the multileader system. The primary advantage of high-planting densities with dwarf trees is to induce early cropping and high yields of excellent quality fruit. Other advantages are reduced labor costs, earlier return on establishment investment, and the feasibility of early renewal of the orchard.

VHDP from 4000 to 6000 trees ha^{-1} is associated with major limitations and can only be recommended in regions with good environmental conditions where the use of MC or Quince Adams rootstocks is possible. Practice has shown that there is no economic justification for increasing planting density above 6000 trees ha^{-1}.

4.4 Tree training systems

Various training systems can be used in pear orchards; the choice is most often determined by the orchard design, in particular the density of planting.

Freestanding Central Leader and Vertical Axe: The Central Leader is a three-dimensional system adapted to LDP, featuring a single leader (McClymont et al., 2021) and three to four tiers of permanent scaffold branches spaced along the trunk and a tree height of 4.0 to 4.5 m (Figure 2.9). The trees need no support and are developed by heading the leader at 90 cm at planting. During the second through the fourth years, the 1-year-old part of the leader is headed back by one-third every year during winter pruning to obtain a strong leader that will be able to bear a high yield without any support. At the end of the second year, a lower tier of scaffold branches is selected and positioned at 30 degrees above the horizontal by tying them to the trunk with a twine (Figure 2.10). A narrower option derived from the Central Leader is the **Vertical Axe** (VA) where trees must be supported by a post or wire system to grow upright. The VA is developed by heading the leader at 120 cm at planting and removing any lateral branches larger than two-thirds the diameter of the leader (Robinson, 2011). During the second through the fourth years, the leader is left intact. At the end of the second year, a lower tier of four scaffold branches is selected and placed in a horizontal position. Beginning in year 4, large

FIGURE 2.9

Central Leader of Packham's Triumph at 1480 trees per hectare with black netting that protects fruit from intensive sunlight.

FIGURE 2.10

At the end of the second year, a lower tier of scaffold branches is placed in a horizontal position or 30 degrees above horizontal by tying them to the trunk with twine.

diameter limbs above the bottom tier of scaffolds (>4 cm) must be removed at the trunk with an angled cut to develop replacement limbs.

Spindle: The Spindle is the most used training system in Europe in MDP or in HDP. It enables the formation of a continuous productive wall, early entry into production, high productivity and fruit quality, easy management of the orchard from the ground, and consequently lower costs. Numerous variants of this system exist which differ for their height and thickness in the lower part of the canopy. Tree size is determined by rootstock, row spacing (0.5-1.5 m), and type of pruning and training (traditional spindle, tall spindle, slender spindle, and super-slender spindle). In all spindle types, the permanent scaffold consists of only one central leader, while the weak fruiting wood is freely distributed along the central leader, with longer branches in the lower part of the tree than in the middle and upper parts (Figure 2.11). The best option for spindle formation is the use of nursery trees with many feathers. Excessive and too vigorous lateral branches (with a diameter greater than one-half of the central leader diameter) are usually removed. The sylleptic shoots of "knip" or 1-year-old nursery trees or the proleptic shoots of 2-year-old nursery trees growing with a small crotch angle are tied from 40 to 45 degrees above the horizontal (Lordan et al., 2017) to a horizontal position shortly after planting to induce cropping and prevent them from developing into substantial lower scaffolds. The requirement for artificial bending depends on the variety. For cultivars such as Comice and Conference, bending the future branches is necessary (Lauri, 2009). With light pruning of the leader or feathers at planting, the spindle tree can be allowed to crop in the second year, which contributes to a rapid return on the initial investment but also induces natural bending of lateral branches, which keeps them weak while favoring leader extension (Robinson, 2011).

From the second to the fourth year, pruning is light and aims to complete the tree structure, to preserve all the developed fruiting wood, and, if necessary, to control

FIGURE 2.11

Slender spindle tree without permanent scaffold branches.

the crop load. Vertical shoot(s) competing with leader must be removed, as well as strong shoots along the leader and vigorous laterals arising from branches, leaving a short-beveled stub at the base to stimulate renewal growth (Figure 2.12). Renewal branches develop and replace the previous side branches when these have grown too large (Schupp et al., 2017). After the tree achieves the desired height, the leader is cut in the dormant season to a weak side branch or a spur at the desired height (Robinson, 2011).

V-system: The V-system is widely used in the Netherlands and Belgium and is a variant of a Y-shaped system with four arms that requires well-feathered 2-year-old nursery trees for tree establishment. Four equally developed feathers are selected and kept as fruiting branches (Deckers and Schoofs, 2001). These branches, which start about 60 cm above the base of the bud union, are not pruned back at planting. Two of them are tied to the left and two to the right perpendicular to the row on a bamboo stick and can be considered as four central leaders on one stem. The productivity in this V-system is based on flower buds on short spurs and on bourse shoots on the four central leaders (Deckers and Schoofs, 2001). This system leads to a lower canopy height and thus enables pruning and harvesting from the ground, provides high yields and large fruit. However, the initial costs are very high, as developing the V-trellis is quite time and material consuming.

Multileader system: The fast development of automatized harvest with robotic technology is driving a transition in the design of the orchard toward narrow-row and two-dimensional planting systems to make fruit more accessible to robotic arms (Musacchi et al., 2023). In addition, such narrow canopies are more adaptable to pruning with shearing machines and easier to harvest with simple harvest aids.

FIGURE 2.12

Pruning a strong branch that arises from the central leader, leaving a short, beveled stub at the base to stimulate renewal growth.

Multileader trees have a natural tendency to form a narrow and dense hedge with little or no secondary structure. Splitting vigor over two or more axes helps to control vegetative growth (Musacchi et al., 2021) because vigor is distributed over more than one leader. In multileader tree orchards, the distance between rows is significantly reduced and leads to an increase in light interception. The minimum distance between rows is limited by the dimensions of the tractor and mechanical implements and the bins used for harvesting pears; thus, the free alley required must be above 1.5 m. Spacing along the row is increased according to the number of leaders to keep them 0.3 to 0.6 m apart, resulting in distance between trees ranging from 1.0 m for 2 stem per tree to 3.0 m for 10 stem per trees (McClymont et al., 2021; Tustin et al., 2022). The simplest two-dimensional canopy consists of two leaders aligned parallel to the row direction, which are developed from preformed nursery trees with two axes (Figure 2.13). The multileader trees are obtained by tying down the leader(s) of the tree provided by the nursery, which is (are) bent horizontally about 80 cm above ground, from which the upright-growing shoots that will form the leaders *(uprights)* are trained. The length of the *uprights* can be reduced to obtain pedestrian-sized trees or can be much taller to increase the yield per hectare (Tustin et al., 2022).

4.5 Tree support and system of anti-hail net

Medium- and high-density planting systems require installation of tree support structures that consist of wooden or concrete posts and three to five horizontal wires. In areas where hail falls frequently, hail protection nets should be installed, taking

FIGURE 2.13

The training system with two leaders parallel to the row direction, formed starting from nursery trees with two axes, results in a flat and narrow tree canopy with a depth of 70 to 80 cm.

FIGURE 2.14

Complete covering of the orchard with nets attached to the front and sides (cage system).

advantage of the support system. White net is most commonly used in pear, except in areas where netting is simultaneously used to protect the fruit from excessive light, in which case black nets are a better option. To provide additional protection against certain insects (oriental stink bug or codling moth), the orchard can be completely enclosed by placing additional nets at the front and sides of the orchard (Figure 2.14). An alternative set-up features nets draped over each individual row (Figure 2.15).

FIGURE 2.15

Each row in the pear orchard is completely covered with netting from bottom to top to protect the pear trees from hail, insects, and intense sunlight.

5. Orchard management

5.1 Pruning

Winter pruning enables the formation of the canopy, control of tree size, improvement of light distribution in the canopy, control of floral bud numbers, and is recommended as the first step in any crop load control program. In combination with fruit thinning, it regulates fruit load. It begins at planting and continues annually. Intensity and methods of pruning mainly depend on cultivar and planting system. In all training systems, vertical and vigorous shoots are removed regularly. Over the years, no other large branches are allowed to remain on the mature tree except the permanent scaffold branches. Permanent lateral branches are kept "simple" or "linear" without any long laterals on the side (Lauri, 2009). In addition, a certain number of older, weaker branches must be removed each year to renew fruiting wood. In the renewal pruning of a spindle system, excess side branches are removed preferentially by size, leaving a short, beveled stub at the base to stimulate renewal growth (Schupp et al., 2017). From this stub, a flat weak replacement branch often grows. If these are left intact, they will naturally bend down with the weight of the crop (Figure 2.16). Renewal pruning is a key to maintaining fruit quality through renewing spurs and creating a favorable light environment (Schupp et al., 2017). In LDP, the same cuts are made on the permanent scaffolds located on the central leader.

Heading back strong branches to develop new shoots is usually avoided because this may lead to an imbalance between vegetative growth and fruiting. However, heading back of weak and moderately strong, 2- or 3-year-old, fruiting branches is regularly applied to reduce the number of spurs (Figure 2.17) and promote renewal on 1-year-old wood. Furthermore, cutting back weak 2-year-old wood prevents excessive flowering and consequently increases fruit set. Comice and Abate Fetel bear mostly on 2- or 3-year old branches, which need to be cut back from

FIGURE 2.16

Flat, weak replacement branches often grow from the stub along the central leader, which are covered with flower buds in the second year and bear fruit in the third year.

FIGURE 2.17

Heading back of weak and moderately strong branches on pear trees is regularly applied to reduce the number of spurs.

one-half to two-thirds to reduce spur numbers, leaving where possible no more than 8 to 10 spurs; otherwise, excessive competition may lead to poor fruit size and quality.

In systems with many axes per hectare, lateral branching of vertical fruiting stems is minimized by complete removal of any vegetative limbs, so that lateral fruiting growth units comprise terminal fruiting bourse shoots and spurs (Tustin et al., 2022).

Summer pruning can be used to control the shape of the tree in the training phase and reduce vegetative growth and transpiration in full production by pruning the water sprouts and strong shoots in the top of the tree in early summer. In the late

FIGURE 2.18

Narrow canopies such as spindle or planar cordon are adaptable to machine pruning with shearing machines.

season, removing the inner and upper suckers promotes better light interception and maturation of the 1-year-old shoots (Musacchi et al., 2021). The mechanization of summer pruning with shearing machines is becoming common practice, but remedial hand pruning is necessary every 2 or 3 years.

Pruning is largely dependent on human labor and is therefore very expensive. **Mechanical and robotic pruning** are two alternative methods to replace manual pruning. Narrow canopies such as spindle or planar cordon are adaptable to mechanical pruning with shearing machines (Figure 2.18). Robotic, selective pruning, which aims to mechanically prune the tree branches to the same quality and level as the human hand, is still forthcoming, but it will enable the complete automation of this complex process.

5.2 Fruit thinning

Fruit thinning may only be necessary with high-yielding varieties such as Bartlett, Conference, Packham's Triumph, etc. Given the difficulties in pollination that pear cultivars suffer, for several of the most commonly grown varieties, PGRs (most often gibberellins) are applied to set parthenocarpic fruit. In some instances, this may result in excessive fruit set, which can cause small size and low commercial value. An excessive number of fruit can also have a negative effect on the formation of flower buds for the next season through their seeds (Wertheim, 2000). Pruning as the only means of adjusting crop load often does not lead to the expected results for many pear varieties. A large number of flowers per tree and good conditions for their fertilization can lead to high fruit set and consequently an excessive number of fruit per tree (Radivojevic et al., 2022).

Thinning out excessive fruit can be done by hand or by application of some chemicals. Hand thinning is recommended in very young pear orchards or as an additional method to remove excess fruit remaining on the trees after chemical thinning, especially small and deformed fruit, to further maximize fruit size and the proportion of first-class fruit in the total yield. Postbloom thinners are generally applied between petal fall and 16-mm fruitlet diameter (Elsysy et al., 2020). Several chemicals are currently available for pear fruit thinning. **1-Naphthaleneacetic acid (NAA)** is applied at a concentration of 15 to 20 mg l^{-1}, while **6-benzyladenine (BA)** is normally used at a concentration of 100 to 150 mg l^{-1}, depending on the variety. **Metamitron** is a relatively new post-bloom thinning agent that inhibits photosystem II (PSII) in the light reaction of photosynthesis, leading to a carbohydrate deficit that promotes fruit abscission (Elsysy et al., 2020). Adequate thinning of Conference can be achieved by single or repeated applications of 175 to 350 mg l^{-1} metamitron at the developmental stage of 8 to 12 mm fruit diameter. A new compound, ethylene-precursor 1-aminocyclopropene-1-carboxylic acid (ACC), is also under testing for its potential as a late thinner. It has been shown to reduce fruit set in Forelle pear at a concentration of 225 to 300 ml l^{-1} (Theron et al., 2018).

5.3 Irrigation

Pear trees require large amounts of water for their growth and development. Irrigation is necessary in all regions where the natural amount of precipitation is not sufficient to meet annual water requirements. Its importance depends on the climate and increases with progression from temperate to drier and arid zones (Naor et al., 2006). Climate change, through rising temperatures and longer periods without precipitation, leads to significantly increased need for irrigation, even in conditions with low evaporative demands. A study conducted under Belgian growing conditions shows that even in that environment (which many would consider moderate), irrigation is necessary to consistently achieve maximum fruit size and yield in Conference (Janssens et al., 2011). To reduce the volumes of water used in irrigation, it is necessary to increase the efficiency of water use through an efficient irrigation system and applying an irrigation schedule that meets the plant water requirements (Naor et al., 2006).

Pear orchards are increasingly using highly efficient microirrigation technologies (drip and microsprinkler systems) that deliver water directly to the root zone under the canopy of the tree. The choice of a type of microirrigation system depends on climatic conditions; soil types, characteristics, and topography; crop water requirements; water availability and quality; and economic parameters. Drip irrigation is usually applied at a very high frequency to directly supply and replenish the water used by the plants. Microsprinkler systems are suitable for a wider variety of soils, but evaporation losses can be significantly higher.

Irrigation scheduling should keep soil moisture within a desired range, usually between field capacity (full point) and a predetermined refill point for optimal growth. The timing and amount of water needed are influenced by prevailing climatic conditions, the stage of growth, soil properties, and the extent of root development. There are different approaches to objectively determine the irrigation schedule, including soil- or plant-based measurement methods, and atmospheric ones (Jones, 2004). Soil water potential can be measured by tensiometers and is related to soil moisture. According to Boland et al., (2002), the soil moisture tension from budburst and flowering to harvest should be between 8 and 40 kPa and after harvest between 8 and 200 kPa.

Atmospheric-based measurements use reference evapotranspiration (ET_o) and crop coefficients (K_c) to estimate crop evapotranspiration (Et_c). This method ($ET_c = ET_o \times K_c$) can be applied to pear orchards, as appropriate K_c values are available. ETo is calculated by the Penman-Monteith equation, which uses meteorological data as inputs, such as temperature, wind speed, radiation, and humidity, which are easily provided by automatic weather stations. K_c accounts for the difference between the evapotranspiration of a standard grass cover and the evapotranspiration for a specific crop in the field and changes with the tree phenological phase during the season. At the beginning of the season, crop factors are generally low and peak just before harvest (Boland et al., 2002). This method is simple and inexpensive, available online in different areas, and provides remote recommendations for irrigation scheduling (Boini et al., 2019). Plant-based systems are more complex and, even though they can accurately determine actual plant water status, they have more experimental than practical applications.

5.4 Plant nutrition

Nutrient management plays an important role in ensuring good tree growth, cropping, and fruit quality. For appropriate mineral nutrition of pear trees, consideration should be given to their being perennials and that the minerals absorbed during the year influence growth and development in the current as well as the following year. Nutrient deficiencies are alleviated by the application of fertilizers and/or adjustment of pH. The fertilization must follow climate characteristics and tree demand. Growers may estimate the nutrient requirements of fruit trees on the basis of leaf and soil analyses, as well as the growth and cropping of the trees. Since pear fruit trees are perennials, foliar analysis is the most accurate way to determine nutritional status of the orchard. Table 2.3 shows standard nutrient values that can be used to interpret the results of the analysis of pear leaf tissue.

Nitrogen (N): Mineral N fertilizers are generally applied annually to the soil surface, but fertigation would be a better option. N fertilizer applied to the soil is prone

Table 2.3 Standard nutrient values used to interpret the results of pear leaf tissue analysis[a]

	Deficiency	Low range	Normal range	High range
Dry matter (%)				
Nitrogen	<1.35	<1.60	1.60-2.40	>2.40
Phosphorus	<0.15	<0.18	0.18-0.26	>0.26
Potassium	<0.16	<0.20	0.20-2.00	>2.00
Calcium	<0.10	<1.30	1.3-3.00	>3.00
Magnesium	<0.05	<0.30	0.30-060	>0.60
mg kg^{-1}				
Manganese	<5	<20	20-200	>200
Iron	<40	<50	50-400	>400
Boron	<5	<35	35-80	>80
Zinc	<5	<20	20-200	>200

[a] *Adapted from Halbrendt et al., (2012).*

to volatilization and runoff losses. Very high individual application rates should be avoided as they lead to low N use efficiency and high nitrate leaching. Therefore, the total nitrogen dose should be divided into at least two parts to increase N use efficiency. The first smaller application should be added just before or after harvest, depending on harvest time. The absorbed N is stored in perennial wood for remobilization in the next season. The second major part, which coincides with the rapid growth of the shoots and the cell division and expansion of the fruit, is added around flowering a few weeks later. In the case of fertigation, the desired amount is applied by a larger number of smaller doses. Foliar applications can be helpful only when tissue N is below normal. The total annual amount of nitrogen should not exceed 100 kg N ha^{-1} for mature orchards with high crop load.

Potassium (K): The time of application of potassium through the soil is not as important as the time of application of nitrogen, as the tree needs potassium from flowering to fruit harvest and its mobility in the soil is generally lower than that of nitrogen. K fertilizer can be broadcast on the soil surface once after harvest on loamy soils especially when larger amounts need to be applied and when the soil is rich in clay or before flowering on sandy soils in humid regions. When potassium is added by fertigation, the nutrient supply is synchronized with the highest plant requirement, so that the application should be aimed at the period from flowering to a few weeks before harvest (Tagliavini and Marangoni, 2002).

Phosphorus (P): Total phosphorus consumption by the pear tree is relatively low, so deficits rarely occur under natural conditions, mainly in soils with inappropriate pH values. By changing the soil pH, the content of easily soluble phosphorus in the soil solution can be increased without having to add fertilizer. P fertilizer can be broadcast on the soil surface after harvest or by fertigation.

Calcium (Ca): The application of calcium in the form of limestone to the soil is usually done before the establishment of orchards. In the mature orchard calcium fertilizers are mainly applied to the canopy to be directly taken up by fruit, thereby increasing the Ca concentration in outer fruit layers (Tagliavini and Marangoni, 2002). $Ca(NO_3)_2$ and $CaCl_2$ applied as a foliar spray are usually recommended to prevent physiological disorders in the fruit.

Magnesium (Mg): Pear trees grown in sandy soils, especially those on which large quantities of fertilizers containing K^+ or NH_4^+ cations have been applied, often show a magnesium deficit. To prevent magnesium deficiency, large supplies of potassium (K^+) or ammonium-nitrogen fertilizer (NH_4^+) should be avoided. Magnesium sulphate or magnesium nitrate is generally used as soil or foliar Mg fertilizer.

Iron (Fe): Pear trees are very sensitive to iron deficiency, which leads to leaf chlorosis (Figure 2.19). Pear trees with chlorosis symptoms at flowering will have reduced fruit set and yield, poor quality, and are likely to only be biennial bearing (Tagliavini and Marangoni, 2002). The most common cause of iron chlorosis is the bicarbonate ion, which occurs in high amounts in calcareous soils. To alleviate Fe chlorosis, pear growers can use iron deficiency tolerant rootstocks, lower the pH in certain parts of the soil by adding acidifiers, use physiologically acidic fertilizers such as potassium sulfate or ammonium sulfate, and use synthetic Fe chelates to cure or prevent Fe deficiency. Fe-EDTA and Fe-DTPA chelates, which are characterized by low stability in alkaline soils, are used for foliar applications. Foliar treatments are only effective in situations with slight or moderate symptoms of iron chlorosis, and the effect is short-lived, requiring repeated applications to maintain the regreening of leaves. Fe-EDDHA, which is characterized by a higher stability in calcareous/alkaline soils, is generally used for soil applications. To avoid leaching of Fe

FIGURE 2.19

Pear trees are very sensitive to iron deficiency, which leads to chlorosis.

chelates, multiple additions of iron chelate in small amounts via fertigation during intensive shoot growth is necessary.

Boron (B): Soil application of boron is effective but requires great care, as the applied amounts can sometimes exhibit a phytotoxic effect (Johnson, 2008). Foliar applications of sodium-tetraborate or boric acid should be preferred, applied before flowering or at the end of summer.

Zinc (Zn): Zinc deficiency is common in high-pH, calcareous soils, as well as leached sandy soils. For foliar sprays, $ZnSO_4$ or a Zn chelate can be used, which are equally effective. The most common spraying period for pears is before flowering or after harvest (Swietlik, 2002).

Manganese (Mn): Manganese deficiency can occur in calcareous soils with high pH. On acidic soils, especially those with a pH below 5.5, the solubility of manganese increases and can reach toxic levels. In case of severe deficiency, foliar applications with manganese chelates or $MnSO_4$ can be carried out before flowering or after harvest.

5.5 Orchard floor and weed management

Orchard floor management is important for preserving favorable physical, chemical, and biological properties of the soil. The orchard floor can be divided into two distinct areas: the one between tree rows (alley row) and that below the trees (tree row). Alley rows can be clean cultivated or under permanent sod cover. Clean cultivation is recommended only in very dry conditions without irrigation, to eradicate weeds and preserve soil moisture. Most frequently, the alley row is in the form of a permanent grass cover, which is sown on purpose, selecting perennial grasses

FIGURE 2.20

Permanent alley row grass covers and herbicide strips in the tree rows.

with a shallow root system (Figure 2.20) mixed with leguminous plants, to improve N soil content without, or minimizing, the use of mineral N.

Such grass cover requires frequent mowing to reduce its transpiration and provide organic mulches beneath tree rows. The number of cuts during the year depends on the amount of precipitation. The tree rows are generally kept free of vegetation. Weed control is normally by herbicide use; more recently, concerns related to soil health are bringing back traditional tillage (Musacchi et al., 2021). Mechanical weed control is usually carried out using equipment such as discs or cultivators. More sustainable techniques that reduce soil disturbance are finger and blade weeders.

Herbicides are the primary tool used to manage vegetation under the tree row. In pear production, pre- and postemergence herbicides can be used together or separately in rotation. Postemergence herbicides are useful for eradication of difficult perennial weeds. Their application carries a risk especially in young pear orchards since some herbicides can be absorbed through green, not yet lignified, bark and damage/kill the young trees. No herbicides should be applied before the ground has settled around the tree and there are no cracks in the soil, to avoid contact with absorbing roots. Postemergence products should be applied when weeds are sufficiently small so that good coverage can be achieved with minimized potential for injury to the young bark of the trees.

6. Main diseases and pests and their management

Commercial pear production often takes place under mild and humid climatic conditions, which favor fungal diseases, such as pear scab *(Venturia pirina)*, brown spot of pear *(Stemphylium vesicarium)*, European fruit tree canker *(Neonectria ditissima)*, and postharvest fruit rots (Wenneker and Thomma, 2020). The use of synthetic fungicides is currently the main means to control diseases. However, despite the routine use of fungicide applications, fruit infections in the orchard are a growing problem. This may be due to the use of noneffective chemicals, ineffective spray application technologies, or inadequate timing of the applications. To develop effective control strategies, it is necessary to assess which are the most important pathogens that occur in a specific region on the crop. Such control methods should focus on reduction of the inoculum pressure of pathogens. Control of the complex diversity of pathogens in orchards is difficult because infections may occur during the entire period from flowering until postharvest (Wenneker and Thomma, 2020).

6.1 Orchard diseases

Pear scab (*Venturia pirina*; syn. *V. pyrina*): European pear scab is an economically important disease in most of the European pear growing areas worldwide (Spotts and Castagnoli, 2010), especially in organic pear cultivation. In Japan/Asia, pear scab on

FIGURE 2.21

(A) Pear scab lesion on Conference shoot (twig scab). (B) Pear scab lesion on Migo fruit. (C) Pear scab lesion on the underside of a Conference leaf. (D) Pear scab lesion on Conference fruit after storage (storage scab).

Asian pears *(Pyrus pyrifolia)* is caused by *Venturia nashicola* (Sutton et al., 2014). On leaves and fruit, the disease appears as olive green to dark brown to black, velvety circular spots (lesions) that consist of sporulating mycelia growing under the epidermis (Figure 2.21). The main impact of pear scab is the loss of fruit quality, while late-season fruit infections may not become visible until fruit have been in storage (Spotts and Castagnoli, 2010). *V. pirina* infects pear leaves, buds, fruit, and twigs. *V. pirina* overwinters on dead leaves as pseudothecia that release airborne ascospores (sexual spores), and it overwinters on twig lesions and mummified fruit as conidia (asexual spores). Ascospores, overwintering conidia, and mycelia are all sources of primary infections in early spring. Secondary (subsequent) infections are caused by conidia dispersed throughout the growing season. Occurrence and severity of twig infections varies among geographical regions (Sokolova et al., 2014). Sustainable pear scab management requires proper sanitation practices, resistant cultivars, and fungicide applications (Porsche et al., 2017). Fungicide applications are the major management strategy applied against pear scab in organic and conventional pear cultivation systems. Disease warning systems are used to predict the risk of infections. There is an urgent need for resistant and competitive pear cultivars. However, breeding programs for pear scab-resistant cultivars are still under development.

Fire blight *(Erwinia amylovora)*: Fire blight is a common and very destructive bacterial disease that can kill whole pear and young apple trees in one growing season of pears (Sutton et al., 2014). The disease can kill blossoms, fruit, shoots, twigs, branches, rootstocks, and entire pear trees (Figure 2.22). The symptoms of fire blight are necrosis with the browning or blackening of dead plant tissue, giving the appearance of burned by fire. Wilting of the shoot tip and the presence of ooze exudates from the stem tissue are typical symptoms and are caused by clogging of the vascular tissue with extracellular polysaccharide (EPS) produced by the bacterium (Sutton et al., 2014). *Erwinia amylovora* can infect and cause severe damage to many plants in the Rosaceae family, including many pome fruit and ornamental plants like apple,

FIGURE 2.22
Pear trees affected by fire blight.

pear, crabapple, mountain ash, hawthorn, and others. *E. amylovora* overwinters in fire blight cankers. In spring, the pathogen produces droplets of orange bacterial ooze that emerge on the canker. Ooze attracts vectors such as insects and birds but is also carried by wind, rain, contaminated pruning tools, and infected plant material to new infection sites. The control of fire blight usually involves planting resistant cultivars, spray application of copper bactericides, antibiotics, biological materials, activators of systemic acquired resistance (SAR), shoot growth retardants, and mechanical removal of infected tree parts by pruning or whole tree removal (Aćimović et al., 2021; Philion and Joubert, 2021). However, among the widespread pear cultivars commonly planted in key production regions, there are no varieties entirely resistant to fire blight (Montanari et al., 2016).

Brown spot of pears *(Stemphylium vesicarium)*: Brown spot of pear (BSP) has a large economic impact in several pear-producing areas located in Italy, Spain, France, Belgium, the Netherlands, and Portugal, and has been lately described in Argentina (Llorente et al., 2012). Symptoms consist of necrotic lesions on leaves, twigs, and fruit, and the fruit can also rot (Figure 2.23). The symptoms in the fruit consist of rounded brown spots, occasionally surrounded by a reddish halo. The symptoms in the leaves consist of necrotic spots that generally develop into a V shape. Saprophytic growth of the fungus on plant debris is common. The species is also able to cause disease in asparagus, onion, and other crops. However, only isolates originating from pear orchards, including isolates from dead grass leaves, are pathogenic on pear leaves or fruit in bioassays. Nonpathogenic isolates of *S. vesicarium* are also present in pear orchards (Köhl et al., 2009). The disease cycle

FIGURE 2.23

Brown spot lesion on Conference fruit.

is characterized by two phases; a pathogenic phase on the aerial pear organs during the growing season and a saprophytic phase on plant debris on the orchard ground during the year (Llorente et al., 2012). In the autumn, pseudothecia are formed on dead pear leaves and fruit on the orchard floor. *S. vesicarium* is able to colonize pear leaf debris and necrotic tissues of herb plants on the ground, where it produces abundant ascospores and conidia that are capable of infecting pear trees (Köhl et al., 2009). Differences in susceptibility to BSP have been observed among pear cultivars (Llorente et al., 2012). Forecasting models have been developed for use in warning systems, and new strategies of control have been proposed for disease management (Llorente et al., 2012). Integrated management of brown spot includes sanitation methods to decrease inoculum potential. These methods consist of leaf litter removal and application of biological control agents based on *Trichoderma* or *Bacillus* spp. to the orchard grounds.

Powdery mildew *(Podospheara leucotricha)*: Powdery mildew is present in most areas were pears are grown, and known as a serious disease of buds and foliage. The net-like russeting reduces the value of susceptible cultivars. Economic loss is related to cultivar susceptibility, climatic conditions, and orchard management practices (Sutton et al., 2014). Infected leaves, shoots, and fruit are covered by a light grey or white powdery coating of spores and mycelium (Sardella et al., 2016; Figure 2.24). *P. leucotricha* is an obligate parasite, thus requires living plant tissue to grow. The fungus overwinters in buds (in apple, not in pear). Germination does not occur in free water. Conidia germinate under conditions of high relative humidity. Abundant sporulation from overwintering shoots and secondary lesions on young foliage

FIGURE 2.24

(A) Powdery mildew on Migo fruitlet and leaves. (B) Powdery mildew on Migo fruitlet.

FIGURE 2.25

Reddening of the foliage of pear trees with pear decline.

leads to a rapid buildup of inoculum (Sutton et al, 2014). Management of powdery mildew of pears requires consideration of cultivar susceptibility. Reducing primary inoculum and protecting leaves and fruit from secondary infections are important. Timely application of effective fungicides is widely used to prevent new infections and reduce the number of spores produced on new lesions.

Pear decline *(Phytoplasma pyri)*: Pear is the main host of pear decline phytoplasma. Pear decline causes significant crop loss as well as the stunting and death of pear trees grown on highly susceptible rootstocks. Pear decline has been reported in several European countries, along the Pacific Coast and in parts of Northeastern United States, Argentina, Australia, and Israel (Sutton et al., 2014). Symptomatology depends on the cultivar and rootstock (Figure 2.25). Generally, the first symptom is seen in autumn when the leaves of affected trees develop a premature red color followed by early leaf fall. The second symptom is seen in the following spring when leaves remain small and pale, with little or no shoot growth and no fruit

FIGURE 2.26

Neonectria cankers on 1-year-old Conference twigs.

production. Pear decline is a disease associated with the presence of phytoplasmas in the sieve tubes (phloem tissue) of diseased pear trees. Pear decline can be transmitted by grafting, budding, and chip budding. The pear psyllids *Cacopsylla pyricola*, *C. pyrisuga,* and *C. pyri* are the vectors of pear decline phytoplasma, and they are mainly responsible for spreading the disease in the field. As a preventive measure, the use of healthy propagation material is a prerequisite. Intensive control of weeds and insect vectors in the orchards is also essential. Resistant rootstocks are the most efficient way to control the disease.

Fruit tree canker *(Neonectria ditissima)*: European canker (Nectria canker) is a common disease in pome fruit orchards, especially in areas with a humid climate and abundant rain, and may cause severe losses, mainly in temperate regions (Wenneker et al., 2017). The fungus has a broad host range and can infect both apple and pear. On pear, large outbreaks are only observed when there are favorable weather conditions but may occasionally cause severe incidences (Sutton et al., 2014). The formation of cankers on woody tissue is the most obvious symptom of *Neonectria ditissima* (Figure 2.26). Cankers can be formed on twigs, branches, and the main stem. *N. ditissima* is a wound pathogen, and all kinds of natural and artificial wounds can be infected. Leaf scars formed during leaf fall are considered to be the major site of infection. Rain is an important vector of the pathogen by aerial rain splash from tree to tree and by rain splash and runoff within infected trees. Sporulation, spore dispersal, and infection of *N. ditissima* are favored by rainfall. The main measures to control European canker are: (i) pruning out infected branches when weather conditions are less conducive to infection; (ii) fungicide applications after pruning, leaf fall, and picking; and (iii) painting of wounds. Control measures are still not sufficiently effective, and more information is needed to refine the management strategies (Harteveld et al., 2023).

6.2 Postharvest diseases

Blue mold (*Penicillium* spp.): Blue mold, known also as soft rot or wet rot, is one of the most destructive rots of stored pears (Yu et al., 2012). It is an important

FIGURE 2.27

Blue mold on Conference pear.

postharvest disease in all pear-producing countries, and its infection may occur during the growing season, at harvest, in storage, or transit (Spadaro et al., 2019). Under modern storage conditions, losses due to this rot should be less than 1%, but if storage problems occur or conditions are less than optimal, losses can be 50% or more (Sardella et al., 2016). Blue mold originates primarily from infection of wounds, such as punctures and bruises, on the fruit. The first symptoms are soft watery brown spots, which undergo rapid enlargement at temperatures between 20°C and 25°C (Figure 2.27). The lesions have a very sharp margin between diseased and healthy tissues, and the diseased tissue can be readily separated from healthy tissue. The surface of older lesions may be covered with masses of blue–green spores. The fungus can spread into neighboring healthy fruit, forming "nests" of decay (Spadaro et al., 2019). *P. expansum* is the most common species related to blue mold (Sutton et al., 2014). Some isolates of *P. expansum* produce the mycotoxin patulin with potential damage to public health, necessitating strict quality control of fruit used for processing (Snowdon, 1990). Conidia surviving on storage bins from year to year are the most important source of inoculum. Infections also arise from waterborne spores in postharvest drench solutions used to float the fruit onto packing lines (Sutton, 2014). Harvesting and handling methods that minimize wounds and bruising are essential for controlling blue mold. Sanitizing bins and storage room walls prior to harvest can significantly reduce the inoculum levels for *P. expansum.* Postharvest applications of fludioxonil or pyrimethanil provide excellent control of blue mold as well as other postharvest diseases. Biocontrol against *P. expansum* might also be a promising method (Wenneker and Thomma, 2020).

Grey mold *(Botrytis cinerea)*: Grey mold is a very common postharvest disease of pear worldwide that can cause heavy losses, especially in cold-stored pears, and develops more rapidly at cold storage temperatures than other postharvest decay (Sardella et al., 2016; Spadaro et al., 2019). If it is left uncontrolled, losses as high as 20% to 60% are not uncommon after an extended period of storage (Jones and Aldwinckle, 1990; Sutton et al., 2014). Grey mold can cause large losses because of the ability to spread from decayed to surrounding healthy fruit through fruit-to-fruit contact ("nesting") during storage. Grey mold originates primarily from infection of wounds, such as punctures and bruises. The disease can take

FIGURE 2.28

Grey mold on Conference pears.

several forms, such as calyx-end rot, stem-end rot, and belly rot (Figure 2.28). Grey mold causes a firm and light brown-to-brown decayed area without sharp margins. The decayed fruit is spongy, and diseased tissue does not separate from the healthy tissue as does blue mold (Spadaro et al., 2019). Under high relative humidity conditions and as storage extends, fluffy white to grey mycelium and greyish spore masses may appear (Sutton et al., 2014). Wet conditions during the growing season can be very conducive for *Botrytis* infections. *B. cinerea* is a fungus that can colonize senescing or dead tissue of many plant species. Conidia produced on infected plant material are the source of inoculum on the orchard. Conidia of the fungus are dispersed mainly by air currents and water splash. The fungus can infect flowers at bloom and fruit during the growing season. (Sutton et al., 2014). Good harvest and handling aimed at minimizing punctures and bruises on the fruit will limit the infections that may occur through wounds. Avoidance of wounds during harvest (special instructions have been made for pickers) reduced *Botrytis* problems in the Netherlands significantly during storage (Wenneker et al., 2021). Applying effective fungicides is key to reducing infections before harvest.

Brown rot (*Monilinia* spp.): Brown rot disease of pear is caused by *Monilinia* spp. *Monilinia fructigena* is especially widespread in Europe, Asia, and South America, and causes serious losses in pear fruit (Snowdon, 1990). Losses can take place preharvest in the orchard and postharvest during cold storage (Sutton et al., 2014). Brown rot on ripening or mature fruit typically develops as a rapidly spreading, firm, brown decay (Figure 2.29). The first symptoms on ripe fruit are small, superficial, circular brown spots that quickly begin rotting. The cream-white to buff sporulating area may have concentric zones of sporulation produced in response to diurnal cycles. The fungus often spreads by growth from diseased fruit to healthy ones in the same cluster or as conidia to other tissues (Xu et al., 2001a,b). The main causative fungi of brown rot are *Monilinia fructicola*, *M. fructigena*, and *M. laxa* (Jones and Aldwinckle, 1990). *M. fructigena* is the most common species on pear in Europe (Sutton et al., 2014). The brown rot fungi overwinter mainly in or on diseased mummified fruit either attached to the tree or on the ground. Other infected tissues on trees, such as twigs, peduncles, and cankers on twigs or branches, can also serve as sources of primary inoculum. Conidia are transported by wind, water, or

FIGURE 2.29

Monilinia rot on Conference pears.

insects to young fruit. Initial infection is via wounds often due to insect damage and especially those wounds made by birds (Snowdon, 1990; Xu et al., 2007). The most important animal vectors are birds, wasps (*Vespula* spp.), beetles, flies (Diptera), and lepidoptera (Holb and Scherm, 2008). Control of brown rot requires an integrated approach, i.e., combining effective control of insect pests and minimizing damage to fruit with close supervision of fruit pickers to ensure only sound fruit are harvested and stored. Removal of infected fruit from the tree and orchard floor during summer and winter reduces the inoculum available for later infections of mature fruit near harvest and overwintering inoculum.

Bull's eye rot (*Neofabraea* spp.): Bull's eye rot pear, previously known as Gloeosporium rot, is an important latent postharvest disease, occurring in major fruit-growing areas (Soto-Alvear et al., 2013). Bull's eye rot appears as singular spherical lesions, with well-defined margins. The diseases have light brown centers and a dark brown border (Figure 2.30). Decayed tissue is firm and acervuli are often present as wet, cream-colored masses in older lesions. Symptoms do not show up in the orchard or until after months of storage (Jones and Aldwinckle, 1990). Bull's eye lesions on apple and pear fruits are generally caused by four *Neofabraea* species: *N. alba*, *N. malicorticis*, *N. perennans,* and *N. kienholzii* (Spotts et al., 2009; Köhl et al., 2018). On pear, *N. alba* has been found in small cankers and pruning stubs, and also consistently detected in leaf litter of apple and pear and in necrotic tissues of dead weeds and grasses (Köhl et al., 2018). Current management practices to control *Neofabraea* spp. in the orchard include pruning of cankers from infected trees to

FIGURE 2.30

Bull's eye rot of pears, with acervuli and spore masses of *Neofabraea alba*.

minimize the buildup of inoculum during the fruit growing season and removal of fallen fruit and dead tree branches from the orchard floor (Creemers, 2014). Furthermore, fungicide application is a common component of bull's eye rot management.

Phytophthora rot (*Phytophthora* spp.): In North America and Europe, several species of *Phytophthora* (mostly *P. syringae* and *P. cactorum*) infect fruit before and during storage (Snowdon, 1990). The disease also occurs in regions with overhead irrigation, when irrigation water is contaminated with *Phytophthora* spp., which is then referred to as "sprinkler-rot" (Sutton et al., 2014). All commonly grown pear varieties are susceptible to the disease. The fungus might also be important as the causal agent of collar rot of pear trees (Snowdon, 1990). The lesions are pale to dark brown in pears, and the boundary between healthy and rotted flesh is firm (Snowdon, 1990). Visible white growth is fungal mycelium that typically appears after several days of incubation at 100% relative humidity (Figure 2.31). On immature pears, the disease is sometimes confused with fire blight. Several *Phytophthora* species are currently known to cause fruit rot of pome fruit, but pear rot is caused mainly by *P. cactorum* (Sanchez et al., 2019). Infection of fruit in the orchard results from rain splash of contaminated soil just before or during harvest, and the lower part of the tree is more often affected (Snowdon, 1990). The infected pears develop rot during the first months of storage. If the soil below the trees is kept free of weeds (generally considered as good orchard practice to conserve soil moisture), soil particles are splashed up by rain, and zoospores come into contact with low-hanging fruit. Zoospores can be present in irrigation water (surface water included); using overtree irrigation provides conditions for disease development by delivering the pathogen to the fruit. Production of fruit near the orchard floor should be avoided.

FIGURE 2.31

Phytophthora rot of Conference pears.

FIGURE 2.32

Side rot of Conference pears.

The fruit should be protected by orchard sprays and a postharvest dip or drench (Snowdon, 1990), and overhead irrigation systems should be avoided to minimize risk of disease (Sutton et al., 2014).

Side rot (*Cadophora* spp.): Side rot of pears is recognized as the cause of major storage losses in the Pacific Northwest (Sugar and Spotts, 1992), the Netherlands, and Belgium (Wenneker et al., 2016; 2021). Typical round to oval, dark brown, and slightly sunken spots (size 0.5 to 1.0 cm in diameter) appeared after 6 or more months of cold storage under CA (Sugar, 2014; Wenneker et al., 2016; Figure 2.32). The primary causal fungus is identified as *C. malorum* (syn. *Phialophora malorum*) in the United States (Sugar and Spotts, 1992). The occurrence of *C. luteo-olivacea* in the Netherlands causing side rot on pear has recently been reported (Wenneker et al., 2016). Results using quantitative PCR and specific primers showed that necrotic pear leaves fallen to the ground, mummies and cankers, and other plant species present on the orchard floor (e.g., weeds and grass) were identified as potential sources of inoculum for *C. luteo-olivacea* (Kohl et al., 2015; Wenneker et al., 2016). Routine application of preharvest fungicides (such as captan and fludioxonil) did not effectively control postharvest side rot in the Netherlands. Good results to control side rot were obtained with postharvest drenches with pyrimethanil. This is the current practice for long storage of pears in the Netherlands.

6.3 Pests

Codling moth *(Cydia pomonella)*: The codling moth is found in most pome fruit production areas around the world. Although apple is the preferred host, it also feeds upon pear, quince, crab apple, and walnut (Sutton et al, 2014). Pears have some natural resistance to attack by codling moth when fruit are small because of their hardness; however, pears can become heavily infected in late summer as they mature. Fruit damage is of two types: (i) a deep entry is where the larva enters the fruit, tunnels into the center, and feeds on the seeds. Brown frass, the feces produced by the larvae, can often be seen extruding from the entry hole; (ii) a sting is a shallow entry where the larva does some feeding but does not actually tunnel into the fruit (Figure 2.33). Effective codling moth control is essential for both integrated and organic fruit growers. The decision support system RIMpro-Cydia was developed for fruit growers and advisors. Precautionary measures, such as removal of wood piles from the orchard and the affected fruit during thinning, are the first step in the effective control of codling moth. Other measures in the strategy involve applications of Cydia pomonella granulovirus (CpGV) and netting (e.g., Alt'Carpo nets) using a single-plot or single-row system. Pheromone mating disruption can also be an effective tool to control codling moth.

Pear psylla (*Cacopsylla* spp.): Several species of pear-feeding psyllids are recognized worldwide, but those identified as the pear psylla, the most important pest of commercial pears, are *Cacopsylla pyricola* in North America and *Cacopsylla pyri* in Europe (Sutton et al., 2014). It has multiple generations per year, a large reproductive capacity, and readily develops resistance to pesticides. Pear psylla adults look like small, dark reddish brown cicadas and overwinter as adults, sheltering on the bark of pear trees (Figure 2.34). The adults become active in spring

FIGURE 2.33

(A) Codling moth adult. (B) Codling moth larva. (C) Codling moth larva entry hole, with brown frass.

FIGURE 2.34

(A) Pear psylla adults. (B) Pear psylla nymphs. (C) Sooty mold on honeydew from pear psylla nymphs.

and start to feed on young leaves and flowers. They begin laying eggs when pear buds begin to swell. Nymphs have sucking mouthparts and feed on plant sap. Depending on the temperature, there are three or more generations per year (Winkler et al., 2007). Pear psylla nymphs produce large amounts of honeydew that encourages the growth of black sooty mold on pear fruit and leaves, resulting in downgraded and rejected fruit. Adult *C. pyri* are a vector of the pathogen "pear decline" *(Candidatus Phytoplasma pyri)*, which reduces shoot and fruit growth and can lead to tree death (Süle et al., 2007). In the past, pear growers relied on synthetic insecticides to control *C. pyri*, while in integrated pest management (IPM) approaches, pesticides with known adverse effects on natural enemies of pear psylla were replaced by selective or nonchemical ones, such as particle film treatment (PFT), against the spring generation of pear psylla and mating disruption against codling moth (Civolani et al., 2023; Puterka et al., 2000).

Pear bud weevil *(Anthonomus pyri)*: The pear bud weevil occurs in Europe and Russia. In autumn, the weevil destroys reproductive and vegetative buds. Since the withdrawal of broad-spectrum insecticides like carbamates and organophosphates and the restricted use of neonicotinoids, pear bud weevils have become an increasing problem in pear fruit in Europe (Beliën et al., 2020). The pear bud weevil forms one generation per year. The weevils principally cause damage because their larvae develop and feed on reproductive or vegetative buds (Figure 2.35). Since the general adoption of IPM and the application of more selective instead of broad-spectrum insecticides, *A. pyri* populations can locally escalate to damaging levels and become a major pest.

Pear midge *(Contarinia pyrivora)*: The pear midge is present in pear orchards in North America and Europe. It is generally controlled by chemical applications for other pests (Sutton et al., 2014). The pear midge overwinters as pupa in the soil. Adults emerge from the soil after spring as small flying midges. Adult females deposit eggs between the petals of unopened flowers. The newly hatched larvae start to feed on the pulp from within the developing fruitlets. Full-grown larvae exit from the fruit (on the tree or the ground), burrow into the soil to pupate later, and emerge the next spring as adult midges. The species has only one generation each year (Sutton et al., 2014, Figure 2.36). Practical recommendation for the control and management

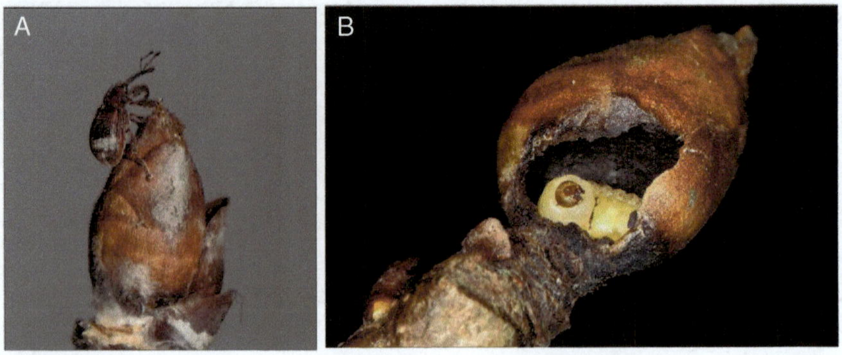

FIGURE 2.35

(A) Pear bud weevil adult. (B) Pear bud larva in feeding on pear bud.

FIGURE 2.36

(A) Distortion of pear fruitlet caused by pear midge infestation. (B) Larvae of pear midge inside pear fruitlet.

of the pear midge are: (i) pest monitoring: the specific pheromone in traps is a sensitive and accurate means of monitoring the short emergence period of adult midges; (ii) removal of young, infested fruit before fruit and larvae fall to the ground; (iii) insecticides, such as pyrethrin, Spinosad and mineral oil; and (iv) mass trapping (Jones et al., 2016).

7. Current and future challenges in main production regions

Climate change will have a significant impact on agriculture, necessitating the development of adaptation measures for risks to agricultural productivity globally. Pear production is considered vulnerable to climate change, including global warming, because pear trees are less adaptable to climate than annual crops. Recent temperature increases have caused various problems in pear production, including incomplete dormancy, flower bud dieback, frost damage due to early flowering, and fruit sunburn. Fire blight, caused by *Erwinia amylovora*, is expanding its infestation. New diseases and pests are also a concern. Pear production areas are expected to move to higher latitudes due to global warming. Adaptation measures must be developed using cultivation technology to utilize the trees that are being cultivated now. It is also necessary to breed new varieties that are resistant to global warming. In the future, it is desirable to build a data-driven smart agriculture system that utilizes advanced technologies such as robots, AI, and ICT (Information and Communication Technology). Specifically, the development of automatic harvesting robots, automatic application of pesticides based on orchard weather and image data, forecasting of disease and insect damage, growth prediction, and harvest season prediction are expected in the future.

References

Aćimović, S. G., Meredith, C. L., Santander, R. D., & Khodadadi, F. (2021). Proof of concept for shoot blight and fire blight canker management with postinfection spray applications of prohexadione-calcium and acibenzolar-SMethyl in apple. *Plant Disease, 105*, 4095–4105.

Bell, R. L., & Itai, A. (2011). In C. R. Kole (Ed.), *Pyrus* in *wild crop relatives: Genomic and breeding resources Vol.6 temperate fruits* (pp. 147–178). Heidelberg: Springer.

Beliën, T., Clymans, R., Bangels, E., Thys, T., & Bylemans, D. (2020). Control of *Anthonomus* spp. weevils in IPM pome fruit orchards. *Acta Horticulturae, 1269*, 209–220.

Boini, A., Manfrini, L., Bortolotti, G., Corelli-Grappadelli, L., & Morandi, B. (2019). Monitoring fruit daily growth indicates the onset of mild drought stress in apple. *Scientia Horticulturae, 256*, Article 108520.

Boland, A. M., Ziehrl, A., & Beaumon, J. (2002). Irrigation Management. *Orchard Crops. Guide to best practice in water management* (pp. 21–50). Knoxfield, Victoria, Australia: Highway Press.

Bower, J. H., Biasi, W. V., & Mitcham, E. J. (2003). Effect of ethylene in the storage environment on the quality of Bartlett pears. *Postharvest Biology and Technology, 28*, 371–379.

Brewer, L., & Volz, R. (2019). Genetics and breeding of pear. *The Pear Genome* (pp. 63–101). Cham, Switzerland: Springer.

Charrier, A., Vergne, E., Dousset, N., Richer, A., Petiteau, A., & Chevreau, E. (2019). Efficient targeted mutagenesis in apple and first-time edition of pear using the CRISPR-Cas9 system. *Frontiers in Plant Science, 10*, 40.

Civolani, S., Soroker, V., Cooper, W. R., & Horton, D. R. (2023). Diversity, biology, and management of the pear psyllids: A global look. *Annals of the Entomological Society of America, 116*, 331–357.

Clayton, M., Biasi, W. V., Southwick, S. M., & Mitcham, E. J. (2000). Retain affects maturity and ripening of Bartlett pear. *HortScience, 35*, 1294–1299.

Creemers, P. (2014). Anthracnose canker and perennial canker. In T. B. Sutton, H. S. Aldwinckle, A. M. Agnello, & J. F. Walgenbach (Eds.), *Compendium of apple and pear diseases and pests* (2nd ed., pp. 51–53). St. Paul, MN: APS Press.

Crisosto, C. H. (2016). Asian pear. In K. C. Gross, C. Y. Wang, & M. Saltveit (Eds.), *The commercial storage of fruits, vegetables, and florist and nursery stocks* (Vol. 66, pp. 205–209). USDA Agriculture Handbook.

Deckers, T., & Schoofs, H. (2001). Management of high-density pear orchards. *The Compact Fruit Tree, 34*, 117–120.

da Silva, G. J., Villa, F., Grimaldi, F., da Silva, P. S., & Welter, J. F. (2018). In J. Al-Khayri, S. Jain, & D. Johnson (Eds.), *Pear (Pyrus spp.) Breeding in Advances in Plant Breeding Strategies: Fruits* (pp. 131–163). Heidelberg: Springer.

Drake, S. R., & Elfving, D. C. (2004). Quality of packed and bin stored Anjou pears as influenced by storage atmosphere and temperature. *Journal of Food Quality, 27*, 141–152.

Einhorn, T. C. (2021). A review of recent *Pyrus, Cydonia* and *Amelanchier* rootstock selections for high-density pear plantings. *Acta Horticulturae, 1303*, 185–196.

El-Sharkawy, I., Jones, B., Gentzbittel, L., Lelievre, J. M., Pech, J. M., & Latche, A. (2004). Differential regulation of ACC synthase genes in cold-dependent and independent ripening in pear fruit. *Plant, Cell & Environment, 27*, 1197.

Elsysy, M. A., Hubbard, A., & Einhorn, T. C. (2020). Postbloom thinning of Bartlett pear with metamitron. *HortScience, 55*, 174–180.

Evans, R. C., & Campbell, C. S. (2002). The origin of the apple subfamily (*Maloideae; Rosaceae*) is clarified by DNA sequence data from duplicated GBSSI genes. *American Journal of Botany, 89*, 1478–1484.

Fischer, T. C., Malnoy, M., Hofmann, T., Schwab, W., Palmieri, L., Wehrens, R., et al. (2014). F1 hybrid of cultivated apple (*Malus* × *domestica*) and European pear (*Pyrus communis*) with fertile F2 offspring. *Molecular Breeding, 34*, 817–828.

Halbrendt, J. M., et al. (2012). Plant Nutrition. *Tree Fruit production guide. Pennsylvania 2012-2013* (pp. 10–16). The Pennsylvania State University.

Holb, I. J., & Scherm, H. (2008). Quantitative relationships between different injury factors and development of brown rot caused by *Monilinia fructigena* in integrated and organic apple orchards. *Phytopathology, 98*, 79–86.

Harteveld, D. O. C., Goedhart, P. W., Houwers, I., Köhl, J., de Jong, P. F., & Wenneker, M. (2023). Detecting the asymptomatic colonization of apple branches by *Neonectria ditissima*, causing European canker of apple. *European Journal of Plant Pathology, 166*, 291–301.

Isuzugawa, K., Murayama, H., & Nishio, T. (2014). Characterization of a giant-fruit mutant exhibiting fruit-limited polyploidization in pear (*Pyrus communis* L.). *Scientia Horticulturae, 170*, 196–202.

Janssens, P., Deckers, T., Elsen, F., Elsen, A., Schoofs, H., Verjans, W., et al. (2011). Sensitivity of root pruned Conference pear to water deficit in a temperate climate. *Agricultural Water Management, 99*, 58–66.

Johnson, R. S., & Handley, D. F. (2000). Using water stress to control vegetative growth and productivity of temperate fruit trees. *HortScience, 35*, 1048–1050.

Johnson, R. S. (2008). Nutrient and water requirements of peach trees. In D. R. Layne, & D. Bassi (Eds.), *The peach: Botany, production and uses* (pp. 303–331). Cambridge, Massachusetts: CAB International.

Jones, A. L., & Aldwinckle, H. S. (1990). *Compendium of apple and pear diseases* (p. 224). Saint Paul, MN: APS Press.

Jones, H. G. (2004). Irrigation scheduling: Advantages and pitfalls of plant-based method.s. *Journal of Experimental Botany, 55*, 2427–2436.

Jones, V. P., Mills, N. J., Brunner, J. F., Horton, D. R., Beers, E. H., Unruh, T. R., et al. (2016). From planning to execution to the future: An overview of a concerted effort to enhance biological control in apple, pear, and walnut orchards in the western US. *Biological Control, 102*, 1–6.

Köhl, J., Groenenboom-de Haas, B. H., Kastelein, P., Rossi, V., & Waalwijk, C. (2009). Quantitative detection of pear pathogenic *Stemphylium vesicarium* in orchards. *Phytopathology, 99*, 1377–1386.

Köhl, J., Wenneker, M., Groenenboom-de Haas, B. H., R. Anbergen, R., Goossen-van de Geijn, H. M., Lombaers-van der Plas, et al. (2018). Dynamics of post-harvest pathogens *Neofabraea* spp. and *Cadophora* spp. in plant residues in Dutch apple and pear orchards. *Plant Pathology, 67*, 1264–1277.

Kaur, S., Gill, M. S., Gill, P. P. S., Jawandha, S. K., & Singh, N. P. (2023). Influence of harvest date on storage quality of Asian pear (*Pyrus pyrifolia*) fruit. *Erwerbs-Obstbau, 65*, 1331–1339.

Kikuchi, A. (1946). Speciation and taxonomy of Chinese pears. *Collected Records of Hort Res Kyoto University, 3*, 1–8.

Kupferman, E. (2003). Controlled atmosphere storage of apples and pears. *Acta Horticulturae, 600*, 729–735.

Köhl, J., Wenneker, M., Groenenboom-de Haas, B. H., Anbergen, R., Goossen-van de Geijn, H. M., der Plas, Lombaers-van, et al. (2018). Dynamics of post-harvest pathogens *Neofabraea* spp. and *Cadophora* spp. in plant residues in Dutch apple and pear orchards. *Plant Pathology, 67*, 1264–1277.

López, A., Zon, K., Dussi, M. C., Reeb, P., Giardina, G., Leskovar, M., et al. (2011). Economic evaluation between chemical thinning vs. hand thinning in Williams pear. *Acta Horticulturae, 909*, 29–37.

Lauri, P. E. (2009). European pear architecture and branch management in France: An INRA research program. *The Compact Fruit Tree, 42*(2), 12–14.

Lelievre, J. M., Tichit, L., Dao, P., Fillion, L., Nam, Y. W., Pech, J. C., & Latche, A. (1997). Effect of chilling on the expression of ethylene biosynthetic genes in Passe-Crassane pear fruits. *Plant Molecular Biology, 33*, 847–855.

Li, J., Zhang, M., Li, X., Khan, A., Kumar, S., Allan, A. C., & Wu, J. (2022). Pear genetics: Recent advances, new prospects, and a roadmap for the future. *Horticulture Research, 9*, Article uhab040.

Lindo-García, V., Larrigaudière, C., Escheverría, G., Murayama, H., Soria, Y., & Ginè-Bordonaba, J. (2019). New insights on the ripening pattern of Blanquilla pears: A comparison between on- and off-tree ripened fruit. *Postharvest Biology and Technology, 150*, 112–121.

Lindo-García, V., Muñoz, P., Larrigaudière, C., Munnè-Bosch, S., & Ginè-Bordonaba, J. (2020). Interplay between hormones and assimilates during pear development and ripening and its relationship with the fruit postharvest behaviour. *Plant Science, 291*, Article 110339.

Llorente, I., Moragrega, C., Ruz, L., & Montesinos, E. (2012). An update on control of brown spot of pear. *Trees, 26*, 239–245.

Lombard, P. B., & Westwood, M. N. (1987). Pear rootstocks. In R. C. Rom, & R. F. Carlson (Eds.), *Rootstocks for fruit crops* (pp. 145–183). New York: John Wiley.

Lordan, S. J., Alegre, S., Montserrat, R., & Asin, L. (2017). Yield and profitability of Conference pear in five training systems in North East of Spain. *Spanish Journal of Agricultural Research, 15*, Article e0904.

Matsuda, N., Ikeda, K., Kurosaka, M., Takashina, T., Isuzugawa, K., Endo, T., et al. (2009). Early flowering phenotype in transgenic pears (*Pyrus communis* L.) expressing the CiFT gene. *Journal of the Japanese Society for Horticultural Science, 78*, 410–416.

McClymont, L., Goodwin, I., Whitfield, D., O'Connell, M., & Turpin, S. (2021). Effects of rootstock, tree density and training system on early growth, yield and fruit quality of Blush pear. *HortScience, 56*, 1408–1415.

Meheriuk, M., Prange, R. K., Lidster, P. D., & Porritt, S. W. (1994). Postharvest disorders of apples and pears. *Agriculture and Agri-Food Canada*, 1737/E.

Montanari, S., Prechepied, L., Renault, D., Frijters, L., Velasco, R., & Horner, M. (2016). A QTL detected in an interspecific pear population confers stable fire blight resistance across different environments and genetic backgrounds. *Molecular Breeding, 36*, 47.

Morgan, D. R., Soltis, D. E., & Robertson, K. R. (1994). Systematic and evolutionary implications of *rbcL* sequence variation in *Rosaceae*. *American Journal of Botany, 81*, 890–903.

Mourgues, F., Chevreau, E., Lambert, C., & Bondt, A. (1996). Efficient Agrobacterium-mediated transformation and recovery of transgenic plants from pear (*Pyrus communis* L.). *Plant Cell Reports, 16*, 245–249.

Musacchi, S., Iglesias, I., & Neri, D. (2021). Training systems and sustainable orchard management for European pear (*Pyrus communis L.*) in the mediterranean area: A review. *Agronomy, 11*, 1765.

Musacchi, S., Sheick, R., Mia, M. J., & Serra, S. (2023). Studies on physiological and productive effects of multi-leader training systems and Prohexadione-Ca applications on apple cultivar WA 38. *Scientia Horticulturae, 312*, Article 111850.

Naor, A., Gal, Y., & Peres, M. (2006). The inherent variability of water stress indicators in apple, nectarine and pear orchards, and the validity of a leaf-selection procedure for water potential measurements. *Irrigation Science, 24*, 129–135.

Nham, N. T., Macnish, A. J., Zakharov, F., & Mitcham, E. J. (2017). Bartlett pear fruit (*Pyrus communis* L.) ripening regulation by low temperatures involves genes associated with jasmonic acid, cold response, and transcription factors. *Plant Science, 260*, 8–18.

Nham, N. T., de Freitas, S. T., Macnish, A. J., Carr, K. M., Kietikul, T., Guilatco, A. J., et al. (2015). A transcriptome approach towards understanding the development of ripening capacity in Bartlett pears (*Pyrus communis* L.). *BMC Genomics, 16*, 762.

Pasquariello, M. S., Rega, P., Migliozzi, T., Capuano, L. R., Scortichini, M., & Petriccione, M. (2013). Effect of cold storage and shelf life on physiological and quality traits of early ripening pear cultivars. *Scientia Horticulturae, 162*, 341–350.

Ou, C., Zhang, X., Wang, F., Zhang, L., Zhang, Y., Fang, M., et al. (2020). A 14 nucleotide deletion mutation in the coding region of the PpBBX24 gene is associated with the red skin of "Zaosu Red" pear (*Pyrus pyrifolia* White Pear Group): a deletion in the PpBBX24 gene is associated with the red skin of pear. *Horticulture research, 7*, 39.

Perchepied, L., Chevreau, E., Ravon, E., Gaillard, S., Pelletier, S., Bahut, M., et al. (2021). Successful intergeneric transfer of a major apple scab resistance gene (Rvi6) from apple to pear and precise comparison of the downstream molecular mechanisms of this resistance in both species. *BMC Genomics, 22*, 1–18.

Philion, V., & Joubert, V. (2021). Use of a growth regulator (prohexadione-Ca) and summer pruning as post symptom rescue treatments following a fire blight infection during bloom. *The Plant Pathology Journal, 103*, 163–174.

Porsche, F. M., Pfeiffer, B., & Kollar, A. (2017). A new phytosanitary method to reduce the ascospore potential of *Venturia inaequalis*. *Plant Disease, 101*, 414–420.

Puterka, G. J., Glenn, D. M., Sekutowski, D. G., Unruh, T. R., & Jones, S. K. (2000). Progress toward liquid formulations of particle films for insect and disease control in pear. *Environmental Entomology, 29*, 329–339.

Radivojevic, D., Fotiric, M., Milivojevic, J., & Oparnica, C. (2022). Chemical thinning induced by single and sequential application of BA, NAA and metamitron on Gala and Golden Delicious apple. *Acta Horticulturae, 1341*, 1–8.

Reynoird, J. P., Mourgues, F., Norelli, J., Aldwinckle, H. S., Brisset, M. N., & Chevreau, E. (1999). First evidence for improved resistance to fire blight in transgenic pear expressing the attacin E gene from *Hyalophora cecropia. Plant Science, 149*, 23–31.

Robinson, T. (2011). High density pear production with *Pyrus communis* rootstocks. *Acta Horticulturae, 909*, 259–269.

Sánchez, E. E., Righetti, T. L., & Sugar, D. (1998). Partitioning and recycling of fall applied boron in Comice pears. *Acta Horticulturae, 475*, 347–354.

Süle, S., Jenser, G., Szita, E., Bertaccini, A., & Maini, S. (2007). Management of pear decline caused by *Candidatus* Phytoplasma pyri in Hungary. *Bull Insectology, 60*, 319–320.

Sanchez, A. D., Sosa, M. C., Lutz, M. C., Carreño, G. A., Ousset, M. J., & Lucero, G. S. (2019). Identification and pathogenicity of *Phytophthora* species in pear commercial orchards in Argentina. *European Journal of Plant Pathology, 154*, 811–822.

Saito, T. (2016). Advances in Japanese pear breeding in Japan. *Breeding Science, 66*, 46–59.

Sansavini, S., Ancarani, V., & Neri, D. (2008). Overview of intensive pear culture: Planting density, rootstocks, orchard management, soil-water relations and fruit quality. *Acta Horticulturae, 800*, 35–50.

Saquet, A. A. (2019). Storage of pears. *Scientia Horticulturae, 246*, 1009–1016.

Sardella, D., Muscat, A., Brincat, J.-P., Gatt, R., Stephen, D., & Valdramidis, V. (2016). A comprehensive review of the pear fungal diseases. *International Journal of Fruit Science, 16*, 351–377.

Sassa, H., Hirano, H., Nishio, T., & Koba, T. (1997). Style-specific self-compatible mutation caused by deletion of the S-RNase gene in Japanese pear (*Pyrus serotina*). *The Plant Journal, 12*, 223–227.

Sawamura, Y., Mase, N., Takada, N., Sato, A., Nishitani, C., Abe, K., et al. (2013). A self-compatible pollen-part mutant of Japanese pear produced by crossing Kosui with pollen from gamma-irradiated Kosui. *The Japanese Society for Horticultural Science, 82*, 222–226.

Schupp, J. R., Winzeler, H. E., Kon, T. M., Marini, R. P., Baugher, T. A., Kime, L. F., et al. (2017). A method for quantifying whole-tree pruning severity in mature tall spindle apple plantings. *HortScience, 52*, 1233–1240.

Snowdon, A. L. (1990). Pome fruits. In , *Vol. 1. A colour atlas of postharvest diseases and disorders of fruits and vegetables* (pp. 170–216). London, UK: Wolfe Scientific.

Shirasawa, K., Itai, A., & Isobe, S. (2021). Chromosome-scale genome assembly of Japanese pear (*Pyrus pyrifolia*) variety 'Nijisseiki' *DNA Research, 28*(2). April 2021, dsab001.

Sokolova, O., & Moročko-Bičevska, I. (2021). Evaluation of *Venturia pyrina* virulence on European pear (*Pyrus communis*) cultivars by an in vitro methodology. *Journal of Phytopathology*, 1–10. https://doi.org/10.1111/jph.13002

Sokolova, O., Moročko-Bičevska, I., & Bankina, B. (2014). Review of the pear scab caused by *Venturia pirina. Research for Rural Development, 1*, 26–33.

Soto-Alvear, S., Lolas, M., Rosales, I. M., Chávez, E. R., & Latorre, B. A. (2013). Characterization of the bull's eye rot of apple in Chile. *Plant Disease, 97*, 485–490.

Spadaro, D., Torres, R., Errampalli, D., Everett, K., Ramos, L., & Mari, M. (2019). Pome fruits. In L. Palou, & J. L. Smilanick (Eds.), *Postharvest pathology of fresh hortcultural produce*. Boca Raton: CRC Press, Taylor and Francis Group.

Spotts, R. A., & Castagnoli, S. (2010). Pear scab in Oregon: Symptoms, disease cycle and management. *Oregon State University Extension Service*. https://catalog.extension.oregonstate.edu/sites/catalog.extension.oregonstate.edu/files/project/pdf/em9003.pdf.

Spotts, R. A., Seifert, K. A., & Wallis, K. M. (2009). Description of *Cryptosporiopsis kienholzii* and species profiles of *Neofabraea* in major pome fruit growing districts in the Pacific Northwest USA. *Mycological Research, 113*, 1301–1311.

Sugar, D., & Basile, S. R. (2013). Integrated ethylene and temperature conditioning for induction of ripening capacity in Anjou and Comice pears. *Postharvest Biology and Technology, 83*, 9–16.

Sugar, D., & Hilton, R. J. (2011). Potential organic methods for management of pear scab. *Acta Horticulturae, 909*, 527–530.

Sugar, D. (2014). Side rot. In T. B. Sutton, H. S. Aldwinckle, A. M. Agnello, & J. F. Walgenbach (Eds.), *Compendium of apple and pear diseases and pests* (2nd ed., pp. 81–82). St. Paul, MN: The American Phytopathological Society.

Sugar, D., & Spotts, R. A. (1993). Dispersal of inoculum of *Phialophora malorum* in pear orchards and inoculum redistribution in pear immersion tanks. *Plant Disease, 77*, 47–49.

Sugar, D., & Spotts, R. A. (1992). Sources of inoculum of *Phialophora malorum*, causal agent of side rot of pear. *Phytopathology, 82*, 735–738.

Sugar, D., & Basile, S. R. (2011). Orchard calcium and fungicide treatments mitigate effects of delayed postharvest fungicide applications for control of postharvest decay of pear fruit. *Postharvest Biology and Technology, 60*, 52–56.

Sutton, T. B., Aldwinckle, H. S., Agnello, A. M., & Walgenbach, J. F. (2014). *Compendium of apple and pear diseases and pests* (2nd ed.). Saint Paul, MN: APS Press.

Swietlik, D. (2002). Zinc nutrition of fruit trees by foliar sprays. *Acta Horticulturae, 594*, 123–129.

Tagliavini, M., & Marangoni, B. (2002). Major nutritional issues in deciduous fruit orchards of Northern Italy. *Hort Technology, 12*, 26–31.

Tamura, F. (2012). Recent advances in research on Japanese pear rootstocks. *Journal of the Japanese Society for Horticultural Science, 81*, 1–10.

Theron, K. I., Lötze, G. F. A., & Reynolds, J. S. (2018). Chemical thinning of Forelle pear with S-abscisic acid and 1-aminocyclopropane-1-carboxylic acid. *Acta Horticulturae, 1206*, 13–20.

Tominaga, A., Ito, A., Sugiura, T., & Yamane, H. (2022). How is global warming affecting fruit tree blooming? "Flowering (dormancy) disorder" in Japanese pear (*Pyrus pyrifolia*) as a case study. *Frontiers in Plant Science, 12*, Article 787638.

Torres, C. A., Valdivia, A., Jorquera, G., & Hernandez, O. (2019). The use of DA meter to assess apple and pear maturity in Chile. *Acta Horticulturae, 1256*, 63–70.

Tustin, D. S., Breen, K. C., & van Hooijdonk, B. M. (2022). Light utilisation, leaf canopy properties and fruiting responses of narrow-row, planar cordon apple orchard planting systems—a study of the productivity of apple. *Scientia Horticulturae, 294*, Article 110778.

Velasco, R., Zharkikh, A., Affourtit, J., et al. (2010). The genome of the domesticated apple (*Malus × domestica Borkh.*). *Nature Genetics, 42*, 833–839.

Villalobos-Acuña, M., & Mitcham, E. J. (2008). Ripening of European pears: The chilling dilemma. *Postharvest Biology and Technology, 49*, 187–200.

Villalobos-Acuña, M., Biasi, W. V., Flores, S., Jiang, C. Z., Reid, M. S., Willits, N. H., et al. (2011). Effect of maturity and cold storage on ethylene biosynthesis and ripening in Bartlett pears treated after harvest with 1-MCP. *Postharvest Biology and Technology, 59*, 1–9.

Wang, C., Tian, Y., Buck, E. J., et al. (2011). Genetic mapping of PcDw determining pear dwarf trait. *Journal of the American Society for Horticultural Science, 136*, 48–53.

Wang, C. H., Li, W., Tian, Y. K., et al. (2016). Development of molecular markers for genetic and physical mapping of the PcDw locus in pear (*Pyrus communis* L.). *Journal of Horticultural Science and Biotechnology, 91*, 299–307.

Wenneker, M., de Jong, P. F., Joosten, N. N., Goedhart, P. W., & Thomma, B. P. H. J. (2017). Development of a method for detection of latent European fruit tree canker (*Neonectria ditissima*) infections in apple and pear nurseries. *European Journal of Plant Pathology, 148*, 631–635.

Wenneker, M., Pham, K. T. K., Lemmers, M. E. C., et al. (2016). First report of *Cadophora luteo-olivacea* causing side rot on Conference pears in The Netherlands. *Plant Disease, 100*, 2162.

Wenneker, M., Pham, K. T. K., Boekhoudt, L. C., et al. (2017). First report of *Neofabraea kienholzii* causing bull's eye rot on pear (*Pyrus communis*) in The Netherlands. *Plant Disease, 101*, 634.

Wenneker, M., Pham, K. T. K., & Köhl, J. (2021). Important postharvest pathogens of pear and their dynamics in plant residues in Dutch pear orchards. In R. Zoppolo, et al. (Eds.), *Acta Horticulturae: Vol. 1303. Proc. XIII international pear symposium* (pp. 501–508).

Wenneker, M., & Thomma, B. P. H. J. (2020). Latent postharvest pathogens of pome fruit and their management: From single measures to a systems intervention approach. *European Journal of Plant Pathology, 156*, 663–681.

Wertheim, S. (2000). Developments in the chemical thinning of apple and pear. *Plant Growth Regulation, 31*, 85–100.

Winkler, K., Helsen, H., & Devkota, B. H. (2007). Predatory bugs show higher abundance close to flower strips in pear orchards. *Proceedings of the Netherlands Entomological Society Meeting, 18*, 31–36.

Wertheim, S. J., Wagenmakers, P. S., Bootsma, J. H., & Groot, M. J. (2000). Orchard systems– conditions for success. *The Compact Fruit Tree, 33*(3), 79–81.

Wu, J., Wang, Z., Shi, Z., Zhang, S., Ming, R., Zhu, S., et al. (2013). The genome of the pear (*Pyrus bretschneideri* Rehd.). *Genome Research, 23*, 396–408.

Xu, X.-M., Butt, D. J., & Ridout, M. S. (1998). The effects of inoculum dose, duration of wet period, temperature and wound age on infection by *Nectria galligena* of pruning wounds on apple. *European Journal of Plant Pathology, 104*, 511–519.

Xu, X. M., Bertone, C., & Berrie, A. (2007). Effects of wounding, fruit age and wetness duration on the development of cherry brown rot in the UK. *Plant Pathology, 56*, 114–119.

Xu, X. M., Guerin, L., & Robinson, J. D. (2001a). Effects of temperature and relative humidity on conidial germination and viability, colonization and sporulation of *Monilinia fructigena*. *Plant Pathology, 50*, 561–568.

Xue, C., Guo, S., Gu, K., Xing, M., & Wu, J. (2023). Optimization of the regeneration and *Agrobacterium*-mediated transformation in pear. *Horticulture Advances, 1*, 13.

Yamamoto, T., & Terakami, S. (2016). Genomics of pear and other Rosaceae fruit trees. *Breeding Science, 66*, 148–159.

Yu, J., & Wang, Y. (2017). The combination of ethoxyquin, 1-methylcyclopropene and ethylene treatments controls superficial scald of d'Anjou pears with recovery of ripening capacity after long-term controlled atmosphere storage. *Postharvest Biology and Technology, 127*, 53–59.

Yu, T., Yu, C., Chen, F., Sheng, K., Zhou, T., Zunun, M., et al. (2012). Integrated control of blue mold in pear fruit by combined application of chitosan, a biocontrol yeast and calcium chloride. *Postharvest Biology and Technology, 69*, 49–53.

Zhang, S. J., Wu, J., Chen, H., Gu, C., Tao, S. T., Wu, J. Y., et al. (2011). Identification of differentially expressed genes in a spontaneous mutant of Naguoli pear (*Pyrus ussuriensis* Maxim) with large fruit. *The Journal of Horticultural Science and Biotechnology, 86*, 595–602.

Zielinski, Q. B., & Thompson, M. M. (1967). *Speciation in* Pyrus*; chromosome number and meiotic behavior* (Vol. 128, pp. 109–112). Bot Gaz.

CHAPTER

Peach

3

Gemma Reig[1], Gregory Reighard[2], Theodore DeJong[3], Luca Corelli Grappadelli[4], James E. Adaskaveg[5], George A. Manganaris[6]

[1]*Department of Agricultural and Forest Engineering, University of Lleida, Lleida, Catalonia, Spain;* [2]*Department of Plant and Environmental Sciences, Clemson, SC, United States;* [3]*Plant Sciences Department, UC Davis, Davis, CA, United States;* [4]*Department of Agricultural and Food Science and Technology, University of Bologna, Alma Mater Studiorum, Italy;* [5]*Department of Microbiology and Plant Pathology, University of California, Riverside, CA, United States;* [6]*Cyprus University of Technology, Department of Agricultural Sciences, Biotechnology & Food Science, Lemesos, Cyprus*

Chapter outline

Temperate Tree Fruits and Nuts. https://doi.org/10.1016/B978-0-443-29137-1.00003-7

1. Origin and spread

Peach [*Prunus persica* (L.) Batsch] was originally domesticated in China and came to Persia (modern-day Iran) via the Silk Trading Route in the 2nd or 3rd century BCE (Manganaris et al., 2023a). From there, it spread into the Mediterranean regions, first Greece and then Rome by the 1st century BC, where Romans spread peach trees throughout their empire, Spain being part of it. Peach trees came to Florida, Mexico, and South America in the mid-1500s via Spanish and Portuguese explorers. It became feral in the Southeastern United States and Mexico and was further spread throughout North America by the Native Americans (Hancock et al., 2008). In the early 18th century, Spanish missionaries introduced the cultivation of peach to California. About the same time, the Russians apparently brought peach seeds or trees by ship to Western North America (Manganaris et al., 2023a). The next phase of introduction in the U.S. took place by direct introduction from China and England in the mid-1850s (Bassi and Monet, 2008). A few Chinese peach cultivars of good quality were intensively used in breeding programs, and most of the current American and European commercial cultivars are their descendants, resulting in a reduction of genetic variability among most current commercial cultivars (Aranzana et al., 2012).

Seed propagation was the main means of distributing this crop until the middle of the 19th century in the United States and Europe, and until the middle of the 20th century in Central and South America. This suggests that there are many peach cultivars that have been subjected to several centuries of regional selection and genetic improvement for adaptation to climate, soils, and availability of water. In addition, the spread of peach trees to Southern China, Taiwan, and Thailand gave rise to low and mid-chill germplasm, which was spread to Florida, throughout the United States and Southern Europe, North Africa, and South America in the 1970s (Manganaris et al., 2023a).

There is controversy about the origin and evolutionary history of this fruit, since the direct wild ancestor of peach remains unknown and is likely extinct, but some results suggest that it emerged between 0.6 and 2.47 million years ago (Mya) (Yu et al., 2018). A study conducted by Zheng et al. (2014) found fossil evidence indicating that peach cultivation and domestication could date back to at least 7500 years ago in the Yangtze River Valley of Southern China. In addition, peach endocarp fossils from 2.6 Mya, found recently in Kunming (Southwest China), are indistinguishable from endocarps of modern peach cultivars, suggesting that peaches may have acquired edible attributes of modern fruits long before domestication, perhaps mediated by frugivorous primates (Su et al., 2015).

2. Economic importance

Peach, including nectarine, is a tree crop widely cultivated in temperate and subtropical zones throughout the world, with an average annual global production of 24.8 million tonnes in the last 5 years. Currently, China accounts for 64% of this production, followed by the European Union (Spain, Italy, Greece, and France) at 12%, Turkey at 4%, the U.S. with approximately 3%, and Iran at 2% (FAOSTAT, 2024). Over the last 10 years, world peach production, including nectarine, has increased more than 20% from 21.2 million tonnes to 26.3 million tonnes in 2022/2023 (Figure 3.1A). This increase appears to indicate an expanding industry; however, some countries have declined in production since 2013/2014, namely the European Union (Figure 3.1C) and the United States (Figure 3.1D). Turkey and Chile contributed to global growth, but the upward trend globally has been overwhelmingly due to expansion in China (Figure 3.1B). Overall, China gains more than offsetting the total combined losses in the rest of the world.

China's average annual production for the years 2018-2022 was 15.4 million tonnes, from more than 863,000 hectares (see Figure 3.1B). China has several peach-producing provinces, but most of the production occurs in the central area. However, the peach-growing area is declining in Northern and Central China but rising in the hillier and mountainous western provinces, with ongoing removal of orchards from traditional farmland (USDA, 2024). Recently, peach greenhouse production and high- density cultivation have emerged in limited areas of Northern China. Due to the addition of these new production areas, combined with improving

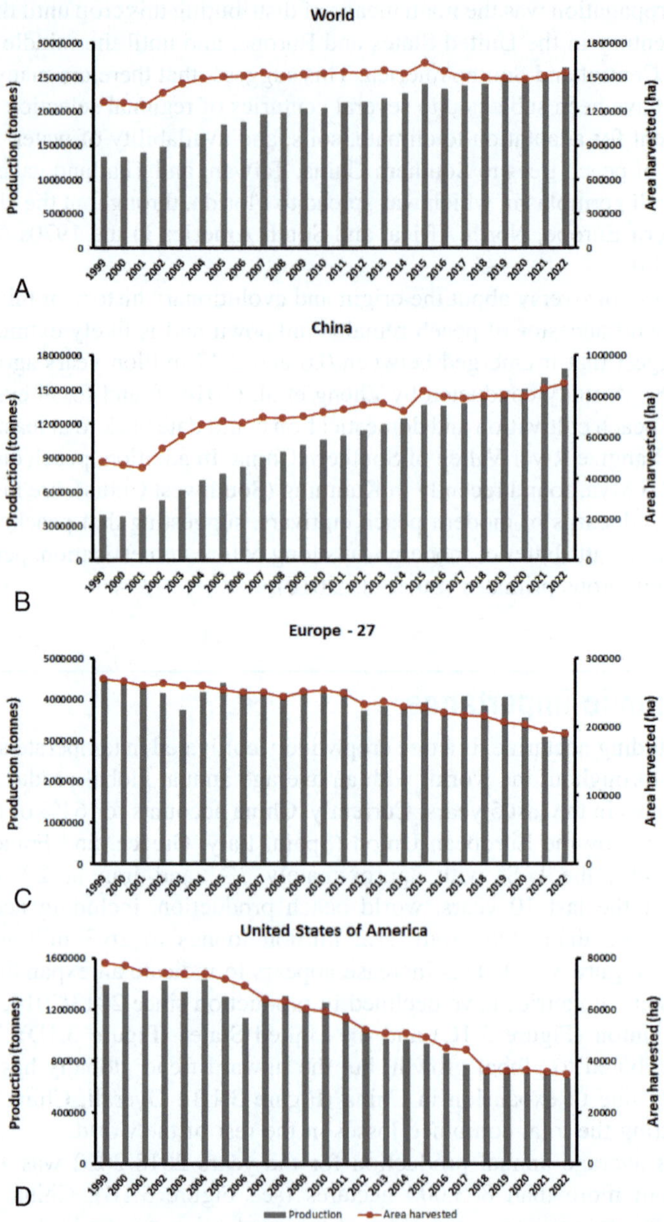

FIGURE 3.1

Evolution of peach production and harvested area (1999-2022) worldwide (A), and for the main peach continent producers, China (B), Europe (C), and United States (D).

From FAOSTAT (2024).

production efficiency, China is steadily increasing volumes of production (Manganaris et al., 2023a).

The European Union (EU) is ranked as the second largest producer with 3.2 million tonnes on approximately 190,000 hectares in 2022 (see Figure 3.1C). Since 2019, peach production has dropped 20%, as damaging rain and frost each year impacted yields, especially in the top growing member state, Spain. During the last decade, total EU peach hectares declined yearly with more rapid reductions in the past 5 years. Oversupply due to declining demand and poor returns for EU growers have prompted tree removals in the top growing countries of Spain and Italy (USDA, 2024).

Within the EU, Spain has led with average peach production of 1.3 million tonnes (Figure 3.2A) over a 5-year period (2018−2022) on 69,700 hectares (Europêch, 2022; FAOSTAT, 2024). The Ebro Valley (Northeastern Spain) continues to be the main producing area, led by Aragon and Catalonia, accounting for around 60% of the total production. This region is followed by the southeastern regions led by Murcia with 30% and Extremadura and Andalusia. Italy with a 5-year average production of 1.1 million tonnes (2018-2022) produced on approximately 56,000 hectares is the second producer within the EU and the third worldwide. Over recent decades, its production has moved south from the traditional northern regions, especially Emilia-Romagna, following the introduction of mid- to low-chill cultivars for both peaches and nectarines, which are better suited to the South. Greece is fifth worldwide and the third EU producer with almost 900,000 tonnes on about 39,000 hectares. The main production is in the northern part of the country with Imathia and Pella being the key producing counties. Peach cultivation has expanded to central Greece and in the southern part with the early-ripening cultivars (Manganaris et al., 2023a). Finally, France is the fourth peach producer in the EU with a 5-year annual average production of 210,000 tonnes (2018−2022) on 11,000 hectares. Its production is mainly located in the Occitania region.

Turkey is the fourth largest worldwide peach producer and the second contributor to growth of world peach production after China for the years 2018-2022. In particular, Turkey produced 1 million tonnes in 2022 on 53,000 hectares (FAOSTAT, 2024). Bursa, Canakkale, Izmir, and Mersin on the Aegean and Mediterranean Seas are the top provinces where peaches and nectarines are produced.

The U.S. ranks as the sixth largest producer with 666,000 tonnes on approximately 35,000 hectares in 2022 (see Figure 3.1D); however, production has declined over 37% in the last decade. The production is mainly confined to California, which produces nearly 80% of the U.S. fresh peach crop, followed by South Carolina, Georgia, Pennsylvania, New Jersey, and Colorado (Manganaris et al., 2023a). California also produces almost 100% of canned peaches, which has dropped 43% since 2013 due to the declining demand (USDA, 2024).

Despite the descending trends in terms of production volumes due to decreasing land planted with peach, the overall production per area planted has grown due to advancements in fruit production systems that are linked to the use of new cultivars, rootstocks, training systems, and modern growing techniques. In addition, the fresh

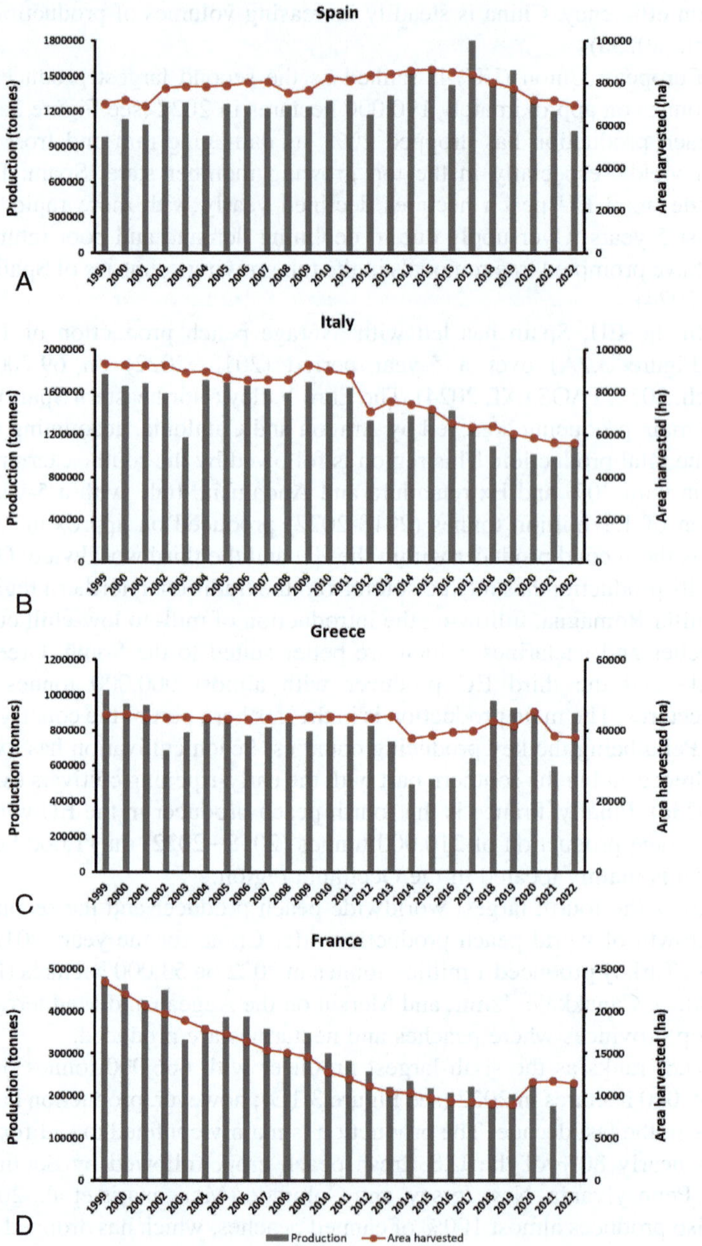

FIGURE 3.2

Evolution of peach production and harvested area (1999–2022) at the European Union level: Spain (A), Italy (B), Greece (C), and France (D).

From FAOSTAT (2024).

and processed sector has developed new products. Indeed, the current approach to increase fresh market consumption, and therefore increase demand, includes releases of cultivars having fruit with enhanced flavor, high total soluble solids (TSS), high aromatic profiles, desirable texture, attractive appearance, increased nutritional value (including high antioxidants and vitamin C contents), and/or cultivars with new fruit types (albino, red-fleshed, flat shape, edible seeds, etc.).

Within the context of increasing production, it is worth mentioning that the introduction and establishment of flat shape types of fruit, both with peach and nectarine, in the main peach growing areas in Spain is a unique case in the world. Its production started from local cultivars at marginal sites, mainly in the Ebro Valley and Murcia. At the end of the 1990s, the introduction of improved cultivars from France and Italy, with fruit that had more red skin color coverage and more regular shapes, fruit size, and stylar cavity closure, together with superior handling and transport conditions, led the Spanish peach sector to develop this fruit type on a commercial scale (Font i Forcada et al., 2022). During the past two decades, breeding programs have improved and increased their cultivars, adapting their goals to climate change challenges and to sector and consumer needs. Currently, Spain is the largest European producer and exporter of flat peaches and nectarines with a total production of 226,000 tonnes (18% of the total peach production in 2022) on more than 10,000 hectares. In 2021, flat nectarine production was 1% of total production; however, growers had attempted to establish them in Spanish orchards one decade earlier with limited success due to the lack of adaptability to climatic conditions, acidic flavor, and poor quality and appearance.

In 2022, Turkey was the top exporter of fresh peaches (25% of the export market share), followed by the European Union (15%), Chile (13%), and the United States (8%) (USDA, 2024). Russia is the top market for Turkey, importing more than 80% of its total production. Currently, half of the total EU exports go to the United Kingdom. Chile is the leading Southern Hemisphere supplier to Europe, United States, and China, while the United States mainly exports to Canada and Mexico.

3. Botany and relevant species

Prunus persica (L.) Batsch belongs to the *Rosaceae* family, subfamily *Prunoideae*, and is grouped within the *Prunus* genus and included in the *Euamygdalus* Schneid section of the *Amygdalus* subgenus. It is a diploid (8 chromosomes, $2n = 16$) with a small genome size (265 M base pairs/1C), self-fertile, and naturally self-pollinating. Peach trees are composed of the root system and the aerial part (trunk, branches, shoots, leaves, flowers, and fruit). The root system, which is usually quite shallow under irrigated conditions (50-60 cm in depth depending on the soil type), is a network of fine roots (orange-white while young, turning dark orange when older with large lenticels) that spread widely, which maximizes and enhances soil contact for efficiently absorbing water and nutrients. Root architectural traits (depth, density, specific length, root/shoot ratio, number of root hairs) determine the temporal and spatial distribution of the root system in the soil, playing a fundamental role in the ability to absorb nutrients (Bassi and Monet, 2008).

The aerial part of peach trees grows to a medium height (4-6 m but up to 8 m) if not pruned, with a smooth dark grey and brown bark that becomes rough with aging. Its growth (or tree habit) is determined by the number and location of lateral shoots produced per year, the strength of apical control, and branch angles. Most commercial peaches have a relatively strong apical dominance, thus producing few lateral shoots in the current year growth (Figure 3.3A). There are distinct growth habits (fastigiate, upright, upright to spreading, spreading and drooping), but the spreading habit is the most common among the current commercial cultivars (Font i Forcada, 2022). Efforts have been made to develop cultivars with upright and drooping growth habits to decrease cost of production but with minimal commercial success (Bassi et al., 2023).

The leaves are lanceolate, glabrous, and serrate, broadest near the middle, often with a glandular petiole. They are usually flat, 5 to 10 cm wide by 10 to 20 cm long, green but turning to a yellow color in the fall before they drop. Leaves are arranged on shoots in an alternating pattern. The flowers are generally pink (Figure 3.3B) but can also be white or red, hermaphroditic, and give rise to fruit, which develops from a single carpel (the ovary, which contains two ovules and is characterized by lack of fuzz in the case of nectarine) and bears one or two seeds (Bassi and Monet, 2008; Manganaris et al., 2023a). Peach cultivars have different fruiting habits, which directly affects pruning. However, most of the current commercial cultivars produce the best fruit quality on 1-year-old shoots (Figure 3.3C), particularly cultivars with

FIGURE 3.3

Peach tree at bloom time trained to an open vase (A), open flower at stage 65 according to BBCH-scale for stone fruit (B), and 1-year old shoots (C).

early and mid-season fruit maturity, whereas some mid- and most of the late-season cultivars produce more fruit on older shoots (Iglesias et al., 2023).

Botanically, the peach fruit is a drupe with a fleshy texture consisting of several parts. It has a thin, edible outer skin (exocarp) derived from the ovary wall, which is a protective layer composed of cuticle, epidermal cells, and some hypodermal cell layers. The epidermis consists of thick-walled flexible cells and surface trichomes or hairs ("fuzz": absent in nectarines) that are extensions of some epidermal cells. An edible flesh of varying thickness beneath the skin (mesocarp) is the main edible portion of the fruit. It consists mainly of storage parenchyma tissue of large and relatively thin-walled cells with high content of water, sugars, acids, and nutrients, including numerous health-promoting compounds. A lignified inner wall (endocarp), commonly referred to as the stone or pit, encloses one or very rarely two seeds. The seed is composed of several parts depending on its maturity. The outer tissue of the seed is the testa or seed coat. Initially, the nucellus occupies the entire inner volume of the seed. With time, this gives way to development of the endosperm, which subsequently gives way to the developing embryo with two cotyledons. Thus, the seed is initially filled with metabolically "cheap" nucellar tissue that is progressively replaced by "higher cost of production" tissues, such as proteins, fats, and carbohydrates (Manganaris et al., 2023a; Reig et al., 2023).

The fruit also presents other discernible morphological characteristics such as skin color (greenish, orange, purple, red, yellow, or white, with many combinations of these colors), flesh color (red, yellow, white, yellow-red, white-red), shape (round, elongated, and flat) (Figure 3.4), size (from 51 mm to more than 90 mm of diameter), flesh texture (melting, slow melting or slow softening, nonmelting, stony hard), flesh acidity (sub-acid [<5 meq 100 mL^{-1}] or sweet [5-9 meq 100 mL^{-1}], balanced [9-12

FIGURE 3.4

Different peach types and flesh colors: Elegant Lady, a yellow-fleshed peach (A); Sweet Regal, a white-fleshed peach (B). Clariss, a yellow-fleshed nectarine (C). Nectarnow, a white-fleshed nectarine (D); Diablotina, a red-fleshed nectarine (E); Ferlot, a yellow-fleshed nonmelting peach (F); Ice peach, a white-fleshed nonmelting peach (G); Mistral-30, a yellow-fleshed flat nectarine (H); Osiris, a yellow-fleshed flat peach (I); and SweetCap, a white-fleshed flat peach.

(Figure adapted from CABI Peach book)

meq 100 mL^{-1}], and acid [12-15 meq 100 mL^{-1}]), and stone adherence (clingstone, freestone, semi-freestone) (Iglesias and Echeverría, 2009; Reig et al., 2023).

Besides *P. persica*, the most widely grown species among the *Prunus* genus, there are four other wild diploid *Prunus* species from China, which are closely related to peach. Table 3.1 includes the main known tree characteristics and uses of each one (Bassi and Monet, 2008; Cao et al., 2022).

4. Breeding and propagation

4.1 Cultivars and selection criteria

Peach as a diploid tree species with eight chromosomes ($2n = 16$) representing eight linkage groups in molecular research. It has many (e.g., 43) tree and fruit traits that are inherited by simple Mendelian inheritance and are relatively easy to incorporate into breeding lines via classical breeding (Monet and Bassi, 2008; Bassi et al., 2023). Peach is grown globally but also has environmental, climatic, and edaphic challenges that breeders must overcome. Quantitatively inherited, adaptive traits targeted by breeders include cold hardiness, chilling requirement, heat requirement, bloom time, fruit development period, and in subtropical regions, high heat tolerance during bloom. Soil characteristics such as high bulk density (clay), excessively drained (sand), high pH (alkaline), and salinity (high salts) have been addressed with rootstock cultivar breeding primarily with interspecific crosses of *Prunus sp.*

Chilling requirement, which is the required minimum cold during winter for flower and vegetative buds to complete rest and emerge from dormancy, is a priority in cultivar development of many breeding programs due to climate change. Warmer winters have been an issue for many older cultivars for blooming normally and setting a peach crop. Traditional cultivars requiring 800 or more chill hours now do not receive enough chill in many warmer production areas such as the Southeastern United States and Mediterranean countries (Figure 3.5). Breeders are creating lower chill cultivars but are also trying to increase the flower bud postdormancy heat requirement to delay bloom and avoid spring freezes. However, many new low chill cultivars tend to bloom earlier than older cultivars and are at increased risk of spring frost events.

Fortunately, chilling requirement is moderately to highly heritable; therefore, breeders can achieve significant genetic gain through selection of parents based on phenotype and recurrent mass selection. Low-chill cultivars have associated problems such as excessive blind nodes, bud abortion, and poor fruit shape that is amplified by inconsistent winter chilling and/or warm spring temperatures in subtropical regions. Breeders have been successful in producing cultivars for regions with mild winters like Southern California, Florida, Brazil, Southeast Asia, Australia, South Africa, and the Mediterranean basin. However, tolerance to or avoidance of freezing temperatures during bloom remains a priority in many production areas. Therefore, selecting for traits such as later phenology, high flower bud

Table 3.1 The most interesting species considered close relatives of peach

***Prunus* sp.**	***P. davidiana* (Carr.) Franch.**	***P. mira* Koehne**	***P. kansuensis* Rehd.**	***P. ferganensis* (Kost. and Rjab)**
Origin	Central China	Qinghai-Tibet plateau in China	Northeast China	Western China
Common name	Mountain peach or Shan Tao	Tibetan peach or Guang He Tao	Wild peach or Kansu Tao	Xinjiang Tao
Estimated genome size (Mb)	237.29	242.94	238.06	237.24
Tree height	Tall (up to 10 m)	Tall (up to 20 m)	-	Tall (up to 8 m)
Bark color	Reddish-brown	-	-	-
Leaves	Ovate-lanceolate, long and glabrous	Lanceolate, villous along the midrib beneath and rounded at base	Villous along the midrib near the base and broadest below the middle	-
Flower	White or light pink	White	-	-
Pit	Small and pitted	Smooth	Furrowed not pitted	-
Uses and special traits	To hybridize with peach, to improve cultivar disease resistance (PPV, PM, LC), or to breed interspecific rootstocks adaptable to marginal soils or replant problems. It is resistant to cold, drought environments, and aphids	Used as seedling rootstock in some regions of India. It shows strong adaptability to cold, high-altitude, and drought, disease, and ultraviolet ray (UV) resistance	It shows high resistance to root-knot nematodes	To improve cultivar disease resistance (PM)

LC, *Leaf curl;* PPV, *plum pox virus;* PM, *powdery mildew.*

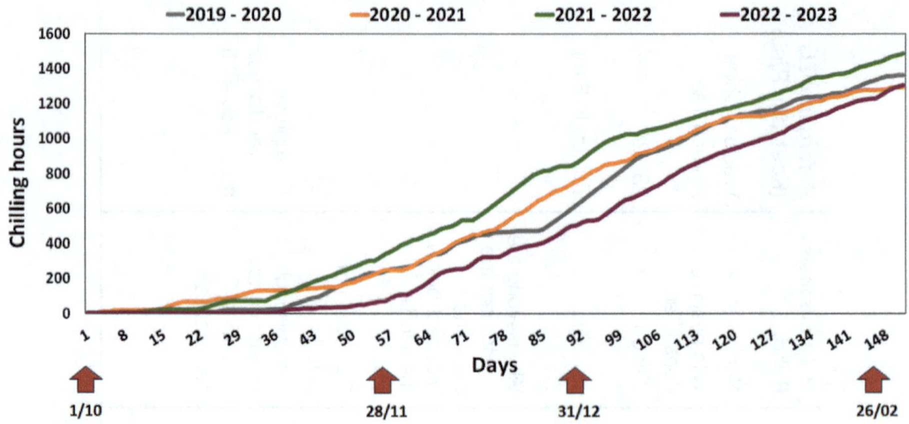

FIGURE 3.5

Chilling hour evolution (from 2019 to 2023) of the main peach growing area in Catalonia, Spain *(From Ruralcat, 2024).*

density, and genetic tolerance to lower freezing temperatures have resulted in late blooming cultivars with high bud set but often have counteracting consequences of indirectly selecting for higher chilling cultivars.

The peach fruit has a diversity of flesh and skin colors, textures, rate of softening, shapes, sizes, and flavors (Manganaris et al., 2022; Bassi et al., 2023; Reig et al., 2023). Flesh color can be white (dominant gene), yellow, or red with different shades within these colors. White flesh peaches were preferred in Asia (country of origin) and some European countries until the late 20th century, while yellow-flesh peaches were, and still are, preferred by consumers in the Americas and most of Europe. However, breeders have focused on breeding more white flesh peaches and nectarines for the markets that traditionally preferred yellow flesh peaches and nectarines. They also have developed new blood-red flesh peaches to be marketed for the health benefits of anthocyanins (antioxidants) (Figure 3.4E). Moreover, several centuries old, red flesh cultivars were only pickled for use due to their tartness and still have a niche market in France and Germany.

Geneticists have mapped four "hot spots" on chromosome 4 that includes fruit quality traits such as stone adhesion to the flesh (freestone or clingstone), flesh texture (melting or nonmelting), fruit weight/size, fruit development period, bloom date (days to bloom), ripening date, maturity, and individual sugars and soluble solids (Gasic and Saski, 2019). Molecular markers for important Mendelian characters such as peach vs. nectarine fruit, flat vs. round fruit, clingstone vs. freestone fruit, and low acidity in fruit have been identified for breeding purposes (Bassi et al., 2023).

Skin color is important for fresh market peaches and nectarines but much less for processing cultivars. Most Western countries prefer a majority red skin overcolor, but Asia and some South American countries and Spain have different or complementary skin color preferences. In Spain, most of the commercial peach cultivars produced and consumed are red-skinned, but there is another type (yellow skin with nonmelting texture) that is consumed both fresh and canned, representing 20% of the total peach production in Spain. Additionally, in some countries, including Spain, yellow fruits are covered with bags, when they are for fresh market, and are promoted as specialty products, mostly during the mid and late harvest season. Red skin color is affected by environment (light) and controlled by several recessive genes that are used in developing new cultivars with a high degree of anthocyanin production in the skin. In contrast, peaches with no anthocyanins have been bred, including an Italian cultivar series trademarked Ghiacco (Ice peach) (Figure 3.4G).

Peaches can have three different flesh textures, which are referred to as melting, nonmelting, and stony hard (Byrne et al., 2012). The melting ("freestone") and nonmelting ("clingstone") flesh types are controlled by four alleles, which allow flexibility to breed for flesh type since the nonmelting trait is recessive to the melting flesh trait. Melting flesh is the main fruit type marketed for fresh consumption, while nonmelting cultivars are largely grown for processing due to their firmness (less bruising and increased flesh integrity), though they are also favored in Asia and several other countries for the fresh market, such as Spain. Currently, new nonmelting cultivars are often sold as fresh market peaches due in part to their increased firmness and unique flavors. The third flesh type, stony hard, is a monogenic recessive trait inherited independently from the other two flesh types. It has been used by breeders to create peach and nectarine cultivars that have very firm flesh, delayed ripening/softening, high sugar content, and increased tree retention of ripe fruit to aid in harvesting operations. A fourth flesh type resembles the stony hard phenotype, but the flesh becomes melting when fully ripe. Due to the long delay until softening occurs, these genotypes are referred to as slow softening cultivars.

Fruit shape is heritable, though multiple genes are involved, and the environment, especially in hot growing regions, can affect it. Peach cultivars are selected for fruit that are roundish and not prone to have prominent tips or sutures, uneven halves, and split pits. The fruit shape of most historical and modern peach cultivars was round and symmetrical, which aided in minimizing bruising during packing and shipping. Other peach shapes are the peento, saucer, doughnut, or flat. These flat-shaped peaches were common in China and eventually became more available in Western markets. One limitation of these peaches is a tendency to tear at the stem when picked and/or develop cracking at the calyx end. These postharvest defects reduce commercial packout and thus require a higher retail price. Fortunately, currently breeders are putting new cultivars on the market with none of these defects, especially those harvested in early and mid-season (Figure 3.6).

Peach fruit size has greatly increased in modern breeding programs. Consumers are attracted to large fruit, and larger-sized fruit fill more baskets and bring higher prices. Selecting for large-fruited genotypes can have trade-offs in quality (i.e., diluted sugars, more bruising, uneven ripening) as well as physiological limitations when selecting for genotypes that require a short fruit development period. Fruit size is also influenced by orchard culture, especially time of flower or fruit thinning and irrigation scheduling.

FIGURE 3.6

Fully red-skinned, flat peach and flat nectarine with improved calyx end.

Peach flavor is a key attribute for the consumers. Peaches are expected to be sweet, and low acid cultivars taste very sweet, but they can be less flavorful than fruit that contain more acidity. Breeders select for genotypes that have fruit with at least 10% TSS that in some cultivars, especially nectarines, can reach 20% (Byrne et al., 2012). The ratio of sugars to acids in peach fruit is important to determine perceived sweetness and flavor profile. There is sufficient genetic heritability to increase sugar content for new cultivars. Acidity levels in peach fruit are controlled by both qualitative and quantitative genes with a dominant gene for low acidity. The dominant nature of the low acid gene has made it feasible to incorporate low acidity into yellow flesh cultivars.

Peaches, like other stone fruits, have desirable phytochemicals that include carotenoids, phenolics, vitamins, and organic acids that can be improved through breeding. However, there is limited effort to increase these compounds, partly due to the difficulty to increase concentrations without negatively affecting flavor and other quality traits. Breeders have been trying to improve postharvest fruit quality to provide consumers with a flavorful peach after it completes its journey through the supply chain. A common complaint and major factor for why peach consumption per capita has declined compared to other fruits is inconsistent fruit quality. This can be attributed to lack of aroma, bland taste, flesh mealiness, and flesh browning, most of which can be a result of poor postharvest handling and storage temperatures (Crisosto, 2006). Much effort has been directed through breeding to develop cultivars that are less affected by longer times spent in the supply chain. Nonmelting cultivars tend to be more tolerant to flesh disorders.

Global concerns about agricultural sustainability have emphasized the reduction of agrochemical inputs in fruit production. Peach is susceptible to several fungal and bacterial diseases that are difficult to control in humid temperate regions even with pesticides. Breeders are focused on incorporating genetic resistance to pathogens such as brown rot *(Monilinia sp.)*, *Taphrina* leaf curl, powdery mildew, and *Xanthomonas arboricola* leaf spot (Bassi et al., 2023). Gains in genetic resistance have been noteworthy thus far for the latter three diseases. Chromosome 4 harbors disease resistance quantitative trait loci (QTLs) that cluster along the chromosome as "hot spots." This region also includes QTLs for resistance to powdery mildew, brown rot, bacterial spot, and peach tree short life (Gasic and Saski, 2019). Resistance to insects that feed on peach fruit has received minimal focus though resistance to green peach aphid, which transmits the plum pox virus (Sharka), has been identified.

Peach tree growth habit is relatively easy to manipulate with gene mutations that include brachytic dwarf, semi-dwarf, weeping, spur type, columnar and intermediate forms such as upright and compact (Scorza et al., 2006). As orchard systems convert to higher density plantings and become intensively mechanized to replace hand labor, new cultivars continue to be developed such as Crimson Rocket with these modified architectural forms. However, few cultivars with modified tree forms have been accepted by commercial producers.

Fortunately, peach has a small genome (n = 265 Mb) for a tree species, and thus it is conducive to molecular research using biotechnology to make genetic advances or improvement sooner than classical breeding for a tree species. The first sequenced peach genome was released in 2013 (http://www.rosaceae.org/peach/genome) and since then gene mapping and annotation has rapidly progressed to where marker-assisted selection (MAS) or breeding (MAB) is now being used to identify and screen for many desirable traits (www.rosbreed.org; https://web.archive.org/web/20150801131947/http://fruitbreedomics.com/) (Gasic and Saski, 2019). Molecular markers linked to traits that are difficult to select by phenotype save considerable time and expense when genes of interest can be found early in the evaluation period of peach progeny. Many of these markers can be found in genomic databases like www.rosaceae.org. Another technology that has been effective to search for and map quantitative traits that are under the influence of many genes with small effects is the genome-wide association study (GWAS). Fruit tree crops with decades of phenotypic data can now be used in GWAS to reconnect traits with their underlying genetics across the genome (Gasic and Saski, 2019).

Unlike row or cereal crops, very few transgenics have been commercialized in tree fruit, including peach. Genetic transformation is a promising tool for genetic improvement of peach, but the main obstacle to genetic engineering is the regeneration of transformed tissues or plantlets (Gasic and Saski, 2019). Technology advancements have now made this option feasible, but no transgenic peach cultivars have been released as of 2024. Similarly, gene editing such as the CRISPR-Cas9 technique, which is considered the most advanced tool for precision genome editing, is in its early stages with tree fruit, and no gene-edited peach cultivars have been released thus far.

4.2 Rootstocks and selection criteria

Peach scion cultivars like many stone fruits are grafted to a rootstock of the same or a related species or hybrid. Rootstocks are used to overcome soil problems associated with high bulk density, pH, salinity, carbonate ions, drought, waterlogging, nematodes, and fungal and bacterial pathogens that injure or kill trees and reduce productivity (Reighard and Loreti, 2008; Reighard 2013; Reig et al., 2016, 2020; Minas et al., 2023a). Rootstocks can also reduce vigor in high-density orchards, improve fruit quality, and increase vigor on replant sites. Traditionally, peach was planted in low-density plantings using vigorous peach seedling rootstocks that were not adapted to many of the soil problems listed previously, nor could they restrict vigor. Therefore, breeding programs have switched the focus of development of new rootstocks from peach seedlings to related *Prunus* species and interspecific hybrids between peach, almond, plum, prune, and apricot. These new rootstocks possess genes to overcome the abiotic and biotic challenges that affect sustainable production. Newly released rootstocks are now available, and some have become commercial standards, replacing peach seedling rootstocks such as Nemaguard, Lovell, GF-305, and Montclar. Classical rootstock breeding in the past took 40 years or more before a rootstock was thoroughly tested and released. Today, breeding programs are fast-tracking this timeline by incorporating molecular technology such as GWAS and MAS to map, identify, and transfer new genes from related non-peach species to create rootstocks with resistance to difficult pathogens such as *Armillaria sp.*

The objectives of current rootstock breeding programs are to develop interspecific hybrids that can be clonally propagated in mass from tissue culture or other means of vegetative propagation to maintain genetic uniformity for complex traits, unlike open-pollinated seedling rootstocks. These rootstocks must be compatible with peach, have nematode and disease resistance, drought or moderate waterlogging tolerance, and impart favorable horticultural characteristics to the scion. Some must also tolerate saline and alkaline soils, while others must be size-controlling. Therefore, selection criteria include graft compatibility; climatic and edaphic adaptation; nematode, fungal, and bacterial resistance; scion vigor control; and imparting better fruit size and quality.

Commonly used peach seedling rootstocks having root-knot nematode *(Meloidogyne sp.)* resistance include Nemaguard, Nemared, Flordaguard, and Guardian (Minas et al., 2023a). *Prunus* hybrids that are vegetatively propagated with similar resistance are Viking, Cadaman, Hansen 536, Isthara, Rootpac 40, Rootpac R, MP-29, and Garnem. Plum species rootstocks such as Adesoto 101 and Empyrean 2 are also resistant. There is little to no resistance in *Prunus* species to ring (*Mesocriconema xenoplax*) and lesion *(Pratylenchus sp.)* nematodes. Ring nematode is especially problematic due to its association with the peach tree short life (PTSL) syndrome and the bacterial canker *(Pseudomonas syringae pv. syringae)* complex. The best ring nematode and PTSL tolerant rootstocks are Guardian, Viking, and MP-29.

Resistance or tolerance to soil fungal pathogens such as *Phytophthora sp.* and *Armillaria/Desarmillaria sp.* has been difficult to find in rootstocks compatible with peach. MP-29, Rootpac R, and Krymsk 86 have resistance or tolerance and are recommended for *Armillaria sp.*-infested sites. For high bulk density soils, Adesoto 101, Empyrean 2, Krymsk 86, Rootpac 20, and Rootpac R are tolerant; whereas, for calcareous soils, peach-almond hybrids such as GF-677, Cornerstone, Garnem,

Rootpac40, Adafuel, Adarcias, Pilowred, and Intensia are recommended as well as Adesoto 101 and Cadaman (Rubio-Cabetas et al., 2017; Reig et al., 2020). Rootpac R, Rootpac 40, and Empyrean 3 exhibited good tolerance to a high pH, *Armillaria*-infested, replant site in Spain, but they have had high mortality on replant sites in North America. Viking and Rootpac 20 have reported tolerance to saline soils (Jimenez et al., 2011). Krymsk 86, Bailey, and Siberian C peach seedling rootstocks are commonly used in colder temperate regions in North America.

An important characteristic for new peach rootstocks is vigor control as high-density plantings utilizing mechanization require smaller, more efficient trees (i.e., more fruit, less wood). Many semi-dwarfing and dwarfing rootstocks are now available. Semi-dwarfing rootstocks (60%–90% of standard) include Controller 6, 7, and 8, Rootpac 20, Empyrean 2, and Ishtara. Dwarfing (<60% standard) rootstocks include MP-29, Adesoto 101, Rootpac 40, Krymsk 1, Pillowred, and Intensia depending on the soil fertility and climate (Iglesias et al., 2023). Unfortunately, these vigor-reducing rootstocks have exhibited variable yield efficiencies (i.e., fruit yield/tree size) in limited testing and thus need more evaluation. In addition, the level of resistance to edaphic, abiotic, and biotic stresses across many different regions is relatively unknown, and with more field testing, many will be replaced with improved releases. A new rootstock must have multiple favorable horticulture and resistance traits and be thoroughly field tested before it becomes widely adopted.

5. Orchard management

5.1 Establishment and fertilization

Before planting a peach orchard, one of the most important decisions is site selection. Critical factors to consider are good air and water drainage, appropriate climate such as winter, spring and summer temperatures, soil classification (texture, bulk density, depth, pH), and site history (soil diseases, replant problems, previous herbicide use) (LaRue, 1989). Once an appropriate site is selected, soil sampling for organic matter, pH, fertility, and, if loamy or sandy soil, nematode sampling to identify root-knot, lesion, and ring species is advisable.

Preplant practices include removing old roots, subsoiling to break up hardpans, and soil fumigation if samples indicate problematic nematodes. Laser leveling on flat ground will help with water distribution and allow for bedding if water drainage is restricted. Depending on the soil analyses, preplant deep incorporation of lime and phosphorus may be necessary.

Planting stock from nurseries are often 1-year-old "June" t-budded or 2-year-old chip-budded trees on seedling or vegetatively propagated rootstocks (e.g., self-rooted cuttings or tissue-cultured plantlets). Grafted or budded trees in containers are also available but often at a higher cost. Whatever nursery tree is selected, the shoot-to-root ratio should be in balance as roots are lost in lifting from the nursery bed or restricted by the container. Also, damaged or broken roots are pruned off before planting. Unless a high-density tree wall is desired, most newly planted trees are cut back to about 50 to 70 cm above ground to encourage lateral buds to break and eventually become the scaffolds to support the fruiting canopy. At planting, the graft union should be approximately 5 cm above the soil surface to prevent scion rooting after the soil settles. With hedgerow plantings, tree rows should be arranged in a north-south direction to maximize light interception.

Planting can be done via tree planter or by hand (shovel) with weed control using herbicides or mechanical means. In the year of planting, N-P-K fertilizer is applied in split applications based on soil analysis and tree growth. In subsequent years, fertilization is based on mid-season or postharvest foliar analyses and estimated nutrient loss from the annual crop removal. The addition of micronutrients such as zinc, boron, or iron might be necessary depending on soil pH. Fertilizer is applied in early spring and/or in small amounts during the growing season to help promote a strong bloom and vegetative growth. If the majority (>50%) of the years' recommended fertilizer is applied in the early spring at pre-bloom and fruit set is adequate (i.e., good pollination and no frost damage), then the remainder of the planned fertilizer amount is applied after fruit set to assist fruit development. However, if there is poor fruit set or significant crop loss, then no additional nitrogen is needed so to avoid excessive vegetative growth. Annual fertilizer amounts (kg/ha) should follow local recommendations for specific growing regions and soil types to maintain the trees within established nutrient sufficiency ranges.

5.2 Irrigation

Water is essential for good peach tree and fruit growth. Irrigation is usually installed prior to the first growing season though wherever summer rainfall is frequent, it can be added later before the first bearing year. Water can be applied by overhead, flood, or ground level drip or microsprinklers. Water sources include canals, wells, or ponds, but water access may be limited due to availability and/or government regulations. Some water has salinity and/or alkalinity properties and must be ameliorated through water treatment or adapted rootstocks if available. Water pH between 6.0 and 7.5 is acceptable with 6.5 to 7.0 preferred. Irrigation scheduling is very important, especially in dry, hot climates and during critical fruit growth stages.

Peach fruit is vulnerable to water stress during stages I and III of fruit growth (Morandi et al., 2023). Stem water potentials of -0.4 to -0.6 MPa are typical during these fruit growth stages. During stage II, shoot and trunk growth are negatively affected by water stress but not the fruit. This stage is often referred to as pit hardening and can be a time to restrict vegetative growth by water restriction. Deficit irrigation can be used to reduce vegetative growth and to save water during stage II of fruit growth and postharvest. Peach trees require 20 to 50 mm of water per week especially in the 2 to 3 weeks before harvest. After harvest, irrigation can be limited or stopped, but during very hot periods in late summer if the stem water potential reaches -1.5 MPa, then irrigation is needed to support flower bud development (i.e., next year's fruit) to avoid the formation of abnormal meristems that produce misshapen fruit (i.e., doubles and deep sutures).

Irrigation scheduling is very important for peach but not easy to do. Climate, soils, rootstocks, and water availability must be considered. Hot dry areas require more water than regions with summer rainfall. Sandy soils drain water faster than clay soils. Almond rootstocks are more drought tolerant than peach rootstocks. Water restrictions are common in some xeric regions. Therefore, irrigation scheduling often relies on estimates of evaporation (ETo) corrected by peach-specific crop coefficients (K_c) that are adjusted for canopy cover. Crop ET changes depending on environmental factors, canopy development and the stage of fruit growth. Peach irrigation needs can be as much as 1000 mm per year in hot, arid climates. Crop coefficients (K_c) for peach have been calculated to range between 0.45 to 1.15 depending on the time of year and stage of growth (Morandi et al., 2023).

Precise measurement of soil water availability is a key component of irrigation scheduling. Some methods to determine timing for irrigation include soil sensors, plant sensors, and/or remote sensing. These provide data that can be incorporated into models that also include crop coefficients, weather, soil characteristics, rootstock, tree density, and orchard architecture to inform the grower when and how much water is needed to produce a marketable crop.

5.3 Pruning and training systems

Pruning is an important cultural practice to remove unwanted shoots and direct new shoot growth in peach trees to create a canopy that intercepts and distributes light efficiently for photosynthesis. Proper and timely pruning improves fruit quality (sugars) and skin color (red blush). Other factors like soil type, climate, rootstock vigor, labor availability, market price, etc., must be considered when deciding how much pruning to do in a specific geographic region. Pruning should direct shoot growth to fill the canopy for the selected training system to optimize light interception, which renews fruiting wood, increases labor efficiency, allows mechanization, facilitates harvest, and improves fruit yield and quality. Both summer and winter (dormant) pruning is necessary to sustain tree productivity.

There are many training systems for peaches, but the three-dimensional (3D) open vase, Palmette and V systems, and the two-dimensional (2D) Tatura systems are the most common (Table 2). Each of these training systems have specific pruning protocols to maintain productivity (Corelli-Grappadelli and Marini, 2008; Anthony and Minas, 2021; Iglesias et al., 2023). The open vase systems are easiest to create, whereas the two-dimensional planar systems require more management. Once the tree canopy has attained the system's desired height and volume, pruning is key to maintaining productivity and fruit quality. Pruning practices to maximize yield differ within each training system, but to maintain orchard sustainability, all methods conceptualize how pruning specific branch types affects future shoot growth and tree structure. Therefore, pruning is part "art" as well as science.

Table 3.2 Training systems characteristics and their mechanization potential for peach orchards

Training system	Canopy/row architecture, volume	Rootstock	Tree × row spacing (m)	Potential mechanization options
Open Vase (e.g., Gobelet)	Three-Dimensional Tree Row, High	Vigorous	2.5 x 5.0 to 4.5 x 6.0	Limited (top hedging)
Palmette, Structured Y Multileader (e.g., Perpendicular V, Quad V, Hex V)	Three-Dimensional Y-Tree Row, High	Vigorous to semi-vigorous	1.8 x 5.0 to 2.2 x 5.0 to 3.0 x 5.4	Partial (top and side hedging, some blossom/fruit thinning)
Central Leader (e.g., Spindle, Fusetto)	Three-Dimensional Tree Row, Medium	Semi-vigorous to semi-dwarfing	2.5 x 4.0 to 3.5 x 5.0	Partial (top and some side hedging, some blossom/fruit thinning)
Planar Multileader (e.g., UFO)	Two-Dimensional Fruiting Wall, Medium-Low	Vigorous to semi-dwarfing	2.0 x 3.0 to 3.0 x 3.5	Extensive (top and side hedging, blossom/fruit thinning)
Planar Trileader (e.g., Trident or Tri-Axis)	Two-Dimensional Fruiting Wall, Medium-Low	Semi-vigorous to semi-dwarfing	1.5 x 3.0 to 2.2 x 4.0	Extensive (top and side hedging, blossom/fruit thinning)
Planar Dual Leader (e.g., Bi-Axis)	Two-Dimensional Fruiting Wall, Medium-Low	Semi-dwarfing to dwarfing	1.0 x 3.0 to 1.5 x 4.0	Extensive (top and side hedging, blossom/fruit thinning)
Planar Central Leader (e.g., Espalier, Super Spindle)	Two-Dimensional Fruiting Wall, Medium-Low	Dwarfing	0.5 x 3.0 to 1.0 x 4.0	Extensive (top and side hedging, blossom/fruit thinning)
Dual Planar Espalier (e.g., Tatura)	Two-Dimensional Y- or V-Dual Fruiting Wall, Medium-Low	Semi-vigorous to semi-dwarfing	0.5 x 3.5 to 1.5 x 4.5	Partial (top and some side hedging, blossom/fruit thinning)

Table adapted from CABI Peach book

There are three types of peach fruiting shoots: proleptic, sylleptic, and epicormic. How each type is treated (pruned) impacts the amount and quality of fruit buds (DeJong et al., 2012). Proleptic shoots originate from visible preformed buds (i.e., vegetative shoot buds) that go through a dormancy phase and produce most of the quality fruiting wood on a peach tree for the following year's crop. Sylleptic shoots originate from actively growing proleptic shoots (i.e., side branches on elongating proleptic shoots) as well as on epicormic shoots (i.e., water sprouts). Epicormic shoots do not arise from visible vegetative buds, but instead originate from meristems that lie quiescent unless apical dominance control has been lost due to heading cuts, branch breakage, or severe pruning. The primary objective of dormant pruning any training system for peach is to retain or maximize quality proleptic and sylleptic fruiting wood each year, while removing epicormic sprouts and any shoots that are inferior or unproductive.

Heading and thinning cuts are the primary ones used to dormant prune peach trees (Corelli-Grappadelli and Marini, 2008; Iglesias et al., 2023). A heading cut removes the terminal section of the shoot's main axis but retains the basal part of the shoot with its existing buds and nodes. A thinning cut removes the entire shoot, leaving no buds or meristems from which to regrow. Heading cuts are important in the early years to form the scaffolds for the training system. Thinning cuts reduce the number of fruiting shoots to eliminate poor quality shoots, increase light penetration, and redirect shoot growth to maintain future crop production. In 3D training systems, after the trees have filled their allotted space and are in full production, dormant thinning cuts are preferred over heading cuts to maintain canopy height and width. In the narrower, less-structured 2D systems where no permanent scaffolds are retained, thinning cuts can be less frequent than heading ones, which are used to renew fruiting wood for the following season. The shoots thus formed will either be left whole, if they are the desirable size for the cultivar; removed entirely ("thinned out") if there are too many; or headed to break apical dormancy and obtain renewal wood for the subsequent growing season.

In 3D training systems like the open vase, the canopy architecture requires correct scaffold placement and summer pruning. Likewise, in 2D training systems, light interception and quality depend on detailed dormant and timely summer pruning, as these planar canopies quickly fill their space. The narrow 2D canopies, whether vertical single planes or V or Y dual planes, facilitate improved cropping precision since the number of fruit-bearing shoots can be quantified and adjusted to modify the crop load via pruning, thereby reducing fruit thinning costs (DeJong et al., 1994). In general, the 2D systems crop earlier with higher yields but require a larger initial investment in trees, trellis materials, and labor.

5.4 Crop load and thinning

Peaches have the potential to set a fruit crop that can damage the tree's woody structure and/or reduce fruit size and quality; thus, fruit thinning is essential most years. Ideal crop load (i.e., fruit per tree) depends on cultivar, tree vigor, age, and location.

A study by Marini et al. (2018) showed a crop load of 2 to 3 fruit per cm^2 of trunk cross-sectional area was the maximum fruit number per tree before fruit size was negatively impacted below the minimal commercial size. Thinning is an expensive cultural practice that traditionally has been done using plastic bats, toilet brushes, or by hand to physically remove flowers or fruitlets. The earlier it is accomplished during the stage I fruit development period, the better the result in larger fruit size. Small fruited and early-ripening cultivars should be thinned preferably within 30 days after bloom if not at bloom.

There have been many thinning methods tested including application of soybean oil, fish oil, surfactants, plant hormones, fertilizers, lime sulfur, caustic chemicals, and water, as well as use of ropes, hand drills, lasers, and specialized mechanical equipment (Marini and Reighard, 2008). Most chemical fruit thinners have been either inconsistent or have not been approved for use on peaches. Currently one compound with promise that is labelled for peach is Accede, an amino acid-based product that is an ethylene precursor. With new size-controlling rootstocks and adoption of high-density training systems, mechanical thinning with machines such as the Darwin string thinner is becoming more commonplace. However, most bloom and fruitlet thinning in traditional open vase trained orchards is done by hand if labor is available. More recently, mechanical thinners have been developed that can be used coupled with the Darwin, or after hand or chemical thinning, such as the "Eclairvale." These thinners are based on a drum fitted with rigid spokes about 80 cm long that enter the canopy as the tractor moves along the row, impacting and causing fruit to drop. This machine is effective at a late stage, typically the end of pit hardening, and can reduce hand-thinning costs, if coupled to the Darwin, to 15% of what would be required by 100% hand-thinning, according to anecdotal reports.

6. Fruit maturation and harvest criteria

6.1 Fruit development and growth

Peach fruit growth in weight (fresh and dry) and diameter of traditional mid- and late-ripening cultivars follows a double-sigmoid model, which defines four growth stages, S1-S4 (Figure 3.7), all expressed as days after full bloom (DAFB). Stage S1 starts at the full bloom phenological stage, and within it occurs the first step of fruit development, fruit set (fertilization of one of the ovules in the ovary). In peach, fruit set is regulated by the orchestrated action of three hormones, which depending on the species/genotype, are auxins and/or gibberellins (GAs), and/or cytokinins (CK) and sustained by the availability of nutrients (calcium, Ca^{2+}; boron, B; potassium, K^+; and nitrogen, N) (Mariotti et al., 2011). S1 and S3 are two fast growth stages, which are sustained mainly by cell duplication and expansion or cell expansion only, meanwhile S2 and S4 are the contrary. In S2, growth slows and pit hardening (PH) occurs. The ripening process occurs after the fruit reaches the maturation stage (end of S3/beginning S4) in which fruit growth is slow and sustained mainly by

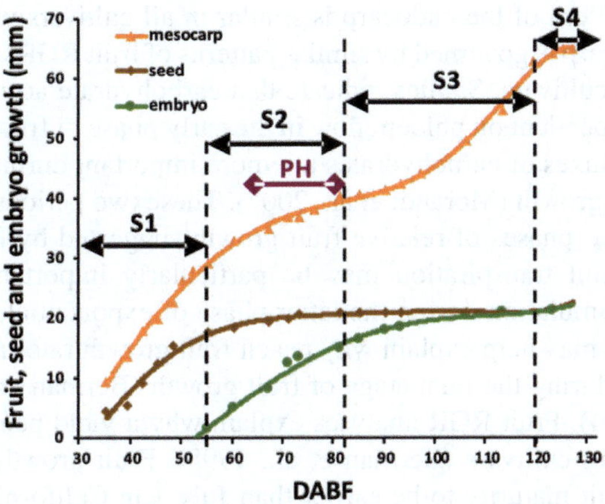

FIGURE 3.7

Fruit and seed growth pattern of cv. Fantasia. Orange triangles represent fruit growth expressed as cross diameter, and brown diamonds and green circles represent seed and embryo growth, respectively, and both expressed as length.

(Figure adapted from CABI Peach book)

cell enlargement (Tonutti et al., 2023). This final S4 stage consists of S4-1, in which fruit reaches its final size, and S4-2, when the fruit ripens (Rodriguez et el., 2019).

Fruit growth or fruit development is genetically controlled, and this allows the selection of cultivars with different harvest dates: early-ripening (60-70 DAFB), mid-ripening (110-130 DAFB), and late-ripening (160-170 DAFB). The developmental events previously described that occur in the four stages of fruit growth of later maturing cultivars occur in all cultivars. However, early-ripening cultivars have a shortened or nondistinct S2 stage compared to mid- and late-ripening cultivars, and late maturing cultivars have an extended S3 stage.

6.2 Understanding what limits peach fruit growth

Theoretically, fruit growth can be limited by "source limitations" (lack of resources to meet requirements for fruit to grow), or "sink limitations" (lack of inherent potential of the fruit to grow). Grossman and DeJong (1995a) used relative growth rates (RGRs) analysis to determine periods when peach fruit growth is primarily source-limited and sink-limited. Fruit of late maturing cultivars are primarily source-limited during stage S1 and S3 of fruit growth, and primarily sink-limited during stage S2 of growth, but duration of limitations is influenced by crop load (DeJong and Grossman, 1995a,b). Early-ripening cultivars tend to be source-limited for a larger portion of their growth period and their sink-limited period is shorter because of a shorter length of development. Additional research indicated that the patterns of dry weight accumulation of peach fruit can more easily be understood by analyzing the fruit as a composite of two organs, the endocarp (and seed), and the mesocarp. The developmental pattern of the endocarp is similar in all cultivars while the development of the mesocarp is governed by similar patterns of fruit RGRs that are truncated in early-ripening cultivars. Studies indicate that carbohydrate accumulation in fruit may be mostly dependent on phloem flow in the early phase of fruit growth, whereas xylem mediated fluxes of carbohydrates are more important during the later phases of expansive fruit growth (Morandi et al., 2007). These two periods of growth correspond with the two phases of relative fruit growth suggested by DeJong and Goudriaan (1989). Fruit transpiration may be particularly important for facilitating carbohydrate accumulation during the latter phase of exponential growth (Morandi et al., 2010). This may help explain why peach fruit growth rates are so sensitive to tree water status during the final stage of fruit growth (Berman and DeJong, 1996; Girona et al., 2004). Fruit RGR analyses explain why a yield penalty is associated with early-ripening cultivars (Berman et al., 1998). Fruit growth models indicate that advancing fruit maturity to be earlier than July 1 in California for nonmelting peach cultivars would result in potential yield penalties of approximately 1.8 t ha^{-1} day^{-1}. Fruit RGR-based fruit growth models also indicate why it is important to thin crop loads to desired levels as soon as it is economically feasible (DeJong et al., 1992; Grossman and DeJong, 1995b, Reig et al., 2022).

6.3 Temperature and hormone effects on peach fruit development and growth

The length of time between bloom and fruit maturity for a given cultivar and year is a function of a genetically determined pattern of fruit development and the temperatures experienced during the first 30 days after full bloom (Marra et al., 2002). The temperature dependence of the length of the fruit development period has been quantified by calculating the cumulative growing degree hours (GDH) during the first 30 days after full bloom (GDH30) for several Californian peach cultivars over numerous growing seasons (Day et al., 2008). Most of the temperature-dependent differences in the time between full bloom and fruit maturity among years for individual cultivars are accounted for by differences in the time between full bloom and reference date (pit hardening +10 days) (Lopez and DeJong, 2007). It has been documented that in the years when early spring temperatures are high (GDH30 > 6000), there is a tendency for fruit sizes to be small, while fruit sizes tended to be larger in years when GDH30 < 6000. This phenomenon may explain why peaches grown in warm, subtropical regions often have small fruit sizes that cannot be substantially improved with early fruit thinning. If global climate change brings increases in early spring temperatures in temperate regions, one of the most pronounced effects of increased spring temperatures, in addition to lack of chilling and fruit set, will likely be shortened fruit development periods and a tendency toward reduced fruit sizes.

Hormones also play an important role in controlling fruit development and growth. CK and gibberellic acid (GA) are involved in the early phases with ethylene and auxins (AUX), which are important also at ripening (Bonghi et al., 2011).

Indeed, peach is a climacteric fruit, therefore the development (S1-S3) and ripening (S4) are oxidative processes, producing reactive oxygen species (ROS), such as hydrogen peroxide (H_2O_2) (Huan et al., 2016). During fruit ripening (S4), an increase in ethylene release and respiration (Zeng et al., 2015) alongside a number of coordinated biochemical and physiological processes result in changes in color, texture, taste, overall flavor, aroma, and nutritional value, making the fruit desirable for fresh consumption at maturity and after cold storage.

Some changes that occur as fruit mature are ethylene dependent, such as flesh softening or the synthesis of specific pigments. Compared to other climacteric fruits, such as tomato, peach at its climacteric peak, which occurs during the ripening stage (S4), has low/moderate respiration rates (15-80 ml CO_2 kg^{-1} h^{-1}) and low ethylene production rates. Ethylene production, though, is highly variable depending on genotype and phenotype and is related to flesh textures (melting, nonmelting, slow melting or slow softening, and stony hard). Cultivars with melting (M) textures exhibit a prominent softening during the late-ripening stage, resulting in a rapid decrease in flesh firmness, concomitantly with an increase of juiciness, a large production of ethylene, and an increase in respiration rate. Nonmelting (NM) types have firm texture when fully mature and soften very slowly, producing high levels of ethylene between fruit growth stages III and IV. This flesh texture may become rubbery when overripe as the flesh never melts. The rubbery texture is related to the loss of endopolygalacturonase (endoPG) activity, as this enzyme is responsible for cleaving pectins from the cell wall. Slow-softening types have firm and crispy texture that, when fully ripe, becomes melting, with slow softening on the tree as they produce ethylene. Finally, stony hard (SH) fruit types have very firm and crispy texture that never softens. They have a complete lack of ethylene production during the late-ripening stage due to the suppression of the key gene in the ethylene enzymatic pathway *(Pp-ACS1)* that has been linked to the low levels of auxins (Tatsuki et al., 2006).

6.4 Fruit maturity changes and harvest traits

It is crucial to understand the evolution of fruit traits to determine optimum harvest time during "on-tree" ripening to assess the correct fruit maturity to ensure proper internal quality and increase packouts and consumer acceptance (Minas et al., 2023b). Consumer acceptance has been in constant decline due to immature, tasteless, and/or overripe fruit resulting from inappropriate harvest decision-making, as well as cultivar textural problems. Fruit maturity traits, which can be external or internal and determined by nondestructive and destructive methods, include both skin and flesh color changes from green to yellow/orange and/or yellow/red, respectively. Color changes are due to accumulation of pigments (e.g., carotenoids or anthocyanins). Flesh softening is due to cell-wall dismantling, sensorial and textural changes resulting from increased carbohydrate accumulation (expressed as dry matter content [DMC] and TSS), and acidity loss due to increased respiration rates, ethylene biosynthesis, and aromatic volatilization (Figure 8, Minas et al., 2023b). However,

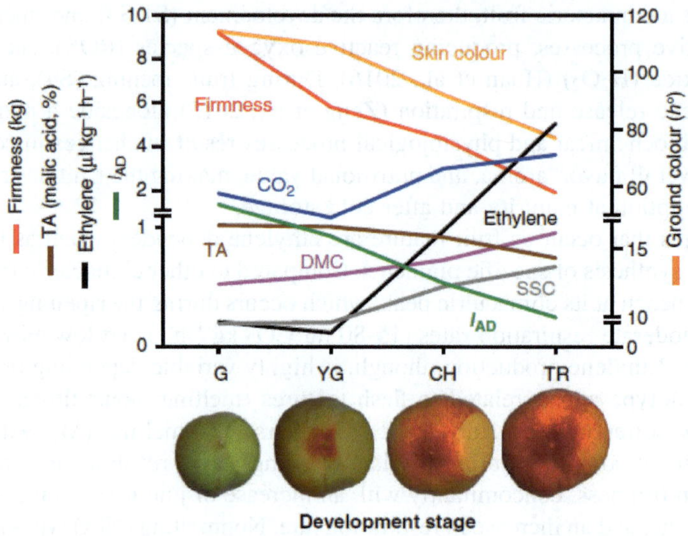

FIGURE 3.8

Peach fruit quality and maturity changes during growth, development, maturation, and ripening "on-tree" for the stages: *G*, green (stage S2); *YG*, yellow green (S3); *CH*, commercial harvest (S4I preclimacteric); *TR*, tree ripe (S4II climacteric), *DMC*, dry matter content; h^o, ground skin hue angle; I_{AD}, index of absorbance difference at 670 and 720 mm; *SSC*, soluble solids content; *TA*, titratable acidity.

(Figure adapted from CABI Peach book)

the increases or the decreases of each trait during growth, development and ripening of the fruit depends on the harvest timing of the cultivar (Figure 3.7).

Similar to other fruit tree crops, the changes in skin color due to the degradation of chlorophyll (green color) and the new synthesis of carotenoids (yellow/orange color), which unmask already existing colors, and/or increases in anthocyanins (red color) are the most visual changes observed during peach ripening (Tonutti et al., 2023). In fact, ground (background) color is the most important trait used for harvest timing decisions, except for the peaches that are fully red-skinned prior to maturation. Highly red-skinned fruit have been the objective of many breeding programs in recent years. Full-red skin obscures visual observation of the ground color, which many times leads to premature harvesting with poor fruit quality for consumers. Red color development is genetically controlled, but environmental factors in the field such as light exposure and temperature also play an important role in this trait (Andreotti et al., 2008). The main anthocyanins accumulating in skin of ripening peaches are cyanidin 3-glucoside and cyanidin 3-rutinoside (Ravaglia et al., 2013).

In parallel changes in skin color, but not visually perceived, flesh color is affected by the same processes as the skin. Peaches can be classified into nine flesh colors (greenish white, snow white, cream white, greenish yellow, light yellow, yellow, orange yellow, orange, and red). Based on the anthocyanin coloration of the flesh, fruit can also be classified into seven categories: absent, weak, only under the skin, under the skin and around the stone, only around the stone, faint in the whole flesh, and intense in the entire flesh (Giovannini et al., 2013); or in three color patterns: whole red flesh, red dots under the skin, and red color around the stone (Zaracho et al., 2023).

The loss of flesh firmness, a genetically programmed ripening-related event that occurs in parallel with ground and skin color hue decline, is a consequence of the decline in both cell-wall strength and cell-to-cell adhesion, together with the turgor pressure decreases that lead to flesh softening and an increase in juiciness. Ninety percent to 95% of the cell wall and middle lamella is composed of cellulose, hemicellulose and pectins, with the mechanical/elastic properties of the cell wall determined by cellulose and hemicellulose. Cell-to-cell adhesion is mainly ensured by pectins, a major constituent of the middle lamella. Indeed, during ripening there is a slight decrease of cellulose content, whereas hemicelluloses are the substrate of hydrolytic endo-1,4-β-glucanase (EG) activity, which increases in the initial phase of peach fruit softening and before the onset of the ethylene climacteric. The endo-PG enzyme is responsible for the depolymerization of the cell wall pectins, especially at the melting stage and the onset of the ethylene climacteric (Tonutti et al., 2023).

TSS should at least be 10% to 11% to ensure that peaches are acceptable to most consumers. This varies depending on the cultivar, but also within the fruit population of the same tree due to inter-fruit competition and location in the tree canopy (Tonutti et al., 2023). The main sugars in peach fruit are sucrose, glucose, fructose, and sorbitol that account for 50% to 80% of the mesocarp dry matter content (DMC) (Reig et al., 2023). Sugar metabolism and accumulation in peaches during ripening is due to the involvement of a number of enzymes responsible for the processes of assimilate import, metabolism, and dilution with relationships among them described in the SUGAR model developed by Genard et al. (2003).

Together with TSS, concentration of organic acids is the main driver of the flavor and sweetness of ripe peaches. High concentrations of sugar are easily hidden by high concentrations of organic acids that dim the perception of sweetness. However, this varies according to the predominant organic acid (e.g., citric acid adds more sourness than malic acid), meaning different organic acids require different concentrations to affect sugar perception (Bento et al., 2020). Malic acid is, in general, the most abundant acid at fruit maturity, followed by quinic, citric, and succinic acid, with traces of shikimic acid. All of them determine the titratable acidity (TA) of the fruit, and their content and profile seem to be quantitatively and qualitatively variable among cultivars. In addition, cultivars can be classified according to the TA content as previously described (Iglesias and Echeverría, 2009).

During the transition from unripe to ripe stages, peach fruit develop a set of volatile organic compounds (VOCs), which varies among cultivars. At the unripe stage, the compounds that impart the green associated aroma are lipid-derived C6 compounds such as n-hexanal, (E)-2-hexenal, (E)-2-hexenol, and (Z)-3-hexenol, which decrease during ripening; whereas esters, lactones, and terpenoids, such as (Z)-3-hexenyl acetate, γ-hexalactone, γ-octalactone, γ-decalactone, δ-decalactone, and linalool provide the ripe peach aroma (Tonutti et al., 2023). Part of the aroma development is controlled by ethylene. In fact, lactone synthesis shows a clear pattern concomitant with the climacteric rise in ethylene production with γ-decalactone being the principal volatile compound at the late-ripening stage (Zhang et al., 2010).

7. Postharvest physiology and technology

The storage potential of peach fruits is limited to a few weeks (1-3) based on the cultivar, the storage regime, and the maturity stage at harvest. Extended storage leads to chilling injury (CI) symptoms that become more evident after removal from cold storage and subsequent maintenance at room temperatures. Thanks to targeted breeding efforts, many new cultivars released into the market are less susceptible to CI disorders. This aspect is particularly important for late-ripening cultivars to increase availability worldwide, offering off-season premium products. To assure the competitiveness and sustainability of the peach industry, mid-season cultivars should follow a protocol for delivery and consumption within a margin that does not exceed 20 days from harvest, and this can be achieved by harvest of cultivars with progressive ripening during the summer period.

Peach storage life is limited due to incidence of postharvest fungal diseases and chilling injury symptoms. The main fungal diseases are brown rot *(Monilinia fructicola)*, gray mold *(Botrytis cinerea)*, Rhizopus rot *(Rhizopus stolonifer)*, and sour rot *(Geotrichum candidum)*. The identification of "reduced risk" fungicides like fludioxonil and fenhexamid led to their introduction as safer postharvest treatments. Recently, natural products produced by fermentation such as natamycin and polyoxin-D are being promoted as agricultural options that have been organically approved (Adaskaveg et al., 2022a).

7.1 Chilling injury

CI is the main physiological disorder limiting export and long-distance peach distribution. It encompasses an array of disorders which mostly refer to fruits which, while maintaining normal water content, lack juiciness and have a highly unappealing texture. Such symptoms have been described using different and sometimes diffuse terms such as mealy, woolly, and leathery (Manganaris et al., 2019). Besides mealiness or woolliness (perception of a dry and wooly texture due to lack of free juice upon consumption) and leatheriness (hard-textured fruit with no juice), internal breakdown evident as flesh browning is a severe disorder, while red flesh pigmentation or bleeding is important for certain cultivars (Figure 3.9). Maturity stage at harvest, storage temperature, atmosphere, and duration are the most relevant factors associated with CI incidence and severity (Manganaris et al., 2019).

FIGURE 3.9

Chilling-damaged nectarine fruit (cv. Caldesi 2000), characterized primarily by dry texture (woolliness) all over the surface and browning symptoms in the flesh near the skin *(left)* and chilling-damaged peach fruit (cv. Andross) with flesh browning symptoms extending all over the fruit, with a zone of about 1 cm below the skin remaining unaffected, characterized by discontinuity and no dry texture, after 4 weeks cold storage (5°C) plus 5 days shelf-life (20°C) *(right)*.

(Figure adapted from Manganaris et al., 2006)

Means to alleviate chilling injury symptoms

1. The elaboration of modern genomics-based breeding programs in peach that use the full potential of genome-wide association mapping and genomic selection appears to provide genotypes resistant/tolerant to CI disorders (Manganaris et al., 2019).
2. **Conditioning** (delayed cold storage) at 20°C and 95% RH followed by forced-air cooling prior to cold storage on fruit harvested at firm-ripe stage can reduce CI susceptibility and assure successful ripening upon removal from cold storage. When these treatments are applied properly, market life increases by up to 2 weeks in the cultivars tested.
3. **Employment of advanced** nondestructive assessment of CI incidence with hyperspectral approaches seems to be a promising tool in the future.
4. **Chemical treatments** (salicylic acid, methyl jasmonate, oxalic acid, γ-aminobutyric acid, and gibberellic acid) have shown promising results at the experimental level but are not yet being widely applied at commercial scales.
5. **Heat treatments** have shown some benefits for CI prevention, yet its efficacy is highly dependent on the cultivar, preharvest factors, and shipping duration. Intermittent warming (IW) has been also reported as a CI delaying strategy. This protocol was tested at a commercial scale in South Africa, yet it was proven to be difficult to apply at commercial scales, while the benefits are modest (Murray et al., 2007).

7.2 Other physiological disorders

7.2.1 Field skin inking or black staining

This is a type of skin discoloration causing fruit rejections. The symptoms appear as brown and/or black spots or stripes that are restricted to the skin (Figure 3.10). The inked areas are normally small but in extreme cases can reach up to 50% of fruit surface. Inking can be prevented or controlled by (1) keeping picking containers clean and avoiding dust contamination on fruits; (2) reducing fruit abrasion damage by treating fruit gently, using air-ride suspension on trailers, and avoiding long hauling; and (3) checking water quality for potential contamination with heavy metals (Fe, Cu, Al) and other measures (Manganaris et al., 2023b; Reig et al., 2022).

7.2.2 Skin burning

This is another type of skin discoloration that has become a frequent problem on specific susceptible peach and nectarine cultivars (Cantín et al., 2011). Research studies over several years indicated that peach and nectarine skin discolorations, field inking and skin burning, are triggered by a combination of physical damage during harvesting-hauling combined with different postharvest stress factors (Cheng and Crisosto, 1994; Cantín et al., 2011; Reig et al., 2022). However, although field inking and skin burn disorders have similar symptoms, they have different triggers and different biological mechanisms of development, and therefore it is important to understand the differences between both cosmetic skin disorders. Skin burning symptoms appear as brown and/or black areas that are restricted to the skin. In contrast to field inking, symptoms are mainly triggered during packing operations, principally at the brushing-washing point, although abrasion that occurs prior to packing may also contribute to its development. Some prevention and mitigation measures to control skin burning are to minimize physical damage or abrasion on the fruit surface during pre- and/or postharvest operations, to apply dry packing (without brushing or a chlorine rinse) for highly susceptible cultivars and to avoid high air velocities during forced-air cooling for skin burning susceptible cultivars.

FIGURE 3.10

Peach fruit skin inking in different peach cultivars.

7.2.3 Corky spot

The symptoms appear as dark sunken spots on the surface of the fruits, especially on fruit sides and blossom end. Internally, flesh initially presents reddish spots that turn brown, corky, and dry as the fruit ripens, making it unsuitable for marketing. This disorder has been attributed to excessive fertilization and some water stress conditions during the growing period that can lead to a nutritional imbalance and a deficiency in fruit Ca content (Peris and Alegre, 2012). Thus, avoiding excess nitrogen and potassium fertilization and water stress conditions is recommended to prevent corky spot.

7.2.4 Skin bronzing

Bronzing refers to patches of skin on the fruit that look bronzed on primarily yellow to light red skin background. Depending on severity, the damage may stretch from a single small patch to most of the peach being covered. Although research shows that many patches are formed prior to harvest, most of the symptoms only appear after storage.

8. Main tree diseases and pests and their management

8.1 Diseases

Leaves, fruit, woody tissues, and roots of peach and nectarine trees may be affected by a range of diseases that are mostly caused by fungi in the phyla Ascomycota, Basidiomycota, and Mucoromycota, but several fungal-like organisms (i.e., Oomycota), bacteria including phytoplasmas and viruses can also be involved. Most of the pathogens can also cause diseases in other crops. Some of the diseases such as brown rot, shot hole, and powdery mildew are of worldwide and common importance, but many others are of more localized significance or occur sporadically. Some diseases are only found in the orchard and are denoted as preharvest diseases, while others continue to cause crop losses after harvest and are referred to as postharvest diseases. Proper disease diagnosis and understanding of pathogen biology and disease epidemiology are a key for effective disease management. The interaction of the pathogen, the host, and the environment is called the disease triangle, which is an important concept in plant pathology that determines the magnitude of disease expression. A list of important fungal diseases is presented in Table 3.3. Some of them are described in the following text, but more detailed information on diseases of peach and nectarine can be found elsewhere (Adaskaveg et al., 2023; Adaskaveg and Förster, 2023).

8.1.1 Brown rot

It is a major fungal disease in most regions of the world that can result in extensive crop losses especially in areas with high rainfall and/or high humidity during the growing season. Three main causal pathogens are involved. *Monilinia fructicola*

Table 3.3 Common fungal diseases of peach and nectarine and primary management practices

Disease	Pathogen	Symptoms of economic significance	Primary management
Blossom, Foliage, and Fruit Diseases			
Brown Rot	*Monilinia fructicola, M. laxa, M. fructigena, M. polystroma, M. mumecola, M. yunnanensis*	Fruit decay	Sanitation, fungicides
Jacket rot and green fruit rot	*Botrytis cinerea*	Fruit decay	Sanitation, fungicides
	Sclerotinia sclerotiorum	Fruit decay	Orchard floor management, fungicides
	Monilinia fructicola, M. laxa	Fruit decay	Sanitation, fungicides
Peach leaf curl	*Taphrina deformans*	Defoliation, fruit off-grades	Fungicides
Powdery mildew	*Podosphaera pannosa (Sphaerotheca pannosa), P. leucotricha, P. clandestina*	Defoliation, fruit off-grades	Cultivar selection, fungicides
Rust	*Tranzschelia discolor, T. pruni-spinosae*	Defoliation, fruit off-grades	Fungicides, (removal of alternate host?)
Scab	*Venturia carpophila*	Fruit off-grades	Fungicides
Shot hole	*Wilsonomyces carpophilus*	Defoliation, fruit off-grades	Fungicides
Trunk and Scaffold Diseases			
Leucostoma (Cytospora) canker (syn. perennial canker)	*Leucostoma cincta (Cytospora cincta), L. persoonii (C. leucostoma)*	Branch dieback, tree decline and death	Sanitation, pruning practices
Silver leaf disease	*Chondrostereum purpureum*	Branch dieback, tree decline and death	Sanitation, pruning practices
Fungal gummosis	*Botryosphaeria dothidea (B. obtusa* and *Lasiodiplodia theobromae)*	Branch dieback, tree decline and death	Sanitation and winter pruning, (removal of alternate host?)

Table 3.3 Common fungal diseases of peach and nectarine and primary management practices—*cont'd*

Disease	Pathogen	Symptoms of economic significance	Primary management
Root and Crown Diseases			
Armillaria root rot	*Armillaria mellea, A. ostoyae, Desarmillaria. tabescens, D. caespitosa*	Root and crown rot, tree decline and death	Site selection (avoidance), sanitation, soil fumigation
Phytophthora root and crown rots	*Phytophthora cactorum, P. cambivora, P. cinnamomi, P. citricola, P. citrophthora, P. cryptogea, P. drechsleri, P. mediterranea, P. niederhauseri, P. parasitica (P. nicotianae), P. syringae,* and others	Root and crown rot, tree decline and death	Irrigation management (e.g., raised beds, sprinkler shields to prevent trunk wetness) rootstocks, fungicides
Verticillium wilt	*Verticillium dahliae*	Wilt	Site selection (avoidance), soil fumigation

is the most virulent and is present in the majority of production regions. *M. laxa* occurs worldwide but is generally less important where *M. fructicola* is also present. *M. fructigena* occurs in Europe mainly on pome fruit crops and occasionally on peach and nectarine. Additional species include *M. polystroma*, *M. mumecola*, and *M. yunnanensis*.

The disease cycle includes blossom blight twig blight, and pre- and postharvest fruit decay (Figure 3.11). Primary inoculum sources in the spring are overwintering, rotted fruit mummies on the tree producing asexual conidia, fruit mummies on the orchard floor producing sexual fruiting structures (apothecia) and ascospores, as well as twig cankers and fruit peduncles. Infections occur over a wide temperature range with an optimum of 22.5°C to 25°C. Minimum wetness durations of 3 to 5 hours are required at 20°C, whereas 18 hours are needed at 10°C.

Blossom blight rarely reduces the crop load, but diseased flowers and shoots provide secondary inoculum for fruit infections. From necrotic anthers and filaments, the infection proceeds to the floral tube, ovary, and peduncle. Flowers wilt, turn brown, and are covered with grayish-tan sporodochia at high humidity or wetness. Flowers typically stick firmly to the twig with a gum drop at the base of the peduncle. Flower infections by *M. laxa* and *M. fructicola* may result in twig cankers, often with extensive gumming. When the twig is girdled by the canker, it dies, but

FIGURE 3.11

Brown rot caused by *Monilinia fructicola:* Blossom blight showing gumming and canker development at the base of the peduncle *(left)* infected peach fruit and shoot dieback *(right)*.

leaves remain attached. Sporulation may continue for several years on large cankers. Infection of fruit from dead flower and canker inoculum is by direct penetration through the cuticle or indirect through injuries or stomata. Susceptibility to infection is high during early fruit development, decreases during green fruit stages, and increases again as fruit mature and ripen (Gradziel, 1994). Quiescent infections that occur on developing fruit in less favorable environments (arid and semiarid climates) may become active when fruit mature prior to or after harvest. Postharvest decay can also be initiated at harvest when airborne and surface-deposited conidia gain entry through small wounds that commonly occur during picking, bulk handling, and transport to a packinghouse. The decay is firm, brownish in color, and develops rapidly. At advanced stages, the fruit surface is covered with cottony, light brown fungal mycelium and conidia. Brown rot decay of fruit in cold storage often appears black with little or no sporulation.

8.1.2 Scab

This is an important fungal disease of peaches and nectarines in warm areas with high rainfall, such as the Southeastern United States, or where tree canopies are wetted by sprinkler irrigation. Scab is caused by the fungus *Venturia carpophila* (syn. *Fusicladium carpophilum, Cladosporium carpophilum*). Infections can occur on twigs, leaves, and fruit, but symptoms are most noticeable on fruit. Twig lesions develop on new succulent growth in the spring and persist throughout the host dormant period. These produce abundant primary inoculum (conidia) after petal fall until mid-spring that is wind- and rain-dispersed. Symptoms on leaves develop in early summer on the lower surfaces as irregular, blotchy lesions slightly darker in color than the surrounding healthy tissue. Lesions turn olive-green once sporulation of the fungus occurs. Infections of leaves rarely cause serious defoliation of trees. On half- to full-sized fruits, small circular spots develop mainly on the upper, exposed surfaces and are first green and turn olive to black once the fungus sporulates. Spots are superficial, slowly enlarge, may coalesce, and make the fruit unmarketable.

8.1.3 Rust

This occurs wherever peaches are grown and causes leaf, fruit, and stem infections. Rust is caused by the biotrophic, macrocyclic, heteroecious species *Tranzschelia discolor* and *T. pruni-spinosae*. Different *formae speciales* of *T. discolor* have been described on almond, peach, and prune with strains on peach being designated as *T. discolor* f. sp. *persicae*. The alternate host of the pathogens are species in the Ranunculaceae, but in mild climates, they can survive on *Prunus* spp. as mycelium in green shoots or as urediniospores (Soto-Estrada and Adaskaveg, 2004). Leaf infections develop as angular, yellow lesions with rusty-brown urediniospore-producing pustules (uredinia) on the lower surfaces (Soto-Estrada et al., 2005). Late in the growing season, uredinia develop into telia that produce dark brown to black teliospores. Heavy leaf infections can result in premature tree defoliation, and fruit infections can cause direct crop loss. Symptoms on immature fruit are green, circular lesions 2 to 3 mm in diameter. On mature fruit, lesions are sunken with yellow halos, and the mesocarp below the lesion is discolored.

8.1.4 Peach leaf curl

It is a cosmopolitan disease and occurs wherever peaches and nectarines are grown. Economic losses can occur after winter and spring seasons with high rainfall, but losses are minor when fungicides are properly timed. The disease is caused by *Taphrina deformans* (Figure 3.12). The fungus overwinters in the asexual, yeast-like stage that contaminates twigs and buds. Newly developing leaves in the spring

FIGURE 3.12

Taphrina deformans causes peach leaf curl *(left)* and fruit deformation *(right)*.

develop discolored areas that thicken and then become wrinkled and puckered, causing the leaves to curl. Infected leaves are light green to yellow, red, and purple. They may be covered with a subtle white layer of sexually produced spore sacs (asci) containing ascospores that become airborne, cause secondary infections, and contaminate buds. Infected leaves eventually turn brown and abscise. Defoliated trees will form new leaves, but fruit set will be sparse in the current and following seasons. Fruit infections are less common and are irregular, raised green or reddish swellings (see Figure 3.12).

8.1.5 Powdery mildew

It occurs worldwide and is most damaging during warm, humid conditions. Excessive durations of wetness will kill conidia of powdery mildew fungi. *Podosphaera pannosa* is the most important pathogen. Other species include *P. clandestina* (syn. *P. oxyacanthae*), *P. tridactyla,* and *P. leucotricha*. Recent morphological and molecular analyses indicate that *P. tridactyla* s. lat. consists of at least 12 species with different specificity to *Prunus* species. Fruit infections cause the most economic damage, but leaf infections are important sources of inoculum. In nurseries, however, leaf infections can cause significant damage to seedlings and small trees (Figure 3.13).

Mildews overwinter on twigs as white, dense, felt-like mycelium, as mycelium in shoots and in the inner bud scales, or as sexual fruiting structures (chasmothecia). The first symptoms on peach leaves and green shoots are small, circular, white, web-like colonies that become powdery once masses of asexual conidia are produced. Leaves may then curl or become stunted. Older infections commonly result in leaf chlorosis and necrosis. Severe infections may result in defoliation. Fruit are susceptible from the early stages of development until pit hardening on peach and nectarine. White circular spots may enlarge, coalesce, and cover large areas of the fruit. Infections usually result in some deformation of the fruit surface with depressed or slightly raised areas. Any fruit with blemishes caused by powdery mildew are generally unmarketable.

FIGURE 3.13

Powdery mildew caused by *Podosphaera pannosa* on peach leaf *(left)* and fruit *(right)*.

8.1.6 Perennial canker (Leucostoma or Cytospora canker)

It is an important disease of stone fruit crops especially in colder climates. It reduces the number of bearing branches, kills twigs, and shortens tree life. Perennial canker is caused by *Cytospora cincta*, *C. leucostoma*, *C. plurivora,* and several other species. On small branches, lesions around dead buds or previous years' leaf scars are sunken, zonate, and discolored, and they may ooze amber gum. On large branches, scaffolds, and trunks, elliptical cankers exude copious amounts of gum, the bark dries and cracks form, exposing the blackened tissue beneath (Gauthier et al., 2021). Cankers typically develop concentric rings from alternation between canker extension and callus formation. Pycnidia erupt through the bark giving the surface a pimpled appearance. These produce conidia, the primary inoculum. Trees stressed by freezing, nutrient deficiency, ring nematode infections, and bacterial canker are predisposed to the disease. Young vigorous trees are less susceptible.

8.1.7 Fungal gummosis

Fungal gummosis can be a serious problem in areas with high rainfall. Tree growth and yield are significantly reduced on susceptible cultivars (Beckman et al., 2003), and trees may ultimately die. The disease is caused by species in the Botryosphaeriaceae including *Botryosphaeria dothidea* (the main pathogen)*, Diplodia seriata*, *Neofusicoccum parvum,* and *Lasiodiplodia theobromae*. All have a broad host range and are wound invaders, but *B. dothidea* can also cause infections at bark lenticels. On tree trunks and large branches, numerous sunken necrotic lesions develop around lenticels that produce copious amounts of gum. The lesions may coalesce and form large cankers. On young branches, small blisters with no gumming occur around lenticels. The pathogens produce abundant conidia in pycnidia that are disseminated by rain splash and ascospores in ascostroma in branch and trunk cankers. Wound infections occur during wetness, and lenticel infections by *B. dothidea* occur primarily in the summer. The pathogens overwinter in diseased bark and woody tissues.

8.1.8 Phytophthora root and crown rots

Phytophthora root and crown rots are destructive soil- and waterborne diseases of stone fruits worldwide. Peach trees can be affected at all ages, and infections often result in tree death. Multiple species of *Phytophthora* in the fungus-like phylum Oomycota of the kingdom Chromista (Stramenopila) are involved including *P. cactorum*, *P. cambivora*, *P. cinnamomi*, *P. citricola*, *P. citrophthora*, *P. cryptogea*, *P. drechsleri*, *P. megasperma*, *P. parasitica (P. nicotianae)*, *P. syringae,* and *P. niederhauserii.* These species differ in their geographic and seasonal occurrence, temperature optima, capability to cause root and/or crown rots, and virulence. *Phytophthora* spp. are mainly soil- or waterborne organisms. They survive as chlamydospores, hyphae, or oospores in root debris in the soil. In the presence of water, they produce abundant sporangia that release numerous motile zoospores that are the main infective propagules.

Infected trees show poor terminal growth, and leaves are small, chlorotic, and sparse. Trees decline progressively over several seasons or die suddenly in late spring or summer. Crown rots develop at the root crown and/or at the base of the trunk. A canker develops that often produces copious amounts of amber gum as a host response. Trunk girdling kills the tree. Cankers may extend up the trunk and occasionally scaffold branches. Aerial cankers develop when inoculum is rain- or sprinkler water-disseminated to upper parts of the tree, especially the scaffold crotches, where water tends to accumulate. Infected roots decay, sometimes extending to the crown. On feeder roots, the outer cortical tissues disintegrate leaving the white stele protruding from the decaying outer tissue.

8.1.9 Armillaria *root rot*

Armillaria root rot (syn. shoestring root rot, oak root rot) affects peaches as well as many other woody plants worldwide. Peach is considered extremely susceptible, and trees can be killed at any age. Severe outbreaks are observed on land planted successively with peaches or where forest trees were recently removed. In contrast to *Phytophthora* root rot, *Armillaria* root rot is also common on well-drained soils.

Early symptoms include poor terminal growth and undersized, curled leaves on all major limbs. Infected trees may collapse suddenly during the summer with most leaves still attached. In orchards with widely spaced trees, the disease may expand in a circular fashion, but in high-density orchards the disease often progresses along rows. White mycelial fans in the cambium between the bark and wood on the crown are highly diagnostic signs of the disease. In advanced stages of colonization, white rot wood decay occurs (Morrison et al., 1991). Black, shoestring-like mycelial strands (rhizomorphs) may be attached to roots, except when *Desarmillaria tabescens* or *D. caespitosa* are involved. In late summer/early fall, clusters of brown mushrooms may develop at the base of infected trees.

Worldwide, several causal pathogens have been identified: *A. mellea* sensu stricto, *A. ostoyae*, *D. tabescens*, and *D. caespitosa* that persist in the soil for many years as mycelium in infected plant tissues or as rhizomorphs. Infections of healthy roots originate from infected root segments in the soil, from root-to-root contact, or from contact with rhizomorphs. Basidiospores from the fruiting bodies are not considered major infection propagules (Rizzo et al., 1998).

8.1.10 Bacterial blast and canker

Bacterial blast and canker affect all *Prunus* spp. and is a major cause of premature peach tree decline and tree death in many production areas. Symptoms first appear as trees emerge from dormancy. Dormant buds are blighted and are often associated with an elliptical canker lesion. Cankers that are reddish-brown underneath the bark are found on twigs and branches and can extend all the way to the trunk. They may have a sour smell in early spring. Branches and scaffold limbs may die, and in severe cases, the entire tree will collapse (Hattingh et al., 1989). Young trees (up to 6-7 years old) are most susceptible to the disease. Cold injury can predispose the tree to infection and by itself cause symptoms similar to bacterial canker.

On peach, two common pathovars of the bacterial pathogen *Pseudomonas syringae*, pv. *syringae* and pv. *persicae*, have been described (Bultreys and Kaluzna,

2010). Pathogenicity and virulence are associated with the production of phytotoxins and plant hormones. The bacteria are ubiquitous in the environment and are disseminated by wind-blown rain, possibly pruning tools, and nursery stock. The primary entry into the peach tree is considered through wounds occurring during leaf drop in fall but possibly also through pruning cuts. Mild temperatures (12°C to 18°C) and wet conditions favor infection (English et al., 1961).

8.1.11 Bacterial spot

Bacterial spot is an important disease of peach and nectarine worldwide and is associated with warm, moist environmental conditions during the growing season. Fruit quality and marketability as well as orchard productivity are reduced. The disease is most severe in production areas with sandy, light soils. The bacterial pathogen *Xanthomonas arboricola* pv. *pruni* is highly specialized to infect *Prunus* species (Stefani, 2010). Numerous small, angular lesions 1 to 3 mm in diameter develop on leaves, fruit, and shoots (Roselló et al., 2012). Lesion centers may become necrotic and abscise creating "shot holes" that can cause the leaf to appear tattered and resemble copper phytotoxicity or shot hole caused by *Wilsonomyces carpophilus*. Severe defoliation can occur on highly susceptible cultivars that results in reduced fruit size, sunburn, and cracking of fruit, as well as reduced tree vigor and hardiness for overwintering. Fruit are most susceptible to infection between shuck split and pit hardening. Lesions often extend into the mesocarp and may exude light to dark amber gum. Infections after pit hardening usually result in a speckled fruit surface appearance. Infected fruit may drop prematurely. Spring cankers develop on shoots of the previous summer's growth from leaf scar infections at the lateral nodes that occurred in late autumn during leaf drop. Summer cankers form on new green shoots and become visible in late spring and summer. The pathogen survives during tree dormancy in symptomless lateral and terminal buds that can develop into twig cankers during late winter and early spring, and epiphytically on symptomless plant surfaces (Shepard and Zehr, 1994). In the spring, the pathogen is carried from between the bud scales with plant growth and is disseminated from spring cankers by wind-driven rain and heavy dew. Multiple cycles of infection can occur with each period of precipitation and heavy dew formation.

8.1.12 Crown gall

Crown gall is a soilborne disease caused by the ubiquitous soil bacterium *Agrobacterium tumefaciens* that attacks plants in many families, especially in temperate regions. The pathogen belongs to the *Rhizobiaceae,* a family of nitrogen-fixing bacteria with several species inducing hyperplastic plant growths. It invades through wounds, and infection is favored by moist conditions and poorly drained soils. Gall development occurs in 2 to 4 weeks at temperatures over 20°C (Fullner et al., 1996). Galls produced by *A. tumefaciens* typically develop at the root crown just below the soil line and on roots. Galls may enlarge to several centimeters in diameter. Newly formed galls are soft and spongy, whereas older ones turn woody and may become infection sites for other organisms including wood decay fungi. Young trees may be stunted, and older trees may exhibit symptoms typical of root problems. Crown gall is most important in the nursery, and symptomatic trees are culled. The bacterium is disseminated through nursery stock and movement of soil. *Agrobacterium* species can survive in the soil for more than 2 years, but survival is shortened by soil temperatures of higher than 34°C and acid conditions.

8.1.13 Plum pox virus (PPV)

Sharka, caused by Plum pox virus (PPV), is causing serious issues to the economic sustainability of stone fruits, including peach. PPV belongs to the genus *Potyvirus* within the family *Potyviridae*, which has gradually spread worldwide (Europe, Middle East, North and South America, and Asia) ranking among the top 10 most studied viruses in the world. PPV has several strains, being identified so far: PPV-Dideron (D), PPV-Marcus (M), PPV-Recombinant (Rec), PPV-El Amar (EA), PPV-Cherry (C), PPV-Winona (W), PPV-Turkey (T), PPV-Cherry Russian (RC), PPV-Ancestor (An), and PPV-Cherry Volga (CV), where M by far is the most virulent one being efficiently transmitted by aphids in peach (Pedrelli, 2024).

The virus spreads between long distance regions through the illegal traffic and insufficiently controlled exchanges of plant material infected, followed by natural and local spread by aphids. PPV is graft transmitted, and the vegetative multiplications of infected plants greatly contributes to the spread of the virus from infected areas if certified virus-free material is not used (García et al., 2014).

Symptoms of Sharka may appear on leaves, shoots, barks, petals, stones, and fruits; these last ones are not dangerous for consumption but unmarketable for the peach industry. Early in the growing season leaves show light-green discoloration, chlorotic spots, bands or rings, vein clearing or yellowing, and leaf deformation. Flower symptoms can occur on petals (discoloration), especially in peach cultivars with pink flowers. Infected fruits show chlorotic spots or lightly pigmented yellow rings or line patterns (especially in Japanese plums and peaches). They may become deformed or show irregular shape (especially cherries and plums) and over time may develop brown or necrotic areas under the discolored rings (Pedrelli et al, 2024).

8.1.14 Postharvest diseases and its management

The most important postharvest decays of peach and nectarine are presented in Table 3.4. The causal pathogens belong to the Eumycota in the phyla Ascomycota and Mucoromycota. They are ubiquitous, heterotrophic, and require water for growth. Most postharvest fungal pathogens need fruit injuries to infect, and these occur during harvest, transportation to packing houses, and packing into boxes for shipping to final markets. Some fungi, such as *Monilinia* or *Colletotrichum* species, can directly infect intact fruit surfaces causing quiescent infections that develop into decays after fruit ripening. Movement of fruit from cold to warmer conditions can cause enough water condensation for infection to occur, whereas fruit tissues with injuries have sufficient moisture to allow fungal growth.

Effective disease management requires detailed knowledge of pathogen biology and epidemiology as well as host physiology. General management strategies

Table 3.4 Postharvest diseases of peach and nectarine and primary management practices

Disease	Pathogen	Primary management
Common		
Brown rot	*Monilinia fructicola, M. laxa, M. fructigena mumecola, M. polystroma, M. yunnanensis*	Minimize injuries, temperature management, sanitation, pre- and postharvest fungicides
Gray mold	*Botrytis cinerea*	Minimize injuries, temperature management, sanitation, pre- and postharvest fungicides
Rhizopus rot, Gilbertella decay, Mucor decay	*Rhizopus stolonifer, Gilbertella persicaria, Mucor piriformis,* other *Mucor* spp.	Minimize injuries, temperature management, sanitation, postharvest fungicides
Sour rot	*Geotrichum candidum*	Minimize injuries, temperature management, sanitation, pre- and postharvest fungicides
Less Frequent		
Alternaria rot	*Alternaria alternata,* other *Alternaria* spp.	Minimize injuries, temperature management, sanitation, fungicides
Anthracnose	*Colletotrichum* spp.	Sorting and grading, temperature management, sanitation, fungicides
Black mold	*Aspergillus niger*, species in the *A. niger* complex	Sorting and grading, temperature management, sanitation, fungicides
Blue mold	*Penicillium expansum,* other *Penicillium* spp.	Sorting and grading, temperature management, sanitation, fungicides
Botryosphaeria fruit rot	*Botryosphaeria dothidea, B. obtusa, Lasiodiplodia theobromae*	Preharvest disease management with fungicides, sanitation
Cladosporium rot	*Cladosporium herbarum*	Minimize injuries, temperature management, sanitation, postharvest fungicides
Sour pit	*Candida inconsipicua*	None available

include a variety of cultural practices and use of chemicals in an integrated management approach. Their relative impact depends on the disease to be managed. Thus, some diseases such as *Armillaria* root rot, perennial canker, fungal gummosis, silver leaf, and *Phytophthora* root and crown rots can often be managed adequately using cultural practices, whereas others such as brown rot, blossom blight, rust, and peach leaf curl are most effectively controlled with the use of fungicides. For most diseases, however, a combination of these two main strategies is the best approach. Other diseases, such as PPV, the management is possible only by prevention, using PPV-free propagation material, contrast to PPV vectors, and field monitoring for PPV presence (Pedrelli et al., 2024).

Cultural practices start with choosing a planting site that is less likely to experience favorable microclimates for pathogen infection or avoiding sites where the soil is known to be infested (e.g., *Armillaria* root rot, *Verticillium* wilt). Using a directional orchard layout with prevailing winds that allows good air circulation and use of clean nursery stock are recommended. Soil water management including an irrigation design that prevents overwatering (e.g., reducing *Phytophthora* root and crown rot) as well as keeping trees vigorous by proper fertilization are important preventative measures. Other practices include properly timed pruning that minimizes large wood-exposing wounds and promotes wound healing (e.g., silver leaf, perennial canker, fungal gummosis) during dry conditions, orchard sanitation such as removing mummified fruit (e.g., brown rot, bacterial spot), infected twigs, and cankers (e.g., brown rot, perennial canker, fungal gummosis) from trees as well as weeds and senescent plant tissues from the orchard floor (e.g., jacket rot and green fruit rot), removal of alternate hosts such as wild *Prunus* spp., avoiding injuries to fruit including those from insects, and use of resistant cultivars if available.

Fungicides or biological treatments need to be applied during the most susceptible stages of host development, during critical times of inoculum production (e.g., rust), and before the occurrence of periods with conducive wetness and temperatures that promote pathogen infection, although some fungicides, especially those in the triazole class, have some postinfection activity. Applications during tree dormancy to reduce inoculum and protect against twig and bud infections can be highly effective for shot hole and peach leaf curl, and bactericides can be used for bacterial blast and canker.

Management of postharvest decays employs a wide range of strategies that begin with appropriate harvest and handling practices to minimize fruit injuries and end with suitable storage conditions of the commodity (Adaskaveg et al., 2022b, 2023, 2025; Adaskaveg and Förster, 2023). Grading of fruit in the packinghouse eliminates most fruit with visible injuries and of low quality. Washing fruit with sanitizers reduces plant as well as human pathogens (Bartz et al., 2022). Low temperatures during storage slow fruit respiration, and high relative humidity prevents water loss. Fungicide and some biological treatments provide a sustainable, decay-free supply of stone fruit commodities and allow marketing of fruit locally and over long distances (Adaskaveg and Förster, 2023). Currently, temperature management that is carefully monitored throughout postharvest handling, transportation, and marketing, and postharvest fungicide treatments are considered the most effective strategies to prevent losses from decay and to maintain postharvest crop life.

8.2 Pests

Leaves, fruit and woody tissues of peach and nectarine trees may be affected by a range of arthropod pests that includes species of Diptera, Lepidoptera, Thysanoptera, Coleoptera, and Hemiptera. Some of the pests such as the Mediterranean fruit fly, the peach twig borer, and San Jose scale are of worldwide and common importance, but others are of more localized significance. However, climate change is removing climatic barriers for the establishment of new pests in new areas where they arrive, as occurred with the brown marmorated stink bug *(Halyomorpha halys)*, which has spread extremely rapidly in many fruit-growing districts all over the world. Proper pest diagnosis and understanding of insect biology and pest epidemiology are a key for effective pest management. A list of important pests and their relative importance is given in Table 3.5. Cultural practices rely on population monitoring with adult trapping devices, both the traditional and the electronic ones, in key hosts where population growth and development occur early in the season, together with the Integrated Pest Management (IPM) program of a specific region. In addition, dropping infested fruit to the ground and collecting and destroying all fruits that remain in the orchard should be considered to reduce propagules for future generations (Papadopoulos et al., 2023).

8.2.1 Mediterranean fruit fly

The Mediterranean fruit fly (medfly), *Ceratitis capitata* (Diptera: Tephritidae), is one of the most important pests for peach production and trading of fresh fruit commodities globally. It is an extremely polyphagous species that can infest fruits of more than 300 plant species, with peaches and nectarines being some of the most preferred hosts reaching up to 100% of yield loss when management is not applied (Papadopoulos et al., 2023). Its current geographic distribution includes almost all Africa, South and Central America, Middle East, Western Australia and islands of the Pacific, Atlantic and Indian oceans, and Europe, especially in the Mediterranean countries, such as Italy, Greece, and Spain (Papadopoulos et al., 2023). It is considered an endemic pest in fruit-growing areas of the Spanish Mediterranean coast (Plá et al., 2021) affecting both fruit production and exportation.

Frequent detection of medfly in California, U.S., is considered to stem from transient populations (McInnis et al., 2017), whereas in North or Central Europe, this pest has also been recorded but only as a limited number of interceptions or short-lived adventive populations. The pest cycle includes many generations per year depending on climatic conditions and the availability of hosts. Under tropical and subtropical conditions there can be up to six or seven generations per year, whereas in the northern areas there can be as few as three. Life history and expectancy of eggs, larvae, and pupae stages for both male and female insects may depend on temperature conditions, the origin of the population, and the host fruit in which immature insects develop. Symptoms on fruits occur after oviposition of the eggs, when the larvae feed on the flesh of the fruit. Fruit often subsequently become infected by fungi and bacteria that intensify the damage and render the fruit

Table 3.5 Main peach and nectarine pests and primary management practices (Papadopoulos et al., 2023).

Order and family	Species	Damage	Main management approaches
Fruit/reproductive organs			
Coleoptera: Curculionidae	*Conotrachelus nenuphar*	Fruit, stems	Cover sprays targeting adults, biological control targeting soil dwelling larvae
Diptera: Tephritidae	*Ceratitis capitata* (Wiedemann)	Ripe and ripening fruit	Bait and cover sprays, mass trapping, sterile insect technique, inundative release of egg-larvae parasitoids
	Bactrocera zonata (Saunders)	Ripe and ripening fruit	Bait and cover sprays, male annihilation with methyl eugenol, mass trapping, aterile insect technique
Lepidoptera: Tortricidae	*Grapholitha molesta* (Busck)	Stem borer, fruit surface and flesh	Adult trapping with pheromone traps, insecticide applications, application of CpGV as a biopesticide, mating disruption
	Thaumatotibia (Cryptophlebia) leucotreta	Fruit	Adult trapping with pheromone traps, insecticide applications, sterile insect technique, mating disruption, biological control (egg parasitoids)
	Platynota idaeusalis (Walker)	Leaves, leaf roller, cosmetic infestation on fruit	Adult trapping with pheromone traps, insecticide applications
Lepidoptera: Gelechiidae	*Anarsia lineatela* Zeller	Stem borer, fruit surface and flesh	Adult trapping with pheromone traps, insecticide applications, mating disruption

Table 3.5 Main peach and nectarine pests and primary management practices (Papadopoulos et al., 2023).—*cont'd*

Order and family	Species	Damage	Main management approaches
Hemiptera: Diaspididae	Pseudalacaspis pentagona (Targioni-Tozzetti)	Shoots, branches, fruit	Cover sprays targeting neonate, crawler larvae, application of winter (not used in Europe anymore) and summer oils, biological control
	Comstockaspis perniciosus (Comstock)	Shoots, branches, fruit	Cover sprays targeting neonate, crawler larvae; application of winter and summer oils; biological control
Hemiptera: Pentatomidae	*Halyomorpha halys* (Stål)	Fruit	Cover sprays or border sprays targeting all life stages, biological control
	Euschistus servus (Say)	Vegetative parts – leaf, buds, fruit	Cover sprays targeting all life stages
Heteroptera: Miridae	*Lygus lineolaris* (Palisot de Beauvois)	Buds, flowers, fruit	Cover sprays targeting all life stages, broadleaf weed management within orchard to minimize alternative hosts
Thysanoptera: Thripidae	*Frankliniella occidentalis* (Pergande)	Mainly fruit, bud, flowers	Weed management within orchard to minimize alternative hosts
Eriophyidae			
Foliage and shoot feeders			
Lepidoptera: Tortricidae	*Adoxophyes orana* Fischer von Röeslerstamm	Leaves, shots, fruit (surface scars)	Adult trapping with pheromone traps, insecticide applications
Hemiptera: Diaspididae	*Pseudaulacaspis pentagona* (Targioni-Tozzetti)	Shoots, branches, fruit	Biological control, crawlers trapping, insecticide application, summer oils

Continued

Table 3.5 Main peach and nectarine pests and primary management practices (Papadopoulos et al., 2023).—*cont'd*

Order and family	Species	Damage	Main management approaches
	Comstockaspis perniciosus (Comstock)	Shoots, branches, fruit	Biological control, crawlers trapping, insecticide application, winter summer oils
Hemiptera: Aphididae	*Muzys persicae* (Sulzer)	Leaf, young shoots	Cover sprays, biological control
Wood borers			
Sesiidae	*Synanthedon exitiosa* (Say)	Wood borer, trunk and roots	Cover sprays, biological control, mating disruption
	Synanthedon pictipes (Grote and Robinson)	Wood borer, scaffolding limbs	Cover sprays, biological control, mating disruption

Table adapted from CABI Peach book

unsuitable for consumption and marketing. In several hosts, the oviposition sting alone can cause fruit to drop from the tree (Papadopoulos et al., 2023).

8.2.2 Peach twig borer

The peach twig borer, *Anarsia lineatella*, is currently globally widespread causing significant damage to young shoots and fruits in many *Prunus* species (almond, apricot, peach, and plum), of which peach and nectarine being the preferred hosts where no management is applied. Its occurrence affects both production and fresh fruit marketing (Papadopoulos et al., 2023).

8.2.3 Brown marmorated stink bug

The brown marmorated stink bug, *Halyomorpha halys* (Stål) (Hemiptera: Pentatomidae), is a pest of global importance. Insecticide management of *H. halys* in fruit trees is complex and primarily relies on broad-spectrum pyrethroids and neonicotinoids (Leskey and Nielsen, 2018). There are currently neither nonchemical control approaches nor resistant peach cultivars (Papadopoulos et al., 2023).

8.2.4 San Jose scale

San Jose scale, *Comstockaspis perniciosus,* has a worldwide distribution with a very high reproductive potential. Scale insects damage plants directly through feeding injury by inserting their piercing/sucking mouthparts into plant tissues and withdrawing nutrients directly from the plant. In peaches, feeding damage can cause leaf chlorosis and twig or limb die-back. In terms of chemical control, insect growth regulators such as buprofezin and pyriproxyfen are effective at managing San Jose scale crawlers, and the systemic nature of spirotetramat can target additional life stages when feeding on plant tissue. In terms of biological control agents, there are several predators and parasitoid wasps that feed on or attack San Jose scale (Papadopoulos et al., 2023).

8.2.5 Green peach aphid

Aphids are soft-bodied insects that use their piercing sucking mouthparts to feed on plant sap. Some of them are very important vectors of plant viruses, such as plum pox virus (PPV). Once PPV becomes established within an orchard, the virus spreads from tree to tree by several species of virus-carrying winged aphids. The efficiency of transmission largely depends upon the virus strain, the host species, the cultivar, and the aphid species. Indeed, aphid vectors *Aphis spiraecola*, *Myzus persicae*, and *Hyalopterus pruni* can transmit PPV efficiently in a nonpersistent manner, while *A. craccivora*, *A. fabae*, *Brachycaudus cardui*, and *B. persicae* are less efficient vectors (Zhou et al., 2021).

Green peach aphid *(Myzus persicae)*, a common pest of peach, is the aphid vector spreading PPV through peach orchard. This insect is native to Europe but now is widely distributed across the United States and Canada. This aphid feeds on the underside of the leaves, curling them badly when the infestation is heavy as well as causing aborting of flowers and newly formed fruit. In addition, on nectarine this aphid feeds directly on young fruit, distorting normal growth and leaving fruit bumpy and unmarketable.

The green peach aphid is attacked by several predators (lady beetles, lacewings, syrphid flies, soldier beetles, and predaceous bugs). Early season establishment of them on peach crop but nectarine, may often suppress green peach aphid populations below damaging levels. Populations can also be suppressed by prebloom sprays, as these aphids are most susceptible just after hatching and before they have crawled into opening buds and flowers, just as buds are beginning to open.

9. Current and future challenges

Like many other tree crops, peach production worldwide is facing many challenges associated with climate change, soil and orchard management (lack of labor, pests and diseases, labor costs, mechanization, etc.), acceptance of gene technology, grower profitability, and consumer preferences.

The increase in global temperatures can adversely affect peach production, affecting bloom time, fruit set and development, ripening process, and fruit quality. Indeed, warmer winters induce insufficient chilling hours, disrupting normal peach fruit growth patterns. Linked to high winter temperatures, and in particular the Ebro Valley basin area in Spain and the Central Valley of California, long periods of foggy days are being reduced. This typical climate event helps the accumulation of sufficient chill hours, especially from mid-November to January. Additionally, the occurrence of extreme weather events during spring and summer, such as heatwaves, untimely spring frosts and hailstorms, and changes in precipitation patterns that increase the risk of droughts or floods, negatively impact peach production and tree health.

Additional challenges of new or chemical-resistant insects and diseases will require the development of reliable algorithms to monitor insect pests to predict their fecundity as well as to predict the development of diseases under different climatic scenarios. Sustainable peach production in the future will be assisted by AI and other technologies to implement effective and economical methods for pest and disease monitoring and surveillance in addition to providing accurate climate forecasting. Furthermore, rapid advances in biotechnology are providing the tools to identify genes to incorporate into cultivars and rootstocks to meet the challenges of climate change, resistant pests/pathogens, labor availability, rising production expenses, and the consumer's expectation of an environmentally friendly grown peach with flavor and nutritive value.

Acknowledgements

The authors would like to thank Dr. Helga Förster for her contribution in the section related to plant disease management.

References

Adaskaveg, J. E., Arpaia, M. L., Förster, H., & Mitcham, E. J. (2025). Postharvest disease and insect control. In A. Kader, J. F. Thompson, & M. Saltveit (Eds.), *Postharvest technology of horticultural crops* (4th ed.). Davis, California: University of California (*in press*). Publication No. 21659.

Adaskaveg, J. E., Chen, D., Cary, D., & Förster, H. (2022a). Overview of management practices with the use of biocontrols, natural products, and conventional fungicides to combat pre- and postharvest diseases of peach and nectarine fruit. *Acta Horticulturae, 1352*, 17–26.

Adaskaveg, J. E., & Förster, H. (2023). Major postharvest diseases of peach and nectarine with a review of their management. In G. A. Manganaris, & C. H. Crisosto (Eds.), *Peach* (pp. 343–365). Wallingford, United Kingdom: CAB International.

Adaskaveg, J. E., Förster, H., & Prusky, D. B. (2022b). Overview of postharvest pathology. In J. E. Adaskaveg, H. Förster, & D. B. Prusky (Eds.), *Postharvest pathology of fruit and Nut crops* (pp. 3–22). St. Paul, Minnesota: American Phytopathological Society Press.

Adaskaveg, J. E., Schnabel, G., Ritchie, D. F., & Förster, H. (2023). Common preharvest diseases of peach and nectarine caused by fungi and bacteria: Biology, epidemiology, and management. In G. A. Manganaris, & C. H. Crisosto (Eds.), *Peach* (pp. 261–342). Wallingford, United Kingdom: CAB International.

Aranzana, M. J., Illa, E., Howad, W., & Arus, P. (2012). A first insight into peach (Punus persica (L.) Batsch) SNP variability. *Tree genetics & Genomes, 8*(6), 1359–1369.

Andreotti, C., Ravaglia, D., Ragaini, A., & Costa, G. (2008). Phenolic compounds in peach (*Prunus persica*) cultivars at harvest and during fruit maturation. *Annals of Applied Biology, 153*, 11–23.

Anthony, B. M., & Minas, I. S. (2021). Optimizing peach tree canopy architecture for efficient light use, increased productivity, and improved fruit quality. *Agronomy, 11*, 1961.

Bartz, J. A., Fatica, M. K., & Schneider, K. R. (2022). Sanitation practices during postharvest handling of fresh fruits and vegetables. In J. E. Adaskaveg, H. Förster, & D. Prusky (Eds.), *Postharvest pathology of fruit and Nut crops* (pp. 23–38). St. Paul, Minnesota: American Phytopathological Society Press.

Bassi, D., Cirilli, M., Foschi, S., Iglesias, I., Giovannini, D., Badenes, M. L., & Gasic, K. (2023). Cultivars. In G. Manganaris, G. Costa, & C. H. Crisosto (Eds.), *Peach* (pp. 92–113). Wallingford, United Kingdom: CAB International.

Beckman, T. G., Pusey, P. L., & Bertrand, P. F. (2003). Impact of fungal gummosis on peach trees. *Hort Science, 38*, 1141–1143.

Bento, C., Gonçalves, A. C., Silva, B., & Silva, L. R. (2020). Peach (*Prunus persica*): Phytochemicals and health benefits. *Food Reviews International, 3*, 1–32.

Berman, M. E., & DeJong, T. M. (1996). Water stress and crop load effects on fruit fresh and dry weights in peach (*Prunus persica*). *Tree Physiology, 16*, 859–864.

Berman, M. E., Rosati, A., Pace, L., Grossman, Y. L., & DeJong, T. M. (1998). Using simulation modeling to estimate the relationship between date of fruit maturity and yield potential in peach. *Fruit Varieties Journal, 52*, 229–235.

Bonghi, C., Trainotti, L., Botton, A., Tadiello, A., Rasori, A., Ziliotto, F., Zaffalon, V., Casadoro, G., & Ramina, A. (2011). A microarray approach to identify genes involved in seed-pericarp cross-talk and development in peach. *BMC Plant Biology, 11*, 107.

Bultreys, A., & Kaluzna, M. (2010). Bacterial cankers caused by *Pseudomonas syringae* on stone fruit species with special emphasis on the pathovars *syringae* and *morsprunorum* race 1 and race 2. *Journal of Plant Pathology, 92*, S21–S33.

Byrne, D., Raseira, M., Bassi, D., et al. (2012). Peach. In M. L. Badenes, & D. H. Byrne (Eds.), *Fruit breeding, Handbook of Pant Breeding 8* (pp. 505–569). U.S.: Springer Science + Business Media.

Cantín, C. M., Tian, T., Xiaoqiong, Q., & Crisosto, C. H. (2011). Copigmentation triggers the development of skin burning disorder on peach and nectarine fruit [*Prunus persica* (L.) Batsch]. *Journal of Agricultural and Food Chemistry, 59*, 2393–2402.

Cao, K., Pemg, Z., Zhao, X., Li, Y., Liu, K., Arus, P., Fang, W., Chen, C., Wang, X., Wu, j., Fei, Z., & Wang, L. (2022). Chromosome-level genome assemblies of four wild peach species provide insights into genome evolution and genetic basis of stress resistance. *BMC Biology, 20*, 139.

Cheng, G. W., & Crisosto, C. H. (1994). Development of dark skin discoloration on peach and nectarine fruit in response to exogenous contaminations. *Journal of American Society Horticultural Science, 119*(3), 529–533.

Community Plant Variety Office. (2015). *Protocol for tests on distinctness, uniformity and stability*. TP/053/2 Rev.

Corelli-Grappadelli, L., & Marini, R. P. (2008). Orchard planting systems. In D. R. Layne, & D. Bassi (Eds.), *The peach: Botany, production and Uses* (pp. 193–220). Wallingford, United Kingdom: CAB International.

Crisosto, C. H. (2006). Peach quality and postharvest technology. *Acta Horticulturae, 713*, 479–488.

Day, K., Lopez, G., & DeJong, T. (2008). Using growing degree hours accumulated thirty days after bloom to predict peach and nectarine harvest date. *Acta Horticulturae, 803*, 163–166.

DeJong, T. M., Andris, H., Beede, R., Day, K., Johnson, R. S., Olson, W., Ramos, D., Yeager, J., & Yoshikawa, F. (1992). Feasibility of increasing cling peach yields by early thinning. *Annual Research Reports of the California Cling Peach Advisory Board*, 1–5.

DeJong, T. M., Day, K. R., Doyle, J. F., & Johnson, R. S. (1994). The Kearney Agricultural Center Perpendicular "V" (KAC-V) orchard system for peaches and nectarines. *HortTechnology, 4*, 362–367.

DeJong, T. M., & Goudriaan, J. (1989). Modeling peach fruit growth and carbohydrate requirements: Reevaluation of the double-sigmoid growth pattern. *Journal of the American Society for Horticultural Science, 114*, 800–804.

DeJong, T. M., & Grossman, Y. L. (1995). Quantifying sink and source limitations on dry matter partitioning to fruit growth in peach trees. *Physiologia Plantarum, 95*, 437–443.

DeJong, T. M., Negron, C., Favreau, R., Day, K. R., Lopez, G., Costes, E., & Guedon, Y. (2012). Using concepts of shoot growth and architecture to understand and predict responses of peach trees to pruning. *Acta Horticulturae, 962*, 225–232.

English, H., DeVay, J. E., Lilleland, O., & Davis, J. R. (1961). Effect of certain soil treatments on the development of bacterial canker in peach trees. *Phytopathology, 51*, 65.

Europêch (2022). https://www.medfel.com (Accessed 6 August 2024).

FAOSTAT. (2024). *Website from Food and Agriculture Organization of the United Nations*. https://www.fao.org/home/en/. (Accessed 8 August 2024).

Font i Forcada, C., Reig, G., Fontich, C., Batlle, I., Alegre, S., Cantín, C. M., Eduardo, I., Carbó, J., Maillard, A., Maillard, L., & Bonany, J. (2022). Flat peach Sublor and Sublim and flat nectarine Perla series. *HortScience, 57*, 1280–1281.

Fullner, K. J., Cano Lara, J., & Nester, E. W. (1996). Pilus assembly by *Agrobacterium* T-DNA transfer genes. *Science, 273*, 1107–1109.

Génard, M., Lescourret, F., Gomez, L., & Habib, R. (2003). Changes in fruit sugar concentrations in response to assimilate supply, metabolism and dilution: A modelling approach applied to peach fruit (*Prunus persica*). *Tree Physiology, 23*, 373–385.

García, J. A., Glasa, M., Cambra, M., & Candresse, T. (2014). Plum pox virus and sharka: A model potyvirus and a major disease. *Molecular Plant Pathology, 15*(3), 226–241.

Gasic, K., & Saski, C. (2019). Advances in fruit genetics. In G. A. Lang (Ed.), *Achieving sustainable cultivation of temperate zone tree fruits and berries Volume 1: Physiology, genetics and cultivation* (pp. 135–162). Cambridge, United Kingdom: Burleigh Dodds Science Publishing Limited.

Gauthier, N., Smigell, C., & Lyons, K. (2021). *Bacterial canker and perennial canker of stone fruit*. University of Kentucky. Plant Pathology Fact Sheet PPFS-FR-T-08.

Giovannini, D., Liverani A, A., Bassi, D., & Lateur, M. (2013). ECPGR priority descriptors for Peach (Prunus persica (L.) Batsch). *European Cooperative Programme for Plant Genetic Resources*, 1–30.

Girona, J., Marsal, J., Mata, M., Arbones, A., & DeJong, T. M. (2004). A comparison of the combined effect of water stress and crop load on fruit growth during different phenological stages in young peach trees. *The Journal of Horticultural Science and Biotechnology, 79*, 308–315.

Gradziel, T. M. (1994). Changes in susceptibility to brown rot with ripening in three clingstone peach genotypes. *Journal of the American Horticultural Society, 119*, 101–105.

Grossman, Y. L., & DeJong, T. M. (1995a). Maximum fruit growth potential and seasonal patterns of resource dynamics during peach growth. *Annals of Botany, 75*(6), 553–560.

Grossman, Y. L., & DeJong, T. M. (1995b). Maximum fruit growth potential following resource limitation during peach growth. *Annals of Botany, 75*(6), 561–567.

Hancock, J. F., Scorza, R., & Lobos, G. A. (2008). Peaches. In J. F. Hancock (Ed.), *Temperate fruit crop breeding: Germplasm to genomics* (pp. 265–298). East Lansing: Springer.

Hattingh, M., Roos, I., & Mansvelt, E. (1989). Infection and systemic invasion of deciduous fruit trees by *Pseudomonas syringae* in South Africa. *Plant Disease, 73*, 784–789.

Huan, C., Jiang, L., An, X., Yu, M., Xu, Y., Ma, R., & Yu, Z. (2016). Potential role of reactive oxygen species and antioxidant genes in the regulation of peach fruit development and ripening. *Plant Physiology and Biochemistry, 104*, 294–303.

Iglesias, I., & Echeverría, G. (2009). Differential effect of cultivar and harvest date on nectarine colour, quality and consumer acceptance. *Scientia Horticulturae, 120*, 41–50.

Iglesias, I., Reighard, G., & Lang, G. (2023). Peach tree architecture: Training systems and pruning. In G. Manganaris, G. Costa, & C. H. Crisosto (Eds.), *Peach* (pp. 17–53). Wallingford, United Kingdom: CAB International.

Jiménez, S., Pinochet, J., Romero, J., Gogorcena, Y., Moreno, M. A., & Espada, J. L. (2011). Performance of peach and plum based rootstocks of different vigour on a late peach cultivar in replant and calcareous conditions. *Scientia Horticulturae, 129*, 58–63.

LaRue, J. (1989). Establishing the orchard. In J. H. LaRue, & R. S. Johnson (Eds.), *Peaches, plums, and nectarines: Growing and handling for fresh market* (pp. 21–30). Cooperative Extension U.C. Davis Publication 3331.

Leskey, T. C., & Nielsen, A. L. (2018). Impact of the invasive brown marmorated stink bug in North America and Europe: History, biology, ecology, and management. *Annual Review of Entomology, 63*, 599–618.

Lopez, G., & DeJong, T. M. (2007). Spring temperatures have a major effect on early stages of peach fruit growth. *Journal of Horticultural Science and Biotechnology, 82*, 507–512.

Manganaris, G. A., & Crisosto, C. H. (2020). Stone fruits. In M. Gil, & R. Beaudry (Eds.), *Controlled and modified atmosphere for fresh and fresh-cut produce* (pp. 311–322). Elsevier. https://www.sciencedirect.com/book/9780128045992/controlled-and-modified-atmospheres-for-fresh-and-fresh-cut-produce#book-description.

Manganaris, G. A., Minas, I., Cirilli, M., Torres, R., Bassi, D., & Costa, G. (2022). Peach for the future: A specialty crop revisited. *Scientia Horticulturae, 305*, 111390.

Manganaris, G. A., Vincente, A. R., Martinez, P., & Crisosto, C. H. (2019). Postharvest physiological disorders in peach and nectarine. In S. Tonetto de Freitas, & S. Pareek (Eds.), *Physiological disorders in fruits and vegetables* (pp. 253–264). Boca Raton: CRC press. https://www.taylorfrancis.com/books/edit/10.1201/b22001/postharvest-physiological-disorders-fruits-vegetables-sergio-tonetto-de-freitas-sunil-pareek.

Manganaris, G., Sansavini, S., Gradziel, T., Bassi, D., & Crisosto, C. H. (2023a). Peach: An introduction. In G. Manganaris, G. Costa, & C. H. Crisosto (Eds.), *Peach* (pp. 1–16). Wallingford, United Kingdom: CAB International.

Manganaris, G., Vicenete, A. R., Echeverria, G., & Crisosto, C. H. (2023b). Postharvest supply chain management protocols and handling of physiological disorders. In G. Manganaris, G. Costa, & C. H. Crisosto (Eds.), *Peach* (pp. 206–225). Wallingford, United Kingdom: CAB International.

Marini, R. P., Fallahi, E., Francescatto, P., et al. (2018). The NC-140 multi-location peach physiology trial: Relationships between peach fruit weight, crop density and early season temperature. *Journal of American Pomological Society, 72*, 231–241.

Marini, R. P., & Reighard, G. L. (2008). Crop load management. In D. R. Layne, & D. Bassi (Eds.), *The peach: Botany, production and Uses* (pp. 289–302). Wallingford, U.K: CAB International.

Mariotti, L., Picciarelli, P., Lombardi, L., & Ceccarelli, N. (2011). Fruit-set and early fruit growth in tomato are associated with increases in indoleacetic acid, cytokinin, and bioactive gibberellin contents. *Journal of Plant Growth Regulation, 30*, 405–415.

Marra, F. P., Inglese, P., DeJong, T. M., & Johnson, R. S. (2002). Thermal time requirement and harvest time forecast for peach cultivars with different fruit development periods. *Acta Horticulturae, 592*, 523–529.

McInnis, D. O., Hendrichs, J., Shelly, T., Barr, N., Hoffman, K., Rodriguez, R., Lance, D. R., Bloem, K., Suckling, D. M., Enkerlin, W., Gomes, P., & Tan, K. H. (2017). Can Polyphagous invasive Tephritid pest populations escape detection for years under favorable climatic and host conditions? *American Entomologist, 63*, 89–99.

Minas, I., Anthony, M., Drogoudi, P., & Costa, G. (2023b). Non-destructive peach fruit maturity and quality assessment. In G. Manganaris, G. Costa, & C. H. Cristos (Eds.), *Peach* (pp. 175–205). Wallingford, United Kingdom: CAB International.

Minas, I. S., Moreno, M. A., Anthony, B. M., Pieper, J. R., & Reighard, G. L. (2023a). Peach rootstock development performance. In G. Manganaris, G. Costa, & C. Crisosto (Eds.), *Peach* (pp. 54–91). Wallingford, United Kingdom: CAB International.

Monet, R., & Bassi, D. (2008). Classical genetics and breeding. In D. R. Layne, & D. Bassi (Eds.), *The peach: Botany, production and Uses* (pp. 61–84). Wallingford, United Kingdom: CAB International.

Morandi, B., Manfrini, L., Losciale, P., Zibordi, M., & Grappadelli, L. C. (2010). The positive effect of skin transpiration in peach fruit growth. *Journal of Plant Physiology, 167*, 1033–1037.

Morandi, B., Rieger, M., & Grappadelli, L. C. (2007). Vascular flows and transpiration affect peach (*Prunus persica* Batsch.) fruit daily growth. *Journal of Experimental Botany, 58*(14), 3941–3947.

Morandi, B., Xiloyannis, C., & Dichio, B. (2023). Irrigation and water management. In G. Manganaris, G. Costa, & C. Crisosto (Eds.), *Peach* (pp. 114–128). Wallingford, United Kingdom: CAB International.

Morrison, D. E., Williams, R. E., & Whitney, R. D. (1991). Infection, disease development, diagnosis, and detection. In C. G. Shaw, & G. Kile (Eds.), *Armillaria root disease* (pp. 62–75). United States Department of Agriculture, Forest Service. Agricultural Handbook 691.

Murray, R., Lucangeli, C., Polenta, G., & Budde, C. (2007). Combined pre-storage heat treatment and controlled atmosphere storage reduced internal breakdown of Flavorcrest peach. *Postharvest Biology and Technology, 44*, 116–121.

Papadopoulos, N. T., Blaauw, B. R., Milonas, P., & Nielsen, A. L. (2023). Biology and Management of insect pests. In G. Manganaris, G. Costa, & C. H. Crisosto (Eds.), *Peach* (pp. 366–420). Wallingford, United Kingdom: CAB International.

Pedrelli, A., Panattoni, A., & Cotrozzi, L. (2024). The sharka disease on stone fruits in Italy: A review, with a focus on Tuscany. *European Journal of Plant Pathology, 169*, 287–300.

Peris, M., & Alegre, S. (2012). Factors affecting corky spot on nectarine fruits in the Ebro valley in Spain. *Acta Horticulturae, 962*, 557–562.

Plá, I., García de Oteyza, J., Tur, C., Martínez, M.Á., Laurín, M. C., Alonso, E., Martínez, M., Martín, Á., Sanchis, R., Navarro, M. C., Navarro, M. T., Argilés, R., Briasco, M., Dembilio, O., & Dalmau, V. (2021). Sterile insect technique Programme against Mediterranean fruit fly in the Valencian Community (Spain). *Insects, 12*, 415.

Ravaglia, D., Espley, R. V., Henry-Kirk, R. A., Andreotti, C., Ziosi, V., Hellens, R. P., Costa, G., & Allan, A. C. (2013). Transcriptional regulation of flavonoid biosynthesis in nectarine (*Prunus persica*) by a set of R2R3 MYB transcription factors. *BMC Plant Biology, 13*, 68.

Reig, G., Cisneros-Zevallos, L., Cosat, G., & Crisosto, C. H. (2023). Peach fruit quality: Components, compositions, and nutritional health benefits. In G. Manganaris, G. Costa, & C. H. Crisosto (Eds.), *Peach* (pp. 226–260). Wallingford, United Kingdom: CAB International.

Reig, G., Garanto, X., Mas, N., & Iglesias, I. (2020). Long-term agronomical performance and iron chlorosis susceptibility of several *Prunus* rootstocks grown under loamy and calcareous soil conditions. *Scientia Horticulturae, 262*, 109035.

Reig, G., Mestre, L., Betrán, J. A., Pinochet, J., & Moreno, M. A. (2016). Agronomic and physicochemical fruit properties of Big Top nectarine budded on peach and plum based rootstocks in Mediterranean conditions. *Scientia Horticulturae, 210*, 85–92.

Reig, G., Torguet, S., Sans, L., Sas, M., & Faro, C. (2022). Inking in Spanish peaches: possible causes and solutions. *Acta Horticulturae, 1352*, 199–206.

Reighard, G. L., & Loreti, F. (2008). Rootstock development. In D. R. Layne, & D. Bassi (Eds.), *The peach: Botany, production and Uses* (pp. 193–220). Wallingford, United Kingdom: CAB International.

Reighard, G. L. (2013). Peach, plum and apricot rootstocks for the 21st Century. In G. Bishop, M. Fountain, R. Harrison, & R. Saville (Eds.), *vol. 119. Aspects of applied biology* (vol. 119) (pp. 59–66). United Kingdom: East Malling centenary conference, East Malling.

Rizzo, D. M., Whiting, E. C., & Elkins, R. B. (1998). Spatial distribution of *Armillaria mellea* in pear orchards. *Plant Disease, 82*, 1226–1231.

Rodriguez, C. E., Bustamante, C. A., Budde, C. O., Müller, G. L., Drincovich, M. F., & Lara, M. V. (2019). A comparative proteomic study between endocarp and mesocarp at very early stages underpins the main differential biochemical processes between these tissues. *Frontiers in Plant Science, 10*, 715.

Roselló, M., Santiago, R., Palacio-Bielsa, A., García-Figueres, F., Montón, C., Cambra, M., & López, M. (2012). Current status of bacterial spot of stone fruits and almond caused by *Xanthomonas arboricola* pv. *pruni* in Spain. *Journal of Plant Pathology, 94*, 15–21.

Rubio-Cabetas, M. J., Felipe, A. J., & Reighard, G. L. (2017). Rootstock development. In R. Socias i Company, & T. M. Gradziel (Eds.), *Almonds: Botany, production and Uses* (pp. 209–227). Wallingford, United Kingdom: CAB International.

Ruralcat. (2024). *Website from Department of climate action*. Food and Rural Agenda of Catalonia's Government (accessed in 20th of October).

Scorza, R., Miller, S., Glenn, D. M., Okie, W. R., & Tworkoski, T. (2006). Developing peach cultivars with novel growth habits. *Acta Horticulturae, 713*, 61–64.

Shepard, D. P., & Zehr, E. I. (1994). Epiphytic persistence of *Xanthomonas campestris* pv. *pruni* on peach and plum. *Plant Disease, 78*, 627–629.

Siew-Wai, C., Shaw, J., Haberle, R., Wen, J., & Potter, D. (2014). Diversification of almonds, peaches, plums and cherries – molecular systematics and biogeographic history of *Prunus* (Rosaceae). *Molecular Phylogenetics and Evolution, 76*, 34–48.

Soto-Estrada, A., & Adaskaveg, J. E. (2004). Temporal and quantitative analysis of stem lesion development and foliar disease progression of peach rust in California. *Phytopathology, 94*, 52–60.

Soto-Estrada, A., Förster, H., DeMason, D. A., & Adaskaveg, J. E. (2005). Initial infection and colonization of leaves and stems of cling peach by *Tranzschelia discolor*. *Phytopathology, 95*, 942–950.

Stefani, E. (2010). Economic significance and control of bacterial spot/canker of stone fruits caused by *Xanthomonas arboricola* pv. *pruni*. *Journal of Plant Pathology, 92*, 99–104.

Su, T., Wilf, P., Huang, Y., Zhang, S., & Zhou, Z. (2015). Peaches preceded humans: Fossil evidence from SW China. *Scientific Reports, 5*, 16794.

Tatsuki, M., Haji, T., & Yamaguchi, M. (2006). The involvement of 1-aminocyclopropane-1-carboxylic acid synthase isogene, Pp-ACS1 in peach fruit softening. *Journal of Experimental Botany, 57*, 1281–1289.

Tonutti, P., Bonghi, C., Drincovich, M. F., & Trainotti, L. (2023). Peach fruit growth, development, ripening and postharvest. In G. Manganaris, G. Costa, & C. H. Crisosto (Eds.), *Peach* (pp. 148–174). United Kingdom: CABI.

USDA. (2024). *Website from U.S. department of agriculture* [Accessed 8 August 2024].

Yu, Y., Fu, J., Xu, Y., Zhang, J., Ren, F., Zhao, H., et al. (2018). Genome re-sequencing reveals the evolutionary history of peach fruit edibility. *Nature Communications, 20*(9), 5404.

Zaracho, N., Reig, G., Kalluri, N., Arús, P., & Eduardo, I. (2023). Inheritance of fruit red-flesh patterns in peach. *Plants, 12*, 394.

Zhang, B., Shen, J., Wei, W., Xi, W., Xu, C. J., Ferguson, I., & Chen, K. (2010). Expression of genes associated with aroma formation derived from the fatty acid pathway during peach fruit ripening. *Journal of Agricultural and Food Chemistry, 58*, 6157–6165.

Zheng, W., Pan, L., Liu, H., Niu, L., Lu, Z., Cui, G., & Wang, Z. (2015). Characterization of 1-aminocyclopropane-1-carboxylic acid synthase (ACS) genes during nectarine fruit development and ripening. *Tree Genetics & Genomes, 11*, 18.

Zheng, Y., Crawford, G., & Chen, X. (2014). Archaeological evidence for peach (*Prunus persica*) cultivation and domestication in China. *PLOS One, 9*, e106595.

Zhou, J., Xing, F., Wang, H., & Li, S. (2021). Occurrence, distribution, and genomic characteristics of plum pox virus isolates from common apricot (*Prunus armeniaca*) and Japanese apricot (*Prunus mume*) in China. *Plant Disease, 105*, 11.

CHAPTER

Sweet Cherry

4

Chita Marlene Ayala Zapata[1], Juan Pablo Zoffoli Guerra[1], Nikos Papadopoulos[2], Jose Quero-Garcia[3] and Victor Blanco[4]

[1]*Pontificia Universidad Católica de Chile, Facultad de Agronomía y Sistemas Naturales, Departamento de Fruticultura y Enología, Vicuña Mackenna 4860, Santiago, Chile;* [2]*University of Thessaly Department of Agriculture Crop Production and Rural Environment;* [3]*INRAE, Biologie du Fruit et Pathologie, Université de Bordeaux, UMR 1332, Villenave d'Ornon, France;* [4]*IRTA, Efficient Use of Water in Agriculture Program, Parc Científic i Tecnològic Agroalimentari de Gardeny (PCiTAL), Fruitcentre, 25003, Lleida, Spain*

Chapter outline

Temperate Tree Fruits and Nuts. https://doi.org/10.1016/B978-0-443-29137-1.00015-3

1. Origin and spread

Sweet cherry (*Prunus avium* L.) originates from areas south of the Caucasus mountains surrounding the Caspian and Black Sea. Archaeological evidence suggests that wild cherries were consumed by prehistoric humans in Western Asia, and its domestication likely began in that region thousands of years ago. Its cultivation was expanded throughout the Mediterranean basin with the ancient Greek and Roman civilizations. During the Middle Ages, cherries continued to be cultivated across Europe, especially in monasteries and royal gardens. European colonists later introduced sweet cherry trees to the Americas in the 1600s, particularly in what is now the United States (Iezzoni et al., 2017). Today, sweet cherries are grown all over the world as they are highly valued by consumers everywhere, reflecting their long history of cultivation, trade, and human enjoyment.

2. Economic importance

Seasonal production: Global sweet cherry production has been increasing exponentially. For example, the 2,765,827 tons produced in 2022 represent a 150% increase since the start of the millennium (FAOSTAT, 2022). The rise in cherry production has been mainly due to the opening of new markets, which has led to increased demand and, consequently, an expansion in the planted area. Additionally, the adoption of new cultivars and rootstocks, novel training systems, higher densities, and plastic covers and greenhouses, as well as microenvironmental monitoring and control, have promoted higher yields and quality in areas with edaphoclimatic limitations and affected by climate change (Lang, 2017).

Production has a marked seasonality between the Northern and Southern hemispheres. In the Northern Hemisphere, harvests begin at the end of April, with greenhouse production in China and a low percentage in Europe. It peaks in June and early July, extending until September, with production in Northern Europe, the U.S., and Canada. In the Southern Hemisphere, production begins with very low volumes in mid-September in South Africa, continuing in mid-October with early cultivars grown in Chile and Argentina, and extending through January to February with fruit produced in New Zealand, Southern Chile, and

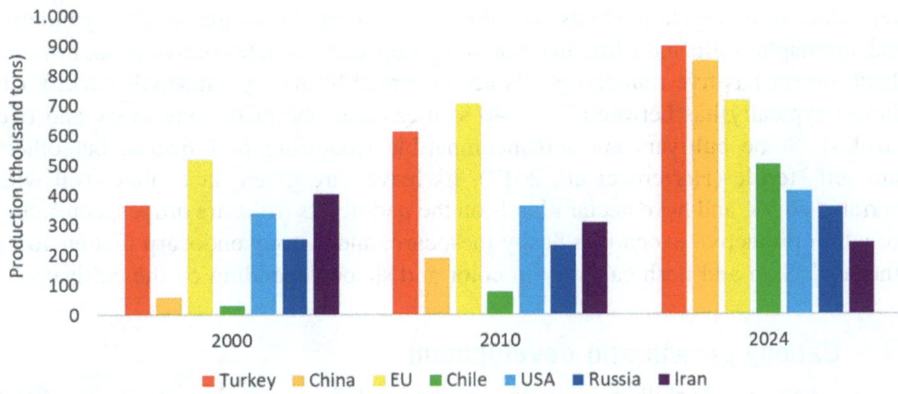

FIGURE 1

Evolution of sweet cherry production in the seven main production countries from 2000 to 2024 (adapted from USDA, 2024).

Australia. It peaks in December and January. The period of lowest global cherry supply is concentrated in March.

Producers: Turkey, China, EU, and Chile were the largest producers worldwide in 2024 (Figure 1). Turkey, followed by China, Uzbekistan, the U.S., Spain, Italy, and Iran are the main cherry-producing countries in the Northern Hemisphere. The production increased in Turkey and China over the last two decades and decreased in France, Italy, Iran, and Germany. The main producer in the Southern Hemisphere is Chile followed far behind by Australia, Argentina, New Zealand, and South Africa.

Exporters: The main exporting countries are Turkey and Chile. During the 2023-2024 season, Chile exported 413,979 tons of cherries, becoming an important factor in the international market.

Importers: The main importing countries include China, Russia, Germany, the U.S., and Canada. China is the world's largest, off—season, importing country, increasing purchases by 8.8 fold between 2012 and 2022. Russia is the second—largest importer, with its main suppliers being Uzbekistan, Turkey, and Azerbaijan.

3. Botany, phenology, and physiology

Sweet cherry belongs to the family Rosaceae, subfamily Amygdaloideae, and the genus Prunus. The sweet cherry is a caducifolious fruit tree with a pronounced apical dominance, naturally reaching up to 30 meters height. The trunk is straight and supports a conical and symmetrical crown that becomes rounded or irregular in mature trees. It naturally develops a root system with deep main roots at its base and abundant secondary roots (Vignati et al., 2022). It forms leaves and flowers from simple reproductive or vegetative buds. The flowers of sweet cherry are radially symmetrical, hermaphroditic and white in color and grouped in fascicles of five to six flowers. Each flower has five rounded petals accompanied by five green sepals each. Each flower typically has between 35 to 40 stamens and one pistil (one ovary and two ovules). Some cultivars are self-incompatible (requiring pollinizers), but others are self–fertile (Herrero et al., 2017). Its leaves are green, lanceolate oval with serrated edges, and have nectar glands on the petiole. Its fruits are drupes, consisting of a thin protective exocarp, a fleshy mesocarp, and a stony endocarp that encloses the seed. Skin and flesh can vary in color and shape depending on the cultivar.

3.1 Canopy growth and development

Sweet cherry needs chill in winter to complete endodormancy. Its reproductive habit is characterized by growing fruit in simple reproductive buds at the base of 1-year-old shoots and spurs on 2-, 3-, and more year-old wood. If the newly formed meristem elongates in the same season, it becomes a lateral extension shoot. The potential fruiting section is comprised of two types of fruit bud populations (Figure 2): nonspur reproductive buds at the base of 1-year-old shoots and single reproductive buds of spurs grown in 2-year-old wood and more. Reproductive buds on spurs (one to eight buds) usually contain one to eight flowers, depending on the cultivar/rootstock combination.

3.2 Fruit growth and development

Fruit growth occurs in a short period (60–90 days), depending on the cultivar. Natural fruit drops occur during either June–July (Northern Hemisphere) or November–December (Southern Hemisphere). The growth pattern of a sweet cherry drupe follows a double sigmoid curve with three developmental stages:

- **Stage I:** The initial phase and high cell division activity occurs in the first half. The number of cells in the mesocarp is genetically determined and remains relatively stable (Olmstead et al., 2007).

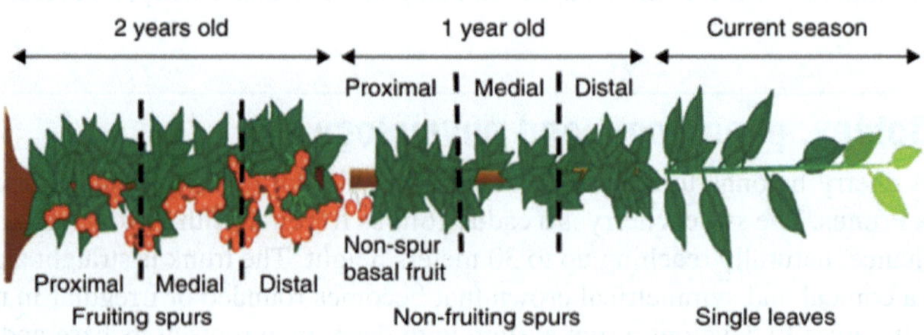

FIGURE 2

A typical sweet cherry productive branch (Ayala & Lang, 2017).

- **Stage II:** This is characterized by endocarp hardening and embryo development, with minimal cell division and increased cell wall thickening. The color changes of the exocarp begin, transitioning from green to light yellow (Einhorn et al., 2013).
- **Stage III:** The fruit shows rapid growth with the enlargement of mesocarp cells, constituting a major sink for carbohydrates in highly productive combinations (Ayala & Lang, 2004). At this phase, the fruit suffers changes that characterize a nonclimacteric type of maturation.

Storage reserves: The storage of carbon reserves (mainly starch) plays a key role during the early stages of spring growth and are accumulated as nonstructural carbohydrates. Reserves are essential for supporting the development of both reproductive and vegetative organs before the new leaves become net exporters (Loescher et al., 1990; Ayala & Lang, 2015).

3.3 Environmental and soil requirements

Sweet cherry requires a range of temperatures for optimal growth (Neilsen et al., 2017; Table 1). During winter, trees must accumulate chill to break dormancy and, during spring, heat to ensure uniform bloom and fruit development (Gratacós & Cortés, 2007). Sweet cherries tend to have high chilling requirements, although they range widely among cultivars (400 to 1600 chilling hours or 30 to 65 chilling portions; Alburquerque et al., 2008). Heat requirements vary between 160 and 680 growing degree days (GDD) from the budburst to harvest (Fadón et al., 2021).

Soil requirements: Soil conditions play a crucial role in root development. An optimal pH range of 6.0 to 7.0 promotes nutrient uptake, acidic soils (pH <5.5) increase aluminum toxicity, and alkaline soils (pH >7.0) restrict micronutrient absorption (Nielsen et al., 2017). Well-drained, aerated soils enhance root elongation, whereas compacted soils limit water infiltration and root growth (Bonomelli et al., 2012).

Table 1 Climatic requirement ranges of sweet cherry

Requirement	Range
Optimal growing temperature	19°C –25°C
Optimal bloom temperature	20°C –25°C
Chilling hours	400–1600
Chilling units	94–1250
Chilling portions	30–65
Growing-degree days (Budburst to harvest)	160–680
Growing degree hours	7000–16,000
Critical temperature	Fruit set < −1.1°C
	Closed flower < −3.9°C
	Open flower < −2.2°C

Adapted from Alburquerque et al., 2008; Fadón et al., 2021; Vignati et al., 2022.

4. Breeding

Numerous reviews and book chapters have been written in the last 30 years dealing with the breeding of sweet cherry cultivars and rootstocks (Iezzoni et al., 2017; Quero-García et al. 2017a, 2019; Hrotkó and Rozpara, 2017; Quero-García, 2019).

4.1 Cultivar breeding

A good, sweet cherry cultivar needs to show a perfect balance between yield and fruit size; both traits are negatively correlated. Yield is the major selection criteria; however, cherry growers will not sell small fruit (< 24–26 mm) at good prices. Therefore, breeders will eliminate those hybrids showing low productivity and/or small fruit size in the first phases of selection. It is highly important to choose precocious cultivars. One of the strategies to increase productivity has been the release of self-fertile cultivars. Another critical factor for profitability is the ability to bear fruit in different environmental conditions. In the context of global warming, chilling and heat requirements are being considered. Indeed, the rise of temperatures in winter reduces chilling accumulation, provoking an irregular and prolonged bloom period and, in the end, a significant loss of productivity. The search for low–chill cultivars has become a top priority for numerous breeding programs. Sources of low–chilling requirements are scarce; one of the most-used genitors for this trait has been the Spanish cultivar "Cristobalina." Hence, new sweet cherry cultivars should combine lower chilling requirements with high or low heat requirements as needed.

One of the most important problems for cherry growers has always been rain–induced fruit cracking (Knoche and Winkler, 2017), and most breeders include cracking tolerance among their major selection criteria. Global warming can also impact fruit quality due to double fruit formation (Wenden et al., 2017). Double fruits are observed after the occurrence of high temperatures during summer, which causes an abnormal cellular differentiation of the floral primordia, giving rise to double pistils.

A very important selection criterion is the extension of the harvest season window. Extra–early harvest cultivars reach very high prices. One strategy is to breed for early bloom cultivars (with a short period from flower to fruit), but these will be more vulnerable to frost damage. Extra–late cultivars are increasingly being created, particularly in Northern European countries such as Germany. Resistance and/or tolerance to pests and diseases are increasingly becoming major selection targets for breeders. Among diseases, one of the most serious is bacterial canker, caused by *Pseudomonas spp*. Among fungal diseases, brown rot *(Monilinia spp.)*, is the most problematic, attacking both flowers and fruits (Børve et al., 2017).

Methodology for cultivar breeding: A classical breeding program will take between 20 to 25 years due to the long juvenility period of sweet cherry. Breeding is based on the creation of new materials, followed by the selection of the best hybrids. During the first evaluation phase (2 or 3 consecutive years), phenotypic evaluation, through visual qualitative observations, allows selection of the most promising hybrids in terms of major traits. In general, depending on the initial number of hybrids, 1% to 5% of the total number of hybrids will be grafted. A minimum of two clonal replicates per hybrid are produced in at least two sites of evaluation, and a higher number of traits will be studied in this phase. After juvenility, the best hybrids will be selected and multiplied. The best selections will be grafted again but multiplied at least 5 to 10 times and established in a higher number of sites. In the end, the most promising hybrids will enter the stage of precommercial evaluation and might simultaneously integrate the scheme of official protection by conducting Distinction, Uniformity, Stability (DUS) trials. Breeders will conduct controlled crosses rather than open pollinations, to combine, in one unique genotype, both parent's favorable characteristics. However, given the relatively low level of success for crosses derived from controlled pollination, numerous sweet cherry breeders have relied on open pollination.

One of the most important fields of research in the last 30 years has been the genetic determinism of traits of agronomic importance. This work has allowed the use of genetic and genomic information, through the development of molecular markers and DNA tests, in the selection process. The different uses and possibilities of "DNA−assisted breeding" (Peace, 2017) have been extensively discussed recently (Herrero et al., 2017; Iezzoni et al., 2017; Quero-Garcia et al., 2017b; 2019; 2022).

The first type of markers to be used were those associated with the self−incompatibility system, controlled by a single locus mapped on linkage group (LG), or chromosome, LG6. The available DNA tests allow breeders to plan their crosses by choosing only intercompatible genitors to verify paternity of hybrids and select hybrids for self−fertility. The second trait with monogenic control for which DNA tests were developed was fruit skin and flesh color, with a major locus mapped on LG3. A molecular marker called Rf, which is a microsatellite or simple sequence repeat (SSR) was developed by Sandefur et al. (2016). Most of the important agronomic traits for sweet cherry show complex polygenic control, traits such as fruit weight/size, fruit firmness, tolerance to rain−induced fruit cracking, fruit composition (sugars, acids, nutraceutical compounds), and also phenology-related traits such as bloom and harvest dates and chilling and heat requirements. By combining phenotypic and genotypic data, Quantitative Trait Locus (QTL) analysis found several genomic regions responsible for the phenotypic variation of these traits. The first trait for which QTL work was later translated into the development of a DNA test was fruit weight. Two SSR markers, which flank the confidence interval of one QTL detected on LG2, help to differentiate between genotypes carrying alleles associated with either small or big fruits. More recently, other molecular markers have been proposed for traits such as bloom date on LGs 1, 2, 4, 5, and 7 (Calle et al., 2021; Branchereau et al., 2022); fruit firmness and fruit cracking tolerance on LGs 3 and 5, respectively (Crump et al., 2023); and fruit productivity, fruit firmness, and fruit cracking on LG1 and 6 and 4 and 5, respectively (Quero-Garcia et al., 2021; Cai et al., 2019).

4.2 Rootstock breeding

Rootstock breeding and the effect of the rootstock on the scion cultivar characteristics such as precocity, productivity, and fruit quality has been an active field of research (Hrotkó and Rozpara, 2017). Sweet cherry rootstocks are chosen from taxa showing good graft compatibility, the main species being *P. avium*, *P. cerasus*, *P. mahaleb*, *P. fruticosa*, as well as their hybrids and some related taxa. The most important target for sweet cherry rootstock breeding has undoubtedly been the search for dwarfing genotypes. The best sources for vigor control have been found in the *Eucerasus* section, *P. fruticosa*, *P. cerasus,* and *P. canescens*.

A fundamental property of any rootstock candidate is a good graft compatibility with a wide range of cultivars. Incompatibility occurs only when using two different species, for instance, *P. avium* grafted on *P. cerasus* or *P. mahaleb*.

The main traits of rootstock adaptation to abiotic stresses are cold hardiness, drought and heat tolerance, adaptability to different soil conditions, and nutrient supply efficiency (Jimenez et al., 2007). The cold hardiest rootstocks are from *P. cerasus* and *P. fruticosa; P. mahaleb* is hardier than *P. avium*. Drought and heat tolerance have been related to root depth, and due to climate change, these traits will become increasingly critical.

As for cultivars, tolerance to pests and diseases is an important goal for sweet cherry rootstock breeders. Resistance to nematodes infestation as well as fungal or bacterial diseases affecting either the root or the foliar system are the main targets of rootstock breeding.

Methodology for rootstock breeding: Contrary to the breeding of sweet cherry cultivars, clonal selection plays a much more important role for rootstock breeding, although hybridization has been largely utilized for the creation of interspecific hybrids. Clonal selection has been carried out on the three major sweet cherry rootstock species, *P. avium, P. cerasus,* and *P. mahaleb*. The first interspecific hybrid created with *P. avium* used as female parent was Colt rootstock *(P. avium* × *P. pseudocerasus)*, on which sweet cherry trees grow vigorously and is particularly well adapted to fertile and humid soils. Within the P—HL series, supposedly derived from crosses between *P. avium* and *P. cerasus*, the dwarf P—HL—A was the most successful. Another promising rootstock was PiKu 1, derived from the cross *P. avium* × *(P. canescens* × *P. tomentosa)*, with moderate or semi-dwarf vigor.

Prunus cerasus has been widely used as a source of dwarfing rootstocks for sweet cherries, although the major disadvantage is variable compatibility and suckering. Hybridization within *P. cerasus* has been little utilized except for the Weiroot series, specifically Weiroot 158, a semi-dwarfing one, as one of the most promising (Bujdoso et al., 2004). Interspecific hybridization using *P. cerasus* as mother genitor was massively conducted at Giessen, Germany, with the cross *P. cerasus* × *P. canescens* obtaining the GiSelA (GI) series. GI.5 has become the standard in countries such as Germany or the Netherlands, reducing growth of trees by 30% to 50%, compared to *P. avium*. However, GI.5 and GiSelA 6 (GI.6) are not well adapted to warm Mediterranean areas. Other programs used species such as *Padus maackii* or *P. subhirtella*, giving rise to rootstocks such as Krymsk 6 or the IP–C series (Hrotkó and Rozpara, 2017), respectively. More recently, the Corette series derived from interspecific crosses, involving *P. cerasus*, *P. avium*, *P. fruticose,* and *P. canescens,* has been released by Michigan State University (Long et al., 2019).

Within species *P. mahaleb*, the first clonal rootstock, Saint Lucie 64 (SL-64), selected in Bordeaux, France (1954), was highly successful thanks to its ease of propagation, compatibility with sweet cherries, and good productivity. Hybridization between *P. mahaleb* and *P. avium* carried out in Oregon, U.S., resulted in the M×M series (for Mahaleb × Mazzard). Out of these, Maxma 14, a semi-vigorous rootstock, is the most popular in the U.S., Chile, and France because of its good adaptation on lime soils (Hrotkó and Rozpara, 2017). Another species that has attracted attention as a potential dwarfing rootstock is *P. fruticosa*, which is a low-growing cherry shrub. Clonal selection was conducted on spontaneous hybrids within central European countries. Finally, out of the *Eucerasus* section, the all-purpose rootstock Adara, a Myrobalan plum *(P. cerasifera).* has been successfully used in Spain to replace SL-64, with higher tolerance to nematodes and asphyxia and a higher longevity.

5. Orchard management

Sweet cherry production systems have evolved toward medium to semi-pedestrian and pedestrian, high density orchards (>1000 up to 2800 trees/ha), considering smaller trees that facilitate labor and increase precocity (Lang, 2001c; Hrotkó et al., 2014). Some important factors influencing orchard practices are the cultivar, the rootstock, the training system, crop load regulation, and mechanization. Such factors are analytically described in the following sections.

5.1 Cultivars

The selection of an optimal cherry cultivar must consider tree structure, self-fertility, fruit quality, storage capacity, and abiotic and biotic tolerance. Cultivars are classified based on the harvest date into early, mid, and late growing season. Early cultivars are harvested at the beginning of the season, have a shorter ripening cycle, and earn the highest prices. Midseason cultivars are harvested in the middle of the growing season and constitute the bulk volume. Late cherries ripen in mid and late summer and predominate in cooler climates. Table 2 provides information for some traditional cherry cultivars, which are described for harvest date relative to Burlat.

5.2 Rootsocks

The selection of an optimal cherry rootstock considers several factors such as vigor, precocity, graft compatibility, and disease tolerance. New precocious dwarfing rootstocks has reduced juvenility and minimized tree volume; although in some

Table 2 Main features of the traditional cultivars worldwide

Cultivar	Harvest date	Fertility	Chilling hours (<7°C)
Brooks	Early	Self-fertile	400–500
Burlat	Early	Cross-pollination	600–800
Santina	Early/Mid-season	Self-fertile	600–800
Regina	Late	Cross-pollination	1000–1400
Lapins	Mid-season	Self-fertile	400–500
Bing	Mid-season	Cross-pollination	1000–1200
Kordia	Late	Cross-pollination	1000–1400
Carmen	Mid-season	Cross-pollination	600–800
Skeena	Mid-season	Self-fertile	800–1000
Rainier	Mid-season	Self-fertile	600–800
Staccato	Late	Self-fertile	1000–1200
Alex	Mid-season	Self-fertile	600–800
Sweetheart	Late	Self-fertile	800–1000
Ferrovia	Mid-season	Cross-pollination	600–900
Ziraat	Mid-season	Cross-pollination	400–500

Adapted from Gratacós & Cortés, 2007; Gonçalves et al., 2008; García et al., 2009; Quero-García et al., 2017; Fadón et al., 2021; Schuster et al., 2014.

FIGURE 3

Range of relative vigor of the most widely used rootstocks worldwide.

conditions, semi-vigorous and vigorous rootstocks are still an option (Lang, 2001b; Hrotkó & Rozpara, 2017; Lang et al., 2017) (Figure 3).

Some of the most widely used rootstocks are briefly described here:

- **Colt** *(Prunus avium × Prunus pseudocerasus)*: It is considered precocious and provides medium to high vigor (Csihon et al., 2018). It does not tolerate light and calcareous soils and low irrigation. It is susceptible to crown gall (Long et al., 2021).
- **Mazzard "Merisier"** *(Prunus avium)*: It is a vigorous rootstock, not suitable for high-density orchards. It adapts to a wide range of soils, but poorly drained soils

should be avoided. It is tolerant to PDV and PNRSV, Nepovirus, and Potyvirus genus (Long et al., 2015; Hrotkó & Rozpara, 2017).

- **Mahaleb** (*Prunus mahaleb* L.): It is resistant to sandy, calcareous, and dry soils but is susceptible to wet conditions. Provides high vigor, but it is less vigorous than Mazzard. It may exhibit incompatibility with some cultivars (Milošević et al., 2014; Hrotkó, 2016).
- **Maxma series:** Includes several interspecific hybrids of *P. mahaleb* × *P. avium.* The best-positioned in the market are Maxma 14 and Maxma 60.
  - **Maxma 14** (*P. mahaleb* × *P. avium* Brokforest): It confers medium vigor and is precocious. It adapts well to various soil types but does not tolerate poorly drained soils. It is moderately susceptible to *Phytophthora spp.* and tolerant to *Pseudomonas spp.* (Hrotkó & Rozpara, 2017; Long & Kaiser, 2010).
  - **Maxma 60** *(P. avium* × *P. mahaleb)*: It confers a high vigor and is mainly used in low-fertile and rocky soils, but it requires well-drainage conditions (Martins et al., 2021).
- **GiSelA (GI) series:** It includes several interspecific hybrids of *P. mahaleb* × *P. canescens.* The best positioned in the market are GiSelA 5 (GI.5), GiSelA 6 (GI.6), and GiSelA 12 (GI.12).
  - **GI.5** *(P. cerasus* × *P. canescens)*: It is a semi-dwarfing rootstock. It shows high precocity, adapts well to cold environments, and tolerates PDV and PNRSV viruses (Franken-Bembenek, 2005). It is not recommended for highly productive and semi-vigorous cultivars.
  - **GI.6** *(P. cerasus* × *P. canescens)*: It is a semi-vigorous rootstock, more vigorous than GI.5.
  - **GI.12** (GI.12, *P. canescens* × *P. cerasus* Leitzkauer): It is a semi−vigorous rootstock less precocious than GI.6.
- **PiKu series:** It includes PiKu 1 *(P. avium* × *[P. canescens* × *P. tomentosa]),* PiKu 3 *(P. pseudocerasus* × *[P. canescens* × *Prunus incisa]),* and PiKu 4 *(P. cerasus* Chatel Morel × *P.* Kursar *[P.kurilensis* × *P. sargentii]).* The best positioned in the market is Piku 1 (Hrotkó & Rozpara, 2017).
- **Krymsk series:** It includes several interspecific hybrids of *P. cerasus, P. fruticose,* and *P. serrulata.* Krymsk rootstocks are known for their good winter hardiness, resistance to bacterial canker, and tolerance to wet soil conditions (Maas et al., 2012). The best positioned in the market are Krymsk 5 and Krymsk 6.
- **Weiroot series:** It is a selection of *P. cerasus* and includes a range of semi-vigorous, semi-dwarfing, and dwarfing rootstocks. Among them, Weiroot 158 is the most used (Bujdosó & Hrotkó, 2005).

5.3 Training systems

They optimize fruit production and quality, facilitate crop management, and maximize access to sunlight and photosynthesis. Training systems are divided into two- and three-dimensional (Lang, 2001a; Stone et al., 2022).

- **Central leader:** From the central axis (trunk) emerge primary or secondary branches that bear fruit. The main branches at the bottom are larger than those at the top, giving a pyramidal shape. It adapts to medium to low vigor rootstocks (Figure 4).

FIGURE 4

Central leader training system.

- **Spanish bush:** It has three to four open axes that dilute vigor and reduce height, facilitating harvest. It has the advantage of bearing without needing support and adapts to medium and high-vigor rootstocks. Fruiting occurs in tertiary branches (Figure 5).

FIGURE 5

Spanish bush training system.

- **Steep leader:** It consists of three or four vertical axes and does not require support. From the main axes emerge secondary and tertiary branches, which bear fruit. This system is adapted to medium to high-vigor rootstocks (Long et al., 2015; Figure 6).

FIGURE 6

Steep leader training system.

- **Kym Green Bush (KGB):** Multiaxe with several temporary vertical branches with spurs only (lateral current season growth is eliminated). It allows non-precocious pedestrian orchards that can be harvested without ladders (Long et al., 2015, Ayala & Valenzuela, 2017; Figure 7).

FIGURE 7

KGB training system.

- **V or Y—Trellis:** It consists of a double fruiting wall over the row with trees arranged in a Y or V shape. In the Y trellis, the trees are arranged (transversally or longitudinally) over the row, and in the V—trellis, the trees alternate in opposite directions between rows (Ayala & Valenzuela, 2019; Figure 8 a,b). The structure requires support to prevent the angle between the main branches or trees from closing. Secondary branches are tied to a horizontal position (wall) and bear fruit (Figure 9). It can be used on semi-vigorous to semi-dwarfing rootstocks. These systems are recommended for semi-pedestrian orchards.

FIGURE 8

(a) Transversal Y-trellis system, (b) V-trellis system.

FIGURE 9

Fruiting wall in a V-trellis training system.

- **Upright Fruiting Offshoot (UFO):** It is considered a precocious semi-pedestrian training system. It allows good light distribution and promotes labor efficiency. It is composed of an inclined trunk (45 degrees) with renewable vertical branches that hold vegetative and reproductive spurs only (lateral shoots are eliminated). It is suitable for semi-vigorous to semi-dwarfing rootstocks (Long et al., 2015; Figure 10).

FIGURE 10

Upright Fruiting Offshoot training system.

- **Tall Spindle Axis (TSA):** It is a narrow and lower central axis, from which 22 to 25 renewable lateral branches of medium vigor form a spiral shape (Ayala & Valenzuela, 2019). Fruit grows on fruiting units composed of 2-year-old wood and at the base of 1-year-old shoots. It allows the establishment of a semi-pedestrian orchard using semi-dwarfing to semi-vigorous rootstocks (Long et al., 2015; Figure 11).

FIGURE 11

Tall Spindle Axis training system.

- **Super Slender Axis (SSA):** It allows the establishment of high–density, semi–pedestrian orchard, requiring dwarfing rootstocks (Musacchi et al., 2015). Trees are formed with a single axis and several renewables on 1-year-old shoots. Requires support and precise pruning (Long et al., 2015; Figure 12).

FIGURE 12

Super Sleder Axis training system.

Table 3 is a summary of most common distances considering different training systems.

Table 3 Most common tree distances used for different training systems

	Distances (m)			
	Tree Spacing		Row Width	
	From	To	From	To
Central Axis	2.0	3.0	3.5	4.0
Spanish Bush	1.8	3.0	4.5	5.5
Steep Leader	3.0	5.0	4.5	5.5
KGB	2,0	2.5	3.5	4.0
Y-trellis	1.2	1.8	3.5	4.0
V-trellis	0.9	1.4	3.5	4.0
UFO	1.2	1.8	3.0	4.0
TSA	1.5	1.8	3.0	4.5
SSA	0.7	1.0	3.0	4.0

From Long et al., 2015; Ayala & Valenzuela, 2019; Stone et al., 2022.

5.4 Crop load regulation

This practice includes pruning and thinning of reproductive structures (buds, flowers, and/or fruits), with the main objective of ensuring an appropriate LA:F ratio to produce high−quality fruit (Ayala & Andrade, 2009).

- **Spring/summer pruning:** It is carried out close or after harvest in more vigorous cultivar/rootstock combinations. Pruning promotes the entry of light to optimize photosynthesis and increase fruit coloration.
- **Winter pruning:** It is common for dwarfing and semi-dwarfing precocious combinations. Its main objective is to regulate the crop load. Additionally, it promotes lateral shoot formation.
- **Bud thinning:** It involves the hand removal of 30% to 50% of reproductive buds to promote an early LA:F ratio adjustment.
- **Flower thinning:** It involves the removal of 30% to 50% of flowers from spurs manually during bloom (Whiting et al., 2006; Rutkowski & Lysiak, 2022).
- **Fruit thinning:** It is the last opportunity to regulate crop load. Hand removal (30% to 60%) of fruits reduces competition for photoassimilates (Figure 13).

FIGURE 13

Fruit thinning at Stage I of fruit development in a self-fertile cultivar.

- **Chemical thinning:** It involves the use of chemical products such as AG_3 and oils (Whiting et al., 2006); however, it has not been widely used since it can lead to over−thinning or have no effect at all.

5.5 Mechanization

In sweet cherry production, mechanization has been successfully implemented in certain tasks to reduce costs associated with manual labor (Seavert & Whiting, 2011). Mechanical pruning has to reduce the height and/or width of the tree in training systems such as UFO, KGB, and Central Axis, which is the most used technology.

5.6 Protected cultivation

One of the most notable technologies adopted in sweet cherry orchards has been plastic covers, which were implemented to reduce the incidence of rain-induced cracking (Ayala et al., 2019). However, they are now also used to modify an orchard's microclimate (Table 4; Blanco et al., 2019a). The use of plastic covers can advance bloom and harvest, requiring a rigorous technical management (Blanke et al., 2017; Janke et al., 2017). Covers are primarily made of high–density polyethylene or low–density polyethylene.

Table 4 Reported responses in physiology, yield, and quality parameters of sweet cherry trees under high–tunnels

Variable	Response	Reference
Bloom and harvest	Earlier	Ayala et al., 2019; Blanco et al., 2019a; Lang, 2014
Yield	Similar	Blanco et al., 2021a; Lang, 2014
	Decreased	Lang, 2014; Palma et al., 2023
Cracking	Decreased	Blanco et al., 2019a; Lang et al., 2016
Fruit Size	Increased	Blanco et al., 2021b; Lang, 2014; Meland et al., 2017b
Firmness	Decreased	Blanco et al., 2021a; Palma et al., 2023
Solid soluble Content	Decreased	Blanco et al., 2019a
	Similar	Blanco et al., 2021b
Vegetative vigor	Increased	Blanco et al., 2021a; Lang, 2014; Palma et al., 2023
	Similar	Blanco et al., 2019a

- **Plastic tents:** These are open, dual–pitch structures with heights between 4 and 6 meters. Their inclined design allows water and hail to slide off easily (Lang et al., 2016). These structures are manually mobile or semi–mobile (Figure 14).

FIGURE 14

Plastic tents (three-wire system).

- **High–tunnels:** These are built with modular galvanized steel frames in an arched shape and are especially useful in areas with extreme weather events (Blanco et al., 2019a; Meland et al., 2017a). Unlike plastic tents, high–tunnels completely cover the top and sides of the orchard (Figure 15).

FIGURE 15

Variants of high-tunnels.

- **Greenhouses:** These structures allow full control of the microclimate, regulating variables such as temperature, humidity, and solar radiation (Ayala et al., 2019; Lang et al., 2016). The cost associated with their construction, maintenance, and operation limits their use.

5.7 Water use and irrigation

Sweet cherry cultivation occurs under both irrigated and rain–fed conditions. However, the species shows low drought tolerance, and it is widely recognized that uncontrolled water stress negatively impacts its growth and yields (Blanco et al., 2022). In warm regions with hot, dry seasons, the lack of irrigation adversely affects vegetative growth, production, and fruit size.

Traditionally, standard sweet cherry irrigation strategies have been based on complementing precipitation to satisfy crop evapotranspiration. However, in highly productive sweet cherry regions where water is limited and atmospheric water demand is high, standard irrigation strategies cannot be followed, and other water management techniques that conserve water need to be adopted (Marsal et al., 2009). In this sense, irrigation strategies like regulated deficit irrigation (RDI), which tailor water supply to the tree's phenology and its sensitivity to water deficit, are a suitable option. During critical phenological stages such as preharvest and floral differentiation, when water deficit can significantly penalize fruit yield or quality, full irrigation is provided to meet the tree's water requirements. In contrast, during drought-tolerant stages such as the midlate postharvest (noncritical periods), water application is reduced but controlling its effects on tree water status (length and intensity of the drought). As a result, the total water used in RDI is less than the optimal amount, increasing water productivity, reducing vegetative growth, and maintaining yield and fruit quality. In sweet cherry cultivation, deficit irrigation is typically implemented during the postharvest phase, after the fruit has been harvested but while vegetative growth is still active (Figure 16). This period coincides with summer, the season of peak crop evapotranspiration. Regarding fruit quality, Marsal et al. (2009; 2010) and Martínez-Hernández et al. (2020) reported an increase in soluble solids concentration in fruit from RDI trees in the combination Summit/SL-64 and Prime Giant/SL-64, in contrast, RDI had no effect in the combination New Star/SL-64.

Several studies have proposed different RDI strategies with deficit during postharvest that can be successfully followed in the long term to achieve significant

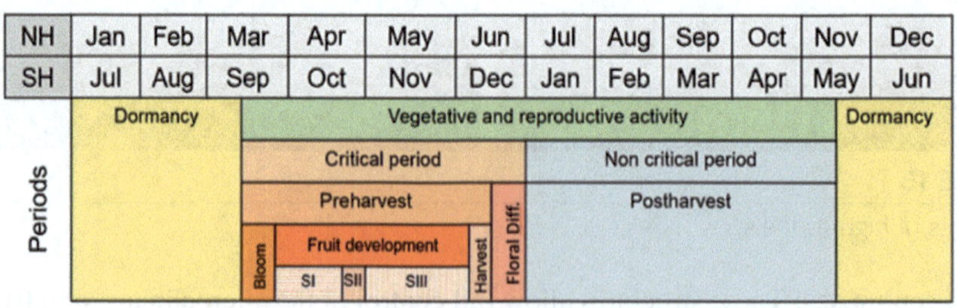

FIGURE 16

Sweet cherry annual calendar in the Northern (NH) and Southern (SH) Hemispheres, and critical period to avoid water deficit.

water savings that range from 760 m^3 ha^{-1} for the combination Sweetheart/Mazzard in Canada (Houghton et al., 2023), to 2700 m^3 ha^{-1} for the combination Prime Giant/SL-64 in Spain (Blanco et al., 2019b). In both examples, the water deficit caused reduced vegetative growth, and as the tree water status was controlled, the intensity of the deficit was not enough to cause negative effects on next year's tree production and fruit quality.

However, when the water deficit was too severe, although greater water savings could be achieved, fruit yield the next year decreased (Hojjati et al., 2024). If water deficit is applied during the whole season including the critical periods (preharvest), as in the sustained deficit irrigation (SDI) strategies, a reduction in the size of the cherry occurs (Blanco et al., 2019b). Thus, for the combination Brooks/Maxma 14 in Chile, a reduction of the 50% of the water applied to the full irrigated treatment saved 2310 m^3 ha^{-1}; however, it significantly decreased fruit yield (Livellara et al., 2011). Consequently, RDI strategies in postharvest are preferred for sweet cherry irrigation management in comparison with SDI strategies. RDI increases water use efficiency, reduces vegetative growth (excessive vigor), and maintains yield without decreasing fruit size (Figure 17).

The application of RDI strategies requires, in addition to an accurate irrigation scheduling, a high knowledge of the water relations in the soil-plant-atmosphere continuum and the use of physiological and physical indicators. Physical indicators such as meteorological variables, soil water potential, and soil water content measure changes in the environment that affect plant water status (Blanco et al., 2022; González-Teruel et al., 2022). Physiological indicators such as leaf, stem, and trunk water potential; stomatal conductance; net photosynthesis; trunk diameter variations; sap flow; leaf temperature; and remote sensing indices assess/quantify plant water status directly or indirectly (Blanco et al., 2019b; Blaya-Ros et al., 2020; Morandi et al., 2018).

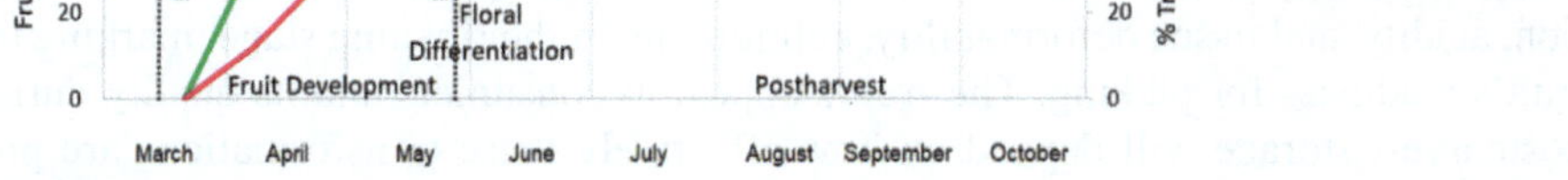

FIGURE 17

Vegetative and fruit growth seasonal patterns for sweet cherry trees and water applied for three irrigation treatments, CTL *(blue)*, RDI (regulated deficit irrigation, *red*), and SDI (sustained deficit irrigation, *yellow*). Based on Blanco et al. (2019b) for the combination Prime Giant/SL-64 in Spain.

5.8 Mineral nutrition

A cherry tree with balanced nutrition will guarantee both good yields and high-quality fruit (Quiroz et al., 2023). The rootstock selection has the greatest influence on tree nutrient absorption, with variations in terms of better absorption of certain mineral nutrients over others (Hrotkó et al., 2014). Another element to consider is the effect of foliar nutrition, which is commonly used in commercial orchards (Nagy et al., 2008). Sweet cherry needs 14 essential elements divided into macronutrients (N, P, K, Ca, and Mg) and micronutrients (B, Zn, Mn, Fe, Cu, S, Mo, Cl, and Ni) (Lemus & Negrón, 2005; Table 5). From bloom to the development of spur leaves and early fruit growth, sweet cherry trees use storage nutrients as a source (Ayala & Lang, 2004). Later, along with vegetative growth and fruit maturation, current root growth and foliage supply nutrients (Artacho & Bonomelli, 2016). The phenology is relevant for a correct fertilization program, giving highlights to the moment of application and the products to be used (Eichert & Fernández, 2012). Micronutrients are applied mainly foliar to complement soil applications (Nagy et al., 2008) throughout the entire growing season, from bloom to leaf fall.

Table 5 Estimated gross annual macronutrient requirements in cherry orchards for adult cherry trees. Ranges depend on different cultivar/rootstock combination and location. Mean values of 12 cultivars.

	Nutrient (kg ha^{-1})				
	N	**P**	**K**	**Ca**	**Mg**
Range	33 - 70	6 - 21	12 - 47	14 - 55	5 - 9
Mean	49	14	32	26	7

From Ughini & Roversi, 2008; Neilsen et al., 2017; Ystaas, 1990.

6. Fruit maturation and harvest criteria

Sweet cherry is a highly perishable fruit whose high value is recognized by its external color and appearance, texture, and flavor. Various physiological and biochemical mechanisms drive the key transformations in the fruit tissue that guide the maturation process. These changes start at the end of SII of fruit development, characterized by alterations in color and a swift evolution in size, sugar concentration, acidity, and tissue deformability, culminating in the ripening stage, marking the fruit's readiness for picking. The fruit's appeal to consumers and its quality during postharvest storage will depend on how effectively these transformations are produced and managed (Correia et al., 2017).

A high-quality cherry should be free of defects, large, firm, and dark in color. Additionally, the presence of a green pedicel and a shiny appearance on the skin are indicators of freshness. Furthermore, factors such as sweetness, the balance between acidity and sugar content, and a firm texture are important sensory attributes that contribute to the overall flavor and acceptability of cherries at the time of consumption.

For optimal quality, sweet cherries must be harvested during a specific time frame ("harvest window"). It corresponds to the stage of development when the fruit achieves its optimal value for the specific trait characteristic of each cherry cultivar. The harvest window considers a physiological stage of the fruit that maximizes storability and postharvest handling. Moreover, factors such as the availability of harvest materials and trained pickers must be aligned with the environmental conditions, fruit physiology and the agronomical management practices to establish the optimal harvest time. Multiple selective harvests (around three), due to variations in bloom times, are typically necessary. These harvests are usually spaced out over 2 weeks, with the exact duration influenced by local climate conditions. During fruit development, changes in skin color are the most closely linked to key indicators of maturity like the build-up of soluble solids, the decrease in fruit firmness (softening), and the emergence of the cherry's unique flavor. Optimal harvest time involves a balance between high soluble solids content and minimal softening. The measure of dry matter, using nondestructive equipment, has been considered as an alternative to quantify sugars and other compounds such as cell wall residues and acids (Toivonen et al., 2017).

Skin color has long been accepted as the best indicator of harvest maturity across several sweet cherry cultivars with dark red/mahogany proposed as the color at which to harvest most cultivars (Crisosto et al., 2003). Harvesting the fruit at an earlier color stage halts the accumulation of soluble solids and fruit mass. Meanwhile, delayed harvests may compromise storage life due to fruit softening and increased sensitivity to mechanical damage (Wang & Einhorn, 2017) (Figure 18).

FIGURE 18

Symptoms of mechanical damage observed after storage, resulting from improper or deficient harvesting practices.

Sweet cherries are harvested at a ripening stage when the respiration is high and fruit are prone to rapid deterioration. Because of this, the cool-down period for fruit must be as short as possible (Toivonen, 2014). Also, exposure of the fruit to the sun should be avoided and physical damage minimized. Ideally, picking must be initiated soon after dawn and concluded around midday, after which fruit temperatures usually peak. The harvest is usually manual, with pickers paid by the weight unit of picked fruit. Hence, the efficiency of the pickers (kg per hour) should be optimized by staff training and selection of the appropriate harvest equipment (basket, bag, or ladder) that align with the orchard training system. Harvest is a critical cost factor, with its efficiency closely tied to the canopy architecture. Pedestrian and semi-pedestrian systems, such as planar architectures (e.g., UFO), and multileader bushes (e.g., KGB), have proven to be almost 1.5 to 2 times more efficient compared to traditional systems like the Central leader or multileader variants (Ampatzidis & Whiting, 2013).

Sweet cherry can be harvested directly into small (3 kg) or large (10 kg) plastic picking baskets and then stacked inside bins for later collection and transport to the packing facility. Alternatively, the fruit can be harvested into a picking bag and then poured gently into a large container (bin). In this case, the incidence of bruising can be higher, but the control of fruit quality is better because the fruit can be seen more clearly by the supervisor.

Sweet cherry fruit is very sensitive to compression and impact damages. The symptoms cannot be easily detected at harvest. Usually, they are evident after 10 days at 0°C, as small indentation (impact) or as deep depression (compression) areas (Zoffoli et al., 2017). By closely monitoring the fruit from harvest through the packing line, it becomes possible to identify the points of injury and implement protective measures, train personnel, or modify specific management practices to reduce these types of damage. The harvesting process also impacts the fruit's appearance in the market; as symptoms of pedicel dehydration. The most effective method to maintain pedicel freshness is using refrigerated transport immediately after hydrocooling, using a mobile hydrocooler, installed directly in the orchard (Figure 19).

7. Postharvest physiology and technology

The sweet cherry fruit is highly perishable due to various forms of physical, pathological, and physiological deterioration during postharvest; however, this deterioration can be mitigated through appropriate postharvest management practices.

7.1 Fruit postharvest deterioration

Sweet cherry belongs to the category of non-climacteric fruits, meaning that the changes during maturation and ripening are not regulated by ethylene. Consequently, the use of 1−MCP, an inhibitor of the ethylene action, is usually not

FIGURE 19

Portable hydrocooler for use as soon as the fruit is harvested.

effective in preventing sweet cherry deterioration during postharvest (Gong et al., 2002). Therefore, the maturation process is driven by its high respiration rate with ABA playing a regulatory role in promoting processes like fruit softening (Ren et al., 2011).

Fruit softening: Fruit firmness is a key trait affecting sweet cherry quality during handling and shipping. Firmness begins to decline at *veraison* and continues through maturation, primarily due to changes in the cell wall and middle lamella. The ripening and softening of fruit involve specific phytohormones, notably ABA and auxins, which have antagonistic roles (Zhai et al., 2022). Postharvest softening is linked to moisture loss during prolonged low-temperature storage. A significant positive correlation exists between moisture content and fruit firmness (skin and flesh) during storage, with moisture decreasing more at 4°C than at 0°C. Temperature fluctuations during refrigerated transport can further soften sweet cherries (Xin et al., 2021).

Dehydration: Postharvest deterioration from dehydration affects both the fruit and pedicel, with the latter being more susceptible due to its less effective cuticle in resisting water loss (Athoo et al., 2015). Browning and shriveling in the pedicel are primarily linked to physical dehydration from transpiration in a nonsaturated atmosphere and associated water pressure deficits (Linke et al., 2010). Another way of water loss is osmotic dehydration, where water moves from the pedicel to the fruit due to differences in osmotic potential in a saturated atmosphere (Knoche et al., 2015). The contribution of the two types of dehydration depends on the water vapor pressure deficit around the fruit during storage. Although MAP combined with low temperatures effectively increase relative humidity inside the bag and reduce dehydration symptoms (Zoffoli & Rodriguez, 2014), they primarily aim to minimize the driving force for transpiration. However, they do not effectively control osmotic dehydration.

Mechanical damage or surface pitting: Mechanical damage occurs when excessive pressure is applied, whether suddenly through impact or gradually through compression. This damage can manifest differently depending on the surface affected; it may appear as pitting, which consists of small depressions measuring 4 to 8 mm, or as bruising, characterized by larger, more pronounced depressions on the fruit (Figure 20). The susceptibility of sweet cherries to mechanical damage varies among cultivars (Param & Zoffoli, 2016). Several factors have been associated with this susceptibility, including firmness, pectin composition and maturity (Wang & Einhorn, 2017). The characterization of the rheological properties of cherries has facilitated the study of how tissue responds to external deformation forces. Strategies that enhance pitting resistance are vital to minimize losses due to this damage. Preharvest treatments of GA_3, methyl jasmonate, salicylic acid, and calcium, as well as fruit thinning, have been linked to decreased mechanical damage (Carrión-Antolí et al., 2024; Zoffoli et al., 2008).

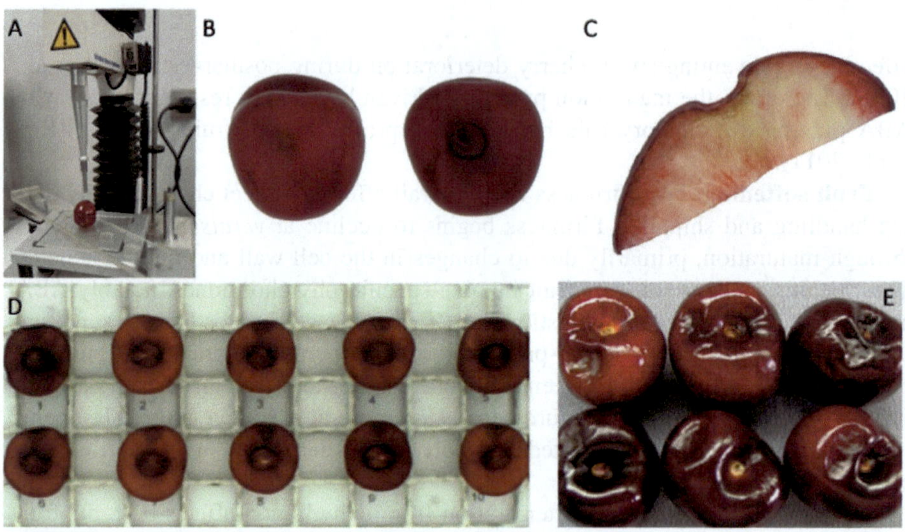

FIGURE 20

Puncture test performed with a texture analyzer (A) Lateral view of fruit before puncture test and after 10 days at 0°C (B), longitudinal (C), and transverse (D) sectioning of the affected mesocarp area and (E) expression of bruising mechanical damage after storage in the shoulder fruit section.

Postharvest cracking: Cracking involves fractures in the cuticle or epidermis of the fruit that do not affect the flesh, resulting from microcracking due to water uptake, which precedes visible macrocracking (Figure 21). Physiological factors contributing to fruit cracking include water uptake, osmotic potential, cuticle properties, and cell anatomy (Correia et al., 2018). While stored, the exposition of cherries to a high-humidity atmosphere and condensed water weakens the cuticle's protective function, allowing water to penetrate hypodermic cells with a more negative osmotic potential than epidermal cells, particularly at the stylar scar end (Winkler et al., 2019). Crack begins with the absorption of water through localized microcracking, affecting only the mesocarp cells in the damaged area without altering cell turgor at this stage (Grimm et al., 2019). As water absorption continues, damage progresses from the initial microfracture to underlying healthy cells, leading to membrane leakage (notably malic acid) and the extraction of Ca^{2+} from the cell wall. This results in increased cell wall thickness and swelling. Ultimately, these processes culminate in macrocracking. This is the so-called Zipper model of cracking (Winkler et al., 2016). The application of calcium chloride ($CaCl_2$) minimizes the thickness and expansion of cell walls, even in areas that are already affected, while also enhancing cell adhesion compromised by crack propagation (Winkler et al., 2024). Preharvest $CaCl_2$ applications have shown potential in reducing postharvest splitting (Matteo et al., 2022). Postharvest $CaCl_2$ applications at concentrations from 0% to 2% during hydrocooling reduced postharvest splitting by increasing Ca uptake, decreasing soluble pectin release, raising the splitting threshold, and reducing the rate of water uptake (Wang & Long, 2015). Fruit sensitivity to postharvest cracking is mainly influenced by cultivar and preharvest conditions, with Brooks and Royal Down being highly susceptible.

FIGURE 21

Types of cracking develop during storage in MAP. Stylar end (A) and lateral-stylar cracking (B).

7.2 Physiological disorders

- **Orange peel, pebbling, or alligator skin** are used to describe a physiological disorder that occurs both on the tree and during postharvest (Coye et al., 2024). This condition is characterized by increased surface roughness. Figure 22 illustrates the typical appearance of these changes. The development of a rough surface during storage is linked to skin dehydration, with water loss resulting from transpiration to the atmosphere and osmotic gradients to the flesh, particularly in water vapor—saturated environments (Schlegel et al., 2018). Certain sweet cherry cultivars, such as Santina, Lapins, Sweetheart, Adriana, and Regina have been identified as more susceptible to this disorder.
- **Internal browning** is a physiological disorder that occurs during prolonged storage at 0°C (over 35 days) and atmosphere of 18 kPa CO_2 and 2 kPa O_2 in certain sensitive sweet cherry cultivars, particularly Regina (Harb et al., 2003) and Skeena. It is characterized by brown pigmentation starting in the mesocarp area near the stone and spreading throughout the flesh. Subepidermal browning may also occur, though less frequently than internal browning (Figure 23). The browning symptoms increase upon ripening after 0°C storage. Alongside these changes, the development of an unpleasant taste diminishes the organoleptic quality, ultimately affecting product acceptance (Zoffoli et al., 2020).

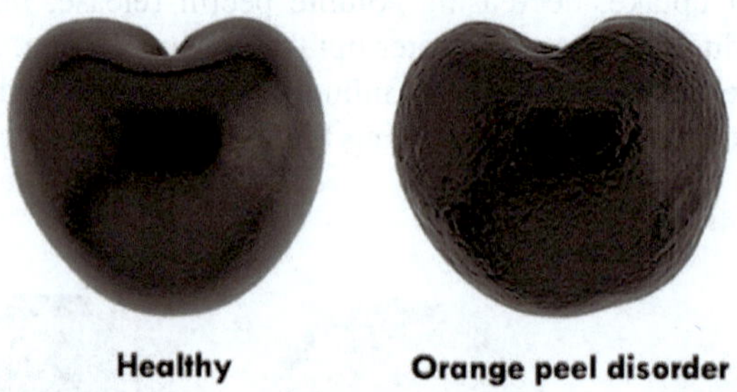

FIGURE 22

Orange peel disorder in sweet cherry cv. Regina after 45 days at 0°C MAP.

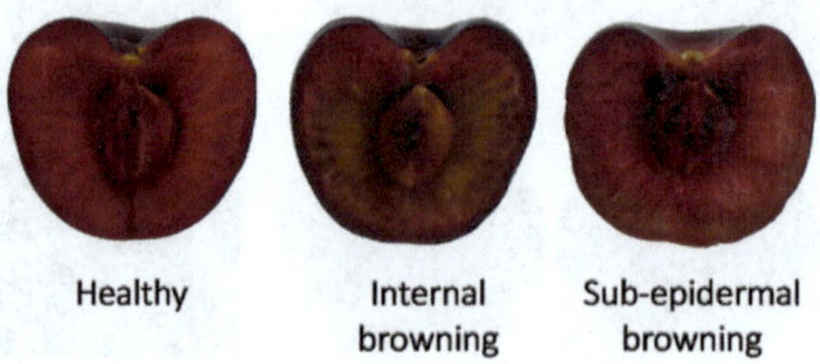

FIGURE 23

Internal browning disorder symptoms in sweet cherry cv. Regina after 45 days at 0°C under modified atmosphere packaging.

7.3 Fruit sorting and packaging process

The process of sorting and packaging cherries begins immediately after harvest and prior to the removal of heat by hydrocooler (Zoffoli, 2022). Cherries reach a pulp temperature of 8°C to 10°C or 0°C to 2°C, depending on whether they will be packaged after 24 to 48 h, respectively. In the packing line, cherries are dumped into sanitized water. Once cleaned, they are lifted out by a belt and continue their movement along the processing line, guided by flowing water. Within the process, the fruit is fed onto a cluster cutter belt where the cherries are separated using a series of precision metal blade saws (Figure 24).

Commercial grading systems for sweet cherries offer advanced capabilities to assess both internal and external fruit quality at high speeds, utilizing a sophisticated automated electronic system (Figure 25). These systems utilize spectral reflectance and imaging processing techniques to provide a 360-degree rotational view, enabling the identification of defects like bruising, cracking, russets, deformations, soft patches, and decayed tissue. Size grading is efficiently conducted using high–definition cameras on high-speed conveyors, ensuring accurate sorting by color and size (Lu et al., 2020). Cutting-edge technology is enhancing the sorting and assessment of internal fruit quality traits. NIR can successfully measure soluble solids in cherries (Escribano et al., 2017).

Selected cherries continue moving along a water flow line where they are treated with fungicide. To ensure a fruit temperature below 5°C during the sealing process

A

B

FIGURE 24

Full view of the passage through a cherry cluster cutting belt (A) and close-up view of the cutter belt (B).

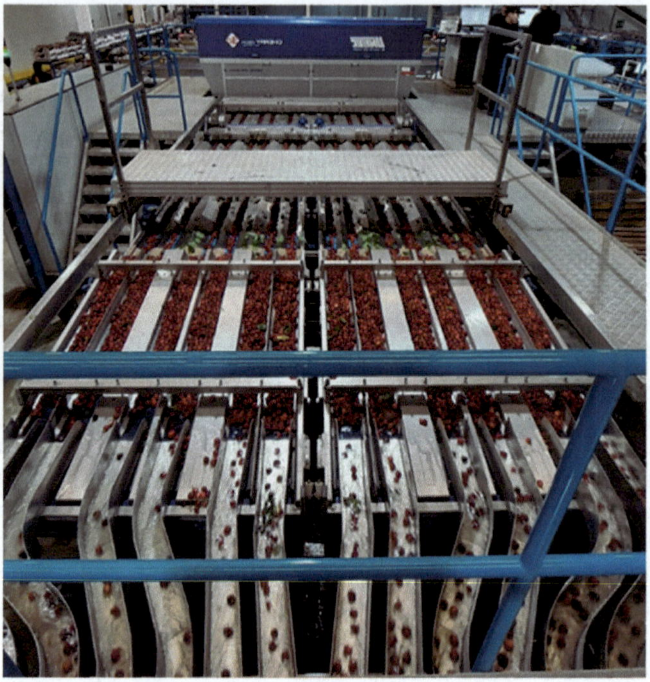

FIGURE 25

Fruit singulation using cascading belts of increasing speed (Unitec machine).

for MAP, the packing line is connected to a chiller that cools the final water flow between 0°C and 1°C. Once sorted, cherries are directed to the appropriate line for automatic packaging into individual bags, clamshells, or tightly filled volume boxes. The bags and clamshells are then placed in either perforated, nonperforated, or modified-atmosphere bags and packed into master cardboard boxes of various sizes (2.5, 5, or 10 kg) and pallets, depending on storage and marketing needs. Perforated and nonperforated bags are used for air transport, while MAP is employed for prolonged sea freight storage (Kargwal et al., 2020). Pallets are then refrigerated using a forced air-cooling method that equalizes the pulp temperature (0°C to 1°C) throughout the fruit, and they are held in a cold storage facility until loading and shipping. Container trucks and vessel chambers are also maintained between 0°C and -1°C to preserve the cold chain until the fruit reaches the market under MAP.

8. Main diseases and pests and their management

8.1 Main diseases and management

There are many pathogens, including fungi, bacteria, viruses, viroids, and phytoplasmas, that infect sweet cherries and lead to damage. They can be grouped by those infecting fruits, foliage, trunk, and woody axes and roots. Among the main preharvest diseases are brown rot *(Monilinia spp.)*, grey mold (*Botrytis cinerea* Pers.) and anthracnose *(Colletotrichum spp., Gloeosporium spp.)*. Foliar diseases include cherry leaf spot (*Blumeriella jaapii* Rehm.), cherry leaf curl (*Taphrina wiesneri* Ráthay) and powdery mildew *(Podosphaera clandestine (Wall.:Fr) Lév.)*. Trunk and root diseases include the Armillaria root rot *(Armilaria mellea Vahl.)*, Verticillium wilt (*Verticilium dahliae* Kleb.), and Leucostoma canker *(Leucostoma spp., Cytospora spp.)*. Bacterial canker (*Pseudomona syringae* pv. Syringae) is a major disease.

8.2 Flower and fruit diseases

Brown Rot *(Monilinia spp.)*: The causal agents are *Monilinia fructicola* and *Monilinia laxa*. High environmental humidity and temperatures above 15°C promote the development of the disease. The presence or damage in inflorescences is visually identified by a blight (necrosis) of petals. Fruits are initially affected by watery rot, which dries out over time, leading to the mummification of infected fruits (Figure 26). Mummified fruits become the most significant source of inoculum. Less frequently, floral whorls or even small twigs, contaminated from previous seasons, can act as sources. The removal of infected structures is one of the key control strategies (Tarbath, et al. 2014).

Gray Mold *(Botrytis cinerea)* is a fungal disease that affects cherries on the tree as well as during storage. Like *Monilinia spp.*, the entry of this pathogen often occurs through microfractures in the fruit's cuticle. A key indicator of this disease is the so-called "watery spot," which evolves into the appearance of grey-colored mycelium (mold) in the same area (Figure 27). Its expression occurs under conditions of high humidity and temperatures between 15°C and 25°C (Serradilla

FIGURE 26

Mummified sweet cherry by *Monilinia spp.*

FIGURE 27

B. cinerea in sweet cherry fruit.

et al., 2021). Control is commonly carried out preventively when favorable conditions arise, even up to 7 days before harvest. The use of various broad-spectrum fungicides can effectively control its incidence. It has been demonstrated that the use of physical barriers, such as plastic covers, reduces the proportion of fruit affected by this disease (Børve et al., 2017).

Anthracnose is a disease caused by fungi of the genera *Colletotrichum spp.*, *Gloeosporium spp.*, and *Sphaceloma spp.* As a fungal disease, anthracnose thrives under high humidity and temperature. It affects buds, leaves, twigs, inflorescences, and fruits (particularly in early stages). When infecting buds, it also impacts production, particularly spurs (Børve et al., 2017). The characteristic symptoms are circular or irregular necrotic lesions on leaves and fruits and aborted fruits, which remain on the tree. These pathogens can also manifest during fruit storage, causing significant losses (Yang et al., 2024). Control measures involve fungicide applications during stage I of fruit growth.

8.3 Foliar diseases

Cherry Leaf Spot *(Blumeriella jaapii)*: The pathogen produces necrotic spots on the leaf blades, which gradually expand, accompanied by chlorosis, ultimately leading to leaf drop. The conditions that favor this disease include the presence of free water, particularly spring rains, which spread the spores to the host leaves. The most used and effective control is the application of fungicides after petal—fall stage. A more modern strategy today involves evaluating cultivars that exhibit resistance to it (Andersen et al., 2018).

Cherry Leaf Curl *(Taphrina wiesneri)*: The fungus primarily resides in buds. The main symptoms include deformities in leaves and shoots, characterized by overgrowth in the affected areas due to the high production of cytokinins by the fungus

(Tsai et al., 2014). This reduces the presence of floral buds and can lead to early defoliation. Its control is through fungicide applications during the growing season. However, preventive control is also possible during the latent stages of the fungus in branches and trunks by removing the infected lignified organs.

Powdery Mildew *(Podosphaera clandestina)* can be easily identified due to a whitish powder that settles on developing leaves and fruit. It causes spots on the fruit's surface, reducing the marketable yield. It is much more frequent in mid-and-late-season cultivars, with infection often coinciding with harvest maturity. The fungus thrives in temperatures between 20° and 27°C and under humid conditions such as that caused by rain, fog, or irrigation (Swamy et al., 2019). The most common control strategy involves the preventive use of fungicides from petal fall to preharvest. However, this approach is not limited to field applications; it also includes postharvest control measures in packing facilities.

8.4 Trunk and root diseases

Armillaria Root Rot *(Armillaria mellea)*: The damage caused by this disease primarily affects the cambium tissue, deteriorating the root and trunk collar area, eventually leading to tree decline and death (Devkota et al., 2020). Conditions that favor pathogen development, aside from its presence, include temperatures ~20°C, with disease development observed even at 28°C. Once the pathogen's presence is identified in the tree, it is immediately considered a loss due to its aggressiveness. Chemical control is not so effective but the removal of all parts of the infected tree from the orchard, and the commercial use of fungi from the genus *Trichoderma spp.* has shown satisfactory results (Devkota et al., 2020).

Verticillium Wilt *(Verticillium dahliae)* is a fungal disease involving multiple species of the genus *Verticillium spp.* This disease causes wilting in tree structures due to vascular damage, as it invades the xylem cavities. It results in stunted growth, leaf chlorosis, wilting of tender structures, and, ultimately, premature senescence. The conditions that promote the development of this disease include soil moisture and temperatures ranging between 21°C and 27°C. Once identified, controlling this disease is ineffective because its presence indicates that the vascular structures have already been colonized. Chemical control is inefficient but management practices (e.g., solarization and crop rotation) before the orchard's establishment have proven effective in preventing infection. The development of resistant rootstocks has also been a strategy employed in woody species (Deketelaere et al., 2017).

Wood cankers: The diseases have various names and causal agents; however, their symptoms are often very similar. *Leucostoma* canker is a fungal disease caused by different species of fungi from the genera *Leucostoma spp. Cytospora spp.*, also known as Cytospora canker, affects the wood of sweet cherry trees with symptoms very similar to those of bacterial canker (caused by the bacteria *P. syringae* pv. syringae), including gummosis (resin-like substance) and branch dieback. For *Leucostoma*, chemical control focuses on the application of fungicides. However, these applications have not shown substantially effective results (Sallato et al., 2021).

FIGURE 28

Bacterial canker symptoms.

In the case of *Pseudomonas spp.*, the use of copper-based compounds has been a traditional strategy during winter with a wide variety of products containing different copper-based active ingredients. Winter pruning during dry weather and applying fungicidal paint to cover pruning cuts reduce the likelihood of *Pseudomonas spp.* entry (Figure 28).

8.5 Main pests and management

Insects and Mites: Three categories of pests stand out: fruit feeders, foliage feeders, and wood feeders. Among the fruit feeders, the most important insects are the European cherry fruit fly, *Rhagoletis cerasi* (L.), the Western cherry fruit fly *Rhagoletis indifferens* Curran, and the two spotted wing drosophila, *Drosophila suzukii* (Matsumura) (Papadopoulos et al. 2017).

Major foliar feeders include the Tortricidae leaf rollers (e.g., *Adoxophyes orana* [Fischer von Rösslerstamm]), the Tenthredinidae *Caliroa cerasi* (L.) and the Aphidide *Myzus cerasi* (Fabricius). There are a few twig and trunk sap-feeders that infest sweet cherries, such as the Coccidae *Sphaerolecanium prunastri* (Boyer de Fonscolombe) and the Diaspididae *Pseudaulacaspis pentagona* (Targioni-Tozzetti). The wood boring insects that infest the sweet cherries include mainly the Buprestidae *Capnodis tenebrionis* L., which is important mainly in young orchards. In addition, the European grape moth *Lobesia botrana* (Denis & Schiffermüller) emerges as an important pest in South America in recent years.

The European red mite (Tetranychidae) may also infest sweet cherries causing significant damage.

In this section, we focus on two major fruit feeders (*R. cerasi*, and *D. suzukii*), one foliage feeder *(C. cerasi)*, one sap-feeder *(S. prunastri)*, the major wood boring pest *(C. tenebrionis)* of sweet cherries, and the European red spider mite *(P. ulmi)*.

European cherry fruit fly *(R. cerasi)*: Adults are rather small (< 4.3 and 3.5 mm long for males and females, respectively), shiny black in color with bright orange–yellow scutellum, well-defined dark patterns on the wings, and green-brown metal eye color (Figure 29).

The pest is rather stenophagous with sweet cherries as the main host (EPPO, 2024b). *Rhagoletis cerasi* completes one generation per year; following a long summer-autumn-winter pupae diapause, adults emerge at approximately the end of blooming (at fruit developmental stage BBCH 71–75). Females lay a single egg in the flesh of ripening cherry fruit and mark it with a host marking pheromone. The hatched maggot feeds in fruit mesocarp and moves close to the pit (Figure 30). Concluding development larvae leave the fruit and drop to the ground, where they pupate, enter a long diapause period and remain there until next spring. *Rhagoletis cerasi* is a key pest for sweet cherries in Europe and West Asia with infestation levels in unprotected fruit exceeding 50% and often reaching 100% (Papadopoulos et al., 2017). In most areas, the infestation levels of very early cultivars are usually low and substantially increase and peak at middle and late ripening cultivars.

Population monitoring is based mainly on the yellow sticky Rebell trap (Figure 31). The addition of ammonia releasing compound can slightly increase

FIGURE 29

Rhagoletis cerasi adult on a ripening fruit (Figure from Paris Prekas).

FIGURE 30

Infested sweet cherry *(right)* and a third instar larva within an infested fruit (Figure from N. Papadopoulos).

FIGURE 31

The yellow sticky Rebell trap deployed in a cherry orchard.

the performance of the yellow sticky traps (Katsoyannos et al., 2000). Current management approaches rely on the application of synthetic pesticides targeting mainly adults or young immatures. Bait spray applications against adults have been tested with good results. The ban of several, mainly systemic pesticides in Europe is associated with a resurgence of *R. cerasi* infestation in many countries. To timely apply pesticides against adults, degree day models have been used together with adult trapping (Kovanci and Kovanci, 2006). Spatial decision-making algorithms have been developed and tested to control the pest in commercial orchards (Ioannou et al.,

2019). Elaborated population growth and dispersion modelling at the orchard level, that has relatively recently developed, may provide a solid foundation for developing effective pest management approaches (Lux et al., 2016). Mass trapping, and the use of host marking pheromone have been tested as well with some promising results. Although there are several organisms feed on the cherry fruit fly and biological control approaches have been tested, their efficacy is considered low.

Spotted Wing Drosophila *(D. suzukii)*: Adults are 2 to 3 mm long with red eyes, a pale brown or yellowish-brown thorax, and black transverse stripes on the abdomen (Figure 32).

Unlike other drosophilids that develop in decaying fruits, *D. suzukii* females bear a serrated sclerotized ovipositor that enables them to lay eggs in intact unripe or ripening fruit causing great damage (EPPO, 2024a). It completes 3 to 13 generations per year. Adults feed on carbohydrate and proteinaceous food to reach reproductive maturing within few days depending on temperature and initiate egg laying on available host fruits (Tochen et al., 2016). In each oviposition event, 1 to 3 eggs are usually laid (approximately 25 eggs/female/day) in the flesh of the fruit, while lifetime female fecundity may reach 400 eggs per female. Larvae feed in the mesocarp, destroying the fruit (Figure 33), while secondary infestations by other drosophilids, fungi, or bacteria intensify damage and lead to fruit collapse (EPPO, 2024a). Pupation takes place either in the soil or within the fruit. Population monitoring with adult trapping and fruit samplings is highly recommended in areas where the pest is present to design control interventions. Pesticide applications are applied mainly against adults and are more efficient when targeting the first generation in spring. The high reproductive potential and short generation time lead often to unpredictable outbreaks, intensified pesticide use, and subsequent, rapid development of resistance to pesticides (Gress and Zalom, 2018). Alternative approaches to control *D. suzukii* include the Sterile Insect Technique (SIT), parasitoids releases, mass trapping, and netting (cover of the crop with net during fruiting and up to harvest) (Rota-Stabelli et al., 2013). Combination of control methods in the frameworks of

FIGURE 32

Drosophila suzukii adults, males *(left)* and females *(right)* (Figure from Vasilis G. Rodovitis and Kostas Zarpas).

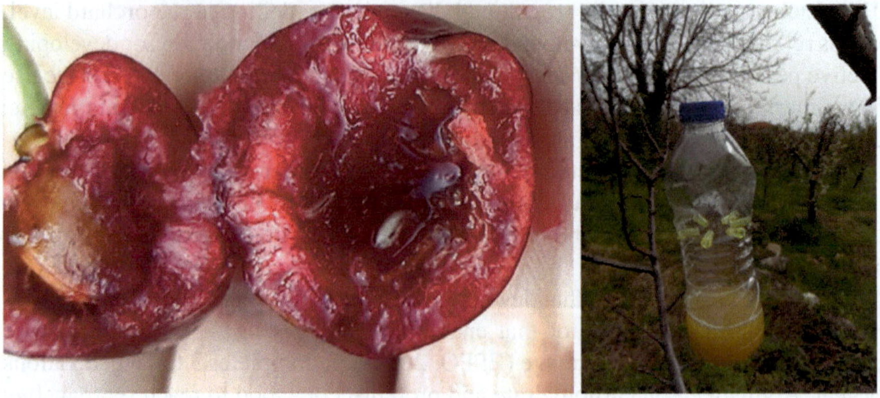

FIGURE 33

Drosophila suzukii larvae within infested sweet cherry (left) and custom-made adult trap *(right)* (Figure Stella A. Papanastasiou and Kostas Zarpas).

a meticulously designed IPM project is highly recommended for the efficient management of *D. suzukii.*

Globose (Plum lecanium) scale *(Sphaerolecanium prunastri)* concludes a single generation per year and overwinters as a second instar nymph. Females (3.5 mm dome-like, dark brown; Figure 34) produce a large number of eggs (>1000) and oviposition may last for weeks. All developmental stages, except adult males, feed on plant sap leading to twig death and a generic weakness of the tree. *Sphaerolecanium prunastri* excretes large quantities of honeydews where sooty mold develops and causes the reduction of the photosynthetic capacity of the tree and impairs the quality of the produced sweet cherries. A list of efficient natural enemies (e.g., Coccinellidae predators and Aphelinidae, Braconidae, and Encyrtidae parasitoid) can regulate *S. prunastri* populations below an economic injury level. Besides biological control by promoting the beneficial native fauna, management of *S. prunastri* populations can be accomplished with dormant oils sprays late in winter, spring applications with insect growth regulators before reaching the adult stage and summer applications against crawlers (Papadopoulos et al., 2017). Special attention should be placed on the long period of crawlers' hatch that is associated with the long oviposition period of females. Overall, to control the plum scale biological control should be considered first, and only if this fails careful and targeted application of insecticides should be implemented.

Flatheaded Wood (peach capnodis) borer *(C. tenebrionis)*: Adults have compact bodies with shiny black coloration, ranging in size from 14 to 28 mm (Figure 35). The life cycle of *C. tenebrionis* lasts up to 3 years, including the adult stage. It overwinters either as an adult in protected locations or as larvae of different instars within tree trunks. In spring, with warmer weather, the adults become active and begin feeding on the upper part of the host plants (Mendel et al., 2003). Females lay eggs near the base of the tree stem preferring weakened trees, when the soil is

FIGURE 34

Ovipositing females of *Sphaerolecanium prunastri* on sweet cherry twig, crawlers and ants feeding on produced honeydews *(top)*, and males trying to mate with young females *(bottom)*.

FIGURE 35

Adult of the Flatheaded Wood borer, *Capnodis tenebrionis*.

dry, and the temperature exceeds 25°C. Newly emerged larvae locate the roots up to 60 cm from the hatching site and penetrate them to feed in the cortex, creating long and large galleries. Larval developmental time ranges from 6 to 18 months and depends on the prevailing temperatures and the rootstock. Pupation typically occurs at the base of the stem, with pupal development taking 3 to 4 weeks. As the infestation progresses, symptoms appear rapidly, leading to tree decline and often death. Key symptoms include yellowing leaves, dry branches, bark injuries, sap ooze, larval frass from cracks, and adult emergence holes on the trunk and branches during spring. Management of *C. tenebrionis* is challenging and based on population monitoring with visual inspections and the collection of adults, since no effective lures are available to be used for trapping. Insecticides (organophosphate and carbamate), although were once the primary control method, are now banned in EU and stem injection with systemic insecticides and the introduction of entomopathogenic nematodes into the soil with irrigation have emerged as potential alternatives (Martínez de Altube et al., 2008).

The European grape fruit moth, *Lobesia botrana* (Denis & Schiffermüller) (synonym: European grapevine moth): *Lobesia botrana* is a major pest for grapevines, and there are no reports of substantial infestation in sweet cherries in the Northern Hemisphere. However, its recent invasion in South America and especially in Chile in 2008 has been associated with severe infestation in sweet cherries (Ayala Zapata, personal observations).

European red mite, *Panonychus ulmi* (Koch) (synonym: European red mite): Females are tiny (<0.4 mm long), dark red with distinct arrays of white spots from where long white hairs arise. Males are smaller in size, and bright red to yellow green. Adults and larvae feed on leaves of cherries, preferably on the underside of leaves. The host range of the pest is quite wide in both families and plant species and may cause considerable damage to sweet cherries. *Panonychus ulmi* concludes 4 to 8 generations per year depending on the prevailing climatic conditions. By sucking the content of leaf cells, it causes mottling of leaves that turn bronze-brown affecting vigor and yield. There are many efficient biological control agents that can keep the population of *P. ulmi* below the economic injury level. However, often in commercial orchards there is a need for systematic application of acaricides. Surveillance is based on sampling dormant spurs during winter and counting overwintering eggs and leaf samples in spring and summer. Dormant spray with winter or summer oils are applied at high density of winter eggs. In case of large summer populations miticides can be used.

9. Current and future challenges in main production regions

Global cherry production faces growing challenges from climate change, pests and diseases, technological limitations, labor shortages, and increasing sustainability demands.

Climate change disrupts the cherry growth cycle through warmer winters, extreme heat, frosts, and droughts, all of which reduce yield and quality. Growers are turning to protective technologies (hail nets, plastic covers, greenhouses) and precision irrigation systems to adapt. Pests like *D. suzukii* and diseases such as *Pseudomonas* spp. are worsening due to climate change, demanding more resilient cultivars and sustainable prevention and control methods. Sustainable practices like integrated pest management (IPM), drip irrigation, and biostimulants are helping reduce chemical use and environmental impact, though they require investment and expertise. Developing new cherry cultivars and rootstocks is essential but slow, taking up to 20 years. These new cultivars/rootstocks must adapt to warmer climates, resist pests, and meet consumer expectations for flavor, size, and shelf-life. Labor shortages, especially for hand-harvesting, are driving up costs. Over 60% of production costs come from harvest in some countries. Solutions include mechanization, improved orchard designs, and better labor conditions. Lastly, environmental regulations are tightening, pushing growers to reduce chemicals and manage waste more sustainably. While this presents challenges, it also creates opportunities to meet growing consumer demand for eco-friendly cherries.

References

Alburquerque, N., García-Montiel, F., Carrillo, A., & Burgos, L. (2008). Chilling and heat requirements of sweet cherry cultivars and the relationship between altitude and the probability of satisfying the chill requirements. *Environmental and Experimental Botany, 64*(2), 162–170.

Ampatzidis, Y., & Whiting, M. (2013). Training systems affects sweet cherry harvest efficiency. *HortScience, 88*(2), 85–94.

Andersen, K. L., Sebolt, A. M., Sundin, G. W., & Iezzoni, A. F. (2018). Assessment of the inheritance of resistance and tolerance in cherry (Prunus sp.) to Blumeriella jaapii, the causal agent of cherry leaf spot. *Plant Pathology, 67*(3), 682–691.

Artacho, P., & Bonomelli, C. (2016). Changes in fine-root production, phenology and spatial distribution in response to N application in irrigated sweet cherry trees. *Tree Physiology, 36*(5), 601–617.

Athoo, T. O., Winkler, A., & Knoche, M. (2015). Pedicel transpiration in sweet cherry fruit: Mechanisms, pathways, and factors. *Journal of the American Society for Horticultural Science, 140*(2), 136–143.

Ayala, M., & Andrade, M. P. (2009). Effects of fruiting spur thinning on fruit quality and vegetative growth of sweet cherry (Prunus avium). *Ciencia e Investigación Agraria, 36*(3), 443–450.

Ayala, M., Blanco, V., & Zoffoli, J. P. (2019). Experiencia chilena en el uso de cobertores plásticos para cerezos. *Revista Frutícola, 41*(3), 18–24.

Ayala, M., & Lang, G. A. (2015). 13C-Photoassimilate partitioning in sweet cherry (Prunus avium) during early spring. *Ciencia e Investigación Agraria, 42*(2), 191–203.

Ayala, M., & Lang, G. A. (2004). Examining the influence of different leaf populations on sweet cherry fruit quality. *Acta Horticulturae, 636*, 85–94.

Ayala, M., & Lang, G. A. (2017). Morphology, cropping physiology and canopy training. In J. Quero-García, A. Lezzoni, J. Puławska, & G. Lang (Eds.), *Cherries: Botany, production and uses* (pp. 269–304).

Børve, J., Ippolito, A., Tanovic, B., Michalecka, M., Sanzani, S. M., Poniatowska, A., et al. (2017). Fungal diseases. In J. Quero-García, A. Lezzoni, J. Puławska, & G. Lang (Eds.), *Cherries: Botany, production and uses* (pp. 338–364). Wallingford, UK: CABI.

Blanco, V., Blaya-Ros, P. J., Torres-Sánchez, R., & Domingo, R. (2022). Irrigation and crop load management lessen rain-induced cherry cracking. *Plants, 11*(23), 3249.

Blanco, V., Torres-Sánchez, R., Blaya-Ros, P. J., Torres-Sánchez, R., Pérez-Pastor, A., & Domingo, R. (2019b). Vegetative and reproductive response of 'Prime Giant' sweet cherry trees to regulated deficit irrigation. *Scientia Horticulturae, 249*, 478–489.

Blanco, V., Zoffoli, J. P., & Ayala, M. (2021a). Eco-physiological response, water productivity and fruit quality of sweet cherry trees under high tunnels. *Scientia Horticulturae, 286*, 110180.

Blanco, V., Zoffoli, J. P., & Ayala, M. (2019a). High tunnel cultivation of sweet cherry (*Prunus avium* L.): Physiological and production variables. *Scientia Horticulturae, 251*, 108–117.

Blanco, V., Zoffoli, J. P., & Ayala, M. (2021b). Influence of high tunnel microclimate on fruit quality and calcium concentration in 'Santina' sweet cherries in a Mediterranean climate. *Agronomy, 11*(6), 1186.

Blanke, M., Lang, G., & Meland, M. (2017). Orchard microclimate modification. In J. Quero-García, A. Iezzoni, J. Pulawska, & G. Lang (Eds.), *Cherries botany, production and uses* (pp. 244–268). Wallingford, UK: CABI.

Blaya-Ros, P. J., Blanco, V., Domingo, R., Soto-Valles, F., & Torres-Sánchez, R. (2020). Feasibility of low-cost thermal imaging for monitoring water stress in young and mature sweet cherry trees. *Applied Sciences, 10*, 5461.

Bonomelli, C., Bonilla, C., Acuña, E., & Artacho, P. (2012). Seasonal pattern of root growth in relation to shoot phenology and soil temperature in sweet cherry trees (*Prunus avium*): A preliminary study in central Chile. *Ciencia e Investigación Agraria, 39*(1), 127–136.

Branchereau, C., Quero-García, J., Echagüe, N. H. Z., Lambelin, L., Fouché, M., Wenden, B., et al. (2022). New insights into flowering date in *Prunus*: Fine mapping of a major QTL in sweet cherry. *Horticulture Research, 9*.

Bujdosó, G., & Hrotkó, K. (2005). Rootstock-scion interactions on dwarfing cherry rootstocks in Hungary. *Horticultural Science, 32*(4), 129–137.

Bujdosó, G., Hrotkó, K., & Stehr, R. (2004). Evaluation of sweet and sour cherry cultivars on German dwarfing rootstocks in Hungary. *Journal of Fruit and Ornamental Plant Research, 12*, 233–244.

Cai, L., Quero-García, J., Barreneche, T., Dirlewanger, E., Saski, C., & Iezzoni, A. (2019). A fruit firmness QTL identified on linkage group 4 in sweet cherry (*Prunus avium* L.) is associated with domesticated and bred germplasm. *Scientific Reports, 9*, 5008.

Calle, A., Grimplet, J., Le Dantec, L., & Wünsch, A. (2021). Identification and characterization of DAMs mutations associated with early blooming in sweet cherry, and validation of DNA-based markers for selection. *Frontiers in Plant Science, 12*, 621491.

Carrión-Antolí, A., Zoffoli, J. P., Serrano, M., Valero, D., & Naranjo, P. (2024). Preharvest gibberellic acid treatment increases both modulus of elasticity and resistance in sweet cherry fruit (cv. 'Bing' and 'Lapins') at harvest and postharvest during storage at 0 °C. *Agronomy, 14*(11), 2738.

Chen, D., Du, X., Zhang, Q., Whiting, M., Scharf, P., & Wang, S. (2012). Performance evaluation of mechanical cherry harvesters for fresh market grade fruits. *Applied Engineering in Agriculture, 28*(4), 483–489.

Correia, S., Schouten, R., Silva, A. P., & Gonçalves, B. (2017). Factors affecting quality and health promoting compounds during growth and postharvest life of sweet cherry (*Prunus avium* L.). *Frontiers in Plant Science, 8*, 2166.

Correia, S., Schouten, R., Silva, A. P., & Gonçalves, B. (2018). Sweet cherry fruit cracking mechanisms and prevention strategies: A review. *Scientia Horticulturae, 240*, 369–377.

Coye, F., Calderón-Orellana, A., Zoffoli, J. P., & Contreras, C. (2024). Influence of preharvest environmental conditions and postharvest relative humidity on the appearance of orange peel disorder in sweet cherry during fruit development and storage. *Chilean Journal of Agricultural Research, 84*, 803–816.

Crisosto, C., Crisosto, G., & Metheney, P. (2003). Consumer acceptance of 'Brooks' and 'Bing' cherries is mainly dependent on fruit SSC and visual skin color. *Postharvest Biology and Technology, 28*(1), 159–167.

Crump, W. W., Peace, C., Zhang, Z., & McCord, P. (2023). Detection of breeding-relevant fruit cracking and fruit firmness quantitative trait loci in sweet cherry via pedigree-based and genome-wide association approaches. *Frontiers in Plant Science, 13*, 823250.

Csihon, Á., Bicskei, D. K., Dremák, P., & Gonda, I. (2018). Performance of sweet cherry cultivars grafted on Colt rootstock. *International Journal of Horticultural Science, 24*(1–2), 7–10.

Deketelaere, S., Tyvaert, L., França, S. C., & Hofte, M. (2017). Desirable traits of a good biocontrol agent against Verticillium wilt. *Frontiers in Microbiology, 8*, 1186.

Devkota, P., Iezzoni, A., Gasic, K., Reighard, G., & Hammerschmidt, R. (2020). Evaluation of the susceptibility of *Prunus* rootstock genotypes to Armillaria and Desarmillaria species. *European Journal of Plant Pathology, 158*(1), 177–193.

Eichert, T., & Fernández, V. (2012). Uptake and release of elements by leaves and other aerial plants parts. In P. Marschner (Ed.), *Marschner's mineral nutrition of higher plants* (3rd ed., pp. 71–84). Academic Press.

EPPO. (2024a). *Drosophila suzukii EPPO datasheets on pests recommended for regulation*. https://gd.eppo.int.

EPPO. (2024b). *Rhagoletis cerasi*. https://gd.eppo.int/taxon/RHAGCE.

Escribano, S., Biasi, W.v., Lerud, R., Slaughter, D. C., & Mitcham, E. J. (2017). Non-destructive prediction of soluble solids and dry matter content using NIR spectroscopy and its relationship with sensory quality in sweet cherries. *Postharvest Biology and Technology, 128*, 112–120.

Fadón, E., Rodrigo, J., & Luedeling, E. (2021). Cultivar-specific responses of sweet cherry flowering to rising temperatures during dormancy. *Agricultural and Forest Meteorology, 307*, 108486.

FAOSTAT - Food and Agriculture Organization of the United Nations Statistics Division. (2022). Available at http://www.fao.org/faostat/en/#data/QC/visualize (Accessed: October 2024).

Franken-Bembenek, S. (2005). Literature review: Reactions of Gisela® rootstocks to pathogens. *Acta Horticulturae, 795*, 303–310.

García, F., Frutos, D., Lopez, G., Carrillo, A., & Cos, J. (2009). Flowering of sweet cherry (Prunus avium L.) cultivars in Cieza, Murcia, Spain. *Acta Horticulturae, 1020*, 191–196.

Gonçalves, B., Correia, C. M., Silva, A. P., Bacelar, E. A., Santos, A., & Moutinho-Pereira, J. M. (2008). Leaf structure and function of sweet cherry tree (*Prunus avium* L.) cultivars with open and dense canopies. *Scientia Horticulturae, 116*(4), 381–387.

Gong, Y., Fan, X., & Mattheis, J. P. (2002). Responses of 'Bing' and 'Rainier' sweet cherries to ethylene and 1-methylcyclopropene. *Journal of the American Society for Horticultural Science, 127*(5), 831–835.

González-Teruel, J. D., Ruiz-Abellon, M. C., Blanco, V., Blaya-Ros, P. J., Domingo, R., & Torres-Sánchez, R. (2022). Prediction of water stress episodes in fruit trees based on soil and weather time series data. *Agronomy, 12*(6), 1422.

Gratacós, E., & Cortés, A. (2007). Chilling requirements of cherry cultivars. *The Compact Fruit Tree, 40*, 7–9.

Gress, B. E., & Zalom, F. G. (2018). Identification and risk assessment of spinosad resistance in a California population of *Drosophila suzukii*. *Pest Management Science, 75*(5), 1270–1276.

Grimm, E., Hahn, J., Pflugfelder, D., Schmidt, M. J., van Dusschoten, D., & Knoche, M. (2019). Localized bursting of mesocarp cells triggers catastrophic fruit cracking. *Horticulture Research, 6*(1), 20.

Harb, J., Streif, J., & Saquet, A. (2003). Impact of controlled atmosphere storage conditions on storability and consumer acceptability of sweet cherries 'Regina. *The Journal of Horticultural Science and Biotechnology, 78*, 574–579.

Herrero, M., Rodrigo, J., & Wünsch, A. (2017). Flowering, fruit set and development. In J. Quero-García, A. Lezzoni, J. Puławska, & G. Lang (Eds.), *Cherries: Botany, production and uses* (pp. 14–35). Wallingford, UK: CABI.

Hojjati, M., Ghanbari Jahromi, M., Abdossi, V., & Torkashvand, A. M. (2024). Enhancing sweet cherry yield and quality under drought stress through melatonin foliar spraying. *Applied Fruit Science, 66*, 2239–2248.

Houghton, E., Bevandick, K., Neilsen, D., Hannam, K., & Nelson, I. M. (2023). Effects of postharvest deficit irrigation on sweet cherry (*Prunus avium*) in five Okanagan valley, Canada, orchards: I. Tree water status, photosynthesis, and growth. *Canadian Journal of Plant Science, 103*, 73–82.

Hrotkó, K., Magyar, L., Borsos, G., & Gyeviki, M. (2014). Rootstock effect on nutrient concentration of sweet cherry leaves. *Journal of Plant Nutrition, 37*, 1395–1409.

Hrotkó, K. (2016). Potentials in *Prunus mahaleb* L. for cherry rootstock breeding. *Scientia Horticulturae, 205*, 70–78.

Hrotkó, K., & Rozpara, E. (2017). Rootstocks and improvement. In J. Quero-García, A. Lezzoni, J. Puławska, & G. Lang (Eds.), *Cherries: Botany, production and uses* (pp. 117–139). Wallingford, UK: CABI.

Iezzoni, A., Wünsch, A., Höfer, M., Giovannini, D., Jensen, M., Quero-García, J., et al. (2017). Biodiversity, germplasm resources and breeding methods. In J. Quero-García, A. Lezzoni, J. Puławska, & G. Lang (Eds.), *Cherries: Botany, production and uses* (pp. 36–59). Wallingford, UK: CABI.

Ioannou, C. S., Papanastasiou, S. A., Zarpas, K. D., Miranda, M. A., Sciarretta, A., Nestel, D., et al. (2019). Development and field testing of a spatial decision support system to control populations of the European cherry fruit fly, *Rhagoletis cerasi*, in commercial orchards. *Agronomy, 9*, 568–587.

Jimenez, S., Pinochet, J., Gogorcena, Y., Beltran, J. A., & Moreno, M. A. (2007). Influence of different vigour cherry rootstocks on leaves and shoots mineral composition. *Scientia Horticulturae, 112*, 73–79.

Kargwal, R., Garg, M. K., Singh, V. K., Garg, R., & Kumar, N. (2020). Principles of modified atmosphere packaging for shelf-life extension of fruits and vegetables: An overview of storage conditions. *International Journal of Current Science, 8*(3), 2245–2252.

Katsoyannos, B. I., Papadopoulos, N. T., & Stavridis, D. (2000). Evaluation of trap types and food attractants for *Rhagoletis cerasi* (Diptera: Tephritidae). *Journal of Economic Entomology, 93*, 1005–1010.

Knoche, M., Athoo, T. O., Winkler, A., & Brüggenwirth, M. (2015). Postharvest osmotic dehydration of pedicels of sweet cherry fruit. *Postharvest Biology and Technology, 108*, 86–90.

Knoche, M., & Winkler, A. (2017). Rain-induced cracking of sweet cherries. In J. Quero-García, A. Lezzoni, J. Puławska, & G. Lang (Eds.), *Cherries: Botany, production and uses* (pp. 140–165). Wallingford, UK: CABI.

Kovanci, O., & Kovanci, B. (2006). Reduced-risk management of *Rhagoletis cerasi* flies (host race *Prunus*) in combination with a preliminary phenological model. *Journal of Insect Science, 6*(1), 34.

Lang, G. A. (2001a). Critical concepts for sweet cherry training systems. *Compact Fruit Tree, 34*(3), 70–75.

Lang, G. A., Early, J. D., Arroyave, N. J., Darnell, R. L., Martin, G. C., & Stutte, G. W. (1985). Dormancy: Toward a reduced, Universal Terminology. *HortScience, 20*(5), 809–812.

Lang, G. A. (2014). Growing sweet cherries under plastic covers and tunnels: Physiological aspects and practical considerations. *Acta Horticulturae, 1020*, 303–312.

Lang, G. A. (2024). Guidelines for the choice of stone fruit rootstocks. *Review on Italus Hortus, 31*, 18–26.

Lang, G. A. (2001b). Intensive sweet cherry orchard systems-rootstocks, vigor, precocity, productivity, and management. *Compact Fruit Tree, 34*(1), 23–26.

Lang, G. A., Sage, L., & Wilkinson, T. (2016). Ten years of studies on systems to modify sweet cherry production environments: Retractable roofs, high tunnels, and rain-shelters. *Acta Horticulturae, 1130*, 83–89.

Lang, G. A. (2017). The cherry industries in the USA: Current trends and future perspectives. *Acta Horticulturae, 1235*, 119–132.

Lang, G. A. (2001c). Underlying principles of high-density sweet cherry production. *Acta Horticulturae, 367*, 325–336.

Lang, G. A., Wilkinson, T., & Larson, J. E. (2017). Insights for orchard design and management using intensive sweet cherry canopy architectures on dwarfing to semi-vigorous rootstocks. *Acta Horticulturae, 1235*, 161–168.

Langille, A. B., Arteca, E. M., & Newman, J. A. (2017). The impacts of climate change on the abundance and distribution of the Spotted Wing *Drosophila (Drosophila suzukii)* in the United States and Canada. *PeerJ, 5*, e3192.

Lemus, G., & Negrón, C. (2005). Nutrición. In G. Lemus (Ed.), *El Cultivo del cerezo* (pp. 97–106). Santiago, Chile: Salesianos. *Boletín INIA N°133.*

Linke, M., Herppich, W. B., & Geyer, M. (2010). Green peduncles may indicate postharvest freshness of sweet cherries. *Postharvest Biology and Technology, 58*, 135–141.

Livellara, N., Saavedra, F., & Salgado, E. (2011). Plant-based indicators for irrigation scheduling in young cherry trees. *Agricultural Water Management, 98*(4), 684–690.

Loescher, W. H., McCamant, T., & Keller, J. D. (1990). Carbohydrate reserves, translocation, and storage in woody plant roots. *HortScience, 25*(3), 274–281.

Long, L. E., & Kaiser, C. (2010). Sweet cherry rootstocks. *Pacific Northwest Extension, 619*(8).

Long, L. E., Kaiser, C., & Thompson, A. (2021). *Sweet cherry rootstocks for the Pacific Northwest, 619*. A PNW Extension Publication.

Long, L., Iezzoni, A., Seavert, C., Auvil, T., Kaiser, C., & Brewer, L. J. (2019). New cherry rootstock and cultivar interactions directly affect orchard profitability. *Acta Horticulturae, 1235*.

Long, L., Lang, G., Musacchi, S., & Whiting, M. (2015). Cherry training systems. *Pacific Northwest Extension, 667*.

Lu, Y., Saeys, W., Kim, M., Peng, Y., & Lu, R. (2020). Hyperspectral imaging technology for quality and safety evaluation of horticultural products: A review and celebration of the past 20-year progress. *Postharvest Biology and Technology, 170*, 111318.

Lux, S. A, Wnuk, A., Vogt, H., Belien, T., Spornberger, A., & Studnicki, M. (2016). Validation of individual-based markov-like stochastic process model of insect behavior and a "Virtual Farm" concept for enhancement of site-specific IPM. *Frontiers in Physiology, 7*, 363. https://doi.org/10.3389/fphys.2016.00363. PMID: 27602000; PMCID: PMC4993809.

Maas, F. M., Balkhoven-Baart, J., & Van Der Steeg, P. A. H. (2012). Evaluation of Krymsk® 5 (VSL-2) and Krymsk® 6 (LC-52) as rootstocks for sweet cherry 'Kordia. *Acta Horticulturae, 1058*, 531–536.

Marsal, J., López, G., Arbones, A., Mata, M., Vallverdu, X., & Girona, J. (2009). Influence of post-harvest deficit irrigation and preharvest fruit thinning on sweet cherry (cv. New Star) fruit firmness and quality. *Journal of Horticultural Science & Biotechnology, 84*(3), 273–278.

Marsal, J., López, G., del Campo, J., Mata, M., Arbones, A., & Girona, J. (2010). Postharvest regulated deficit irrigation in Summit sweet cherry: Fruit yield and quality in the following season. *Irrigation Science, 28*(3), 181–189.

Martínez de Altube, M.d. M., Strauch, O., Fernandez De Castro, G., & Martinez Peña, A. (2008). Control of the flat-headed root borer *Capnodis tenebrionis* (Linné)(Coleoptera: Buprestidae) with the entomopathogenic nematode *Steinernema carpocapsae* (Weiser)(-Nematoda: Steinernematidae) in a chitosan formulation in apricot orchards. *BioControl, 53*(3), 531–539.

Martínez-Hernández, G. B., Blanco, V., Blaya-Ros, P. J., Torres-Sánchez, R., Domingo, R., & Artés-Hernández, F. (2020). Effects of UV-C on bioactive compounds and quality changes during shelf life of sweet cherry grown under conventional or regulated deficit irrigation. *Scientia Horticulturae, 269*, 109398.

Martins, V., Silva, V., Pereira, S., Afonso, S., Oliveira, I., Santos, M., et al. (2021). Rootstock affects the fruit quality of 'Early Bigi' sweet cherries. *Foods, 10*(10), 2317.

Matteo, M., Zoffoli, J. P., & Ayala, M. (2022). Calcium sprays and crop load reduction increase fruit quality and postharvest storage in sweet cherry (*Prunus avium* L.). *Agronomy, 12*(4), 829.

Meland, M., Frøynes, O., & Kaiser, C. (2017a). High tunnel production systems improve yields and fruit size of sweet cherry. *Acta Horticulturae, 1161*, 117–124.

Meland, M., Maas, F., & Kaiser, C. (2017b). Effects of irrigation and gibberellic acid on yield and fruit quality of sweet cherry produced in high tunnels. *Acta Horticulturae, 1177*, 307–312.

Mendel, Z., Assael, F., & Ben-Yehuda, S. (2023). Host selection and root colonization of cyanogenic stonefruit species by *Capnodis* spp. (Coleoptera: Buprestidae). *Annals of the Entomological Society of America, 96*, 127–134.

Morandi, B., Manfrini, L., Boini, A., Ponzo, F., & Corelli Grappadelli, L. (2018). Effects of mild water shortage on water relations, leaf gas exchanges, fruit growth and vascular flows of two different cherry cultivars. *Acta Horticulturae, 1197*, 127–132.

Musacchi, S., Gagliardi, F., & Serra, S. (2015). New training systems for high-density planting of sweet cherry. *HortScience, 50*(1), 59–67.

Nagy, P. T., Thurzó, S., Kincses, I., Szabó, Z., & Nyéki, J. (2008). Effect of foliar fertilization on leaf composition, sugar and organic acid contents of sweet cherry. *International Journal of Horticultural Science, 14*, 45–48.

Neilsen, D., Neilsen, G., & Forge, T. (2017). Environmental limiting factors for cherry production., pp. 189-222. In J. Quero-García, A. Lezzoni, J. Pulawska, & G. Lang (Eds.), *Cherries: Botany, production and uses* (pp. 189–222). Wallingford, UK: CABI.

Olmstead, J. W., Iezzoni, A. F., & Whiting, M. D. (2007). Genotypic differences in sweet cherry fruit (Prunus avium L.) size are primarily a function of cell number. *Journal of the American Society for Horticultural Science, 132*, 697–703.

Palma, M., Sepúlveda, Á., & Yuri, J. A. (2023). Effect of plastic roof and high tunnel on microclimate, physiology, vegetative growth and fruit characteristics of 'Santina' sweet cherry. *Scientia Horticulturae, 317*, 112012.

Papadopoulos, N. T., Lux, S. A., Köppler, K., & Beliën, T. (2017). Invertebrate and vertebrate pests: Biology and management. In J. Quero-García, A. Iezzoni, J. Puławska, & G. Lang (Eds.), *Cherries: Botany, production and uses* (pp. 305–337). Wallingford, UK: CABI.

Param, N., & Zoffoli, J. P. (2016). Genotypic differences in sweet cherries are associated with susceptibility to mechanical damage. *Scientia Horticulturae, 211*, 410–419.

Peace, C. P. (2017). DNA-informed breeding of rosaceous crops: Promises, progress and prospects. *Horticulture Research, 4*, 17006.

Quero-García, J., Branchereau, C., Barreneche, T., & Dirlewanger, E. (2022). DNA-informed breeding in sweet cherry: Current advances and perspectives. *Italus Hortus, 29*, 14–37.

Quero-García, J., Campoy, J. A., Castède, S., Pitiot, C., Barreneche, T., Lerigoleur-Balsemin, E., et al. (2017b). Breeding sweet cherries at INRA-Bordeaux: From conventional techniques to marker-assisted selection. *Acta Horticulturae, 1161*, 1–14.

Quero-García, J. (2019). Cherry breeding in the world: Current analysis and future perspectives. *Italus Hortus, 26*, 9–20.

Quero-García, J., Letourmy, P., Campoy, J. A., Branchereau, C., Malchev, S., Barreneche, T., et al. (2021). Multi-year analyses on three populations reveal the first stable QTLs for tolerance to rain-induced fruit cracking in sweet cherry (*Prunus avium* L.). *Horticulture Research, 8*, 136.

Quero-García, J., Schuster, M., López-Ortega, G., & Charlot, G. (2017a). Sweet cherry varieties and improvement. In J. Quero-García, A. Iezzoni, J. Puławska, & G. Lang (Eds.), *Cherries: Botany, production and uses* (pp. 60–94). Wallingford, UK: CABI.

Quiroz, M. P., Blanco, V., Zoffoli, J. P., & Ayala, M. (2023). Study of mineral composition and quality of fruit using vascular restrictions in branches of sweet cherry. *Plants, 12*(10), 1934.

Ren, J., Chen, P., Dai, S. J., Li, P., Li, Q., Ji, K., et al. (2011). Role of abscisic acid and ethylene in sweet cherry fruit maturation: Molecular aspects. *New Zealand Journal of Crop and Horticultural Science, 39*(3), 161–174.

Rota-Stabelli, O., Blaxter, M., & Anfora, G. (2013). Drosophila suzukii. *Current Biology, 23*(1), R8–R9.

Rutkowski, K., & Łysiak, G. P. (2022). Thinning methods to regulate sweet cherry crops—a review. *Applied Sciences, 12*(3), 1280.

Sallato, B., Grove, G., & Johnson, A. (2021). *Bacterial canker in Washington sweet cherries.* WSU Tree Fruit. https://treefruit.wsu.edu/bacterial-canker-in-washington-sweet-cherries/.

Sandefur, P., Oraguzie, N., & Peace, C. (2016). A DNA test for routine prediction in breeding of sweet cherry fruit color, Pav-Rf-SSR. *Molecular Breeding, 36*, 33.

Schlegel, H. J., Grimm, E., Winkler, A., & Knoche, M. (2018). Orange peel disorder in sweet cherry: Mechanism and triggers. *Postharvest Biology and Technology, 137*, 119–128.

Schuster, M., Grafe, C., Wolfram, B., & Schmidt, H. (2014). Cultivars resulting from cherry breeding in Germany. *Erwerbs-obstbau, 56*(2), 73–82.

Seavert, C., & Whiting, M. D. (2011). Comparing the economics of mechanical vs. hand harvest of sweet cherry. *Acta Horticulturae, 903*, 725–730.

Serradilla, M. J., Moraga, C., Ruiz-Moyano, S., Tejero, P., Córdoba, M. de G., Martín, A., et al. (2021). Identification of the causal agent of aqueous spot disease of sweet cherries (*Prunus avium* L.) from the Jerte Valley (Cáceres, Spain). *Foods, 10*(10), 2411.

Stone, C. H., Close, D. C., Bound, S. A., & Hunt, I. (2022). Training systems for sweet cherry: Light relations, fruit yield and quality. *Agronomy, 12*(3), 643.

Swamy, P., Probst, C., & Grove, G. G. (2019). Incidence of *Podosphaera clandestina* on sweet cherries (*Prunus avium*) and the influence of postharvest handling practices on the survival of conidia on harvested fruit. *Postharvest Biology and Technology, 156*, 110936.

Tarbath, M. P., Measham, P. F., Glen, M., & Barry, K. M. (2014). Host factors related to fruit rot of sweet cherry (Prunus avium L.) caused by Botrytis cinerea. *Australasian Plant Pathology, 43*(5), 513–522.

Teribia, N., Tijero, V., & Munné-Bosch, S. (2016). Linking hormonal profiles with variations in sugar and anthocyanin contents during the natural development and ripening of sweet cherries. *New Biotechnology, 33*(6), 824–833.

Tochen, S., Walton, V. M., & Lee, J. C. (2016). Impact of floral feeding on adult *Drosophila suzukii* survival and nutrient status. *Journal of Pest Science, 89*(3), 793–802.

Toivonen, P. M. A., Batista, A., & Lannard, B. (2017). Development of a predictive model for 'Lapins' sweet cherry dry matter content using a visible/near-infrared spectometer and its potential application to other cultivars. *Canadian Journal of Plant Science, 97*(6), 1030–1035.

Toivonen, P. M. A. (2014). Relationship of typical core temperatures after hydrocooling on retention of different quality components in sweet cherry. *HortTechnology, 24*(4), 457–462.

Tsai, I., Tanaka, E., Masuya, H., Tanaka, R., Hirooka, Y., Endoh, R., et al. (2014). Comparative genomics of *taphrina* fungi causing varying degrees of tumorous deformity in plants. *Genome Biology and Evolution, 6*(4), 861–872.

Ughini, V., & Roversi, A. (2008). Estimation of sweet cherry fertilizer requirements by the Szűcs method varies by cultivar. *Acta Horticulturae, 795*, 733–738.

USDA. (2024). *U.S. Department of Agriculture. Foreign Agricultural Service. Data and analysis*. Retrieved from USDA Open Data (Accessed: January 2025).

Vignati, E., Lipska, M., Dunwell, J. M., Caccamo, M., & Simkin, A. J. (2022). Fruit development in sweet cherry. *Plants, 11*(12), 1531.

Wang, Y., & Einhorn, T. C. (2017). Harvest maturity and crop load influence pitting susceptibility and postharvest quality deterioration of sweet cherry (*Prunus avium* L.). *Acta Horticulturae, 1161*, 613–620.

Wang, Y., & Long, L. E. (2015). Physiological and biochemical changes relating to postharvest splitting of sweet cherries affected by calcium application in hydrocooling water. *Food Chemistry, 181*, 241–247.

Wang, Y., & Long, L. E. (2014). Respiration and quality responses of sweet cherry to different atmospheres during cold storage and shipping. *Postharvest Biology and Technology, 92*, 62–69.

Wenden, B., Campoy, J. A., Jensen, M., & López-Ortega, G. (2017). Climatic limiting factors: Temperature. In J. Quero-García, A. Lezzoni, J. Puławska, & G. Lang (Eds.), *Cherries: Botany, production and uses* (pp. 166–188). Wallingford, UK: CABI.

Whiting, M. D., Ophardt, D., & McFerson, J. R. (2006). Chemical Blossom Thinners vary in their effect on sweet cherry fruit set, yield, fruit quality, and crop value. *HortTechnology, 16*(1), 66–70.

Winkler, A., Bunger, P., Lang, P. M., Schumann, C., Bruggenwirth, M., & Knoche, M. (2024). Mode of action of calcium in reducing macrocracking of sweet cherry fruit. *Journal of the American Society for Horticultural Science, 149*, 61–74.

Winkler, A., Grimm, E., & Knoche, M. (2019). Sweet cherry fruit: Ideal osmometers? *Frontiers in Plant Science, 10*, 164.

Winkler, A., Peschel, S., Kohrs, K., & Knoche, M. (2016). Rain cracking in sweet cherries is not due to excess water uptake but to localized skin phenomena. *Journal of the American Society for Horticultural Science, 141*(6), 653–660.

Xin, Y., Liu, Z., Zhang, Y., Shi, X., Chen, F., & Liu, K. (2021). Effect of temperature fluctuation on colour change and softening of postharvest sweet cherry. *RSC Advances, 11*(37), 22969–22982.

Yang, Y., Li, A., Guo, M., Kong, Y., Zhang, J., Wang, J., et al. (2024). Improving the storage quality and aroma of sweet cherry by postharvest 3-phenyllactic acid treatment. *Scientia Horticulturae, 338*, 113661.

Ystaas, J. (1990). Nutritional requirement of sweet cherries. *Acta Horticulturae, 274*, 521–526.

Zhai, Z., Xiao, Y., Wang, Y., Sun, Y., Peng, X., Feng, C., et al. (2022). Abscisic acid-responsive transcription factors PavDof2/6/15 mediate fruit softening in sweet cherry. *Plant Physiology, 190*(4), 2501–2518.

Zoffoli, J. P., & Rodríguez, J. (2014). Effect of active and passive modified atmosphere packaging of sweet cherry. *Acta Horticulturae, 1020*, 115–119.

Zoffoli, J. P., Muñoz, S., Valenzuela, L., Reyes, M., & Barros, F. (2008). Manipulation of Van sweet cherry crop load influences fruit quality and susceptibility to impact bruising. *Acta Horticulturae, 795*, 877–881.

Zoffoli, J. P., Naranjo, P., Rodríguez, J., Jara, A., Contreras, C., & Param, N. (2020). Características de poscosecha de cereza cultivar Regina. *Revista Frutícola Copefrut, 42*(2).

Zoffoli, J. P. (2022). New post-harvest technologies for high quality sweet cherries. *Italus Hortus, 29*(3), 68–81.

Zoffoli, J. P., Toivonen, P., & Wang, Y. (2017). Postharvest biology and handling for fresh markets. In J. Quero-García, A. Iezzoni, J. Pulawska, & G. Lang (Eds.), *Cherries: Botany, production and uses* (pp. 460–484). Wallingford, UK: CABI.

CHAPTER

Sour Cherry

5

Mirko Schuster[1], Zsuzsanna Békefi[2], Geza Bujdoso[2], Daina Feldmane[3], Milica Fotirić Akšić[4], Marek Szymajda[5]

[1]*Julius Kühn-Institute, Federal Research Centre for Cultivated Plants, Institute for Breeding Research on Fruit Crops, Dresden, Germany;* [2]*Hungarian University of Agriculture and Life Sciences, Institute of Horticultural Siences, Fruit Growing Research Centre, Budapest, Hungary;* [3]*Institute of Horticulture, Graudu str. 1, Cerini, Krimūnu p., Dobele, Latvia;* [4]*Department of Fruit Science, Institute of Horticulture, Faculty of Agriculture, University of Belgrade, Belgrade, Serbia;* [5]*Department of Horticultural Crop Breeding, The National Institute of Horticultural Research, Skierniewice, Poland*

Chapter outline

1. Origin and distribution

The sour cherry, *Prunus cerasus* L., is traditionally grown in the temperate climate zone of the Northern Hemisphere. The sour cherry is an important fruit crop in Eastern and Central Europe and parts of Southwest Asia, as well as in the north-eastern parts of the United States. The sour cherry is native from Central and Southern Europe to Northern India, Iran, and Kurdistan. The region from the southern border of the Black Sea along Anatolia and the Caucasus to Iran has been described as the center of origin. These regions in Eastern Europe and Minor Asia are also the native distribution areas of landraces of sour cherry (Faust and Suranyi, 1997). Some local races of sour cherry are Pandy (syn. Köröser Weichsel, Crisana) and Cigany

Temperate Tree Fruits and Nuts. https://doi.org/10.1016/B978-0-443-29137-1.00005-0

(syn. Zigeunerkirsche, Gypsy cherry) in Hungary, Mocanesti in Romania, Oblačinska in Serbia, Marasca in Croatia, Strauchweichsel and Weinweichsel in Germany, Vladimirskaja in Russia/Moldova, and Kütahya in Turkey. Most of the common local cultivars are selections or hybrids of these sour cherry races.

Currently, sour cherry growing is dominated through a small collection of cultivars. In most cases, these cultivars are landrace or clone selections of regional cultivars. In Middle Europe, the main sour cherry cultivar is Schattenmorelle with its appropriated local synonyms, for example, Łutowka in Poland, Griotte du Nord or Griotte Noir Tardive in France, and Benelux and occasionally English Morello in Great Britain. This cultivar is self-compatible and highly productive with dark red fruits and juice. The origin for this cultivar is probably the Chateau de Moreille in France. In the United States, the cultivar Montmorency dominates sour cherry growing. The origin of this nearly 400-year-old cultivar is France. Montmorency is self-compatible and highly productive with bright red-colored fruits with clear juice. The landrace cultivar Pandy (syn. Crisana, Köröser) or related cultivars are most popular in Hungary and Romania. Pandy is self-sterile, has an excellent fruit quality, and is tolerant to diseases. In Turkey, several types of the Kütahya landrace are the main cultivated sour cherries. Kütahya has very good fruit characteristics and is self-fertile.

2. Economic importance

Nowadays, sour cherry cultivation is facing a number of challenges in all countries where it is considered an economically important crop. The main problem is the market situation with prices being below production costs for several years. Due to stable low prices, cultivation of sour cherries is not profitable, which determines the future and economic importance of this fruit species. This phenomenon is leading to a decline in production in many sour cherry-producing countries, with the exception of the Russian Federation and Iran. In Hungary and Belarus, the area has remained stable.

Another important issue is that the processing industry is owned by a few large companies across Europe, with only a few local and small factories. The production of sour cherries for consumption was very important in the past; in the 2000s, Russia was the main market. After the embargo against that country, the market almost completely collapsed, as it was only possible to export to a few countries.

As a result of these developments, sales prices and resulting income descended in relation to the general economic development. As a result, sour cherry orchards are growing older and older, and no one is planting new orchards, except for a few growers in a few countries. This process can affect everything from research to nursery production to commercial fruit production. In addition, labor and other input costs are rising, so there is no price stability for either producers or consumers. This situation can only be changed by increasing purchase prices and improving the productivity of sour cherry production.

3. Botany and relevant species

The classification of cherries—like the case of other *Prunus* species—is a matter of continuous revision. According to the widely accepted classification, sour cherry (*P. cerasus* L.), also called sour or tart cherry, together with sweet cherry (*P. avium* L.) and ground cherry (*P. fruticosa* Pall.) belongs to the Rosaceae family in the genus *Prunus*, subgenus *Cerasus* Pers., and section *Eucerasus* Koehne.

The sour cherry is an allotetraploid species with $2n = 4x = 32$ chromosomes. It originated as a hybrid of the tetraploid ground cherry (*P. fruticosa*, $2n = 4x = 32$) and unreduced pollen of sweet cherry (*P. avium*, $2n = 2x = 16$). Olden and Nybom (1968) confirmed the allotetraploid origin of sour cherry by morphological and chemotaxonomic studies with artificial tetraploid hybrids between *P. fruticosa* and unreduced gametes of sweet cherry. Further analysis by using isozyme markers, genomic *in situ* hybridization, karyotype studies, and chloroplast DNA markers suggested the hybrid origin of *P. cerasus* (Schuster and Schreiber, 2000; Dirlewanger et al., 2007). Recent molecular studies have described the sour cherry genome as a segmental allotetraploid consisting of two subgenomes, one derived from the sweet cherry, *P. avium*, and the other from the ground cherry, *P. fruticosa* (Goeckeritz et al., 2023; Wöhner et al., 2023).

P. cerasus is divided into two subspecies: subsp. *cerasus* and subsp. *acida*. *P. cerasus* subsp. *cerasus* (syn. *P. cerasus* var. *caproniana* L.) is a tree with a strong trunk, upright branches, a circular stone, and called English Amarelle or Kentish cherry (Scholz and Scholz, 1995). Among the *P. cerasus* subsp. *cerasus,* two varieties were distinguished: var. *cerasus* and var. *austera* (L.) EHRH. Cherries belonging to var. *cerasus* have red, sweet-sour fruits with coloring juice. The stone is easily separated from the stalk. The var. *austera* types have light red, glassy, and sour fruits with noncoloring juice. The separation of the stone is difficult from the stalk. The members of the second subspecies, *P. cerasus* subsp. *acida,* are shrubs or small trees with long, thin, drooping branches. The stones are elliptically shaped. The fruit is dark red to black, sour, with coloring juice. These sour cherries are called also Morello cherries. This subspecies also includes the var. *marasca* (Reichb.) Schneid., the Marasca cherry. This type of cherry is native in Croatia and has small, dark red fruits with a deep red juice, while the taste is bitter and very acidic.

Hybrids between *P. avium* and *P. cerasus* are called *P.* × *gondouinii* (Poit. & Turpin) Rehder or Duke cherries. Generally, the appearance of the trees and fruits resemble sweet cherries, but the taste of the fruit is rather acidic. The fruit skin and juice color show great variability among them.

Sour cherry cultivars are frequently considered to be self-compatible, although self-incompatible and partially self-compatible cultivars also exist. Self-incompatibility in sour cherry is caused by a gametophytic self-incompatibility system. According to a one-allele-match model, self-compatible sour cherries must contain a minimum of two nonfunctional *S* alleles (Hauck et al., 2006).

The most commonly used rootstocks for sour cherry are *P. mahaleb* and *P. avium*. All dwarfing sweet cherry rootstocks, such as Gisela 5, Piku 1, and Piku 4, can also be used successfully as sour cherry rootstocks. New research has shown that the use of dwarfing rootstocks increases fruit set in partially self-fertile sour cherries.

4. Breeding and propagation

The cultivation of sour cherries dates back to the Neolithic period. The first cultivars were named and described in the 17th century. The first breeding activities in sour cherries go back to the conscious selection of valuable genotypes in sour cherry-growing areas. As a result, a large number of sour cherries have been selected. However, the characteristics of the first sour cherries could not meet the new requirements of the developing fruit growth and consumption. As a result, several sour cherry-breeding programs were initiated in Europe and North America in the second half of the last century. In the breeding programs in Germany and Hungary, the landrace Köröser Weichsel (syn. Köröser, Pandy) has been used mainly as one parent part to improve the fruit characteristics and tolerance to biotic stress.

In Poland, most new cultivars have been selected from artificial crosses with Łutówka (syn. Schattenmorelle). In Hungary, Romania, Serbia, and Denmark, many new cultivars have resulted from regional clonal selection of the landraces Pandy, Mocanesti, Oblačinska, and Stevnsbaer, respectively, or are hybrids between landraces and cultivars. In the United States (Michigan), the improvement of Montmorency sour cherry was implemented by selecting more productive accessions (Faust and Suranyi, 1997). Interspecific hybrids of sour cherry with *P. fruticosa* have been developed to increase cold hardiness in breeding programs in Russia and Canada. Similar to the Canadian cultivars, the new U.S. cultivars are also derived from European germplasm, as the sour cherry is not native to North America. Currently, sour cherry cultivation is still dominated by a few main cultivars. The new sour cherry cultivars, with their improved fruit characteristics, have not yet become established, partly because they do not meet the requirements of the processing industry.

The breeding goals in sour cherry are strongly determined by the predominant processing of the fruits. That is why the selection is focused on genotypes with high fruit quality and good processing characteristics. These include a fruit size of 21 to 24 mm for processing, only larger than 24 mm for the fresh market, with small round stones that separate well from the flesh, high sugar and acid content, and strong juice coloring. Characteristics for mechanical harvesting, such as high fruit firmness, easy and dry removal of the fruit from the stalk, and an upright growth of the tree are also important goals. Additionally, breeding work is focused on tolerance to biotic and abiotic stress. These mainly include a high tolerance to *Monilia* blossom blight, *Monilinia laxa*, and to Cherry leaf spot, *Blumeriella jaapii,* as well as to winter and spring frost. The prerequirements for a successful new cultivar are its self-compatibility and a high and regular fruit set. Sour cherries are often self-compatible, although self-incompatible and partly self-incompatible cultivars exist. To extend the ripening period of sour cherries, early ripening cultivars are being sought.

Breeding sour cherries requires a long time, up to 20 years or more. In three stages of selection—seedling selection, first clone testing, and second clone testing—the different characteristics are evaluated with varying degrees of intensity and selected accordingly. A future cultivar should combine as many of the desired breeding objectives as possible. The most common breeding technique used is intraspecific hybridization. Other techniques such as interspecific hybridization, clonal selection, and mutation breeding are less widely applied.

Cultivars, landraces, and promising breeding clones have been the main source of breeding material. In some breeding programs around the world, wild *Prunus* species have also been used as donors to enlarge the genetic base for different breeding objectives. Progeny of different interspecific crosses between *P. cerasus* with *P. maackii*, *P. fruticosa*, *P. padus*, *P. serotina*, and *P. spinosa* have been used as new donors for resistance to diseases and frost tolerance in Russia, United States, and Germany.

The main sour cherry cultivars of global importance are summarized in Table 5.1, and new cultivars or of local importance are summarized in Table 5.2.

Table 5.1 Sour cherry cultivars with global importance

Cultivar	Parentage	Origin	Fertility
Érdi Bőtermő (syn. Danube)	Pandy × Nagy Angol	Hungary	Self-compatible
Fanal (syn. Heinrichs Konservenkirsche)	Unknown	Germany	Self-compatible
Kelleriis 16 (syn. Morellenfeuer)	(Ostheimer Weichsel × Früheste der Mark) × open pollinated	Denmark	Self-compatible
Montmorency	Unknown	France	Self-compatible
Oblačinska	Unknown	Serbia	Self-compatible
Pandy (syn. Köröser, Pandy Meggy, Köröser Weichsel, Kereska, Crişana)	Unknown	Carpathian basin	Self-incompatible
Schattenmorelle (syn. Große Lange Lotkirsche, Łutówka, Griotte du Nord, Moreillska, Skyggemorel)	Unknown	Germany and other countries in Central and Eastern Europe	Self-compatible
Stevnsbaer	Unknown	Denmark	Self-compatibility
Ujfehértói Fürtös (syn. Balaton, Ungarische Traubige)	Unknown	Hungary	Partly self-compatible

Table 5.2 New sour cherry cultivars and cultivars with local importance

Cultivar	Parentage	Origin	Fertility
Achat	Köröser × B7,2,40 (Fanal × Kelleriis 16)	Germany	Self-compatible
De Botoşani	Unknown	Romania	Partly self-compatible
Bucovina	Unknown	Romania	Self-compatible
Carmine Jewel	Kerr's Easy Pick (*P. cerasus* × *P. fruticosa*) × Northstar	Canada	Self-compatible
Cigány (syn. Cigány Meggy, Gypsy Cherry, Zigeunerkirsche)	Unknown	Carpathian basin	Self-compatible
Crişana 2	Unknown	Romania	Self-incompatible
Csengődi	Unknown	Hungary	Partly self-compatible
Debreceni Bőtermő	Unknown	Hungary	Partly self-compatible
Kántojánosi 3	Unknown	Hungary	Partly self-compatible
Kütahya	Unknown	Turkey	Self-compatible
Gerema	Kelleriis 14 × open pollinated	Germany	Self-compatible
Ilva	Unknown	Romania	Self-compatible
Iskra	Köröser Weichsel × Heimanns Rubin	Serbia	Self-compatible
Jachim	Köröser Gierstädt × Safir	Germany	Self-compatible
Jade	Köröser × Röhrigs Weichsel	Germany	Self-compatible
Latvijas Zemais (syn. Lietuvas Zemais, Žagarvyšné, Läti Madalkirss)	Unknown	Baltic states	Partly self-compatible
Lenka	Unknown	Serbia	Self-compatible
Lyubskaya	Unknown	Russia	Self-compatible
Marasca	Unknown	Croatia	Partly self-compatible
Mocăneşti	Unknown	Romania	Self-incompatible
Morina	Köröser × Reinhards Ostheimer	Germany	Partly self-compatible
Nana	Crişana × open pollinated	Romania	Self-compatible
Nefris	Łutówka × open pollinated	Poland	Self-compatible

Table 5.2 New sour cherry cultivars and cultivars with local importance—*cont'd*

Cultivar	Parentage	Origin	Fertility
Nevena	Köröser Weichsel × Heimann's Konserven Weichsel	Serbia	Self-compatible
Northstar	English Morello × Serbian Pie No. 1	United States	Self-compatible
Petri	Unknown	Hungary	Partly self-compatible
Rival	Griot Moscovski × Nana	Romania	Partly self-compatible
Sabina	Łutówka × Schirpotreb	Poland	Self-compatible
Safir	Schattenmorelle × Fanal	Germany	Self-compatible
Sophia	Čačanski Rubin × Heimanns Konserven Weichsel	Serbia	Self-compatible
Spinell	Köröser × B7, 2, 40 (Fanal × Kelleriis 16)	Germany	Partly self-compatible
Šumadinka	Köröser Weichsel × Heimanns Konserven Weichsel	Serbia	Self-compatible
Tarina	Anglaise Hâtive × Visin Tufa	Romania	Partly self-compatible
Taurus	Köröser Weichsel × Vowi	Germany	Self-compatible
Tiki	Stevnsbaer × Fanal	Denmark	Self-compatible
Turmalin (syn. Coralin)	Kelleriis 16 × P10,17,5 (Köröser × Schattenmorelle)	Germany	Self-compatible
Vladimirskaja	Unknown	Russia	Self-compatible
Zhivitsa	Griotte d'Ostheim × Dönissens Gelbe	Belarus	Self-incompatible

5. Orchard management

Establishment and location: Sour cherries are grown in the sites with the wide range of soil fertility and moisture, from semiarid regions (Iran and Turkey) and skeletal, rocky soils (Croatia) to humid regions and alluvial soils (Poland and Baltic states). This indicates the wide adaptation of sour cherries to different environments. Standard requirements for cherry orchard location are: plain land or upper and middle part of gentle slope with a groundwater level not higher than 1.50 m, soil properties: organic matter at least 1.5% to 3.0%, and medium or high content of plant-available phosphorus and potassium content. Cherries best absorb nutrients in weakly acidic to almost neutral soil (pH_{KCl} 6–6.8), but they can also adapt to a larger amplitude of soil reaction (pH_{KCl} 5.5–7.2). In the Northern Hemisphere, southeast-facing slopes are preferable with the best light and temperature conditions, whereas south- and west-facing slopes are subject to larger temperature fluctuations. The lowest parts and north-facing slopes should be avoided; here, cold air masses increase the damages in winter and spring frosts. The orchard microclimate can be improved by vigorous tree rows, which protect the cherries from the coldest winds, but the orchard should be open to airflow in other directions.

When renewing cherry plantations, it is not always possible to apply crop rotation, but replantation in the same field leads to poor growth and productivity. Green manure, preplant fertilizing, mulching, and creating raised beds can improve the situation, but it will not completely mitigate unfavorable replanting effect. A dwarfing rootstock like Gisela 5 suitable for sour cherry growing is less susceptible to replantation and can increase the fruit set of partially fertile sour cherries.

Pruning and training systems: The sour cherry is mainly used for the industrial purposes; therefore, their growing system should be adapted to mechanization to be cost-effective, with a canopy suitable for mechanical harvest, a rootstock with good anchorage, and a trunk tolerant to microdamage. The rootstocks developed from *P. mahaleb* and *P. avium* are being used widely.

Self-propelled and trailed harvesters have been developed for mechanical harvesting of sour cherries. The most popular pruning systems in sour cherry orchards for mechanical shaking of the trunk and primary branches are the open vase or the delayed open vase canopy. The density of these orchards is between 285 and 667 trees/ha. When using full-row ("over the row") mechanical sour cherry harvesters with vertical shaking heads along the row and above the tree, the canopy must consist of a straight central trunk and elastic, thin, productive shoots, avoiding strong branches between the rows. The planting distances and trunk parameters must be adapted to the mechanical harvester. This harvesting method is suitable for different tree densities, starting from a row spacing of 2.8 m, but the tree height is limited. Similarly, a straight, vertical trunk must be maintained for mechanical pruning of side branches. Some sour cherry cultivars are prone to deflection of the central leader, in which case supports may be necessary for young trees (Figures 5.1 and 5.2).

Cultivars of sour cherries that taste good when eaten fresh are still grown in relatively small quantities and harvested by hand or in "pick yourself" orchards. Here, the use of vigor-reducing rootstocks and nonvigorous sour cherry cultivars is recommended.

Pruning is an important part in sour cherry production to maintain tree growth and health. Sunny weather and low relative air moisture facilitate the drying of pruning wounds and diminish the risk of trunk diseases. In the warmest part of the sour cherry-growing area, the pathogens that attack the woody tissues spread their spores during the dormant period, so pruning should not be carried out from defoliation until the appearance of new leaves or flowers. The optimal period for dormant pruning runs from March until the end of the blossoming season. This is a very short period,

FIGURE 5.1

Mechanical sour cherry tree harvester—shaking machine. *(From Mirko Schuster.)*

FIGURE 5.2

Mechanical sour cherry tree harvester—full-row trailed harvester. *(From WEREMCZUK FMR.)*

therefore just the thick branches can be removed from the canopies. In the coolest parts of the sour cherry-growing area, the sunny and dry periods before bud burst are suitable for spring pruning, but later, during leaf development, air humidity and pathogen spread increases. Winter pruning of sour cherries has been applied in areas with suitable conditions (snowless soil, dry air).

The optimum window for summer pruning is between mid-July and the end of August, with dry and sunny weather during and after pruning. The maximum percentage of the "removable" part must not exceed 30% of the total canopy. The aim of this treatment is to increase the light penetration in the canopy; therefore, broken, too weak, and/or infected branches and those that prevent more light in the canopy must be removed. In the summer pruning, a couple of centimeter-long remaining branches (pins) should be left in the canopy to avoid wounds on the main branches. It is typical of sour cherry to produce blind wood. On the surface of blind wood, there were flower buds; these buds do not contain shoots, so after the fruit ripens, there is nothing left. During summer pruning, blind wood can be cut back to allow new shoots to form. To obtain high-quality sour cherry fruit, the balance between crop load, vegetative growth, and leaf area must be maintained. If the length of annual shoots of the previous season is less than 30 cm, the vegetative growth must be advanced by more intense pruning in the spring and by ensured uptake of nitrogen. The direct thinning of flowers or fruits is recognized as unprofitable for sour cherries. For cultivars prone to overabundant production, the crop load has been adjusted by thinning of producing branches, primarily by removing the weakest ones.

Irrigation: Both scion and rootstock have a strong influence on the drought tolerance of grafted sour cherry trees. The effect of irrigation on sour cherry growth and productivity is mostly beneficial, but the magnitude of this effect varies greatly depending on growing conditions, cultivars, and rootstocks. In semi-arid growing sites, adult sour cherries on the Mahaleb rootstocks performed well when the irrigation compensated at least 30% of crop seasonal evapotranspiration with 240 to 526 mm water used annually (Papenfuss and Black, 2010). Under Hungarian conditions, sour cherries had a high productivity in seasons with annual precipitation of 700 to 800 mm, whereas in dry seasons, sour cherry plantations needed 200 to 250 mm of additional water (Kovács et al., 1999).

In European countries with humid climates, the yield increase of irrigated sour cherries varied between 17% and 31%. The decisions for the necessity of irrigation were made based on the compensation of daily evaporation (from 30% to full), soil water potential (no less than -0.01 MPa to -0.05 MPa), and soil volumetric moisture (no less than 20%). In general, the annual water consumption for drip or sprinkler irrigation was 28 to 152 mm ha^{-1}, estimating the total area of the orchard, not just the wetted strip.

The time of fruit growth and ripening in sour cherries coincides with the formation of next year's buds, which require balanced water conditions; the trees should avoid severe water stress in the preharvest period. Looking at sour cherry tree development over the years, irrigation has the most pronounced effect on young tree growth (Feldmane et al., 2019). In the first years, the establishment of a strong trunk in adequate moisture conditions or growth inhibition due to drought may have a long-lasting effect on further tree growth and productivity.

Fertilization: The ability of trees to store nutrients in roots, trunks, and other perennial organs can help overcome nutrient deficit for a period of time. However, balanced nutrition of fruit trees is a prerequisite not only for yield but also for yield stability over the years, tree vitality, and stress tolerance. The analysis of nutrient content in sour cherry leaf and soil has been used to determine the need for fertilization. Rather insignificant variation in leaf nutrient content was observed among sour

Table 5.3 Macroelement content in sour cherry leaf dry mass

Macroelement	Optimum content in sour cherry leaf dry mass	Reference
N	2.5%–3.2%	Sadowski and Jadczuk, 2001
P	0.15%–0.30%	Borowy et al., 2018
K	1.6%–2.0% 1.2%–1.8%	Friedrich, 1993 Sadowski et al., 1990
Ca	0.7–3.0 ppm	Hanson and Proebsting, 1996
Mg	0.4–0.9 ppm	Hanson and Proebsting, 1996

cherry cultivars (Pole et al., 2017), while the rootstock effect was significant (Milošević et al., 2023). The optimum contents of macroelements in sour cherry leaf are given in Table 5.3. Public opinion insists on a decrease of industrially produced mineral fertilizer treatments due to high energy consumption for their production and environmental issues. By decreasing industrial fertilizer rates, nutrient uptake should be advanced by optimal soil and air moisture, temperature, pH level, microbiological activity, and leaf fertilizing.

Nitrogen (N) is one of the major plant building elements, being a component of amino acids and chlorophyll. Sour cherries are considered N-demanding plants, especially during shoot and fruit growth. Annual shoots of 30 to 50 cm in length indicate a sufficient supply of N. Considering the risk of N leaching from the soil into the groundwater and to support soil enzyme activity, a rate of 40-60-80 kg N/ha per year was recognized as most suitable for mature sour cherry trees. Promoting soil organic matter content and microbiological activity can meet the N requirements of sour cherry without the use of mineral fertilizers, at least in humid regions (Sanchez et al., 2003).

Phosphorus (P) has many essential functions for fruit trees, playing a role in energy transfer, flower bud development, and others. However, the amount of P required to perform these functions in sour cherries is relatively low. The amount of P naturally present in the soil often can be sufficient for sour cherry growing (Mercik et al., 1990). However, at least for newly planted sour cherry trees, soil P availability should be monitored and supported.

Potassium (K) has a remarkable role in osmotic regulation, synthesis, and transport of sugars and acids in plants. Sour cherry fruits are known to be one of the sources of K in human nutrition, and the amount of K in the whole cherry plant is high compared to other mineral elements. K availability has the greatest effect on sour cherry growth and yield among most macronutrients (N, P, K, Ca). Different approaches and strategies are used to ensure K uptake. Higher rates of fertilizer (up to 166 kg K or 200 K_2O per ha) have been used to increase soil K to optimum levels (Rutkowski and Łysiak, 2022); moderate rates compensate for K removal by harvest and pruning. At least on some alluvial soils, leaf K content was stabilized and yield increased by applying 40 to 80 kg K/ha (Roeva et al., 2022).

Among many important functions, the role of **calcium (Ca)** in membrane permeability and cell wall structure and **magnesium (Mg)** in chlorophyll synthesis should be noted. The supply of Ca and Mg in the soil was provided by natural minerals or improved by liming and foliar treatments. Foliar sprays with Ca fertilizers reduced fruit cracking and fruit rot, as well as juice leakage from the stem scar in mechanically harvested sour cherries (Wójcik and Wawrzyńczak, 2014).

The levels of **trace elements** considered to be optimal in sour cherry leaf dry matter (DM) were: 25 to 200 mg/kg iron (Fe) and manganese (Mn), 5 to 20 mg/kg copper (Cu), 15 to 70 mg/kg zinc (Zn), and 25 to 60 mg/kg boron (B) (Hanson and Proebsting, 1996). In more recent sour cherry studies, the B content was lower: 21 to 33 mg/kg in Latvia (Feldmane et al., 2017) and 13.7 to 20.4 mg/kg in Serbia (Milošević et al., 2023). Considering the role of B in fruit set, leaf treatments with B fertilizer in the second half of the growing season or at leaf fall were tested in sour cherries. Postharvest B sprays increased B status in flowers (Wójcik and Morgaś, 2015), improved pollen viability and germination capacity, and reduced interannual variation in fruit set (Feldmane et al., 2017). Fruit set was most consistently increased when leaf B content was 19 to 23 mg/kg (Hanson, 1991) or 21 to 26 mg/kg (Feldmane et al., 2017).

6. Fruit maturation and harvest criteria

After pollination and fertilization, during the first growth stage, all parts of the sour cherry fruit increase in size until the seed approaches its maximum size. During the second growth stage, the endocarp hardens and forms the stone, while the growth of the mesocarp, the flesh of the fruit, slows down. After stone hardening, during the third growth stage, fruit ripening begins with a faster increase in mesocarp. During ripening, dry matter accumulation and changes in physical characteristics and biochemical composition occur in the fruit. Visible fruit characteristics such as size, color, and shape develop into their final fruit characteristics. Once the fruit is fully ripe, senescence follows with a gradual loss of quality.

The time between flowering and harvesting of sour cherries varied considerably between cultivars and locations, i.e., 60 to 70 days in Latvia, 60 to 75 days in Poland (Jadczuk-Tobjasz and Bednarski, 2007), 72 to 77 days in Serbia, and 82 to 104 days in Denmark (Vittrup Christensen, 1986). Different approaches have been used to characterize the maturity of cherries. In the international BBCH scale of phenological development, ripening begins with the onset of fruit coloring (BBCH 81) and progresses through advanced coloring (BBCH 85) to fruit ripe for consumption; the fruit has typical taste and firmness (BBCH 89). In several studies, sour cherry ripening was characterized by four or five ripening stages based on fruit color and size, starting with fruit of incomplete size, lightly colored red, and ending with dark red, full-sized fruit (Miloševič and Miloševič, 2012). The cessation of fruit growth in dimensions and weight is one of the signs that ripening has ended and senescence has begun.

Sour cherry is nonclimacteric fruit and does not continue to ripen after harvest. Therefore, the taste, aroma, and biochemical properties of the fruit must meet the requirements of the consumer or processor at the time of harvest. Soluble solids content (SSC), which is related to sweetness, and titratable acid content (TA) have been recognized as key flavor components. They were previously estimated separately or combined in the ripening index (RI) or maturity index (MI) calculated as the ratio of SSC and TA (Miloševič and Miloševič, 2012; Wojdyło et al., 2014).

The range of SSC for different sour cherry cultivars spans from 11% to 25% (Yarilgac et al., 2005, Grafe and Schuster, 2014, Wojdyło et al., 2014). During ripening, SSC increases in sour cherries, reaching a maximum at full ripeness, but decreases in senescent fruit. SSC is strongly correlated with fruit dry matter (Grafe et al., 2009). The influence of growing conditions on the quality of sour cherries is strong. However, fruit characteristics such as SSC and TA are genetically determined. Therefore, the typical level of SSC can be used as one of the harvesting criteria. The TA content in sour cherry cultivars ranges from 1.28% to 3.25% (Yarilgac et al., 2005; Grafe and Schuster, 2014; Wojdyło et al., 2014). The decrease of TA content in fruit fresh weight was mostly observed during ripening of sour cherries (Poll et al., 2003; Miloševič and Miloševič, 2012). The decrease in TA content in fruit fresh weight is more related to cell expansion and dilution effect rather than acid degradation (Famiani et al., 2020). However, there are cases where the TA content was increased during sour cherry fruit development, as in Turkey, cv. Kütahya from 0.5% at the stage of advanced coloring to 1.2% in fully colored fruit (Aslantas et al., 2016). Similarly, acidity increased in ripe sour cherries from day 1 to day 11 after ripening and decreased slightly thereafter (Kaack, 2017). The range of SSC/TA among cultivars was relatively large: 4.3 to 15.8 (Papp et al., 2010; Wojdyło et al., 2014; Grafe and Schuster, 2014). A significant part of the soluble solids consists of sugars, mainly glucose, fructose, sorbitol, in some cultivars also sucrose, and in a few cultivars, galactose and xylose (Sokół-Łetowska et al., 2020). The ratio of total sugar content to TA is also used as a fruit quality indicator, ranging from 4.17 to 7.95 in sour cherry cultivars grown in Poland (Sokół-Łetowska et al., 2020).

The content of ascorbic acid in sour cherry cultivars varied from 5.45 to 22.18 mg/100 g (Wojdyło et al., 2014), 5.45 to 22.11 mg/100 g (Sokół-Łetowska et al., 2020), and 8.23 to 10.44 mg/100 g (Ivanova et al., 2022). Despite the different results on the total TA content, researchers agree that the ascorbic acid content decreases during ripening.

The color of sour cherries develops by increasing the anthocyanin content during sour cherry ripening. The anthocyanin content from 1st to 18th maturity day increased on average from 86.7 to 124.8 mg/100 g fruit weight (FW) in cultivars grown in Denmark (Kaack, 2017). In Marasca sour cherry ecotypes, nonuniform changes in anthocyanin content were observed, with higher levels at the last stage of ripening, from 3.18 to 19.75 g/kg DM (Pedisić et al., 2010).

Total anthocyanin content among cultivars ranged from 57.7 to 226.5 mg/100 g FW in cultivars grown in Denmark (Kaack, 2017); approximately 400 to 2000 mg/L

FW in Hungary (Sass-Kiss et al., 2005); and 844.77 to 994.00 mg/100 g DM in cultivars grown in Poland (Sokół-Łetowska et al., 2020).

During cherry ripening, fruit redness increases and yellowness decreases. In general, L*, C*, and H° values decrease linearly or in a curvilinear pattern in both skin and flesh of sour cherries during ripening, and lower H° and C* values were associated with higher anthocyanin content (Pedisić et al., 2009). However, in dark-colored fruits of cv. Cigančica, blue color was also present at a high level, and H° increased in the last stage of ripening (Pedisić et al., 2009). For mature sour cherry fruits, the following average colorimetric values were detected in various studies: L* 17 to 28, C* 9 to 10, H° 32 to 315 for cultivars Kelleriis and Cigančica (Pedisić et al., 2009); L* 20, C* 10, H° 12 to 15 for cultivar Kütahya (Aslantas et al., 2016); L* 21 to 23, a* 6 to 7, b* 3, C* 7 to 8, H° 22 to 53 for cultivar Łutówka (Rutkowski and Łysiak, 2023).

Decreases in the volatiles benzaldehyde and hydrogen cyanide, which contribute to the typical sour cherry flavor, were observed in sour cherry cv. Stevnsbaer clone 23' during ripening (Kaack, 2017). However, the influence of the growing year on the volatile content was predominant, and no clear tendencies of volatile content changes during maturation were found in another study (Poll et al., 2003).

Fruit firmness and fruit retention force from stalk are important characteristics for cultivar suitability for mechanical harvesting, which is widely used for sour cherries. For mature cherries, fruit firmness in the German sour cherry collection ranged from 81.5 to 146.2 g/mm, with the widespread cultivars Montmorency and Schattenmorelle having fruit firmness of 94.6 and 118.4 g/mm, respectively (Grafe and Schuster, 2014). In Poland, fruit firmness of mature cv. Łutówka varied from 1.94 to 2.19 N (Rutkowski K., Łysiak, 2022). Fruit firmness decreased during ripening under the influence of many processes, such as the increase of vacuoles, cells, and intercellular spaces; the decrease of the cell wall area per fresh weight of fruit flesh; and changes in enzyme composition. In senescent fruit, firmness decreases further due to osmotic changes, cell wall degradation, and loss of cell-cell adhesion under the influence of enzymatic activity. Sufficiently low fruit retention force is one of the most important criteria for mechanical harvesting. The decrease in fruit retention force from the stalk was related to the formation of the abscission layer; it started 12 to 15 days before fruit ripening when the cell walls became thinner and less rigid (Stösser et al., 1969). The abscission layer was not formed by the vascular bundles, and final fruit retention occurred by mechanical detachment of the bundles. During ripening, the fruit retention force decreased from 4.25 to 1.78 N (Aslantas et al., 2016). The fruit retention force below 3.0 N was acceptable for mechanical harvesting.

7. Postharvest physiology and technology/processing

Physiologically, sour cherry is a nonclimacteric fruit, which means that it must be harvested as ripe as possible. During ripening, color changes due to chlorophyll degradation, firmness decreases as complex carbohydrates are converted to sugars, while pectin and hemicelluloses are degraded and solubility increases. Acidity decreases and sugar content increases, while volatile compounds are synthesized. All fleshy *Prunus* fruits have a short shelf-life, so cold storage is essential.

The sour cherry is a light or dark red fruit with a high dry matter content, a sour taste due to the dominant L-malic acid, and a bitter-sweet aroma. Sour cherry fruits have a high respiration rate during the harvesting period, making it difficult to maintain their quality. The fruit undergoes a rapid aging process and is therefore very susceptible to microbial spoilage, which reduces shelf-life after harvest. The most common microorganisms found in sour cherries during storage are *Penicillium, Alternaria,* and *Fusarium.*

Sour cherries are generally processed into jams, toppings, purées; juices, alcoholic beverages; frozen, dried, or canned fruit; vinegar; confectionery; fermented products and concentrates; or used as natural colorings, flavors, and food additives, rather than for consumption. The fruit is a rich source of polyphenols, especially anthocyanins, which are responsible for its specific color. The anthocyanins contained in sour cherries are cyaniding derivates, which are distinguished by a sugar unit bound to the aglyconic part of the molecule, with cyanidin-3-glucosilrutinoside being present in the highest concentration (Pedisić et al., 2010; Alrgei et al., 2015). Other anthocyanins found in sour cherry include cyanidin-3-glucoside, cyanidin-3-rutinoside, cyanidin-3-sophoroside, peonidin-3-glucoside, malvinidin-3-glucoside, delphinidin-3-glucoside, pelargonidin-3-glucoside, cyanidin-3-arabinosylrutinoside, cyanidin-3-gentiobioside, and peonidin-3-rutinoside (Papp et al., 2010; Toydemir et al., 2013; Woźniak et al., 2016). In addition, sour cherry fruits contain a considerable amount of other colorless polyphenols, flavan-3-ols such as catechin, epicatechin, and proanthocyanidins, and hydroxycinnamic phenolic acids such as neochlorogenic, chlorogenic, p-coumaric, and caffeic acid (Bonerz et al., 2007; Alrgei et al., 2015). The presence of quercetin, kaempferol, and isorhamnetin glycosides are also found in sour cherry fruits (Alrgei et al., 2015; Wozniak et al., 2016).

Controlled atmosphere (CA) or modified atmosphere packaging (MAP) with lower oxygen content and increased carbon dioxide (CO_2) was used to extend the storage period of sour cherries in cold stores. In this way, the respiration rate, chemical oxidation rate, and growth of aerobic microorganisms are reduced, while decline in soluble solid content, retention of firmness, stem color and brightness, and the reduction of surface pitting and microbial decay in the fruit are reduced. Mihály et al. (2020) determined that sour cherry fruits packaged under modified atmosphere could be stored at 0°C for 42 days, with improved quality and weight loss, and a significant reduction in the percentage of spoilage. Ravanfar et al. (2014) immersed sour cherry fruits in hot water and found that the treatment delayed dehydration and preserved the visual appearance of the fruit without adversely affecting taste, aroma, or flavor.

As the sour cherry is rich in bioactive compounds, it is extremely important to preserve those valuable ingredients as much as possible during processing. Sour cherry juice has a limited shelf-life due to microbial deterioration and enzymatic activity, so techniques must be used to successfully inactivate vegetative microorganisms and enzymes to extend the shelf-life of the juice. Pasteurization is the most commonly used technique, but lately, the application of nonthermal plasma has been shown to be efficient for the inactivation of microorganisms, with moderate heating of the treated sample resulting in a higher concentration of anthocyanins and phenolic acid in sour cherry juice than pasteurization (Elez Garofulić et al., 2015). In addition, microfiltration (Bagger-Jørgensen et al., 2007), blanching (Jia et al., 2012), boiling treatment, microwave treatment (Yıldız et al., 2022), high pressure processing (Tenuta et al., 2023), and combined thermal treatment with ultrasound and oxygen (Alifakı and Gök, 2024) are used to preserve sour cherry juice. High hydrostatic pressures, pulsed electric fields, cold plasma, and osmode hydration are used to obtain products of good quality and medium-long shelf-life without performing high-impact heat treatments (Dalla Rosa, 2019). To achieve the stability of anthocyanins in sour cherry juices, various color fixation techniques are used, such as the addition of ascorbic acid antioxidants or acerola juice (Navruz et al., 2016). To extend their shelf-life, sour cherries have recently been preserved dry. Drying allows food to be preserved by removing water and prevents the growth of microorganisms and decay. Sour cherries are dried by hot air drying, vacuum drying, osmotic dehydration, or osmotic—microwave vacuum drying to preserve fruit quality, flavor, texture, and crispness.

Sour cherry byproducts: The processing of sour cherry fruits produces stalks, leaves, peels, and seeds as byproducts, which have received much more attention, especially in the last decade. These byproducts are used as animal feed but are mostly considered waste (Kołodziejczyk et al., 2013). Sour cherry pomace consists of peel and flesh and is mainly obtained after juice pressing. It is rich in phenolic compounds (anthocyanins, flavan-3-ols, phenolic acids, and flavonols), carbohydrates, vitamins, carotenoids, fibers (lignin, cellulose, hemicellulose, and pectin), and minerals (Kołodziejczyk et al., 2013; Greiby et al., 2014; Kosmala et al., 2009; Woźniak et al., 2016). Studies by Ferretti et al. (2010) and Kirakosyan et al. (2009) reported the presence of melatonin in this byproduct, which plays a role in the sleep-wake cycle. To date, the extraction of phytochemicals from sour cherry pomace (a process of recovering bioactive compounds from food industry waste) and the encapsulation of phytochemicals (coating the bioactive compounds with a biopolymer material to protect them from damaging factors such as oxygen, water, and light and to increase their stability) are processes used in the food industry. Due to the high content of bioactive components, sour cherry pomace has potential application in snacks, breakfast cereals, beverage mixes, pet foods, bakery products, confectionaries, packaging films as a slow-release antioxidant, yogurts, and many other value-added products. Sour cherry pomace powder can also be used as a nutraceutical and colorant. Sezer et al. (2021) used microwave, high-pressure, enzymatic, and heat treatments to improve the yield of soluble fiber from sour cherry pomace by converting the insoluble fiber into a soluble form.

Another byproduct of sour cherry juice production, which is sold canned or frozen, is sour cherry stone that consists of stone and kernel. Many conventional (shaking, extraction, thermal treatment, and Soxhlet extraction) and green methods (ultrasound-assisted extraction, microwave-assisted extraction, enzymatic hydrolysis, and high hydrostatic pressure) have been used to extract all the bioactive compounds from sour cherry (Natić et al., 2020). While the stone is mostly used as fuel or considered as waste, the kernel is used as a source of oil, protein, fiber, lipids, vitamins (B1, B3, B5, and B6), minerals (Ca, Fe, Mg, Cu, Zn), and polyphenols (flavonoids, anthocyanidins, stilbenes, catechins, flavins, procyanidins). The kernels contain an average of 17% to 36% oil, making them a potential source for oil production (Popa et al., 2011). The oil contains fatty acids, tocopherols, carotenoids, sterols (β-sitosterol, Δ5-avenasterol, sitostanol), tocochromanols, and squalene, but no amygdalin is detected (Kazempour-Samak et al., 2021). The dominant unsaturated acids are oleic acid and linoleic acid, while palmitic acid and stearic acid are the dominant saturated acids (Górnaś et al., 2016). Başyiğit et al. (2020) microencapsulated sour cherry seed oil to increase its oxidative stability, while Altuntas et al. (2023) encapsulated sour cherry seed oil for use as a bioactive food ingredient. Gheju et al. (2016) found that sour cherry stone is also a good source of high porosity carbons, and they succeeded in using it as an adsorbent to remove Cr (VI) from polluted waters.

8. Main diseases and pests and their management

8.1 Fungal and bacterial diseases

***Monilia* blossom blight and brown rot, *Monilinia* ssp.:** *Monilia* blossom blight and brown rot—caused by three species of the genus *Monilinia*: *Monilinia laxa* (Aderh. and Ruhland) Honey, *Monilinia fructicola* (G. Wint.) Honey, and *Monilinia fructigena* Honey ex Whetzel—are major diseases of sweet and sour cherries throughout the world. Generally, *M. laxa* is present worldwide, while *M. fructicola* is found mostly in North America, Australia, and New Zealand. *M. fructigena* is more common in Europe and the Middle East. *Monilia* blossom blight infections during flowering are most commonly caused by the species *M. laxa*, which can infect flowers at temperatures slightly above 5°C. The fungus attacks the flowers and enters the tree through the stigma, anthers, and petals. The fungal mycelium grows down the stem into the branch and then spreads upward. The leaves on the infected branches die and the shoot dries up (Figure 5.3). The brown rot of fruit can occur at any stage of fruit development (Figure 5.4). All three species of *Monilinia* spp. can be responsible for fruit root, with mixed infections on the same fruit. The conidia penetrate the half-ripe and ripe fruit through small openings. Later, spore layers develop in a circle around the infection site. The timing of fungicide application is crucial for controlling blossom blight. Flowers must be protected before, during, and after blooming. Regarding fruits, the critical phase occurs just before harvest. Cultural practices, such as pruning infected twigs and the removal of rotten fruit or mummies, can diminish inoculum levels but do not eliminate the disease.

FIGURE 5.3

Sour cherry trees severely affected by *Monilia* bloosom blight. *(From Mirko Schuster.)*

FIGURE 5.4

Fruit root on sour cherry fruits infected by *Monilinia* spp. *(From Mirko Schuster.)*

Cherry leaf spot diseases, *Blumeriella jaapii* (Rehm) v. Arx: Cherry leaf spot (CLS) diseases, caused by the fungus *Blumeriella jaapii* (Rehm) v. Arx, is one of the most important diseases in all humid cherry-growing regions worldwide, especially in sour cherry growing. Heavy infections can lead to serious damage by leaf defoliation (Figures 5.5 and 5.6). Symptom development begins in late May to early June. Small, round, reddish-purple to dark brown to reddish-purple spots form on

FIGURE 5.5

Sour cherry trees infected by cherry leaf spot diseases, *Blumeriella jaapii. (From Mirko Schuster.)*

FIGURE 5.6

Sour cherry leaves infected by cherry leaf spot diseases, *Blumeriella jaapii (From Mirko Schuster.)*

the upper surface of the leaf, merging especially along the midrib. On the underside of the leaf, small white-reddish spots first appear, bounded by the leaf veins, which merge to form larger complexes. After rainfall, conidia multiply rapidly in the acervuli. These can be seen as a whitish coating on the underside of the leaf, and when dry, take on the appearance of white, scattered spots. Infected leaves turn yellow prematurely and then fall off early in the season, which may result in severe defoliation. As the biology of the fungus is similar to apple scab, the same or similar fungicides can be used. Primary infection must be prevented during the shoot growth period from May onward. The disease can usually be safely controlled with three to four treatments at 10- to 14-day intervals in wet weather.

Anthracnose, *Colletotrichum gloeosporioides*: Anthracnose (bitter rot) on sour cherry fruit caused by *Glomerella cingulata* (Stoneman) Spauld. & H. Schrenk (anam.: *Colletotrichum gloeosporioides* [Penz.] Penz. & Sacc. in Penz.). However, recent investigations have shown that *C. acutatum* is responsible for the epidemics (Damm et al., 2012; Sundelin et al., 2015). Anthracnose in sour cherry is an economically important disease. *C. acutatum* inoculum overwinters on cherry buds and is present in orchards throughout the year. Anthracnose infections are characterized by circular, tan to brown necrotic lesions with concentric rings. Lesions grow rapidly and usually occur on ripe or near-ripe fruit (Figure 5.7). Serious yield losses can occur under epidemic conditions. The recommended strategy for controlling anthracnose is to spray synthetic chemicals from petal fall for the following 6 weeks and to remove infected plant material from orchards to reduce the inoculum for ripening fruit.

Bacterial canker, *Pseudomonas syringae*: Bacterial canker is caused by bacteria belonging to the *Pseudomonas syringae* species complex, a group of polyphagous bacteria that can infect more than 180 plant species. Bacterial canker is the most common cause of disease in cherry and occurs in stone fruit growing areas throughout the world. The disease causes severe yield losses and sometimes leads to the death of entire trees. The bacteria enter through wounds and leaf scars in

FIGURE 5.7

Sour cherry fruit infected by *Colletotrichum gloeosporioides*. *(From Mirko Schuster.)*

winter and lie dormant until spring, when black necrotic lesions develop. Flowers, fruit, and leaves are all affected by the disease and show a range of necrotic symptoms. From dying flowers, which are a source of secondary infection, bacteria can spread to shoots and branches, leading to necrotic dieback and canker formation. Three main periods of copper treatment are recommended for sour cherry orchards: the leafless period, the flowering period, and the leaf fall period. However, the most effective protection against bacterial canker is prevention. When establishing new plantations, it is important to use healthy nursery stock in addition to well-adapted cultivars. In orchards, pruning during dry weather to remove infected shoots is an important treatment. In cases of severe infestation, the entire tree must be removed.

8.2 Insect Pests

Black cherry aphid, *Myzus cerasi* (Fabricius): The black cherry aphid, *Myzus cerasi* (Fabricius) (Hemiptera: *Aphididae*), is one of the main pest species on cherry trees in Europe, Middle East, and Asia and has recently expanded to Australia, New Zealand, and North America. This aphid causes deformation of the young leaves, impacts the photosynthesis rate, and induces inward curling and drying of the upper shoots of sour cherries (Figure 5.8). *Myzus cerasi* is also an important transmitter of many plant viruses. The body of apterae individuals of *M. cerasi* on *Prunus* trees is shiny, and body color varies from dark brown to black. *Myzus cerasi* show a holocyclic and heteroecious life cycle, migrating between primary host (i.e., trees) and secondary hosts (i.e., herbaceous plants) during the year. Wingless females emerge from eggs laid on young shoots and bud bases of cherry trees during the winter and

FIGURE 5.8

Leaf damage caused by black cherry aphid, *Myzus cerasi. (From Mirko Schuster.)*

feed and develop on leaves on young shoots of cherry trees in spring period. Later, winged individuals migrate to secondary hosts in early summer. From the start of colonization, insecticide control is necessary if the damage threshold is exceeded. Natural enemies such as ladybugs and their larvae and hoverfly larvae can significantly reduce existing aphid colonies.

Cherry fruit fly, *Rhagoletis cerasi* (L.) and *Rhagoletis cingulata* Loew.: The European cherry fruit fly, *Rhagoletis cerasi* (L.), can attack sour cherries to varying degrees from year to year. The American cherry fruit fly, *Rhagoletis cingulata* Loew., which arrives a little later, has also been recently observed at various locations in Europe. Because of its later appearance, it may become a major pest of sour cherries.

Both cherry fruit flies are about 4 to 5 mm long, shiny black, with a yellowish scutellum. The wings are crystal clear with dark cross-bands (Figure 5.9). Wing banding is one of the features used to distinguish the two species. While *Rhagoletis cerasi* has a shortened, inwardly displaced cross-band, *Rhagoletis cingulata* has it on the outside at the end of the wing, and the inwardly directed cross-bands are often joined. The larva is about 6-mm long and yellowish-white in color. The flight period of *Rhagoletis cerasi* begins in mid-May. The flies that have pupated in the previous year hatch in May. After a maturation period of 8 to 10 days, the flies begin to lay their eggs. The yellowish eggs, about 2-mm long, are laid individually under the skin of the fruit, which turns red. Usually only one larva develops per fruit, but multiple infestations are possible. The larvae feed in the fruit for about 3 to 4 weeks and then leave the fruit through a borehole to pupate in the soil, where they overwinter as chrysalises. *Rhagoletis cingulata* emerges in early summer, about 4 weeks later than *Rhagoletis cerasi*.

FIGURE 5.9

Cherry fruit fly, *Rhagoletis cerasi* (L.) on sour cherry fruits. *(From Mirko Schuster.)*

Orchards must be protected against insects every year and require insecticide sprays several times per season, when adult fruit flies start flying. Since insecticides for protection are being restricted, environmentally friendly methods of pest control should be considered, such as applying nematodes to the soil, spraying trees and fruits with biopesticides containing the entomopathogenic fungus *Beauveria bassiana*, or applying a dense mesh on the undergrowth soil thus preventing emerging adults from getting to the host plants and fruit. Another method of control is the use of traps with attractants for adult mass trapping.

Spotted-wing drosophila, *Drosophila suzukii*: The spotted-wing drosophila is an invasive pest that has been present in Europe since 2008. The species belongs to the family of fruit flies (Diptera: *Drosophilidae*) and is native to Asia. Unlike the common species, *D. melanogaster*, it feeds on healthy fruit that is about to be harvested, preferably stone fruits and berries. The large number of host spores, the short generation cycle, and the high reproductive rate make *D. suzukii* a very dangerous pest that can cause enormous crop losses.

The fly is a few millimeters in size and has red eyes. The body is yellow-orange in summer and dark brown to black in winter. A typical feature of the males is a black spot on the wing tips. Females have no black spot on their wings (Figure 5.10). They are characterized by their ovipositor, which is equipped with strong, saw-like dark teeth, which they use to penetrate the intact skin of the fruit to lay their eggs inside. Females begin to lay eggs when the fruit changes color. Many generations can be produced each year. The lifespan of the flies is about 6 weeks in summer and up to 8 weeks in winter until the following year. The use of fine-mesh netting (mesh size 0.8 to 1 mm, max. 1.2 mm) is a very effective protective measure to minimize infestation by *D. suzukii*. Control with insecticides is difficult. The high reproduction rate and rapid succession of generations make repeated insecticide applications necessary. This harbors the risk of developing resistance to individual active substances. Furthermore, waiting periods must be observed before harvesting. Insecticides are nevertheless an important component in a combination of protective measures.

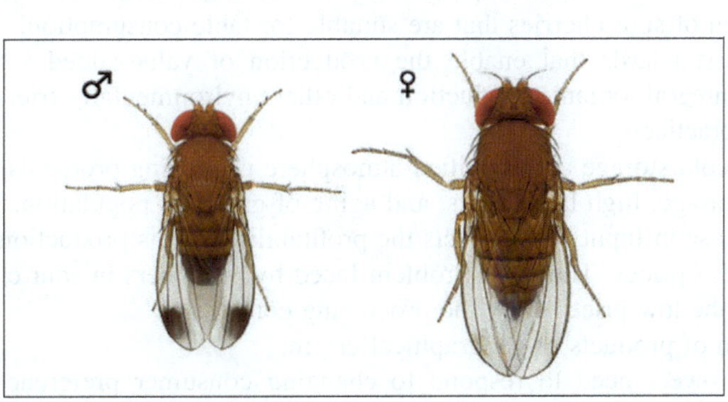

FIGURE 5.10

Spotted-wing drosophila, *Drosophila suzukii. (From Agroscope CC BY-ND 2.0.)*

9. Current and future challenges in main production regions

Sour cherry production is concentrated in the traditional growing areas of Central and Eastern Europe and North America. The majority of sour cherry production is for processing, with only a small proportion sold as fresh fruit. To compensate for low purchase prices, greater mechanization of sour cherry production will be required in the future. This includes techniques for mechanical harvesting and tree training. The key to successful sour cherry production is the selection of suitable cultivars that are well adapted to local conditions and market requirements. Historically, only a limited number of cultivars have dominated the global sour cherry industry. Only a few sour cherry-breeding programs in the world are breeding new cultivars that are well suited for future sour cherry cultivation. The breeding traits focus on selecting sour cherries that are suitable for mechanical harvesting and processing, have high fruit quality, and are well adapted to biotic and abiotic stresses. As with other fruit species, sour cherries present a number of challenges. These include:

- Most production is still conducted using traditional methods and small farms (up to 1–2 hectares), and home gardens lack the capacity to compete in the current market. It is imperative that governments provide financial support to farmers to enable them to invest in new, modern sour cherry orchards.
- Unfavorable climatic conditions – spring frost, rain, high temperatures, and/or winds during blooming period. A seasonal water deficit can lead to reduced fruit size and quality. The lack of irrigation systems in some regions can be a significant constraint on tree productivity.
- Adaptation to climate change – utilization of adapted cultivars and rootstocks, finding new growing areas, or producing sour cherries in high tunnels in regions with unsuitable ecological conditions.
- Introduction of precision farming protocols toward sustainable use of resources, improving of programs and management of pests and invasive species, reducing of pesticide use, optimization of soil management and nutrition.
- Development of certification programs for virus-free plants.
- Production of sour cherries that are suitable for table consumption.
- Establish standards that enable the production of value-added sour cherries through integral, organic production and other environmentally friendly farming practices.
- Improve cold storage and modified atmosphere packaging protocols.
- Labor shortage, high labor costs, and aging of growers' population.
- The increase in input costs affects the profitability of this production.
- Inconsistent prices. The main problem faced by producers in sour cherry cultivation is the low price set by the processing companies.
- Protection of products of geographical origin.
- Cherry growers need to respond to changing consumer preferences that are pushing the industry toward more sustainable and responsible practices.
- Opportunities to invest in new processing programs.

- Disruption of the supply chain. The horizontal and vertical linkage of stakeholders within the chain from producer to consumer should be organized. The activities of all stakeholders (producers, processors, exporters, consumers, etc.) in the development of market chains.

References

Alifakı, Y.Ö., & Gök, S. B. (2024). Optimization of minimal thermal treatment conditions for sour cherry (*Prunus cerasus* L.) juice using ultrasound-ohmic combination treatment. *J Food Saf, 44*, Article e13111.

Alrgei, H. O. S., Dabić, D.Č., Natić, M. M., Rakonjac, V., Milojković-Opsenica, D., Tešić, Ž., et al. (2015). Chemical profile of major taste- and health-related compounds of Oblačinska sour cherry. *J Sci Food Agric, 96*, 1241–1251. https://doi.org/10.1002/jsfa.7212

Altuntas, U., Altin-Yavuzarslan, G., & Ozcelik, B. (2023). Enhanced oxidative stability and bioaccessibility of sour cherry kernel byproducts encapsulated by complex coacervates with different wall matrixes by spray- and freeze-drying. *ACS Omega, 8*, 23782–23790.

Aslantaş, R., Angin, I., Boydas, M. G., & Ozkan, G. (2016). Fruit characteristics and detachment parameters of sour cherry (*Prunus cerasus* L. cv. Kütahya) as affected by various maturity stages. *Erwerbs-Obstbau, 58*(2), 127–134. https://doi.org/10.1007/s10341-016-0270-1

Başyiğit, B., Sağlam, H., Kandemir, Ş., Karaaslan, A., & Karaaslan, M. (2020). Microencapsulation of sour cherry oil by spray drying: Evaluation of physical morphology, thermal properties, storage stability, and antimicrobial activity. *Powder Technology, 364*, 654–663.

Bagger-Jørgensen, R., Casani, S., & Meyer, A. S. (2007). Microfiltration of red berry juice with thread filters: Effects of temperature, flow and filter pore size. *Journal of Food Process Engineering, 25*(2), 109–124.

Bonerz, D., Wurth, K., Dietrich, H., & Will, F. (2007). Analytical characterization and the impact of ageing on anthocyanin composition and degradation in juices from five sour cherry cultivars. *European Food Research and Technology, 224*, 355–364.

Borowy, A., Chrzanowska, E., & Kapłan, M. (2018). Comparison of three sour cherry cultivars grown in Central-Eastern Poland. *Acta Sci. Pol. Hortorum Cultus, 17*(1), 63–73.

Dalla Rosa, M. (2019). Advanced technologies for cherry processing and packaging. *Italus Hortus, 26*(1), 51–58.

Damm, U., Cannon, P. F., Woudenberg, J. H. C., & Crous, P. W. (2012). The *Colletotrichum acutatum* species complex. *Studies in Mycology, 73*, 37–113.

Dirlewanger, E., Claverie, J., Wünsch, A., & Iezzoni, A. F. (2007). *Cherry, fruits and nuts* (pp. 103–118). Berlin, Heidelberg: Springer.

Elez Garofulić, I., Režek Jambrak, A., Milošević, S., Dragović-Uzelac, V., Zorić, Z., & Herceg, Z. (2015). The effect of gas phase plasma treatment on the anthocyanin and phenolic acid content of sour cherry Marasca (*Prunus cerasus* var. Marasca) juice. *LWT-food Science and Technology, 62*, 894–900.

Famiani, F., Bonghi, C., Chen, Z. H., Drincovich, M. F., Farinelli, D., Lara, M. V., et al. (2020). Stone fruits: Growth and nitrogen and organic acid metabolism in the fruits and seeds - a review. *Frontiers in Plant Science, 11*, Article 572601.

Faust, M., & Suranyi, D. (1997). Origin and dissemination of cherry. *Horticultural Reviews, 19*, 263–317.

Feldmane, D., Butac, M., Militaru, M., Kalva, E., Grotuze, S., Missa, I., et al. (2019). Response of adult sour cherry trees to woodchip mulch and drip irrigation. *Acta Hortic, 1235*, 337–344. https://doi.org/10.17660/ActaHortic.2019.1235.46

Feldmane, D., Madalina, M., Butac, M., & Pole, V. (2017). The effect of foliar boron application on flower bud winterhardiness, fruit set and fruit quality of sour cherries in Latvia. *Fruit Growing Research, XXXII*, 47–52. http://publications.icdp.ro/publicatii/lucrari%202017/II.2.%20Feldmane.pdf.

Ferretti, G., Bacchetti, T., Belleggia, A., & Neri, D. (2010). Cherry antioxidants: From farm to table. *Molecules, 15*, 6993–7005.

Friedrich, G. (1993). *Handbuch des Obstbaus* (p. 621). Eugen Ulmer Verlag.

Górnaś, P., Juhneviča-Radenkova, K., Radenkovs, V., Mišina, I., Pugajeva, I., Soliven, A., et al. (2016). The impact of different baking conditions on the stability of the extractable polyphenols in muffins enriched by strawberry, sour cherry, raspberry or black currant pomace. *LWT-food Sci Technol, 65*, 946–953.

Gheju, M., Balcu, I., & Jurchescu, P. (2016). Removal of hexavalent chromium from aqueous solutions by use of chemically modified sour cherry stones. *Desalin Water Treat, 57*, 10776–10789.

Goeckeritz, C. Z., Rhoades, K. E., Childs, K. L., Iezzoni, A. F., VanBuren, R., & Hollender, C. A. (2023). Genome of tetraploid sour cherry (*Prunus cerasus* L.) Montmorency identifies three distinct ancestral *Prunus* genomes. *Horticulture Research, 10*(7), Article uhad097.

Gonçalves, B., Silva, A. P., Moutinho-Pereira, J., Bacelar, E., Rosa, E., & Meyer, A. S. (2007). Effect of ripeness and postharvest storage on the evolution of colour and anthocyanins in cherries (*Prunus avium* L.). *Food Chem, 103*, 976–984.

Grafe, C., Höfer, M., & Schuster, M. (2009). Evaluation of dry matter in sour cherry (*Prunus cerasus* L.). *Acta Hortic, 839*, 281–286.

Grafe, C., & Schuster, M. (2014). Physiochemical characterization of fruit quality traits in a German sour cherry collection. *Scientia Horticulturae, 180*, 24–31.

Greiby, I., Mishra, D. K., & Dolan, K. D. (2014). Inverse method to sequentially estimate temperature-dependent thermal conductivity of cherry pomace during nonisothermal heating. *J. Food Eng, 127*, 16–23.

Hanson, E. J., & Proebsting, E. L. (1996). Cherry nutrient requirements and water relations. In L. Webster (Ed.), *Cherries: Crop physiology, production and uses* (pp. 243–257). CAB International.

Hanson, E. J. (1991). Sour cherry trees respond to foliar boron applications. *Hort Science, 26*, 1142–1145.

Hauck, R. N., Yamane, H., Tao, R., & Iezzoni, A. F. (2006). Accumulation of nonfunctional *S*-haplotypes results in the breakdown of gametophytic self-incompatibility in tetraploid *Prunus*. *Genetics, 172*, 1191–1198.

Ivanova, L., Serdyuk, M., Malkina, V., Tonkha, O., Tsyz, O., Mazur, B., et al. (2022). Cultivar features of polyphenolic compounds and ascorbic acid accumulation in the cherry fruits (*Prunus cerasus* L.) in the Southern Steppe of Ukraine. *Agronomy Research, 20*, 588–602.

Jadczuk-Tobjasz, E., & Bednarski, R. (2007). Preliminary evaluation of the growth and yielding of 10 sour cherry cultivars. *Zeszyty Naukowe Instytutu Ogrodnictwa, 15*, 17–27.

Jia, G. A. O., Baogang, W. A. N. G., Xiaoyuan, F. E. N. G., & Yongqing, Z. H. U. (2012). Effect of blanching on quality of sour cherry (Prunus cerasus L. CV. CAB) juice American-Eurasian. *J Agric & Environ Sci, 12*(1), 123–127.

Kaack, K. (2017). Maturation and picking time for sweet cherries (*Prunus avium*) and sour cherries (*Prunus cerasus* L.). *Eur Food Res Technol, 243*, 539–546.

Kazempour-Samak, M., Rashidi, L., Ghavami, M., Sharifan, A., & Hosseini, F. (2021). Sour cherry (*Cerasus vulgaris* Miller) kernel oil as the novel functional edible oil: Sensory evaluation and antioxidant and physicochemical properties. *Journal of Food Quality*. Article ID 5529613.

Kirakosyan, A., Seymour, E. M., Llanes, D. E. U., Kaufman, P. B., & Bolling, S. F. (2009). Chemical profile and antioxidant capacities of tart cherry products. *Food Chem, 115*, 20–25.

Kołodziejczyk, K., Sojka, M., Abadias, M., Vnas, I., Guyot, S., & Baron, A. (2013). Polyphenol composition, antioxidant capacity, and antimicrobial activity of the extracts obtained from industrial sour cherry pomace. *Industrial Crops and Products, 51*, 279–288.

Kosmala, M., Milala, J., Kołodziejczyk, K., Markowski, J., Mieszczakowska, M., Ginies, C., et al. (2009). Characterization of cell wall polysaccharides of cherry (*Prunus cerasus* var. Schattenmorelle) fruit and pomace. *Plant Foods Hum Nutr, 64*, 279–285.

Kovács, J., Nyéki, J., Szabó, Z., Ligetvári, F., & Soltész, M. (1999). The necessity and possibilities of irrigation in fruit growing under conditions of Hungary. *International Journal of Horticultural Science, 5*, 93–94.

Mercik, S., Slowik, K., & Wasowski, T. (1990). Effect of 65-year static soil fertilization on the growth, fruiting and leaf nutrient status of a tart cherry trees. *Acta Hortic, 274*, 331–338.

Mihály, K., Sándor, E., Kovács, C., & Takács, F. (2020). Effect of different storage methods for the shelf-life and the fungal populations of the Hungarian sour cherry cultivars. *Acta Hortic, 1275*, 427–432.

Milošević, T., & Milošević, N. (2012). Fruit quality attributes of sour cherry cultivars. *ISRN Agronomy, 2012*. article ID 593981.

Milošević, T., Moreno, M. A., Milošević, N., & Milinković, M. (2023). Regulation of yield, fruit size, and leaf mineral nutrients of the šumadinka sour cherry cultivar with help of rootstocks. *Journal of Plant Growth Regulation, 42*, 5587–5599.

Natić, M., Dabić Zagorac, D., Ćirić, I., Meland, M., Rabrenović, B., & Fotirić Akšić, M. (2020). Chapter 56 - cold pressed oils from genus *Prunus*. In M. F. Ramadan (Ed.), *Cold pressed oils: Green technology, bioactive compounds, functionality, and applications* (pp. 637–658). Academic Press.

Navruz, A., Türkyılmaz, M., & Özkan, M. (2016). Colour stabilities of sour cherry juice concentrates enhanced with gallic acid and various plant extracts during storage. *Food Chem, 197*, 150–160.

Olden, E. J., & Nybom, N. (1968). On the origin of *Prunus cerasus* L. *Hereditas, 59*, 327–345.

Papenfuss, K. A., & Black, B. L. (2010). Regulated deficit irrigation of montmorency tart cherry. *Hort Science, 45*, 1437–1440.

Papp, N., Szilvássy, B., Abrankó, L., Szabó, T., Pfeiffer, P., Szabó, Z., et al. (2010). Main quality attributes and antioxidants in Hungarian sour cherries: Identification of genotypes with enhanced functional properties. *Int J Food Sci Technol, 45*(2), 395–402.

Pedisić, S., Dragovic-Uzelac, V., Levaj, B., & Škevin, D. (2010). Effect of maturity and geographical region on anthocyanin content of sour cherries (*Prunus cerasus* var. *marasca*). *Food Technology and Biotechnology, 48*(1), 86–93.

Pedisić, S., Dragovic-Uzelac, V., Levaj, B., Škevin, D., & Skendrović Babojelić, M. (2009). Color parameters and total anthocyanins of sour cherries (*Prunus cerasus* L.) during ripening. *Agriculturae Conspectus Scientificus, 74*(3), 259–262.

Pole, V., Feldmane, D., Ruisa, S., & Butac, M. (2017). The effect of some soil properties on nutrient movement in sour cherry (*Prunus cerasus*) leaves. *Acta Hortic, 1150*, 311–316.

Poll, L., Petersen, M., & Nielsen, G. S. (2003). Influence of harvest year and harvest time on soluble solids, titrateable acid, anthocyanin content and aroma components in sour cherry (*Prunus cerasus* L. cv. Stevnsbær). *European Food Research and Technology, 216*, 212–216.

Popa, V. M., Misca, C., Bordean, D., Raba, D. N., Stef, D., & Dumbrava, D. (2011). Characterization of sour cherries (*Prunus cerasus*) kernel oil cultivars from Banat. *Journal of Agroalimentary Processes and Technologies, 17*(4), 398–401.

Ravanfar, R., Niakousari, M., & Maftoonazad, N. (2014). Postharvest sour cherry quality and safety maintenance by exposure to Hot- water or treatment with fresh Aloe vera gel. *J Food Sci Technol, 51*(10), 2872–2876. https://doi.org/10.1007/s13197-012-0767-z

Roeva, T., Leonicheva, E., Leonteva, L., & Stolyarov, M. (2022). Potassium dynamics in orchard soil and potassium status of sour cherry trees affected by soil nutritional conditions. *Journal of Central European Agriculture, 23*(1), 103–113. https://doi.org/10.5513/JCEA01/23.1.3313

Rutkowski, K., & Łysiak, G. P. (2023). Influence of mulching on replantation disease in sour cherry orchard. *Agriculture, 13*(8), 1587. https://doi.org/10.3390/agriculture13081587

Rutkowski, K., & Łysiak, G. P. (2022). Weather conditions, orchard age and nitrogen fertilization influences yield and quality of Łutówka sour cherry fruit. *Agriculture, 12*(12), 2008. https://doi.org/10.3390/agriculture12122008

Sadowski, A., & Jadczuk, E. (2001). Results of 11-year N-fertiliser trial in a sour cherry orchard. *Acta Hortic., 564*, 278–294.

Sadowski, A., Nurzyński, J., Pacholak, E., & Smolarz, K. (1990). Określenie potrzeb nawożenia roślin sadowniczych. *II Zasady, Liczby Graniczne i Dawki Nawożenia. Instrukcja Upowszechnieniowa nr 3. SGGW, s. 25.*

Sanchez, J. E., Edson, C. E., Bird, G. W., Whalon, M. E., Willson, T. C., Harwood, R. R., et al. (2003). Orchard floor and nitrogen management influences soil and water quality and tart cherry yields. *Journal of the American Society for Horticultural Science, 128*(2), 277–284.

Sass-Kiss, A., Kiss, J., Milotay, P., Kerek, M. M., & Toth-Markus, M. (2005). Differences in anthocyanin and carotenoid content of fruits and vegetables. *Food Research International, 38*, 1023–1029.

Scalisi, A., O'Connell, M. G., Islam, M. S., & Goodwin, I. (2022). A fruit colour development index (CDI) to support harvest time decisions in peach and nectarine orchards. *Horticulturae, 8*(5), 459. https://doi.org/10.3390/horticulturae8050459

Scholz, H., & Scholz, I. (1995). Prunus. In H. Scholz (Ed.), *Gustav hegi illustrated flora of central Europe. Volume IV, Part 2B Spermatophyta: Angiospermae: Dicotyledones 2(3* (pp. 446–510). Blackwell Wissenschafts-Verlag Berlin – Wien.

Schuster, M., & Schreiber, H. (2000). Genome investigation in sour cherry, *P. cerasus* L. *Acta Hortic, 538*, 375–379. https://doi.org/10.17660/actahortic.2000.538.66

Sezer, D. B., Ahmed, J., Sumnu, G., & Sahin, S. (2021). Green processing of sour cherry (*Prunus cerasus* L.) pomace: Process optimization for the modification of dietary fibers and property measurements. *Journal of Food Measurement and Characterization, 15*, 3015–3025.

Sokół-Łetowska, A., Kucharska, A. Z., Hodun, G., & Gołba, M. (2020). Chemical composition of 21 cultivars of sour cherry (*Prunus cerasus*) fruit cultivated in Poland. *Molecules, 25*, 4587.

Stösser, R., Rasmussen, H. P., & Bukovac, M. J. (1969). A histological study of abscission layer formation in cherry fruits during maturation. *J Amer Soc Hort Sci, 94*, 239–243.

Sundelin, T., Stromeng, G. M., Gjaerum, H. B., Amby, D. B., Orstad, K., Jensen, B., et al. (2015). A revision of the history of the *Colletotrichum acutatum* species complex in the Nordic countries based on herbarium specimens. *FEMS Microbiolgy Letters, 362*(16), fnv130. https://doi.org/10.1093/femsle/fnv130

Tenuta, M. C., Artoni, E., Fava, P., Bignami, C., & Licciardello, F. (2023). Shelf life extension and nutritional quality preservation of sour cherries through high pressure processing. *Foods, 12*, 342.

Toydemir, G., Capanoglu, E., Kamiloglu, S., Boyacioglu, D., De Vos, R. C., Hall, R. D., et al. (2013). Changes in sour cherry (*Prunus cerasus* L.) antioxidants during nectar processing and *in vitro* gastrointestinal digestion. *J Funct Foods, 5*, 1402–1413.

Vittrup Christensen, J. (1986). Evaluation of characteristics of 18 sour cherry cultivars. *Danish Journal of Plant and Soil Science, 90*, 339–348.

Wöhner, T. W., Emeriewen, O. F., Wittenberg, A. H. J., Nijbroek, K., Wang, R. P., Blom, E.-J., et al. (2023). The structure of the tetraploid sour cherry Schattenmorelle (*Prunus cerasus* L.) genome reveals insights into its segmental allopolyploid nature. *Front Plant Sci, 14*, Article 1284478.

Wójcik, P., & Morgaś, H. (2015). Impact of postharvest sprays of nitrogen, boron and zinc on nutrition, reproductive response and fruit quality of Schattenmorelle tart cherries. *Journal of Plant Nutrition, 38*, 1456–1468.

Wójcik, P., & Wawrzyńczak, P. (2014). Effect of preharvest sprays of calcium on cracking and Schattenmorelle sour cherry fruit quality harvested mechanically. *Journal of Plant Nutrition, 37*(9), 1487–1497.

Woźniak, Ł., Marszałek, K., & Skąpska, S. (2016). Extraction of phenolic compounds from sour cherry pomace with supercritical carbon dioxide: Impact of process parameters on the composition and antioxidant properties of extracts. *Separation Science and Technology, 51*, 1472–1479.

Wojdyło, A., Nowicka, P., Laskowski, P., & Oszmiański, J. (2014). Evaluation of sour cherry (*Prunus cerasus* L.) fruits for their polyphenol content, antioxidant properties, and nutritional components. *Journal of Agricultural and Food Chemistry, 62*, 12332–12345.

Yıldız, D., Benzer Gürel, D., Çağındı, Ö., & Kayaardı, S. (2022). Heat treatment and microwave applications on homemade sour cherry juice: The effect on anthocyanin content and some physicochemical properties. *Current Plant Biology, 29*, Article 100242.

Yarilgac, T., Balta, M. F., Ozrenk, K., & Muradoglu, F. (2005). Indigenous sour cherry (*Prunus cerasus* L.) germplasm of Lake Van basin. *Asian Journal of Plant Sciences, 4*, 558–561.

CHAPTER

Apricot

6

Juan Alfonso Salazar[1], Germán Ortuño Hernández[1], Jill Stanley[2], Pavlina Drogoudi[3]

[1]*Department of Plant Breeding, Centro de Edafología y BiologíaAplicada del Segura (CEBAS-CSIC), PO Box 164, E-30100 Espinardo (Murcia) Spain;* [2]*The New Zealand Institute for Plant & Food Research Ltd, 990 Earnscleugh, RD 1, Alexandra 9391, New Zealand;* [3]*Hellenic Agricultural Organization (ELGO)—"DIMITRA," Institute of Plant Breeding and Genetic Resources, Department of Deciduous Fruit Trees, 59200 Naoussa, Greece*

Chapter outline

Temperate Tree Fruits and Nuts. https://doi.org/10.1016/B978-0-443-29137-1.00006-2

1. Origin and spread

Apricot *(Prunus armeniaca* L.*)* is one of the oldest fruits cultivated by humans, with a history spanning thousands of years. Although it is now widely cultivated across the globe, its origins are deeply rooted in the mountainous regions of Central Asia, encompassing parts of China, Kazakhstan, Kyrgyzstan, and Uzbekistan. These areas, characterized by extreme continental climates with hot summers and harsh winters, provided ideal conditions for the apricot's adaptation. As a result, the apricot is remarkably resilient, capable of thriving under climatic fluctuations.

The exact origin of the apricot has been a subject of debate for many years. While Armenia has historically been associated with the fruit—an association reflected in its scientific name—more recent botanical and genetic research points to Central Asia as the true center of origin for the apricot. Studies suggest that the apricot was likely domesticated in this region over 4000 years ago, and from there, it spread in multiple directions: westward, across the mountains into Europe, and eastward, reaching China. In China, the apricot has played an important role not only in agriculture but also in traditional medicine, where its use has been documented since the Zhou Dynasty (1046–256 BCE). It was valued for its purported digestive benefits and ability to alleviate respiratory conditions.

In addition to its historical and cultural significance, the apricot is now an economically important fruit on a global scale. The major global producers include countries such as Turkey, Uzbekistan, and Iran, which lead the world in apricot cultivation. However, to fully understand the contemporary relevance of this fruit, it is crucial to examine the pathways through which its cultivation spread from its origin, primarily via ancient trade and migration routes. This dispersal not only underscores the apricot's remarkable adaptability to diverse climates and terrains but also highlights the rich cultural exchanges between the civilizations that have cultivated and consumed it throughout history.

1.1 Evolutionary origin and spread of the apricot

Apricot (*Prunus armeniaca* Linnaeus syn. Armeniaca vulgaris Lamarck) belongs to the family Rosaceae, subfamily Amygdaloideae, within the genus Prunus, subgenus Prunus, and section Armeniaca (Lam.) Koch (Figure 1). This species is commercially cultivated worldwide (Faust et al., 1998).

The number of species within the Armeniaca section varies between 3 and 12, depending on the taxonomist. However, seven distinct species are generally recognized: *P. armeniaca* L., the common European apricot; *P. ansu* (Maxim.) Kost., the Ansu apricot; *P. holoserica* Batal., the Tibetan apricot (originating from Northern China and known for its frost tolerance); *P. brigantina* Vill., the Briançon apricot or alpine plum; *P. mume* (Sieb.) Sieb. & Zucc., the Japanese apricot (native to humid regions of China and resistant to fungal diseases); *P. mandshurica* (Maxim.), the Manchurian apricot; and *P. sibirica* L., the Siberian apricot (from Northern China, distinguished by its cold tolerance). Two additional species are frequently recognized: *Prunus × dasycarpa* Ehrh., known as the black or purple apricot, and *P. sibirica var. davidiana* (Carrière), which appears to be of hybrid origin (Shi et al., 2013).

Several phylogenetic studies have explored the dispersal pathways of apricot germplasm from its centers of origin to the regions where it is currently cultivated. Vavilov (1951) identified three key centers of apricot origin (Figure 2): two primary centers of diversification in China (Gansu Province, the mountainous regions of Northeastern and Central China, and Northeastern Tibet) and Central Asia (Afghanistan, Northwestern India and Pakistan, Kashmir, Tajikistan, Uzbekistan, Xinjiang Province in China, and the western Tian-Shan Mountains). Additionally, the Near East was proposed as a secondary center of diversification, encompassing the mountains west of the Caspian Sea, including the Caucasus, Georgia,

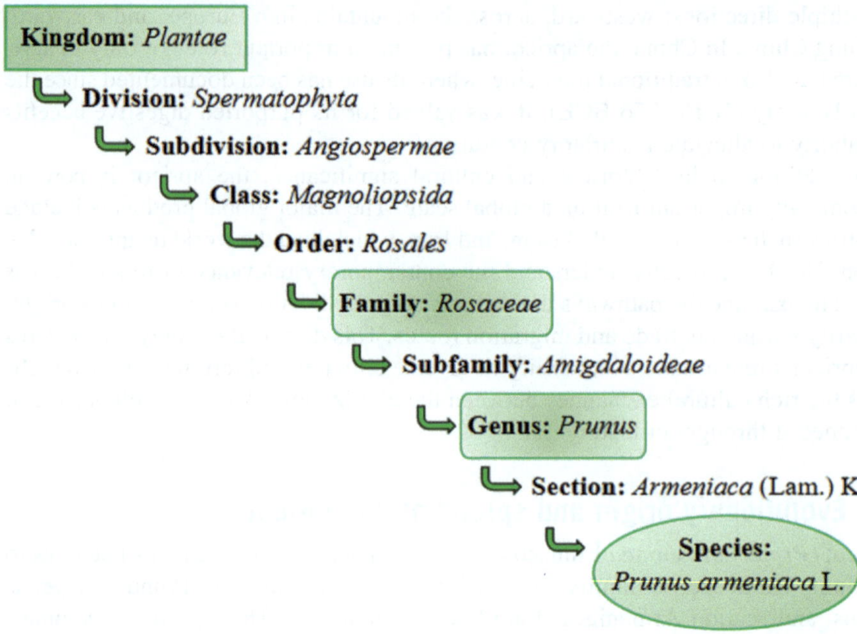

FIGURE 1

Taxonomic classification of the apricot.

Azerbaijan, Armenia, Turkey, and Northern Iran (Vavilov & Dorofeev, 1992). In China, apricot was among more than 30 temperate fruit crops listed, and two subgroups were proposed as additional centers of diversification: the Northern Chinese group (Siberian and Manchurian apricots) and the Eastern Chinese group (Ansu apricot).

Cultivation of *Prunus armeniaca* in China dates back over 5000 years to the Tertiary period. Central Asia is considered the primary center of apricot propagation, particularly in the mountainous regions extending from the Southern Tian-Shan to the Hindu Kush and Kashmir, covering Afghanistan, Tajikistan, and Uzbekistan. This region was crucial for the early domestication of the apricot, while a secondary center emerged in the mountains of Northeastern Dagestan, in the Near East, where the species was widely cultivated. This geographical location, serving as an intermediary between the area of domestication and the distribution of wild species, played a pivotal role in the species' expansion (Miller, 2014).

Spread of apricot to Europe was closely tied to ancient trade routes, with the Silk Road being a significant factor in its dissemination. This extensive network connected Central Asia with Persia, the Mediterranean, and Europe, facilitating the exchange of goods, including apricots, as well as cultural and agricultural knowledge. Along this route, apricot spread westward, reaching the Mediterranean

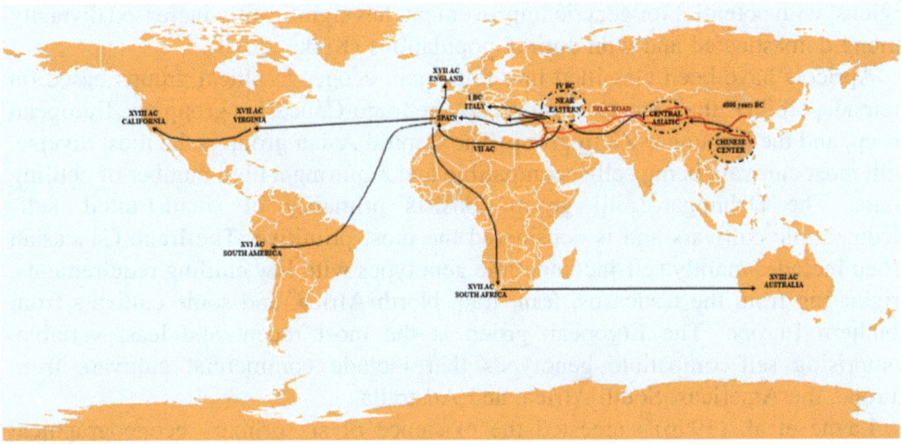

FIGURE 2

Historical spread of apricot through the Silk Road and other global trade routes (Silk Road highlighted in *red*).

Source: Vavilov, 1951.

basin from Iran to ancient Greece and Rome. It became an important crop in Syria and Turkey, where it was known as the "Armenian apple." Additionally, Roman-Persian military campaigns and the expeditions of Alexander the Great into the Turkestan region and the Fergana Valley in the 4th century BCE further contributed to its expansion.

In a subsequent wave of diffusion, apricot was introduced to Western Europe primarily from Syria, Iran, and Armenia by the Arabs in the 1st century BCE, who brought it to the Iberian Peninsula. From there, its cultivation spread across Europe. By the 16th century, apricot had been introduced to South America via maritime and overland trade routes. In the 17th century, its cultivation expanded to England, South Africa, and the United States (Virginia). Finally, in the 18th century, it was introduced to California by Spanish missionaries, and from South Africa, it reached Australia (Vavilov, 1951; Faust et al., 1998). Global spread of apricot, facilitated by trade routes and colonization, underscores the significant role of cultural and commercial interactions in the dissemination of agricultural crops.

1.2 Domestication of the apricot

The development of new apricot cultivars has primarily occurred on the fringes of the species' early growth areas, resulting from the hybridization of indigenous genotypes from different groups. The interaction between cultivars from different regions, with potential for genetic improvement, has significantly increased diversity among domesticated and wild apricot populations (Krška, 2018).

Apricots have been classified into four main ecogeographical groups based on their adaptability: the Central Asian group, the Irano-Caucasian group, the European group, and the Dzhungar-Zailij group. The Central Asian group is the most diverse, with most cultivars being self-incompatible and requiring a high number of chilling hours. The Dzhungar-Zailij group consists primarily of small-fruited, self-incompatible cultivars and is considered the most primitive. The Irano-Caucasian group includes mainly self-incompatible genotypes with low chilling requirements, originating from the Caucasus, Iran, Iraq, North Africa, and some cultivars from Southern Europe. The European group is the most recent and least variable, comprising self-compatible genotypes that include commercial cultivars from Europe, the Americas, South Africa, and Australia.

Layne et al. (1996) suggested the existence of six primary ecogeographical groups: Central Asian, East Chinese, North Chinese, Dzhungar-Zailij, Irano-Caucasian, and European. However, this classification has become more complex due to the introduction of new cultivars resulting from the hybridization of genotypes across these groups (Faust et al., 1998).

Maghuly et al. (2006) further subdivided the European group into Eastern and Western Europe and introduced a new North American group. Although the European and North American groups are of recent origin and exhibit lower variability, most new apricot cultivars developed through breeding programs have arisen from crosses within these cultivar groups (Alburquerque et al., 2008).

Genetic diversity studies support the hypothesis that species related to Asia could be the source of resistance to the plum pox virus (PPV) found in North American cultivars (Badenes et al., 1996). Meanwhile, morphological, phenological, and fruit quality traits studied in the European ecogeographical group indicate that native cultivars from Spain and Tunisia are valuable for early ripening traits, although the genetic variability was lower than anticipated (Badenes et al., 1998). Despite these similarities, human selection has influenced different regions of the genome in apricots cultivated in Europe and China, meaning that distinct genomic modifications may produce similar adaptive phenotypes. In both groups, genes affected by domestication are associated with perennial plant life cycles, fruit quality, and disease resistance. The large population sizes, long life cycle of perennial crops, and resulting gene flow between wild and cultivated populations have allowed domesticated apricots to maintain high genetic diversity, with relatively low proportions of the genome affected by selection (0.42% in European apricots and 0.22% in Chinese apricots) (Groppi et al., 2021).

Genetic diversity analysis, geographic origin, and agronomic traits have been used to define four new groups based on these characteristics (Hagen et al., 2002): Diversification, Geographic Adaptability, Continental Europe, and Mediterranean Basin. For apricots, each cultivar is typically confined to a specific geographic region with unique ecological conditions. Apricot cultivation spans temperate and subtropical climates, from the cold winters of Siberia to the subtropical regions of North Africa, and from the desert areas of Central Asia to the humid zones of Japan and eastern China. In temperate regions, the expansion of apricot production is often restricted by spring frost damage during flowering and the lack of cultivars adapted to local chilling requirements. High precipitation and humidity, especially during flowering or harvest periods, pose significant challenges due to fungal diseases that can destroy flowers and shoots or cause fruit rot and cracking (Bird et al., 1995).

In the context of climate change, apricot cultivation areas are expected to shift significantly in the coming years. Traditional growing regions may face less favorable conditions due to rising temperatures and reduced water availability, pushing farmers to either adapt their agricultural practices or relocate to new areas. This geographical redistribution of apricot cultivation will be critical for ensuring long-term sustainability and production in the face of a warming climate.

2. Economic importance

The world apricot production ranks eighth among the rest of the deciduous fruit tree crops, being lower than apples, grapes, peaches, pears, plum, and strawberries, and similar to walnuts (95.8, 74.9, 26.4, 26.3, 12.4, 9.6, 3.9 MT, respectively), according to FAOSTATS for 2022.

The mean total world production of apricot is ~3.8 million tons, and the harvesting area is ~580,000 ha (FAOSTAT, mean for 2020-2022). Half of the world's apricot harvesting areas are in countries bordering the Mediterranean Sea, with Turkey being the largest producer mainly for dried fruit (Table 1). Apricot is also produced in the Middle East, in the Caucasus, Central Asia (with Uzbekistan as the second largest producer), and China. In Europe, Italy is the largest apricot producer. Over the last 10 years, the biggest change in plantings was an important increase in the planting areas in Afghanistan, Greece, and Armenia (169%, 68%, and 38%, respectively) (data compared between 2020-2022 and 2010-2012).

The apricot trade remains far lower compared to apples, pears, peaches, and nectarines. The main exporting countries of apricot are Turkey, followed by Spain and Italy. Dried apricots are exported mainly by Turkey (57%), followed by Tajikistan (16%) and Afghanistan (12%). The main importing countries of apricot are Russia (29%), followed by Germany (19%) and Austria (5%).

3. Botany and relevant species

Apricots are cultivated for their fruits and flowers, and sometimes their kernels. They belong to the Armeniaca section of the family Rosaceae, subfamily Prunoideae. *Prunus armeniaca* L. refers to both the wild progenitor and the cultivated species (also called "common apricot").

Table 1 Heatmap showing the mean harvested areas (ha) and production (million tons-MT) of the main apricot producing countries for the years 2020-2022 and evolution (% change) compared with the years 2010-2012 (FAOSTAT).

	Area harvested (Ha) 2020-2022	2020-2022 vs. 2010-2012 (%)	Production (MT) in 2020-2022	2020-2022 vs. 2010-2012 (%)
Turkey	136,493	23	812	31
Uzbekistan	39,934	3	468	24
Iran	50,715	-3	291	-16
Italy	17,667	-9	198	-22
Algeria	28,787	-41	194	-23
Afghanistan	22,404	169	170	134
Pakistan	14,667	-49	153	-18
Spain	19,217	4	108	14
Greece	10,187	68	105	50
Japan	13,800	-17	91	-6
France	11,773	-11	89	-43
Armenia	10,987	38	85	63
Morocco	10,137	-18	80	-43
Russian Federation	13,820	26	79	41
China	21,048	0	71	-20
Egypt	4413	-29	68	-29
Ukraine	6900	-26	59	-32
Kyrgyzstan	8329	8	26	16

Natural populations of *P. armeniaca* still occur but only in Central Asia. Closely related species is *P. mume* Sidb. and Zucc., which is primarily cultivated for its flowers and secondly for its fruits, consumed as salted and smoked. There are four other closely related species, such as *P. sibirica* L., *P. mandshurica* (Maxim.) Koehne, *P. holosericea* and *P. brigantina*; the first three are endemic in Eastern Asia (mostly China), while the more distant *P. brigantina* Vill. occurs in the French and Italian Alps (Croppi et al., 2021). All apricot related species are diploid ($2n = 16$, $\times = 8$), with relatively small genome sizes (~220–230 Mbp).

Species within the *Prunus* genus exhibit a high degree of synteny (Dondini et al., 2007), enabling interspecific hybridization. In recent years, new interspecific hybrids have been developed using apricot species as progenitors through artificial cross-pollination. Hybrids of apricot are as follows (Lichou and Jay, 2012):

- **Aprium:** apricot x plum, with dominant traits being from apricot by three-quarters
- **Plumcot:** plum x apricot, with appearance closer to plum

- **Pluot:** plum complex hybrid
- **Nectarcot:** nectarine x apricot
- **Peacotum:** peach x apricot x plum

Major apricot botanical traits and their implications for cultivation are discussed below:

- Apricots are small- to medium-sized deciduous trees, and cultivars differ in their branch habit, producing growth upright or spread. Therefore, the distance between trees in a planting design need to consider the cultivar's natural branch habit; when having upright branch habit, the distance between trees can be closer. In apricots, a multibranch vase system usually has five or six branches, whereas in peach it is limited to three or four branches.

Flowers are solitary, 2 to 3 cm, pink to white, with deep red sepals appearing before leaves. They are formed on 1-year-old wood and more often on short spurs. Spurs are productive for several (3—5) years. The highest-quality fruit is borne on younger spurs. Pruning needs to take into consideration the cultivar's natural habit of producing flowers on **1-year-old wood or on short spurs.**

Apricot cultivars have chill and heat requirements for **breaking dormancy**, and vegetative buds require less chilling than reproductive (flower) buds. There are major differences in chill requirements among commercial cultivars that need to be taken into consideration for successful cultivation in an area (Pantelidis and Drogoudi, 2023). Differences in chill requirements among cultivars is the major reason that apricot can thrive in diverse agroclimatic conditions.

Cultivars are **self-fertile, semi-fertile, or auto-sterile,** and there are also **self-incompatible cultivars**. Bloom needs to be synchronized between the cross-pollinating cultivars, and the time of bloom is influenced by winter temperatures and the cultivars chill and heat requirements for breaking dormancy. Therefore, in specific agroclimatic conditions a major factor affecting adaptability is the choice of cultivar combination. Moreover, the introduction of honeybees or bumblebees in orchards is highly recommended, especially in years when less favorable climate conditions occur during bloom. Bumblebees fly in lower temperatures than honeybees (less than 10°C).

Apricots produce more flowers than are needed to ensure the production of an adequate crop. Therefore, apricot usually requires **thinning** to ensure adequate fruit size. Floral initiation occurs in summer (late May or June) and most of the flowers which set into fruit are produced on spurs. **Cultivation practices and weather conditions in the previous season** may affect the next year's flower production.

There is a great variation in apricot **fruit appearance and chemical traits**. Fruit from apricot cultivars differ in size, shape, skin and flesh color, traits described in the Community Plant Variety Office (CPVO) for apricot. The fruit skin color close at maturity can vary as: (a) the skin ground color (from white to dark orange), (b) amount of over color, (c) the hue of over color (from orange red to purple), and (d) the intensity and distribution of coloration. The adherence of stone to flesh is a trait influencing the fruit postharvest performance. Among cultivars, apricot also differ in sweetness, acidity, and skin thickness. Consumers demand large fruit size, high coloration, high sweetness, and balanced acidity. Cultivars with early ripening time usually produce fruit with less sweetness due to the short developmental period from bloom to harvest (Drogoudi et al., 2008). Thus, in very early cultivars, fruit usually lack optimum organoleptic traits, having low sweetness and high acidity. The apricot kernels may produce cyanide depending on the cultivar. There is a claim that bitter apricot kernels can help cure cancer, however, this claim has not been scientifically proven.

4. Breeding and propagation

4.1 Background

The genetic plant breeding has its roots in ancient times as early as 8000 BCE tracing back to the earliest stages of agriculture (Fabian et al., 2001). It was probably after the first domestications of cereal grains, when agriculture started to be able to discern among interesting characteristics of plants, saving seeds from the best candidate plants for future new crops. Subsequently, directed pollination crosses with selected pollen in plants began to be carried out at the end of the 19th and beginning of the 20th century, marking the onset of the era of plant genetic improvement. Pioneers in this field, such as Gregor Mendel and later other researchers, laid the groundwork for the systematic application of genetics in crop improvement. As the principles of genetic inheritance became better understood, controlled crosses were initiated to select and enhance specific traits in plants, such as yield, disease resistance, and fruit quality. This approach has evolved and been refined over time, becoming an integral part of modern agriculture. Therefore, from the last century the best plant material was selected according to its expressed phenotype; used in fruit trees, this pollen as a parent for new crosses obtained new selections including the desired traits, which nowadays is known as classic breeding (Badenes et al., 2000).

Novel approaches to enhance *Prunus* species, encompassing advancements in germplasm, molecular marker development, and propagation techniques have arisen in recent times. Within germplasm enhancement, efforts are directed toward integrating genes from related *Prunus* species to address several traits such as self-compatibility, growth habit, resistance to PPV, drought resistance, and fruit quality.

Promising propagation methods encompass both in-vitro techniques for plant material assessment and in-vivo micrograft techniques enabling the early propagation of genotypes at high risk. Additionally, nurturing seedlings in controlled environments, which includes inducing an artificial rest period in cold chambers, presents a viable strategy for obtaining robustly growing plants throughout the year. On the other hand, molecular markers have emerged as indispensable tools in *Prunus* breeding endeavors. These molecular markers have been utilized for genetic profiling of germplasm, delineating genetic relationships among cultivars and species, and constructing genetic maps. Analytical methodologies for marker-assisted selection entail employing mapping populations with desired traits and conducting bulk segregant analysis (Martínez-Gómez et al., 2003; Zhebentyayeva et al., 2012).

4.2 Principles of apricot breeding and significance of propagation

To meet the current needs of farmers, industry and consumers, specific traits are the aim of breeding programs in apricot. For farmers, the trait of self-fertility is of paramount importance as it is related to productivity. Resistance to diseases (mainly PPV and others) is also a trait with high importance for introgression into new varieties due to devastating damage in Europe. Due to the recent climate change challenges, low chilling and high heat requirements are also the aims of breeding programs worldwide. Farmers also require traits such as high-quality fruit that can stand handling and reduced input cost.

It is crucial to consider covering the breadth of the current production calendar in the Northern Hemisphere (May-June) and even extending it (April-July), which is an important trait for growers and the industry. The industry also requires good postharvest performance. For the consumer, traits related to fruit quality, such as size, high coloration, and good organoleptic features related to high sweetness, balanced acidity, thin skin, and high aroma, are important.

Many apricot breeding programs nowadays focus on traits related to resistance to abiotic stresses such as drought or frost risk (Gao et al., 2024); biotic stresses (fungal diseases, viruses, bacteriosis, etc.) linked to the plant health (Rubio et al., 2023); phenological traits (chilling requirements, flowering, fruit development, etc.) linked to early or late character (Campoy et al., 2012); pomological traits (physical, biochemical, sensory traits, etc.) related to fruit quality (Ruiz et al., 2005ab), and also postharvest conditions (Ortuño-Hernández et al., 2024ab). In this context, breeding is also oriented toward producing rootstocks on one hand and cultivars on the other.

Obviously, plant propagation means are evidently necessary to implant and/or extend certain commercially interesting cultivars in different geographical areas according to need. The most natural propagation method for apricot trees is seeds. The interior of the seed contains the embryo resulting from the fertilization of the ovule (female) by pollen (male) from the same flower (self-pollination) or from flowers of other apricot trees, giving rise to new intraspecific hybrids naturally. It could also happen that pollen from other *Prunus* species closely related phylogenetically is present, leading to interspecific hybrids (Nicolás-Almansa et al., 2023). This way, the species ensures a certain genetic diversity naturally.

4.3 Propagation: methods and role of propagation

Propagation in apricots *(Prunus armeniaca)* involves the reproduction of new plants, and it plays a crucial role in ensuring the availability, distribution, and preservation of desirable apricot cultivars. Several methods of propagation can be employed for apricots, each with its own advantages and considerations. The primary methods include seed propagation, vegetative propagation methods including grafting or budding, and in-vitro propagation. Let's explore these methods and the role of propagation in more detail (Figure 3).

4.3.1 Seed propagation

Apricots can be grown from seeds, usually extracted from mature fruits. The seeds are cleaned, dried, stratified at low-temperature (5°C), and then planted in a suitable growing medium (Seif El-Yazal et al., 2021). The main aim of this propagation is to introduce genetic variability considering the genetic background of the parents, and this method is the most suitable for genetics breeding programs to introduce the most desirable traits in future seedlings to obtain new improved varieties.

FIGURE 3

Summary of different propagation methods. (A) seed propagation, considering seed stratification, seeding in pots, and plantation on the field; (B) vegetative propagation by buds or grafted scions; (C) in-vitro cultivation by plant explants. (*Bottom left corner:* Pérez-Caselles et al., 2023).

4.3.2 Vegetative propagation

In this kind of propagation, we obtain identical genotypes as clones. One method would be layering, where a low-hanging branch is bent down to the ground, wounded, and covered with soil. Once roots develop, the new plant is separated from the parent. Another way would be through cuttings; this method involves planting cut segments of branches into the soil, where they develop roots and grow into new plants. However, these methods are not recommended due to its very low success as both difficult and inefficient. For this reason, the most useful is grafting or budding (Mendelné et al., 2023).

Grafting involves attaching a scion (a part of the desired apricot cultivar) onto a rootstock (a different apricot plant with good root characteristics). Budding is a similar process, but it involves grafting a bud rather than a scion. Therefore, the main aim of grafting and budding is to look for the best combination of rootstock and cultivar to ensure the desired qualities of both varieties (rootstock cultivar-grafted). These vegetative propagation methods provide clones of the parent plant, ensuring the retention of specific traits, particularly when a genetically identical plant is desired.

4.3.3 In-vitro propagation

The apricot is a species that is difficult to propagate and root vegetatively. Therefore, as we have previously seen, it has so far been propagated by grafting onto heterogeneous rootstocks. Thus, this technique could be suitable to generate an apricot clonal rootstock (Nas et al., 2023). Micropropagation in apricot through in-vitro culture is a method used to propagate apricot plants using tissue culture techniques in a controlled laboratory environment (Yıldırım et al., 2011). It involves taking explants from a selected mother plant and culturing them in a growth medium containing nutrients and plant hormones in sterile culture vessels.

Micropropagation represents the primary and predominant commercial utilization of tissue culture methodologies. It is extensively applied across a diverse range of herbaceous and woody plant species, utilizing techniques such as enhanced axillary bud formation, organogenesis, and/or somatic embryogenesis. Propagation in-vitro of pathogen-free, elite, selected, or difficult-to-propagate genotypes progresses through four distinct phases: (1) establishment of explants, (2) regeneration and proliferation, (3) acclimatization and rooting, and (4) final transplantation *ex-vitro*.

The explants are selected to contain meristematic cells, which are cells that can differentiate into different types of tissues and therefore regenerate complete plants. These explants are sterilized to remove any microbial contaminants and placed in the appropriate growth medium, where their growth and development are stimulated.

During the micropropagation process, environmental conditions such as temperature, light, and humidity are carefully controlled to optimize the growth of the explants and minimize contamination. Over time, the explants develop shoots that can be separated and transplanted to another growth medium for rooting. Once the seedlings have adequate roots, they can be transferred to soil for acclimatization and growth in greenhouse or open field conditions. This method allows for the rapid production of a large number of plants that are genetically identical to the mother plant, which is useful for mass propagation of selected cultivars and germplasm conservation.

Having said that, each propagation method that we use depends on our priorities. Therefore, if we pretend to preserve a bank germplasm in-vivo of our cultivars, we will use grafting or budding on a rootstock. We will use the same method with different purpose, when we try to obtain an orchard cultivated with a specific cultivar to ensure a desired fruit or product with a commercial goal. On the other hand, in-vitro would be the only way to replicate an identical rootstock, which would be beneficial to ensure certain interesting characteristics that may favor the grafted cultivar.

4.4 Rootstocks and selection criteria

It is important to note that the selection of the most suitable rootstock is a very important decision, because the success of the grafted cultivar depends on it and definitely the success of the cultivation in commercial terms. Some of the commercial rootstocks and their characteristics are shown in Table 2.

4.4.1 Rootstock-graft compatibility

Ensuring rootstock-graft compatibility is paramount. The rootstock must be compatible with the apricot cultivar intended for grafting, as not all rootstocks are universally suitable. Anatomical compatibility is crucial, with two main types of graft incompatibility identified: "translocated" and "localized." Translocated incompatibility manifests through yellowing leaves and leaf loss, while localized incompatibility arises from anatomical irregularities at the graft union. Both issues impede normal growth and can lead to plantation failure. Therefore, rigorous testing of new rootstocks is essential to confirm suitability for specific purposes (Reig et al., 2018). Typically, seedling rootstocks of the same genus and species are preferred due to their favorable genetic compatibility and ability to impart similar vigor to the cultivar.

4.4.2 Edaphoclimatic conditions

Soil adaptability and climatic conditions are also important; the rootstock must be suitable for the type of soil where the tree will be planted. This includes factors such as soil texture (sandy, clayey, calcareous, acidic, etc.), pH, and drainage capacity, as well as tolerance to frost, drought, and high temperatures in the climatic area. For this reason, it is important to note that each region, and even within the same region, there are different soils and climatic conditions that influence the choice of a suitable rootstock. An example of a rootstock resistant to difficult soil conditions is the Myrobalan rootstock *(P. cerasifera)*. However, this type of rootstock can induce strong growth, leading to excessive vigor (Remzi et al., 2022).

Table 2 Summary of some of the most commercial rootstocks and their characteristics available in the market.

Rootstock	Species	Origin	Propagation	Characteristics
Real Fino	*Prunus armeniaca*	CSIC, Spain	Seed	Suitable for heavy and calcareous soils with a high pH.
Montclar	*Prunus persica*	INRA, France	Seed	Rapid fruit production, quite resistant to chlorosis, and very uniform.
GF-305	*Prunus persica*	INRA, France	Seed	Well-developed root system, and vigorous growth. It poorly tolerates heavy and poorly aerated soils and is sensitive to nematodes.
Lovell	*Prunus persica*	G.W. Thissell, U.S.	Seed	Suffers in waterlogged soils. Poor tolerance to calcareous soil. Susceptible to root-knot nematode, resistant or tolerant to bacterial canker, susceptible to *Phytophthora* and *Armillaria*.
Torinel	*Prunus domestica*	INRA, France	Seed	Medium vigor, less compared to Myrobalan, and an earlier production start compared to the latter. It adapts well to clayey soils and areas prone to waterlogging.
Adesoto	*Prunus insititia*	CSIC, Spain	Seed	It confers good vigor, uniformity, good productivity, increased fruit size, and early ripening to the plant. Versatile and adapts to water-scarce, clayey, and calcareous soils. Tolerant to *Armillaria*. Produces suckers.
Nemaguard	*P. persica x P. davidiana*	USDA, U.S.	Seed	Excellent for well-drained soils. In poorly-drained soil, plant on a hill - very susceptible to wet feet. Poor tolerance to calcareous soil.

Continued

Table 2 Summary of some of the most commercial rootstocks and their characteristics available in the market.—*cont'd*

Rootstock	Species	Origin	Propagation	Characteristics
Nemared	*P. persica* x *P. davidiana*	USDA, U.S.	Seed	Resistant or tolerant to root-knot nematode, partly resistant to lesion nematode. Susceptible to ring nematode, bacterial canker, *Armillaria*, and *Verticillium wilt*. Partly resistant to crown gall.
Myrobalan B	*Myrobolan (Prunus cerasifera)*	EM, UK	Seed	Wide adaptability and is tolerant to wet soil. Good crown gall resistance, fair phytophthora resistance, low bacterial canker resistance, low root-knot nematode resistance, low oak root fungus resistance.
GF677	*P. persica* x *P. dulcis*	INRA, France	Vegetative	Very susceptible to *Phytophthora*, very susceptible to bacterial canker, some susceptibility to root-knot nematode, some susceptibility to ring nematode, very resistant to plum pox virus, some susceptibility to *Verticillium Wilt* (Blackheart).
Citation	*P. salicina* x *P. persica*	Zaiger, U.S.	Vegetative	Semivigorous, some resistance to *Phytophthora*, some susceptibility to bacterial canker, and some resistance to root-knot nematode.
Mirared	*P. cerasifera* x *Nemared*	Agromillora, Spain	Vegetative	The vigor is approximately 10-15% less than GF677. It is resistant to nematodes and tolerant to *Armillaria* and *Phytophthora*. It is more tolerant of heavy soils than GF677.

Table 2 Summary of some of the most commercial rootstocks and their characteristics available in the market.—*cont'd*

Rootstock	Species	Origin	Propagation	Characteristics
Myrobalan 29-C	*Myrobolan (P. cerasıfera)*	Marion E. Gregory, U.S.	Vegetative	High vigor, resistant to nematodes. Adapts to compact soils, tolerates active limestone and drought.
Rootpac R	*Myrobolan (P. cerasıfera)*	Agromillora, Spain	Vegetative	Vigor similar to Nemanguard. Highly tolerant to root-knot nematodes, resistant to *Rosellinia necatrix*. Tolerant to root asphyxia and lightly susceptible to root-lesion nematodes.
Marianna 26/24	*P. cerasifera x P. munsoniana*	W.L. Howard, U.S.	Vegetative	It is resistance to many soil pathogens. It has very high vigor and is the most resistant rootstock to root asphyxia. Moderately resistant to limestone and nematodes.
Marianna GF 8-1	*P. cerasifera x P. munsoniana*	INRA, France	Vegetative	It adapts well to soil conditions and winter cold. It shows sensitivity to *Verticillium wilt.*
GF-31	*P. cerasifera* Ehrh. *x P. salicina* Lindl	INRA, France	Vegetative	It thrives in shallow, calcareous, and rocky soils but is sensitive to root asphyxiation. Resistant to nematodes of the *Meloidogyne* group.

4.4.3 Diseases and pests

As for resistance to diseases and pests, it is important to choose a rootstock that is resistant to common diseases and pests in the area where the apricot tree will be planted. This may include resistance to nematodes, soil fungi, and other pathogens. For example, some of the most important pests of rootstocks are nematodes. *Meloidogyne spp.* is the most prevalent in the Mediterranean apricot-producing region, with some common rootstocks, mainly of peach and almond, being susceptible. Meanwhile, there exist other plum rootstocks such as Myrobalan B, Marianna

GF 8-1, and GF 31 that are tolerant to this pest (Mendelné Pászti et al., 2023). In terms of fungal diseases, *Armillaria mellea* and *Phytophthora spp.* are among the most significant pathogens, primarily affecting the roots and causing root rot and leaf wilting. Some plum hybrid rootstocks, such as Marianna 2624, which originated in Davis, California, seem to be more resistant to *Armillaria mellea* compared to others, although they are not immune. Regarding *Phytophthora spp.*, findings suggest a widespread susceptibility among peach and almond rootstocks, while there is a general trend toward resistance among plum hybrid rootstocks (Browne et al., 2017).

On the other hand, bacterial canker, caused by *Pseudomonas syringae*, is another significant threat to apricot trees that produces lesions that lead to cracks, resulting in gummy exudations, which can cause leaf loss. Using resistant rootstocks can be an effective strategy to manage this disease. Some rootstocks known for their resistance or tolerance to bacterial canker for apricots would be the rootstock from Myrobalan *(Prunus cerasifera)* and Marianna 2624 *(Prunus cerasifera x Prunus munsoniana)*, which offer some resistance, or Citation *(Prunus salicina x Prunus persica)*, which provides moderate resistance to this disease.

4.4.4 Vigor and tree size

In terms of vigor and tree size, the vigor of the rootstock will influence the final size of the tree. Depending on the available space and cultivation system, a rootstock that produces smaller or larger trees may be preferred (Atkinson, 2001).

For example, a vigorous rootstock can have certain advantages such as allowing rapid tree establishment, leading to early fruit production, greater tolerance to abiotic stress with deeper and more extensive root systems that enhance drought tolerance or adaptability to poor soils, and potentially higher yields. However, they are also characterized by certain disadvantages: they require more space, reducing planting density; their management is more challenging due to increased pruning and more costly phytosanitary treatments; and despite vigorous growth, they may sometimes take longer to reach full productivity compared to less vigorous rootstocks. In contrast, less vigorous rootstocks offer advantages such as easier management in terms of pruning, harvesting, and application of phytosanitary treatments, enabling higher-density orchards that can increase production per unit area. They may also yield commercially viable fruit earlier than vigorous rootstocks if managed properly. However, like their counterparts, they have their own drawbacks: they may have lower tolerance to water stress due to shallower root systems, making them more susceptible to drought conditions; they might produce less fruit in terms of quantity; and generally, they require more favorable climate and soil conditions (Mendelné Pászti et al., 2023).

Dwarfing rootstocks, therefore, are a good option in terms of management and production efficiency but also present challenges related to stress tolerance and potential need for additional support. The choice of dwarfing rootstock should be based on a careful assessment of local soil and climate conditions, as well as specific production goals of the farmer.

The selection between vigorous and less vigorous rootstocks depends on various factors including desired management practices, soil and climate conditions, and production objectives. Both types have advantages and disadvantages that should be carefully considered to optimize apricot production. Other aspects that are no less important, but more difficult to determine, are the influence that these rootstocks may have on the productivity and fruit quality of the apricot tree. It is important to choose a rootstock that does not have negative effects on these aspects; therefore, sometimes it is crucial to have information on other cultivars grafted onto certain rootstocks to understand their behavior before making such an important decision, although it is true that not all cultivars behave the same way with a given rootstock. In a study conducted in recent years, the effect of different rootstocks with varying vigor on Serbian apricot genotypes was compared. These rootstocks were Myrobalan (vigorous), St. Julien A (semi-vigorous), and Pumiselect (dwarfing), tested under various conditions of dry, sandy-loam, and acidic soil. Apricots on Myrobalan had strong growth, lower yield, and the poorest fruit quality characteristics in most cases compared to St. Julien A and Pumiselect. Pumiselect showed a good capacity to reduce tree growth and increase yield, fruit weight, flesh ratio, soluble solids content, and ripening index. St. Julien A exhibited similar influences as Pumiselect (Milošević et al., 2019a).

Finally, another equally important aspect, and no less significant to be taken into account, would be the commercial availability of the rootstock based on the characteristics we need for our growing area and the specific cultivar (Table 2).

4.5 Cultivars and selection criteria

As we have seen, choosing the right rootstock is no small matter, as it represents the vital support for the cultivar we want to graft and the chances of success for our crop. However, similarly, the choice of cultivar is an even more important decision, as it will determine the type of fruit we want to offer to the market (Figure 4). Thus, aside from ensuring good rootstock-graft compatibility, this cultivar must also have good adaptability to its climatic environment, just like the rootstock. Additionally, we will address the main considerations to consider to choose a suitable cultivar.

4.5.1 Early or late character of a cultivar

The decision to choose an earlier or later cultivar will depend on the climatic area of the crop. The phenology of fruit trees, such as the apricot tree, is closely related to their chilling requirements and the timing of flowering, and therefore, the harvest (Ruiz et al., 2007). The chilling requirement is the number of chill hours (generally below 7°C) that a fruit tree needs during the winter to break dormancy and allow normal bud development in the spring. Thus, a specific number of chill hours is necessary for the end of winter dormancy, which varies among cultivars. Depending on the accumulation of chill hours in a region, we will implement cultivars with either low or high chill requirements (Campoy et al., 2012). This way, we ensure that the chill hours are met and that we have a good bloom.

FIGURE 4

Apricot typology during fruit development and ripening in the orchard (CEBAS-CSIC, Spain, Murcia).

Additionally, the choice of cultivars with either early or late flowering is important to sometimes avoid potential frost risks. It is also important to consider the fruit development cycle (Salazar et al., 2016). For example, early flowering cultivars may not necessarily result in an early harvest if the fruit has a long growth cycle. Therefore, both flowering and fruit development need to be considered. Selecting cultivars with different ripening times can help extend the harvest season and ensure a continuous supply of fresh fruit.

Finally, the chilling requirement is a critical factor in the phenology of apricot trees and other fruit trees. It directly influences the timing of flowering and harvest, affecting the productivity and quality of the fruit. Understanding and managing these requirements in relation to local climatic conditions is essential to optimize fruit production, especially in the current context of climate change (Egea et al., 2022).

4.5.2 Selection of self-compatible apricot trees

The selection of self-compatible apricot trees is crucial for pollination, production, and various aspects related to successful crop management. Self-compatible apricot cultivars can self-pollinate, meaning that their flowers can be pollinated by their own pollen (Herrera et al., 2018). This ensures more efficient pollination, especially in conditions where the presence of pollinators is limited or variable. However, self-incompatible apricot trees require pollen from a different cultivar for successful pollination, but even then, they must be intercompatible (Herrera et al., 2022). Definitely, the choice of self-compatible apricot trees has a significant impact on pollination efficiency, production consistency, and ease of orchard management. Opting for self-compatible cultivars can ensure more reliable and high-quality production, reducing the dependence on external pollinators and simplifying agronomic management practices. This is particularly beneficial in areas with limited pollinator populations or in climatic conditions that do not favor pollinator activity. Today, there are molecular tools that enable the selection of self-compatible individuals since previous studies have identified both compatibility and incompatibility alleles (Halász et al., 2010; Muñoz-Sanz et al., 2020). This represents a significant advancement, allowing the certification of plants as self-compatible or not.

4.5.3 Selection of apricot cultivars based on uniformity

Other agronomic criteria include the occurrence of deformities in the fruit, such as irregular shapes and excessive skin roughness, which are unattractive to consumers. Therefore, morphological traits such as fruit weight or shape play an important role in consumer acceptability. The appearance of double fruits that develop fused together, resulting in anomalous fruit, as well as protrusions or cheeks that are more developed than others. The appearance of scars in the form of lines or marks on the skin caused by mechanical damage from pests and diseases. Skin spots of different colors result from pest damage or nutritional deficiencies. Another significant defect is cracking, characterized by the appearance of cracks or splits in the skin due to fluctuations in water availability (Ding et al., 2018). Defects in size can increase the size variability of the fruits on the same tree, which can complicate the sorting and packaging process, or cause very small fruits that do not reach the appropriate commercial size, reducing their market value. Therefore, considering the choice of cultivars without these defects is important to avoid unexpected rejection from consumers.

4.5.4 Selection of apricot cultivars based on fruit quality

The selection of new apricot varieties in terms of fruit quality must consider several criteria to ensure not only an attractive and tasty product but also one that is nutritious and beneficial for health (Fatima et al., 2018). Some of these characteristics are related to sensory attributes such as flavor, which is one of the most important criteria. An adequate balance between sugar and acidity is crucial for good taste. Sensory tests, especially with trained panels, are essential to evaluate this aspect, as well as quantitative evaluations that allow us to quantify the percentage of total acidity or soluble solids content. Aroma is another parameter indicative of higher organoleptic quality. An intense aroma in apricots adds extra value to this type of fruit (Solís-Solís et al., 2007). Aroma can also be evaluated sensorially and/or by assessing aromatic compounds through chromatography.

Fruit color is another important trait to consider. An attractive color is essential when selecting a variety, and it does not necessarily have to be a specific color, as it depends on consumer demand, which can range from yellow, orange, and reddish tones to pink. Some fruits may show uniform yellow or orange color with a striking blush that can range from reddish to pinkish tones and even completely reddish cultivars. Additionally, the presence of shine on the skin enhances the visual appeal for consumers.

Another aspect related to fruit quality is fruit texture. Excessively soft fruits indicate over ripeness, while overly firm fruits may be unripe and lack juiciness, although this depends on the harvest time. However, it is necessary to differentiate between texture and firmness. Firmness allows us to know the hardness of the fruit at the moment of harvest, while texture is more of a sensory aspect related to the fruit's fibrousness and the internal state of the flesh. Both characteristics can be measured and quantified using a texturometer through parameters such as maximum firmness, Young's modulus, bioyield, etc. (Contador et al., 2015).

Nowadays, foods with health benefits are increasingly appreciated. In this regard, the nutraceutical quality of apricots is becoming more important, with a focus on fruits with high antioxidant content, such as carotenoids (especially beta-carotene) and polyphenols (Ruiz et al., 2005ab; Zhou et al., 2020). Other compounds include vitamins and minerals. The levels of vitamins, especially vitamin A (derived from beta-carotene) and vitamin C, as well as minerals such as potassium and magnesium, are crucial for nutraceutical quality. The presence of bioactive compounds such as flavonoids and phenolic acids should be evaluated for their contribution to health and their anti-inflammatory and anticancer properties. All these compounds can be determined through various chemical analyses using high performance liquid chromotagraphy (HPLC).

Finally, selecting apricot trees based on fruit quality involves a comprehensive evaluation encompassing sensory attributes, nutraceutical quality, and commercial value. This holistic approach ensures that fruits are not only pleasant for consumers and profitable for producers but also beneficial for health.

4.5.5 Selection of apricot cultivars based on postharvest handling of the fruit

The shelf-life and postharvest handling of apricots are crucial factors, especially from the perspective of product commercialization (Salazar et al., 2022). Apricots should be harvested at their optimal ripeness, typically when the fruit has reached the characteristic color of the cultivar, is firm but yields slightly to pressure, and separates easily from the tree. Therefore, manual harvesting is preferable to avoid mechanical damage. Care must be taken not to squeeze the fruit too hard and to avoid excessive damage and dropping.

Immediately after harvest, apricots should be sorted, selected, and even cleaned to remove damaged fruit, dust, pesticides, and other residues. This can be done by washing with water or using dry cleaning techniques. Generally, storage for preservation is achieved by controlling temperature and humidity. Apricots should be stored at low temperatures (0°C–2°C) and high relative humidity (90%–95%) to minimize dehydration and loss of firmness. Proper storage conditions can extend the shelf-life to at least 1 to 3 weeks, and in the best cases, up to 4 weeks (Wani et al., 2018).

Other storage methods, sometimes in combination with cold storage, include controlled atmospheres. Storing apricots in controlled atmospheres with reduced oxygen levels and increased carbon dioxide levels can help reduce the respiration rate and delay ripening (Dorostkar et al., 2022). Therefore, the selected cultivar should, as far as possible, address all the issues mentioned, choosing those cultivars that have fewer damages or losses during harvesting, as well as those that withstand long postharvest periods.

4.5.6 Selection of apricot cultivars based on disease resistance

Selecting apricot trees based on disease resistance is crucial for maintaining healthy orchards and ensuring high-quality fruit production. Disease-resistant cultivars can significantly reduce the need for chemical treatments, lower production costs, and minimize environmental impact. Therefore, it is important to select disease-resistant cultivars to common diseases such as brown rot caused by *Monilinia spp*., powdery mildew caused by *Podosphaera pannosa,* bacterial canker caused by *Pseudomonas syringae,* PPV (Sharka), or apple chlorotic leaf spot virus (ACLSV) among others.

As for fungal diseases, brown rot produces brown, sunken lesions on fruit, blossom blight, and twig cankers (Psota 2012). Another fungal disease would be powdery mildew producing white, powdery fungal growth on leaves, shoots, and fruit (Hassan et al., 2019). In terms of bacterial diseases, bacterial canker is the most significant, producing dark, sunken cankers on trunks and branches, gumming, leaf spots, and bud death (Akbaba et al., 2023). As for disease caused by virus, Sharka produces symptoms such as chlorotic mottling and rings in leaves and irregular spots and deformations in fruits reducing the commercial quality; it could even reduce the vigor of the plant and fruit production. As for ACLSV, this produces chlorosis, mottling, and deformities in leaves as well as spots and deformations in fruits, in some cases, even causing a general tree weakening and reduced fruit production (Rubio et al., 2023).

Some managements to prevent fungal diseases include pruning to improve air circulation, removing infected plant material, and applying fungicides at critical times such as during bloom and before harvest. Additionally, using genetically resistant cultivars is recommended, along with maintaining proper spacing and pruning for enhanced air circulation. If the disease appears, it is advisable to prune infected branches, apply copper-based bactericidal treatments, and avoid pruning during wet conditions. Sometimes, for diseases like Sharka, it's best to completely remove the plant.

Definitely, to select suitable cultivars according to diseases criteria, it is important to obtain Disease-Free Certification, as acquiring planting material that is certified as disease-free can prevent the introduction of pathogens into the orchard. This information is often available from nurseries or agricultural extension services. Currently, there are efficient molecular markers available for selecting genotypes with resistance to specific diseases, such as identifying alleles that are resistant or susceptible to Sharka (Zuriaga et al. 2018; Gürcan et al. 2019). All these molecular markers are being applied to improve the efficiency of the current apricot breeding programs.

4.6 Future prospects and challenges

Finally, as we have seen, breeding and propagation in horticulture encompass two main aspects: rootstocks and cultivars, each with distinct selection criteria and challenges. On one hand, rootstocks are crucial as they influence the vigor, disease resistance, and adaptability of grafted plants. Selection focuses on traits like tolerance to adverse conditions, enhancement of fruit yield, and compatibility with various soil types and climates. On the other hand, cultivar development targets improving fruit quality traits such as taste, texture, and appearance to meet consumer preferences and market demands. Emphasis is placed on genetic resistance to diseases and pests, adaptability to diverse environmental conditions, and the integration of cutting-edge breeding technologies.

However, if we explore the challenges of rootstocks and cultivars separately, a good rootstock should have good plant-microorganism interaction, avoiding infection by pathogens to optimizing nutrient uptake efficiency and prevent pest and diseases. For this reason, the interaction between rootstock soil is primary important irrespectively if the rootstock is resistant or not to certain pests and diseases. On the other hand, the compatibility between rootstock and cultivar is also important since the success of this association depends on it, as well as a suitable interaction between the cultivar and the environment to ensure a good photosynthesis balance with a suitable foliage.

Definitely, although the interaction must work as a perfectly joined set, the plant-soil interaction depends only on the root's rootstock while the plant-air interaction depends mainly on the cultivar. For this reason, once each interaction is known separately, we have to study the rootstock-cultivar interaction more in depth. Therefore, breeding challenges involve navigating genetic and molecular complexity to explore better understanding about the biological mechanisms linked to the growth, yield, adaptability, fruit quality, and resistance to diseases among others and ensuring sustainability amidst changing climate patterns tailored to meet evolving agricultural and market needs.

5. Orchard management

5.1 Site selection and planting

The establishment of an apricot orchard begins with site selection. A well-drained site with nutrient-rich soil that has good aeration is ideal for apricot trees, although soil amendments can improve less suitable soils (Kaska, 2006). Apricots grow well in soils with a pH of around 6.5 to 7.5. Apricot trees do not grow well in wet soils; however, they can tolerate drought better than many other deciduous fruit species (Kaska, 2006). Planting on land previously occupied by apricots or other *Prunus* species can result in replant disorder, which requires soil treatments similar to those for other fruit tree species, e.g., fumigation, biofumigation, or soil amendments (Foschi et al., 2010). The alternative strategy is to select rootstocks that are tolerant, e.g., Penta (Remorini et al., 2015), or resistant, e.g., Rootpac 20 (Reighard et al., 2022) to replant disorder.

Other site considerations include sufficient winter chilling for the cultivars to be planted (Fadón et al., 2020), frequency and degree of frosts, particularly during flowering and early fruitlet stages (Rodrigo et al., 2006), and sufficient growing degree days over summer for fruit development and maturation (Ruml et al., 2010). Waterlogging results in anaerobic conditions that affect tree health, and therefore soils prone to waterlogging should be avoided or suitable drainage work undertaken before planting. Some rootstocks are more tolerant to waterlogging (Hassan et al., 1986). Planting generally takes place between leaf fall and bud break but will depend on the winter conditions (Kaska, 2006).

5.2 Nutrition, soil and root health

It is important that the soil has sufficient nutrients available for plant uptake. Regular monitoring of soil and leaf nutrient content is recommended (Kaska, 2006) to assist in making decisions on rates of fertilizer and ratios of the various nutrients. Applying balanced fertilizers with suitable nitrogen, phosphorus, and potassium ratios is crucial for good yields and fruit quality, including affecting the incidence of disorders (Bussi et al., 2003; Radi et al., 2003). Fertilizers can be applied to the soil (Khasawneh et al., 2022), through fertigation systems (Kaska, 2006), or as foliar applications (Moale et al., 2021). Supplying high nitrogen rates can result in undesirable rates of vegetative growth (Rettke et al., 2006). Recommendations of the best timing of applications varies; however, applications in autumn appear important for nutrient supply in the spring (Bussi et al., 2003; Weinbaum et al., 1980).

Incorporating organic matter into the soil enhances soil structure and water retention, and is also important for root health and growth. Use of cover crops or living mulches can increase the biodiversity of microbial communities in the soil, as well as improve organic matter and yields (Demir, 2019). Weed suppression in apricots orchards can be undertaken using herbicides, mechanical weed control, or through the establishment of cover crops (Tursun et al., 2018).

5.3 Irrigation

Apricot orchards typically use irrigation to ensure adequate water for plant and fruit requirements. Drip irrigators or microsprinklers are efficient methods for delivering water directly to the root zone; however, in some apricot-growing regions, overhead sprinklers supply water for irrigation as well as frost protection.

Efficient water management practices are becoming more important as water scarcity is of greater concern, and temperatures are increasing as a result of climate change (Torrecillas et al., 2018). Apricot trees respond to water stress primarily using avoidance mechanisms, such as stomatal regulation, reduced photosynthesis, and leaf shedding (Torrecillas et al., 2018). Young trees could be preconditioned to endure drought by watering at lower rates compared with infrequent watering to field capacity (Ruiz-Sánchez et al., 2000). Regulated Deficit Irrigation (RDI) has been shown to be effective in reducing water requirements with little effect on yield and fruit quality (Temnani et al., 2023; Torrecillas et al., 2018). Sensitive stages to water deficit include the second rapid fruit growth stage and immediately postharvest, as this affects performance in the following season (Temnani et al., 2023; Torrecillas et al., 2018). There are differences among cultivars in the extent of impact of timing and degree of RDI (Ezzat et al., 2021).

Shaded leaves of apricot have lower water use efficiency than exposed leaves (Xiloyannis et al., 2000). Therefore, training systems that increase leaf exposure or pruning to improve light penetration into canopies are strategies to improve overall water use efficiency.

5.4 Training systems

There is a wide choice of training systems for apricots, with tree densities spanning from 300 up to 2300 trees per hectare (Stanley et al., 2024). Traditional low-density, three-dimensional training systems of vase and center leader (Figure 5) on non-dwarfing rootstocks have wider spacings. With the development of lower vigor rootstocks and observations of trends in apple training systems, more intensive plantings on different training systems have been introduced in the past 20+ years. These

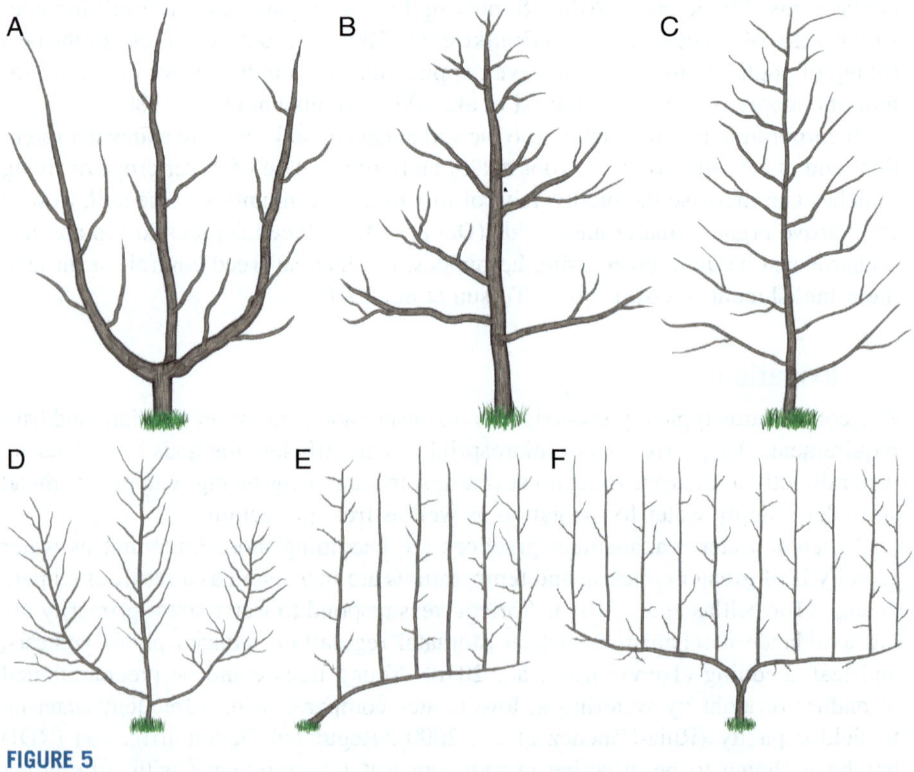

FIGURE 5

Examples of different training systems for apricot. Three-dimensional training systems: (A) vase, (B) centre leader, (C) spindle; two-dimensional training systems: (D) trident, (E) upright fruiting offshoot (UFO), (F) planar cordon. Drawings: Tony Corbett, The New Zealand Institute for Plant & Food Research Ltd, New Zealand.

include other three-dimensional systems such as spindle axe (Figure 5), with planting densities between 1000 and 1700 trees per hectare, and two-dimensional systems (Figure 5), both sloping systems (such as vee or Tatura trellis) and vertical systems (such as parallel-U, trident, upright fruiting offshoots [UFO] and planar cordon), with planting densities between 600 and 2300 trees per hectare (O'Connell, 2022; Scofield et al., 2020).

The decision on planting distances depends on rootstock and scion combinations, soil type, climate, and training system. Higher density systems require greater initial investment, with higher tree numbers and, in some cases, installation of trellises. However, they usually result in faster establishment of canopy area and earlier yields (Southwick et al., 1999).

A disadvantage of large three-dimensional training systems is the challenge of maintaining good light penetration into the lower canopy zones as trees mature, as fruit yield, size, and quality can be lower in these lower inner zones (Stanley et al., 2014, 2018). Higher density two-dimensional systems can reach a higher light interception than three-dimensional systems yet still have good light penetration into the canopy and fruit quality (Scofield et al., 2020; Stanley et al., 2018). They are also proving to have greater labor efficiency for pruning, thinning, and harvest, and are easier systems for uptake of mechanization and automation.

5.5 Crop load and thinning

Adjusting the crop load is essential for consistent yields, larger fruit size, and fruit quality traits such as soluble solids content, which can affect consumer satisfaction (O'Connell, 2022; Stanley et al., 2014, 2015). Reducing crop load also helps reduce the risk of limb breakage due to heavy fruit loads. Crop load adjustment usually reduces yield; however, it may be necessary to achieve the fruit size and quality required by the consumer (O'Connell, 2022; Stanley et al., 2015). Earlier thinning results in larger fruit size and weight, with thinning at flowering producing higher yields than other thinning times (Rettke and Dahlenburg, 1999).

Reduction of crop load can be achieved by reducing the numbers of flowers being produced in spring, by preventing flowers from setting fruit or by fruitlet thinning (Webster and Spencer, 2000).

Reducing the numbers of flowers being produced in spring is very unpredictable. Reductions in flower numbers can be achieved by removing some of the potential flower bud sites during pruning, which could be in winter or else after bud burst when flower bud are evident (Webster and Spencer, 2000). Another method is to apply gibberellic acid the previous year, which reduces flower numbers in spring (Southwick and Fritts, 1995).

Options for thinning at flowering include mechanical thinning (Lei et al., 2023) and application of thinning chemicals such as dinitro-o-butylphenol (although use has been withdrawn in many countries), ammonium thiosulfate, and 6-benzyladenine (Looney and Killick, 1979). Apricot flowering occurs much earlier in spring than many other deciduous fruit crops, and frost can have a major effect of flower damage and subsequent fruit numbers (Kaya et al., 2018). Many growers would prefer to have a greater cost for thinning fruitlets than risk having insufficient flowers from thinning at flowering combined with frost damage.

Early chemical thinning of fruitlets, for example, with 1-naphthalenacetic acid (NAA) and ethephon, is often carried out just after full bloom when initial fruit set is still unknown and is also considered too risky by many growers (Webster and Spencer, 2000). Chemical thinning around the time of pit hardening is also an option; however, for apricot there can be some phytotoxicity or ineffectiveness (Webster and Spencer, 2000). Mechanical thinning has been tested, in association with hand-thinning (Assirelli et al., 2021).

Hand-thinning is very labor intensive but enables selective thinning that is not possible from chemical or mechanical thinning methods. Small, misshapen, and damaged fruit can be removed first, and multiple fruit on a single spur can be thinned down (O'Connell, 2022). Fruit on 1-year-old wood except at the base tend to be smaller and can therefore be removed (Stanley, 2016). Fruit should then be uniformly thinned out, taking cultivar and local knowledge into account (O'Connell, 2022). Differential thinning, to remove more fruit from lower light regions of the tree and retain more fruit in more illuminated zones, may also be beneficial.

5.6 Pruning

Pruning aims to shape the tree to the desired training system, improve light penetration, and ensure appropriate wood remains that is suitable for fruit production the following season. Dead or diseased wood should be removed, and some limbs may need to be taken out to maintain an open canopy (Kaska, 2006).

Architectural and fruiting habits of different cultivars also need to be considered, with cultivar habits falling into at least three different groupings. Cultivar habits can also affect the best timing for pruning. Susceptibility to bacterial canker *(Pseudomonas syringae)* can be higher in autumn and winter, so in regions with high bacterial canker pressure, pruning in spring or late summer may be preferable.

Summer pruning can be used as way to control vigor, increase light into the canopy during fruit maturation and increase flower bud differentiation; however, the cultivar habit and maturation timing alter the effect (Musacchi et al., 2014). Summer pruning also increased carbohydrate accumulation in shoots. Summer pruning in conjunction with winter pruning on young trees was found to be particularly beneficial in increasing the carbohydrate content in shoots and flower buds. The combination of summer and winter pruning can also affect some fruit quality parameters.

6. Fruit maturation and harvest criteria

6.1 Background

In stone fruits like peaches, plums, cherries, and apricots, fruit maturation takes place in sequential stages starting with fruit set after blooming, following by fruit development or growing, maturation, harvesting, and senescence. In this sense, paying attention to fruit maturation, we have to highlight different concepts concerning harvest time: physiological maturation, commercial maturation, and eating ripeness, some of which can coincide depending on our purpose. Each stage has different characteristics that influence fruit quality, handling, and postharvest. The following is an explanation of the differences:

Physiological maturity is reached when the fruit has completed its internal development and has accumulated maximum reserves (sugars, carotenoids, phenolic compounds, etc.), but it may not be ready for consumption. In this sense, the fruit has reached the end of its growth on the plant; it may still be firm or lack the flavor and aroma suitable for eating. This is a key point for fruit quality, but it is not yet fit to be consumed. This maturation stage can coincide with commercial maturation, and it is important for growers as fruit can be harvested at this stage if it needs to be transported or stored for extended periods.

Commercial maturity is the stage at which fruit is harvested for distribution and sale, with the expectation that it will continue ripening off the tree. In this sense, commercial maturity encompasses a broader window that lies between physiological maturity and consumer maturity, depending on the destination market. Therefore, in commercial maturity, the fruit has developed the size needed and the sufficient external color for the market, but it may still be firmer than at the consumer maturity stage, which helps with handling and transportation, as it continues to ripen during storage or transport. Consumers generally do not purchase it at this stage for immediate consumption. Therefore, commercial maturity prioritizes durability and external appearance, as the fruit will continue ripening in postharvest conditions (Manganaris and Crisosto, 2020).

As for **eating ripeness**, this is the optimal stage at which the fruit has the best organoleptic characteristics (taste, aroma, texture) ensuring greater sensory acceptability for consumption. The fruit is soft, juicy, and has the best flavor, aroma, and external and internal colors. It is not suitable for prolonged transport or storage due to its high perishability. This is the moment when the fruit is considered ready to eat. Therefore, fruit at this stage must be consumed quickly, as it is highly sensitive to handling and deteriorates easily (Aubert and Chanforan 2007; Salazar et al., 2022).

Finally, in stone fruits such as apricots, their climacteric nature makes them the most perishable, in contrast to plums and peaches. Therefore, at the commercial level, the key factor is the timing of harvest. Depending on whether the fruit is intended for export to distant countries or not, it will be harvested earlier or later. Fruits destined for longer distances will be harvested closer to physiological maturity, while those traveling shorter distances will be harvested later, closer to eating ripeness.

6.2 Fruit development and maturation

Commonly, in stone fruits such as apricot, fruit development is characterized by a double sigmoid curve. This growth curve starts after flowering and is distinguished by four stages: the first stage (S1), takes place as an exponential rapid growth (cell division and elongation); the second stage (S2) takes place as the hardening of the endocarp (pit hardening); the third stage (S3) takes place again as an exponential growth (cell division and elongation); and finally, in the last stage (S4), the fruit achieves its final size and starts the fruit maturation (Rodriguez et al., 2019). These processes of fruit growth and maturation, therefore, involve a series of processes and/or changes at the molecular and biochemical levels in the fruit, where a complex network of metabolites and proteins is significantly altered (Lombardo et al., 2011).

It is important to highlight those fruits like apricots that have a climacteric nature, which allows them to continue ripening after harvest. Ethylene is a hormone that is not only related to ripening but also to seed germination, cell elongation, cell differentiation, senescence, etc. Moreover, this hormone is a key regulator in all types of abiotic stresses. Other important hormones related to the ripening process include abscisic acid (ABA). In climacteric fruits, there is an accumulation of ABA content preceding climacteric ethylene production, reaching its maximum in fully ripe fruit and activating ethylene biosynthesis genes. Auxins, which are compounds present in plant growth, also play a role. The effect of auxin differs between climacteric and nonclimacteric fruits. In climacteric fruits, auxin seems to accelerate the ripening process, while in nonclimacteric fruits, auxin can negatively regulate ripening. On the other hand, gibberellins (GAs) play a stimulatory role in fruit development by enhancing cell division and enlargement and delaying fruit ripening by decreasing ethylene release. Cytokinins (CKs), with their endogenous levels, induce fruit growth by stimulating cell division and may act by inhibiting auxin responses, at least partially, during fruit set and growth to increase fruit size. Other hormones such as jasmonates (JAs) have shown a delay during the ripening (García-Gómez et al., 2020).

Ultimately, we can find several genes encoding proteins related to fruit growth and maturation, color change, or compounds associated with the nutraceutical properties of the fruit. Therefore, in the case of growth and maturation, we can find genes encoding proteins such as polygalacturonase (PG), pectin methylesterase (PME), auxin-sensitive protein (IAA), ferredoxin (PET), myo-inositol-1-phosphate synthase (INO1), NAC domain-containing protein (NAC), MADS-box protein gene (*MADS-BOX*), or amino-cyclopropane-1-carboxylate synthase (ACS). Regarding color-related genes, we can highlight the carotenoid cleavage dioxygenase 4 (CCD4) gene, beta-carotene-3-hydroxylase (CRTZ), zeta-carotene desaturase (ZDS), flavonoid 3-O-glucosyltransferase (UFGT), chalcone synthase (CHS), flavonoid-3'-monooxygenase (CYP75B1), and anthocyanidin synthase (ANS). Finally, some of the most notable genes directly related to the nutraceutical properties of the fruit, along with those already mentioned related to carotenoids, flavonoids, and anthocyanins, are sucrose synthase (SUSY), GDP-L-galactose phosphorylase (VTC2-5), 3-hydroxyisobutyryl-CoA hydrolase (HIBCH), and lipoxygenase (LOX2) (Ortuño-Hernández et al., 2024a).

Finally, all these hormonal regulation processes influence organoleptic quality, which is strongly affected by the balance between organic acids and sugars (Borsani et al., 2009). Therefore, the evolution of these compounds will significantly affect the level of consumer acceptance. On the other hand, there are also other compounds, such as volatile aldehyde and alcohol compounds (Dabbou et al., 2016), that influence sensory perception related to aroma, which can undoubtedly make fruits like apricots more or less appealing.

6.3 Harvest criteria and methods

As we have seen, the perception of organoleptic quality is the result of an intricate network of processes related to genes, proteins, and metabolites that affect the final quality of the fruit. Therefore, it is necessary to establish methodologies and harvesting parameters based on the intended use of the fruit. During the ripening process, hundreds of genes regulate important traits such as the acidity-sugar balance, fruit color, firmness, and even chlorophyll degradation. In the next section, the methodologies and equipment commonly used to measure the mentioned traits, which can serve as harvesting criteria, are described (Figure 6).

6.3.1 Firmness

Fruit firmness is usually the most used criterion for harvesting fruits like apricots, as this characteristic is the most linked to their shelf-life. Firmness can be measured directly in the field using a manual penetrometer with an 8 mm puncture, and values between 10 to 15 Newtons (N) are optimal for consumption maturity. In a laboratory, a texture analyzer (TA.XT plus) can further measure firmness through compression, with the fruit being compressed up to 5 mm in both cases (Salazar et al., 2022). In this latter case, values around 25 N usually indicate proximity to consumption maturity. This is due to the use of different plungers and methodologies, as well as the fact that, in compression testing, the skin is not removed, meaning that the skin's elasticity itself contributes to the higher values. For commercial harvesting, it is

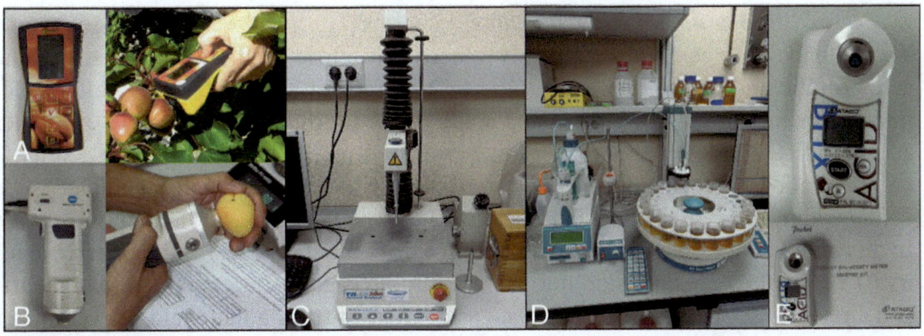

FIGURE 6

Details of some of the measuring equipment used in stone fruit: (A) DAmeter (I_{AD}), (B) Minolta colorimeter (Cie Lab), (C) Texturometer (TA.XTplus), (D) Automatic acid-base titrator, and (E) handheld refractometer (oBrix and total acidity).

recommended to aim for values between 30 and 40 N, while for compression, these values typically exceed 40 N (Stanley et al., 2013), with the optimal value varying according to the type and texture of the fruit. It is important to note that all these values depend on whether the fruits are being sent to nearby or more distant markets.

6.3.2 Skin color and chlorophyll degradation

Skin color is also a common indicator of harvest time, assessed visually in the field. Typically, the onset of skin color variation coincides with chlorophyll breakdown and the stage of eating ripeness, characterized by a fully orange color (Velardo-Micharet et al., 2021). As chlorophyll diminishes, carotenoid content increases, making the orange color more intense. Another aspect is the "blush" color, which is characterized by an intense red hue that partially or fully covers the fruit's skin surface. Consequently, today there are cultivars with fruit ranging from light yellow to orange and finally fully red skin. It is important to note that orange skin color is due to the presence of carotenoids, whereas red skin color is related to anthocyanin content (Xi et al., 2019). Precise color measurements can be made by MINOLTA colorimeters measuring the CIELAB color parameters L*, a*, and b*, being indicators for the evolution of chlorophyll and carotenoid contents (Ayour et al., 2016). Therefore, when monitoring fruit maturity in crops like apricots, color is a key indicator of fruit ripeness. With a colorimeter, it is possible to objectively measure these color changes, helping to determine the optimal harvest time. Currently, these devices are being designed for handling in the field, allowing producers and technicians to take quick measurements on crops.

As for chlorophyll degradation, the most well-known device used especially in peach is the DA-Meter, which measures the Differential Absorption Index (I_{AD}). This index is calculated through two wave lengths (670 and 720 nm) near the chlorophyll-a absorption peak (Costa et al., 2010). This device is practical and portable, allowing farmers and technicians to take quick and accurate measurements directly in the field without damaging the fruit.

6.3.3 Soluble solid-acidity balance

The balance between sugars (mainly glucose, fructose, and sucrose) and organic acids (malic and citric) is crucial (Gómez-Martínez et al., 2021). This balance defines the perception of sweetness and acidity in fruit, directly influencing consumer acceptability. Analyzing soluble solids (measured in °Brix) and titratable acidity is recommended to determine the optimal eating ripeness. Generally, soluble solids content above 12° Brix and balanced acidity indicate good organoleptic quality. In this context, apricots with high soluble solids content (>10%) and moderate acidity (0.7%−1.0%) are considered to have desirable organoleptic quality (Manganaris and Crisosto, 2020). To stimate the sugar content in fruit, a handheld refractometer can be used to measure the refractive index of soluble solids in °Brix. Additionally, some devices can measure both sugar (°Brix) and total acidity (%) simultaneously, determining sugar levels through refractometry and acidity levels through electroconductivity. Additionally, titratable acidity can be measured with a titration device, which uses a base such as sodium hydroxide to perform an acid-base titration up to pH 8.1 and thus estimate acidity (Salazar et al., 2022). Techniques such as ion chromatography (IC) and HPLC analysis can also be implemented to measure the specific content of sugars and organic acids (Fan et al., 2017).

6.4 Conclusions

Understanding how traits like firmness, color, chlorophyll index, soluble solids, and acidity change is essential when making harvesting decisions for apricots or other stone fruit species. Therefore, it is important to study the maturation curves of each of these traits at different harvest times to gather useful information that helps us determine the optimal harvest moment based on our preferences. This means identifying the most commercially favorable time to harvest the fruit at its maximum shelf-life potential without excessively compromising its organoleptic quality. Thus, if we prioritize shelf-life, which is closely related to fruit firmness, we must consider that we will lose organoleptic quality. In this case, we need to assess the level of quality we are sacrificing in terms of texture, sugars, and acidity, as well as the balance between the latter two. In the specific case of apricot trees, it would be advisable to conduct more in-depth studies for each commercial variety, allowing farmers to select the most suitable variety based on their interests and the preferences of consumers in each region. Therefore, it is beneficial to create records for each variety at different harvest times, indicating all these values to observe their evolution and make decisions based on nondestructive harvest criteria, such as fruit color and chlorophyll index.

Moreover, there are now more and more modern imaging techniques that aim to monitor fruit ripening to develop models that can predict firmness or soluble solids (Mozaffari et al., 2022). The development of these models could undoubtedly help us predict not only these traits but also those mentioned previously.

7. Postharvest physiology and technology

Apricot is a highly perishable fruit due to its climacteric nature, meaning that its ripening process continues even after harvest. This postharvest phenomenon makes fruit highly vulnerable to rapid quality degradation, particularly in terms of texture, flavor, color, and nutritional value. The physiology of apricots and the postharvest technologies applied to them are critical factors for extending shelf-life and reducing economic losses, which can reach up to 40% in certain production regions. Postharvest management of apricots involves a range of practices and technologies aimed at slowing the physiological processes that lead to spoilage. These include refrigeration, controlled atmosphere storage, chemical treatments, and advanced technologies such as edible coatings and controlled dehydration. This section discusses both the key physiological changes occurring in apricots after harvest and the principal technologies employed to preserve their quality during storage and transportation.

7.1 Postharvest physiology of apricot

Climacteric nature of apricot: Apricot is a climacteric fruit, which means that after harvest, it experiences an increase in respiration rate and ethylene production. This gaseous plant hormone plays a crucial role in regulating various ripening processes, such as fruit softening, color changes, and the development of volatile aromatic compounds. Furthermore, ethylene is a key regulator of plant responses to various biotic stresses, including drought, flooding, wounding, cold, pathogen infection, and chemical damage (Katz et al., 2004).

During the ripening of climacteric fruits, the ethylene-sensitive system II becomes active and operates in an autocatalytic manner. In this process, the increase in ethylene production triggers a positive feedback loop that accelerates ripening, significantly shortening the fruit's shelf-life if proper handling measures are not taken. The presence of this system suggests that climacteric fruits have evolved to synthesize ethylene autocatalytically, utilizing a positive feedback mechanism to regulate ethylene production during ripening (Liu et al., 2015; Manganaris et al., 2011).

Ethylene biosynthesis is regulated by a well-defined enzymatic pathway involving three reactions: (a) the precursor S-adenosyl-L-methionine (AdoMet) is converted into S-adenosyl methionine (SAM) by S-adenosyl-L-methionine synthetase (MAT); (b) 1-aminocyclopropane-1-carboxylic acid (ACC) synthase (ACS) converts SAM into ACC; and (c) ACC oxidase (ACO) oxidizes ACC to release ethylene. Notably, the conversion of ACC to ethylene, catalyzed by ACO, is oxygen-dependent. However, under anaerobic conditions, ethylene formation is completely suppressed. In addition to ACC, ACS also produces 5'-methylthioadenosine (MTA), which is recycled for methionine synthesis, thereby ensuring high rates of ethylene biosynthesis (García-Gómez et al., 2021). The main challenge in managing apricots postharvest is to delay this physiological cycle without compromising the fruit's organoleptic quality. Under inadequate storage conditions, apricots can become overripe within a few days, making them prone to mechanical damage, loss of firmness, and the development of postharvest diseases.

Postharvest biochemical changes in apricot ripening: The postharvest ripening of apricots triggers a sequence of biochemical changes that directly impact fruit quality, including pulp softening, color transition, the sugar-to-acid ratio, and the profile of volatile compounds. During this stage, cell wall degradation is critical to softening, governed by specific hydrolytic enzymes whose gene expressions regulate this process. The main modifications occur in cell wall polysaccharides, such as pectins, hemicelluloses, and cellulose, controlled by key genes: the *PG* gene encodes polygalacturonase, which depolymerizes pectins, while *PME* encodes pectin methylesterase, facilitating their breakdown (Ortuño-Hernández et al., 2024a). Additionally, *XTH* regulates the production of xyloglucanases, which degrade hemicelluloses, and *CEL* encodes cellulases responsible for breaking down cellulose microfibrils, contributing to fruit softening.

This degradation process is tightly regulated by ethylene, whose action is mediated by the ethylene receptor encoded by the *ETR* gene. This gene regulates the fruit's response, activating a signaling cascade that induces *PG*, *PME*, *XTH*, *CEL*, and *EXP* genes, accelerating cell wall degradation and promoting loss of firmness (Xu et al., 2021). Coordinating these processes is essential for apricot ripening, impacting postharvest quality, texture, and shelf-life.

Another fundamental aspect is the color change, resulting from chlorophyll breakdown and carotenoid accumulation, which imparts the characteristic orange hue. Chlorophyll degradation, mediated by the *SGR* (Stay-Green) gene, initiates this process, while the *PAO* (Pheophorbide a Oxygenase) gene converts chlorophyllide to colorless compounds, enhancing visual chlorophyll disappearance. Carotenoid accumulation, facilitated by the *PSY* (Phytoene Synthase) gene, produces the vibrant orange tones of mature apricots, a key visual maturity indicator. In terms of chemical composition, there is an increase in soluble sugars and a reduction in organic acids, regulated by genes like *SPS* (Sucrose Phosphate Synthase) and *INV* (Invertase). These changes, along with malic acid reduction facilitated by *MDH* (Malate Dehydrogenase), result in a sweeter sugar-to-acid ratio, a critical parameter for sensory quality (García-Gómez et al., 2020).

Finally, the synthesis of volatile compounds enriches apricot's aromatic profile, enhancing sensory acceptance. Genes such as *AAT* (Alcohol Acyltransferase) promote the synthesis of fruity esters, while *LOX* (Lipoxygenase) and *ADH* (Alcohol Dehydrogenase) contribute to green notes (García-Gómez et al., 2022). These volatile compounds are essential for the commercial value of apricots and their sensory appeal.

7.2 Postharvest technologies

The primary factors affecting the postharvest physiology of apricots are temperature, relative humidity, and the atmospheric composition of the storage environment. To enhance shelf-life and preserve the quality of apricots, various postharvest technologies are applied, each addressing these factors in a targeted manner. Key techniques include refrigeration, controlled and modified atmospheres, chemical treatments, and drying. Each of these methods is designed to slow down deterioration processes, maintain fruit firmness, and retain organoleptic properties, tailored to meet specific market demands.

Refrigeration: It is the most widely used technology to delay apricot ripening, effectively reducing both respiration rate and ethylene production. Storing the fruit at temperatures between 0°C and 5°C significantly extends its shelf-life by slowing down the metabolic activities leading to deterioration. However, prolonged storage at temperatures near 0°C can induce "chilling injury," which manifests as water-soaked areas in the pulp, thereby diminishing the fruit's sensory quality.

Studies indicate that refrigerated storage can extend apricot shelf-life by up to 2 to 3 weeks, depending on the variety and initial ripeness conditions (Liu et al., 2019). Careful handling is essential to avoid temperature fluctuations during transport and storage, as such variations can accelerate deterioration and compromise fruit quality.

Controlled and modified atmospheres: The use of controlled (CA) and modified atmospheres (MA) is an increasingly common practice for apricot storage, aimed at reducing fruit respiration rates to extend shelf-life. These conditions involve adjusting oxygen levels to 2% to 3% and increasing carbon dioxide to 5% to 10%. When combined with low temperatures, these atmospheres can effectively extend apricot shelf-life to 4 to 6 weeks (Dorostkar et al., 2022).

- Controlled Atmospheres (CA): In CA systems, gas levels are continuously monitored and adjusted throughout storage using specialized equipment that maintains optimal gas concentrations in the storage environment.
- Modified Atmospheres (MA): The use of plastic packaging to passively generate a modified atmosphere is a simpler, cost-effective alternative. Through fruit respiration, oxygen is consumed and CO_2 accumulates within the package, creating a modified environment that slows down apricot metabolism.

Chemical treatments: The application of 1-methylcyclopropene (1-MCP) has transformed postharvest technology for climacteric fruits such as apricots. This compound acts as an ethylene inhibitor by blocking ethylene receptors in fruit tissues, thereby delaying the ripening process. The use of 1-MCP in sealed chambers immediately after harvest has proven effective in slowing fruit softening and reducing disease incidence during storage (Ortuño-Hernández et al., 2024b). Furthermore, antifungal treatments are essential to prevent decay caused by pathogens such as *Botrytis cinerea* and *Monilinia spp.*, which are primary culprits of postharvest diseases in apricots. Fungicides such as thiabendazole and iprodione have shown efficacy in reducing these infections during storage (Aydin, S. & Ulvi, M., 2019). However, rising concerns over the use of chemical agents have stimulated research into biological and natural alternatives.

Drying and dehydration: Drying is one of the oldest and most effective postharvest technologies for preserving apricots and remains widely used in the industry (İncedayi et al., 2016). In Turkey, apricot drying is primarily carried out in the Malatya region, renowned for producing the majority of the world's dried apricots. Two main drying methods are employed:

- Natural Sun Drying: Apricots are laid out directly in the sun, which imparts a dark brown color and a sweeter flavor. This traditional method uses no additives, though exposure to the open-air increase's susceptibility to contaminants and potential physical damage.
- Sulfur Drying: To retain the bright orange color and extend shelf-life, apricots are pretreated with sulfur dioxide (SO_2). This treatment inhibits microbial growth and maintains an attractive market color. SO_2 content varies based on export regulations, with levels set at 2000 ppm for Europe and 2500 to 3000 ppm for the U.S. and other destinations.

Modern drying methods, such as hot air drying and lyophilization, also offer improved preservation of the fruit's organoleptic and nutritional properties. Lyophilization, in particular, involves freezing the fruit and removing water by sublimation, nearly completely preserving nutrients and texture, although it is significantly more costly than traditional methods.

7.3 Recent innovations in postharvest technology

Edible coatings: Among the most promising innovations in postharvest technology is the use of edible coatings. These coatings are thin films applied to the fruit's surface, creating a semipermeable barrier that reduces water loss and oxygen exposure, thereby slowing the ripening process and maintaining freshness. Common materials used for edible coatings include starch, carnauba wax, and chitosan, all of which are biodegradable and nontoxic, making them a sustainable alternative to plastic packaging. Recent studies have shown that edible coatings, specifically chitosan grafted with gallic acid, can extend the shelf-life of apricots while effectively preserving both sensory quality and nutritional value (Han et al., 2024).

Biotechnology and genetic treatments: The development of apricot varieties with enhanced resistance to postharvest conditions represents a major advancement in fruit preservation technology. Genetic improvement programs focus on selecting varieties with reduced respiration rates and greater resistance to postharvest diseases. Through biotechnology, genes can be introduced to enhance these characteristics without compromising the fruit's flavor or texture. Collectively, innovations in refrigeration, controlled atmospheres, chemical treatments, edible coatings, and biotechnology have significantly enhanced apricot shelf-life and quality during storage and transport. In a global context of rising demand for high-quality fresh produce, these technologies present new opportunities for the sustainable management of apricots in the future.

8. Main diseases and pests

Apricot trees, like other fruit trees, can be affected by diseases and pests, which may lead to very low production, reduced yield, lower fruit quality, or even the death of the tree, resulting in a complete loss of the crop and significant economic consequences. For this reason, it is essential to consider the various diseases and pests that can impact these trees. Among the diseases, we can differentiate those caused by bacteria, fungi, and viruses. Notable bacterial diseases include bacterial canker and crown gall. Important fungal diseases include Brown rot, root rot, rust, shot hole disease, and verticillium. Among viral diseases, there are Viruela, Prunus Necrotic Ringspot Virus, and perhaps the most devastating for major fruit trees, Sharka. In terms of pests, these include aphids, beetles, moths, and mites. In the following sections, each will be described, along with how certain symptoms can be devastating for the fruit or even the entire tree.

8.1 Diseases

8.1.1 Bacterial

Bacterial blast: This disease, also called bacterial canker, is caused by *Pseudomonas* spp. bacteria, most commonly *Pseudomonas syringae* pv. *Syringae* (Pss). This is a major issue in most apricot-growing countries around the world. Symptoms include cankers on branches and trunk, with amber-colored gum oozing out (Figure 7), blossom blight, spotting on fruit and leaves, and dead buds (Akbaba et al., 2023). When cut away, the inner bark and wood is brown. Limbs die back, and entire trees can die. The bacteria can move into plants from water splash (rain or irrigation) or pruning equipment and enter the plant via pruning wounds, buds, flowers, and leaves but not leaf scars (Wimalajeewa, 1987). The greatest susceptibility is in late autumn and winter for buds and in late winter and spring for stems. Freezing and thawing can cause damage to the bark, which can allow entry of Pss. Irrigation should be delivered under the trees rather than overhead sprinklers. Pruning is best carried out in late summer when susceptibility is lower, and rainy weather should be avoided. Wood should be pruned off at least 100 mm below the infection and removed from the orchard. Large pruning wounds can be painted to reduce entry of the pathogen. Pruning equipment should be sterilized between trees. Copper is often applied several times during the season, however, Pss can become resistant. Some scions and rootstocks are more tolerant to Pss, and higher grafting points, e.g., 1.8 to 2.5 m above the ground, can reduce infection.

Crown gall: Crown gall *(Agrobacterium tumefaciens)* is a bacterial disease that affects many types of fruit trees, including apricot (McLaren et al., 1999). The bacterium causes galls to form on the roots, crown, or trunk, which can impede the flow of water and nutrients within the tree. This can lead to stunted growth, poor fruit production, and tree decline. The bacteria enter the tree through wounds in the roots or crown, often caused by planting, pruning, or insect damage. The bacteria can survive in the soil for many years, making it difficult to eliminate once introduced to an orchard. The use of *Agrobacterium radiobacter*, a closely related, nonpathogenic

FIGURE 7

Amber-colored gum, caused by *Pseudomonas syringae* pv. *Syringae*, oozing from an apricot branch.

bacterium, has proven to be the most effective control. Sanitation and certified disease-free material can also assist. Certain rootstocks have some resistance, e.g., Mahaleb, Marianna 2624, and Myrobalan 29C.

8.1.2 Fungal diseases

Brown rot: It is caused by *Monilinia* spp., particularly *M. laxa* and *M. fructicola*. Brown rot is a common fungal disease that affects blossoms, twigs, and fruit. This disease is particularly problematic in wet and humid conditions and can lead to significant crop loss. Brown rot initially infects flowers from ascospores, causing them to turn brown and wilt (Strickland et al., 2019). The fungus can spread to shoots and fruit. In shoots and twigs, a sunken canker may form. In fruit, a brown, circular spot develops, which expands, and fungal mycelia bearing spores appear, causing fruit to rot. Infected fruit can serve as a source of further infection. The fungus overwinters in mummified fruit and cankers on infected trees. Spores are spread by wind, rain, and insects, especially during wet conditions in the spring. Management should focus on removing overwintering inoculum, by removing dead wood, mummified fruit, and infected twigs. During summer, infected wood and fruit should be removed as it is observed. Fungicide application should be targeted around rain events and are particularly important in early spring. Fungicides need to be managed carefully to avoid resistance. Postharvest treatments can also assist with control of brown rot.

Powdery Mildew: This disease is caused by *Podosphaera pannosa* (Wallr.) var. *Persicae*. The most visible symptom is the appearance of circular white spots, especially on leaves and shoots; however, when the infection is severe, fruits also show characteristic symptoms of mildew attack, often associated with necrosis, deformities, and premature fruit drop, leading to unavoidable economic losses (Salazar et al., 2016). The highest incidence of the disease occurs in nurseries. On young leaves, blisters typically appear and become covered with grayish-white fungal spots. This leads to leaf deformation and premature leaf fall. In addition, green shoots and flowers may be directly attacked, showing white fungal spots, resulting in buds not opening properly and flower discoloration, which may eventually dry up. Fruits often develop circular spots that are initially pink and later turn dark brown, reducing the fruit quality. Controlling this disease often requires substantial amounts of pesticides, which can lead to residues in fruits, soil, water, etc. (Sharkawy Sahar et al., 2023). An efficient way to address this disease could be to cultivate, as much as possible, plants that are resistant or have low susceptibility to the disease by using parent plants with these characteristics in breeding programs for this species (Salazar et al., 2016).

Root rots: Root rot in apricot trees is caused by several soil-borne pathogens, with the most common being *Phytophthora* spp., *Armillaria* spp., and *Fusarium* spp. These fungi and oomycetes thrive in poorly drained soils and can infect the tree's roots, leading to a gradual decline in health. Symptoms of root rot include yellowing of leaves, stunted growth, wilting, and premature leaf drop. Infected trees may exhibit poor fruit production or no fruit at all. In severe cases, the tree may die. Cankers can form on the bark of infected crown and root tissue, and the tissue under the bark may appear dark brown. Treatment options include improving soil drainage and avoiding overwatering, which exacerbates the disease and selection of resistant rootstocks, e.g. Marianna 2624 and Myrobalan (McLaren et al., 1999). Fungicide treatments, such as those containing phosphorous acid, can help control *Phytophthora* infections, while soil fumigation or treatment with biocontrol agents may manage *Fusarium* or *Armillaria*. Additionally, removing and destroying infected plant material and practicing crop rotation can help reduce the spread of root rot pathogens.

Rust: Rust is caused by the fungal pathogen *Tranzschelia discolor*. It is characterized by orange or yellowish spots on the upper surfaces of leaves, and orange powdery spore masses on the undersides and spots on the fruit can follow (McLaren et al., 1999). As the disease progresses, it causes premature defoliation. Rust is typically more problematic in humid climates or during wet weather. Management includes removing and destroying infected leaves to reduce the spread of spores. Fungicide applications can be effective in preventing rust, especially during the early stages of infection.

Shot hole disease: Shot hole disease, also called Coryneum blight, is a fungal disease caused by *Wilsonomyces carpophilus* that damages apricot leaves, fruit, and shoots. It is particularly prevalent in wet spring conditions (Muhammad et al., 2022). Infected leaves develop small, dark purple or brown spots that eventually fall out, leaving holes that resemble a "shotgun" pattern. Many sunken, brown lesions with purple, water-soaked margins develop, mainly on the upper parts of the fruit. Severe infestations lead to defoliation, reduced fruit yield, and weakened trees. The fungus overwinters as hyphae and conidia in infected plant material (e.g., bud scales and in cankers on twigs) and is spread by rain and irrigation water. Removal of infected wood from the orchard and avoiding overhead irrigation are standard management practices. Fungicides are applied in late autumn before rains start and in spring.

Verticillium wilt: Verticillium wilt caused by the soil-borne fungus *Verticillium dahliae*. This pathogen invades the tree's vascular system, blocking water and nutrient transport, leading to wilting, yellowing, and curling of leaves. Verticillium wilt is difficult to treat once established, but managing soil health through proper irrigation, mulching, and improving drainage can help reduce stress on the trees. Removing infected trees can also reduce spread.

Silverleaf: Silverleaf in apricot trees is caused by the fungal pathogen *Chondrostereum purpureum*. The disease gets its name from the silvery, metallic sheen that appears on the leaves of infected trees, particularly on branches with vascular damage. There are no chemical treatments for silverleaf, so management typically involves avoiding unnecessary wounds, properly sterilizing pruning tools, and removing and destroying infected branches.

8.1.3 Viruses

Viruela: This disease is caused by the Apple Chlorotic Leaf Spot Virus (ACLSV). The name originates from Spain, where it is known as apricot pseudo-pox ("Viruela"), and in Italy, it is referred to as "butteratura." This disease significantly impacts not only apricot-producing regions but also other *Prunus* species. Typically, visible symptoms do not appear on leaves, although in peaches, a "dark green fleeting sunken mottle" may occasionally be seen. However, the most noticeable symptoms are found on the fruits, which develop severe deformities, making them unmarketable (Rubio et al., 2019). Recently, two apricot cultivars have been identified as resistant to this virus: the heirloom Spanish cultivar Mauricio and the French cultivar Bergeron (García-Ibarra et al., 2012). To manage this disease effectively, it is essential first to apply cultural practices by carefully managing propagation plant material to avoid infections. Additionally, in breeding programs, it is advantageous to use ACLSV-resistant parents to develop new descendants with this resistance trait (Rubio et al., 2023).

Prunus Necrotic Ringspot Virus: Prunus Necrotic Ringspot Virus (PNRSV) is a widespread viral pathogen in the *Prunus* genus, caused by a member of the *Tymoviridae* family. The virus is typically transmitted through infected rootstocks, grafting, and mechanical means, as well as seed and pollen. Symptoms include necrotic ring spots, chlorosis, and reduced growth, particularly in the leaves. The virus can also cause fruit deformities and lower yields. While there is no chemical cure for PNRSV, management strategies include using certified virus-free planting materials and removing and destroying infected trees to prevent further transmission. Some rootstocks show resistance to PNRSV.

Sharka: This disease is considered one of the most destructive diseases of apricots. It is caused by the PPV and is prevalent in most apricot growing countries around the world. Sharka significantly reduces tree growth and yield and thus is an economically important issue for growers (Milošević et al., 2019b). It also reduces flesh:stone weight ratio, and in severe cases it reduces fruit weight. It can also change the fruit chemical composition due to premature ripening, including increasing soluble solids content and titratable acidity whilst reducing secondary metabolites. This can have a negative effect on flavor, which is often reported as bland or tasteless, and fruit become unsuitable for consumption. Symptoms are not always visible, like for plum, but often rings or spots can be observed on stones and possibly on fruit skin. Presence can be detected using PCR. PPV is transmitted by aphid vectors or grafting and has a wide host range among Prunus species. There is no treatment for Sharka. The use of certified virus-free material is recommended and, most importantly, growing virus-resistant cultivars (Vávra et al., 2023).

8.2 Pests

Aphids: A number of aphids can infest apricot trees including green peach aphid *(Myzus persicae)* and the black cherry aphid *(Myzus cerasi)*. These small, soft-bodied insects feed on plant sap and can multiply quickly, especially under favorable conditions. Aphids infest blossoms and new leaves but are less common on apricots than other stone fruit. Mineral oils can be applied to smother eggs, and insecticides can control active stages. Biological control, such as the introduction of natural predators like ladybugs or parasitic wasps, can assist in keeping populations low.

Giant-Headed Worm: *Capnodis tenebrionis* L. belongs to the order *Coleoptera*, family *Buprestidae*, and is a devastating pest of *Prunus* species, primarily in Mediterranean areas. The main damage is usually caused to the roots by the larval stage. The larvae damage the vascular bundles, leading to reduced photosynthetic activity, which can result in the death of the tree within just 2 years. Symptoms tend to appear very late, including yellow leaves, dry branches, bark injuries, and sap oozing due to larval activity (Arapostathi et al., 2024). Managing this pest is challenging, mainly due to the lack of efficient monitoring tools, effective insecticides, resistant rootstocks, and entomopathogens targeting this unique pest. Therefore, further research is needed to develop monitoring devices and semiochemical-based control methods for *C. tenebrionis* (Kokici et al., 2020). Chemical controls, such as organophosphate and carbamate insecticides, have been commonly used, except for those now banned (Ben-Yehuda et al., 2000). Biological control is still in the research phase, as the currently available predators and parasitoids have minimal impact on this pest.

Peach twig borer: The peach twig borer *(Anarsia lineatella)* is a significant pest of apricot trees, causing damage to both shoots and fruit. This pest is common in warm, dry climates and can cause substantial economic losses if not managed effectively. They usually overwinter as second instar larvae (Ivanova, 2006). The larvae attack buds and new shoots, with around three generations per year. Larvae can also bore into fruit. Insecticides should be applied at key periods. Parasitic wasps and predators, such as green lacewings, can help reduce populations. Pheromone traps are useful for monitoring and controlling adult moths.

Oriental fruit moth: The oriental fruit moth (*Cydia molesta* Busck.) is a major pest of apricot, particularly in temperate regions. The larvae feed on shoots and fruit, causing substantial damage (McLaren et al., 1999; Ivanova, 2006). Oriental fruit moths typically have four generations per year, depending on temperature. They overwinter as larvae in cocoons within the bark or soil. Insecticides should be applied at appropriate times to target young larvae. Parasitic wasps and other natural predators can help suppress oriental fruit moth populations. The best time to control this moth is before the young larvae enter the shoot, with pheromone traps for monitoring populations to determine application timing.

Leafroller: Some leafroller larvae *(Tortricidae)* cause damage by feeding on leaves and fruit (McLaren et al., 1999; Ivanova, 2006). They use webbing to build a shelter and can roll leaves up or bind them together. Two to four generations occur per year, and they overwinter as second to fourth instar larvae. Pheromone traps can provide a tool to target insecticide sprays.

Brown marmorated stink bug: The brown marmorated stink bug (BMSB), *Halyomorpha halys*, is an invasive species from the Pentatomidae family that poses a significant threat to economically important crops. Native to East Asia, it has spread widely across Europe and the United States. In apricot trees, BMSB infestation results in damage such as discoloration, necrotic areas on both fruit and leaves, and the development of sunken spots and lesions on the fruit. While pesticide applications can help control the pest, additional management strategies are necessary to effectively mitigate damage and prevent further spread.

Mites: A number of mites can attack apricots, including European red mite (*Panonychus ulmi* [Koch]) and two-spotted mite (*Tetranychus urticae* Koch) (McLaren et al., 1999). Mites feed on leaves, causing spotting and mottling. Severe infestations can result in defoliation. They can also damage fruit of some cultivars. A number of generations can occur in 1 year. Predatory mites are a good control for mites; however, miticides can be used cautiously to avoid resistance development.

9. Current and future challenges in main production regions

In recent years, climate change has imposed the most devastating challenges in apricot production worldwide. Failure to satisfy dormancy due to increase in winter temperature may cause flower bud drop, while flowers may not be fully developed inducing yield failure. Moreover, the conditions of "false spring" during the winter months cause flowering during the winter months and damaging frosts. Cultivar traits related to medium chill, high heat requirements are most needed to avoid frost damages and satisfy chill requirements. Agroclimatic characterization in the apricot growing regions and careful planning with cultivars adapting to a specific area are required. Effective techniques such as monitoring the chill accumulation during the winter and spring months and spraying with chemicals or nutrients for breaking dormancy are needed. Phenotyping and characterization for chill requirements and phenology of cultivars are needed before commercial cultivation.

Scarce availability of water resources and soil salinization problems have resulted due to climate change. Breeding for cultivars with resilience to biotic and abiotic stresses are needed, for example, traits such as resistance to PPV-Markus strain, being more damaging and having spread in many regions of the world. Various breeding programs are worldwide, and a range with cultivars with optimum appearance, taste, and high yield are available. Yet it is important that cultivars are evaluated in local pedoclimatic conditions before commercial use, especially nowadays.

The increase in the cultivation costs and the uncertain prices are weak points in apricot production. Mechanization is needed to reduce production cost and also due to a lack of field-workers (both in availability and specialization). It is also important to increase apricot consumption by choosing cultivars with optimum fruit quality. Moreover, the consumer needs to be able to recognize the cultivar and have consistency of taste during the season. There is a need to expand and plan the productive calendar. In the future, for apricot production, we also need to consider sustainability in the production systems.

References

Akbaba, M., Hürkan, K., & Ozcan, O. (2023). Characterization of causal agents of bacterial canker on apricot plantations and risk mapping using GIS in Aras Basin (Türkiye). *Journal of Phytopathology, 171*, 517–536.

Alburquerque, N., García-Montiel, F., Carrillo, A., & Burgos, L. (2008). Chilling and heat requirements of sweet cherry cultivars and the relationship between altitude and the probability of satisfying the chill requirements. *Environmental and Experimental Botany, 64*(2), 162–170. https://doi.org/10.1016/j.envexpbot.2008.01.003.

Arapostathi, E., Panopoulou, C., Antonopoulos, A., Katsileros, A., Karellas, K., Dimopoulos, C., et al. (2024). Early detection of potential infestation by Capnodis tenebrionis (L.) (Coleoptera: Buprestidae), in stone and pome fruit orchards, using multispectral data from a UAV. *Agronomy, 14*, 20. https://doi.org/10.3390/agronomy14010020.

Assirelli, A., Caracciolo, G., & Roccuzzo, G. (2021). New tools for mechanical thinning of apricot fruitlets. *Agriculture-Basel, 11*(11), 1138. https://doi.org/10.3390/agriculture11111138.

Atkinson, C., & Else, M. (2001). Understanding how rootstocks dwarf fruit trees. *Compact Fruit Tree, 34*, 46–49.

Aubert, C., & Chanforan, C. (2007). Postharvest changes in physicochemical properties and volatile constituents of apricot (*Prunus armeniaca* L.). Characterization of 28 cultivars. *Journal of Agricultural and Food Chemistry, 55*(8), 3074–3082. https://doi.org/10.1021/jf063476w.

Aydin, S., & Ulvi, M. (2019). Residue levels of pesticides in nuts and risk assessment for consumers. *Quality Assurance and Safety of Crops & Foods, 11*(6), 539–548. https://doi.org/10.3920/QAS2018.1405.

Ayour, J., Sagar, M., Alfeddy, M. N., Taourirte, M., & Benichou, M. (2016). Evolution of pigments and their relationship with skin color based on ripening in fruits of different Moroccan genotypes of apricots (Prunus armeniaca L.). *Scientia Horticulturae, 207*, 168–175. https://doi.org/10.1016/j.scienta.2016.05.027.

Badenes, M. L., Asíns, M. J., Carboxell, E. A., & Glacer, G. (1996). Genetic diversity in apricot, Prunus armeniaca, aimed at improving resistance to Plum Pox Virus. *Plant Breeding, 115*(2), 133–139. https://doi.org/10.1111/j.1439-0523.1996.tb00888.x.

Badenes, M. L., Martínez-Calvo, J., & Llácer, G. (1998). Analysis of apricot germplasm from the European ecogeographical group. *Euphytica, 102*(1), 93–99. https://doi.org/10.1023/A:1018332312570.

Badenes, M. L., Pastor, I., Martínez-Calvo, J., & Llácer, G. (2000). Improved efficiency in apricot breeding: Earlier assessment of seedling progeny for resistance to plum pox virus. *J Hort Sci & Biotechnol, 75*(4), 459–464.

Ben-Yehuda, S., Assael, F., & Mendel, Z. (2000). Improved chemical control of Capnodis tenebrionis and C. carbonaria in stone-fruit plantations in Israel. *Phytoparasitica, 28*(1), 27–41. https://doi.org/10.1007/bf02994021.

Bird, G. W., Ogawa, J. M., Ritchie, D. F., Urio, K., Uyemoto, J. K., & Zehr, E. I. (1995). *Compendium of stone fruit diseases*. St. Paul: The American Phytopatological Society.

Borsani, J., Budde, C. O., Porrini, L., Lauxmann, M. A., Lombardo, V. A., Murray, R, et al. (2009). Carbon metabolism of peach fruit after harvest: Changes in enzymes involved in organic acid and sugar level modifications. *Journal of Experimental Botany, 60*(6), 1823–1837. https://doi.org/10.1093/jxb/erp055.

Browne, G. T. (2017). Resistance to Phytophthora species among rootstocks for cultivated Prunus species. *HortScience, 52*(11), 1471–1476. Retrieved from: Jun 14, 2024. https://doi.org/10.21273/HORTSCI10391-17.

Bussi, C., Besset, J., & Girard, T. (2003). Effects of fertilizer rates and dates of application on apricot (cv Bergeron) cropping and pitburn. *Scientia Horticulturae, 98*(2), 139–147. https://doi.org/10.1016/S0304-4238(02)00203-0.

Campoy, J. A., Ruiz, D., Allderman, L., Cook, N., & Egea, J. (2012). The fulfilment of chilling requirements and the adaptation of apricot (Prunus armeniaca L.) in warm winter climates: An approach in Murcia (Spain) and the Western Cape (South Africa). *European Journal of Agronomy, 37*, 43–55. https://doi.org/10.1016/j.eja.2011.10.004.

Contador, L., Shinya, P., Díaz, M., Hernandez, E., & Infante, R. (2015). Evaluation of textural properties of peach and nectarine through texture profile analysis. *Acta Horticulturae, 1079*, 633–636. https://doi.org/10.17660/ActaHortic.2015.1079.86.

Costa, G., Noferini, M., Fiori, G., Ziosi, V., Berthod, N., & Rossier, J. (2010). Establishment of the optimal harvest time in apricot (Orangered and Bergarouge) by means of a new index based on vis spectroscopy. *Acta Hortic, 862*, 533–539. https://doi.org/10.17660/ActaHortic.2010.862.84.

Croppi, et al. (2021). Population genomics of apricots unravels domestication history and adaptive events. *Nature Communications, 12*, 3956.

Dabbou, S., Lussiana, C., Maatallah, S., Gasco, L., Hajlaoui, H., & Flamini, G. (2016). Changes in biochemical compounds in flesh and peel from *Prunus persica* fruits grown in Tunisia during two maturation stages. *Plant Physiology and Biochemistry, 100*, 1–11. https://doi.org/10.1016/j.plaphy.2015.12.015.

Demir, Z., Tursun, N., & Isik, D. (2019). Effects of different cover crops on soil quality parameters and yield in an apricot orchard. *International Journal of Agriculture and Biology, 21*(2), 399–408. https://doi.org/10.17957/IJAB/15.0000.

Ding, G. X., Wang, B. M., Fu, B. C., Cang, G. Y., Niu, X. Y., & Wen, P. F. (2018). Water adhered on the fruit surface during maturation period is the main cause of apricot cracking. *Acta Hortic, 1214*, 137–144. https://doi.org/10.17660/ActaHortic.2018.1214.23.

Dondini, L., Lain, O., Geuna, F., Banfi, R., Gaiotti, F., Tartarini, S., et al. (2007). Development of a new SSR-based linkage map in apricot and analysis of synteny with existing Prunus maps. *Tree Genetics and Genomes, 3*(3), 239–249. https://doi.org/10.1007/s11295-006-0059-8.

Dorostkar, M., Moradinezhad, F., & Ansarifar, E. (2022). Influence of active modified atmosphere packaging pre-treatment on shelf life and quality attributes of cold stored apricot fruit. *International Journal of Fruit Science, 22*, 402–413. https://doi.org/10.1080/15538362.2022.2047137.

Drogoudi, P. D., Vemmos, S., Pantelidis, G., Petri, E., Tzoutzoukou, C., & Karayiannis, I. (2008). Physical characters and antioxidant, sugar and mineral nutrient contents in fruit from 29 apricot (*Prunus armeniaca* L.) cultivars and hybrids. *Journal of Agricultural and Food Chemistry, 56*, 10754–10760. https://doi.org/10.1021/jf801995x.

Egea, J. A., Caro, M., García-Brunton, J., Gambín, J., & Egea, J. (2022). Agroclimatic metrics for the main stone fruit producing areas in Spain in current and future climate change scenarios: Implications from an adaptive point of view. *Frontiers in Plant Science, 13*, 842628. https://doi.org/10.3389/fpls.2022.842628.

Ezzat, A., Salama, A.-M., Szabó, S., Yaseen, A. A., Molnár, B., & Holb, I. J. (2021). Deficit irrigation strategies on tree physiological and chemical properties: Treatment effects, prediction based model analyses and inter-correlations. *Agronomy, 11*(7), 1361. https://doi.org/10.3390/agronomy11071361.

Fabian, J. R., & Keene, N. H. (2001). *Transmission genetics. 139-144, 4* 2008–2016. Elsevier. https://doi.org/10.1016/B978-0-12-374984-0.01568-0.

Fadón, E., Herrera, S., Guerrero, B. I., Engracia Guerra, M., & Rodrigo, J. (2020). Chilling and heat requirements of temperate stone fruit trees (*Prunus* sp.). *Agronomy, 10*, 409. https://doi.org/10.3390/agronomy10030409.

Fan, X., Zhao, H., Wang, X., Cao, J., & Jiang, W. (2017). Sugar and organic acid composition of apricot and their contribution to sensory quality and consumer satisfaction. *Scientia Horticulturae, 225*, 553–560. https://doi.org/10.1016/j.scienta.2017.07.016.

Fatima, T., Bashir, O., Gani, G., Bhat, T., & Jan, N. (2018). Nutritional and health benefits of apricots. *International Journal of Unani and Integrative Medicine, 2*, 5–9. https://doi.org/10.33545/2616454X.2018.v2.i2a.25.

Faust, M., Suranyi, D., & Nyujto, F. (1998). Origin and dissemination of apricots. In I. Wiley, & J. & sons (Eds.), *Horticultural reviews* (Vol. 22, pp. 225–260). Wesport then New York.

Foschi, S., Spotti, C., Minuto, A., & Minuto, G. (2010). Shank application of chloropicrin as an alternative to methyl bromide for Prunus tree replant. *Acta Horticulturae, 883*, 157–162. https://doi.org/10.17660/ActaHortic.2010.883.17.

Gómez-Martínez, H., Bermejo, A., Badenes, M. L., & Zuriaga, E. (2021). Nutraceutical profiles of apricots (Prunus armeniaca L.) as a source of fruit quality traits for breeding. *Spanish Journal of Agricultural Research, 19*(4), Article e0703. https://doi.org/10.5424/sjar/2021194-18331.

Gürcan, K., Çetinsağ, N., Pınar, H., & Macit, T. (2019). Molecular and biological assessment reveals sources of resistance to Plum pox virus - Turkey strain in Turkish apricot (Prunus armeniaca) germplasm. *Scientia Horticulturae, 252*, 348–353. https://doi.org/10.1016/j.scienta.2019.04.003.

Gao, J., Guo, W., Liu, Q., Liu, M., Shang, C., Song, Y., et al. (2024). The physiological response of apricot flowers to low-temperature stress. *Plants, 13*, 1002. https://doi.org/10.3390/plants13071002.

García-Gómez, B. E., Ruiz, D., Salazar, J. A., Rubio, M., Martínez-García, P. J., & Martínez-Gómez, P. (2020). Analysis of metabolites and gene expression changes relative to apricot (*Prunus armeniaca* L.) fruit quality during development and ripening. *Frontiers in Plant Science, 11*, 1269. https://doi.org/10.3389/fpls.2020.01269.

García-Gómez, B. E., Salazar, J. A., Egea, J. A., Rubio, M., Martínez-Gómez, P., & Ruiz, D. (2022). Monitoring apricot (*Prunus armeniaca* L.) ripening progression through candidate gene expression analysis. *International Journal of Molecular Sciences, 23*, 4575. https://doi.org/10.3390/ijms23094575.

García-Gómez, B. E., Salazar, J. A., Nicolás-Almansa, M., Razi, M., Rubio, M., Ruiz, D., et al. (2021). Molecular bases of fruit quality in *Prunus* species: An integrated genomic, transcriptomic, and metabolic review with a breeding perspective. *International Journal of Molecular Sciences, 22*, 333. https://doi.org/10.3390/ijms22010333.

García-Ibarra, A., Dicenta, F., Martínez-Gómez, P., & Rubio, M. (2012). Evaluation of apricot (*Prunus armeniaca* L.) resistance to Apple chlorotic leaf spot virus in controlled greenhouse conditions. *Eur J Plant Pathol, 133*, 857–863.

Groppi, A., Liu, S., Cornille, A., et al. (2021). Population genomics of apricots unravels domestication history and adaptive events. *Nat Commun, 12*, 3956. https://doi.org/10.1038/s41467-021-24283-6.

Hagen, L., Khadari, B., Lambert, P., & Audergon, J.-M. (2002). Genetic diversity in apricot revealed by AFLP markers: Species and cultivar comparisons. *Theoretical and Applied Genetics, 105*(2), 298−305. https://doi.org/10.1007/s00122-002-0910-8.

Halász, J., Pedryc, A., Ercisli, S., Yilmaz, K. U., & Hegedűs, A. (2010). S-Genotyping supports the genetic relationships between Turkish and Hungarian apricot germplasm. *J. Amer Soc Hort Sci., 135*(5), 410−417. https://doi.org/10.21273/JASHS.135.5.410.

Han, Y., Liu, Y., Li, Z., Sun, J., Song, F., & Chen, J. (2024). Effects of chitosan grafted with gallic acid treatment on the postharvest physiology and biochemistry of Saimaiti apricots. *Journal of Stored Products Research, 108*, Article 102390. https://doi.org/10.1016/j.jspr.2024.102390.

Hassan, M., Monir, G., & Shoala, T. (2019). Biological and chemical control of powdery mildew (*Sphaerotheca pannosa* (wallr.) var. persicae) in apricot. *International Journal of Sustainable Development, 2*, 1−19.

Hassan, M. M, Salem, A. T., & Fayek, M. A. (1986). Response of apricot seedling lines to waterlogging. *Scientia Horticulturae, 29*(1−2), 125−130. https://doi.org/10.1016/0304-4238(86)90038-5.

Herrera, S., Lora, J., Hormaza, J. I., Herrero, M., & Rodrigo, J. (2018). Optimizing production in the new generation of apricot cultivars: Self-incompatibility, S-RNase allele identification, and incompatibility group assignment. *Front Plant Sci, 9*, 527. https://doi.org/10.3389/fpls.2018.00527. PMID: 29755489. PMCID: PMC5935046.

Herrera, S., Lora, J., Hormaza, J. I., & Rodrigo, J. (2022). Self-incompatibility in apricot: Identifying pollination requirements to optimize fruit production. *Plants (Basel), 11*(15), 2019. https://doi.org/10.3390/plants11152019.

İncedayi, B., Tamer, C. E., Sinir, G.Ö., Suna, S., & Çopur, Ö. U. (2016). Impact of different drying parameters on color, β-carotene, antioxidant activity and minerals of apricot (*Prunus armeniaca* L.). *Food Science and Technology, 36*, 171−178. https://doi.org/10.1590/1678-457X.0086.

Ivanova, L. (2006). Key pests of apricot and control. *Acta Hortic, 701*, 651−654. https://doi.org/10.17660/ActaHortic.2006.701.115.

Kaska, N. (2006). Orchard management in apricots. *Acta Horticulturae, 717*, 287−293. https://doi.org/10.17660/ActaHortic.2006.717.58.

Katz, E., Lagunes, P. M., Riov, J., Weiss, D., & Goldschmidt, E. E. (2004). Molecular and physiological evidence suggests the existence of a system II-like pathway of ethylene production in non-climacteric *Citrus* fruit. *Planta, 219*, 243−252. https://doi.org/10.1007/s00425-004-1228-3.

Kaya, O., Kose, C., & Gecim, T. (2018). An exothermic process involved in the late spring frost injury to flower buds of some apricot cultivars (*Prunus armenica* L.). *Scientia Horticulturae, 241*, 322−328. https://doi.org/10.1016/j.scienta.2018.07.019.

Khasawneh, A. R., Alsmairat, N., Othman, Y. A., Ayad, J. Y., Al-Hajaj, H., & Qrunfleh, I. M. (2022). Controlled-release nitrogen fertilizers for improving yield and fruit quality of young apricot trees. *Scientia Horticulturae, 303*, Article 111233. https://doi.org/10.1016/j.scienta.2022.111233.

Kokici, H., Laterza, I., Bari, G., Meneghini, M., Addante, R., & De Lillo, E. (2020). A preliminary report on the effects of the rootstocks on the postembryonic development of Capnodis tenebrionis on semi-artificial substrates. *Bull Insectology, 73*(1), 95−101.

Krška, B. (2018). Genetic apricot resources and their utilisation in breeding. In *Breeding and health benefits of fruit and nut crops*. InTech. https://doi.org/10.5772/intechopen.77125.

Layne, R., Bailey, C., & Hough, L. F. (1996). Apricots. *Fruit Breeding, 1*, 79−111.

Lei, X. H., Yuan, Q. C., Xyu, T., Qi, Y. N., Zeng, J., Huang, K., et al. (2023). Technologies and equipment of mechanized blossom thinning in orchards: A review. *Agronomy, 13*(11), 2573. https://doi-org.ezpakl.plantandfood.co.nz/10.3390/agronomy13112753.

Li, K., DiLegge, M. J., Minas, I. S., Hamm, A., Manter, D., & Vivanco, J. M. (2019). Soil sterilization leads to recolonization of a healthier rhizosphere microbiome. *Rhizosphere, 12*, Article 100176.

Lichou, J., & Jay, M. (2012). *Monographie abricot*, 978-2-87911-311-1 (p. 567). Paris [FRA]: CTIFL Centre Technique Interprofessionnel des Fruits et Légumes | 2012. Frenche.

Liu, B., Zhao, H., Fan, X., Jiao, W., Cao, J., & Jiang, W. (2019). Near freezing point temperature storage inhibits chilling injury and enhances the shelf life quality of apricots following long-time cold storage. *Journal of Food Processing and Preservation, 43*, Article e13958. https://doi.org/10.1111/jfpp.13958.

Liu, M. C., Pirrello, J., Chervin, C., Roustan, J. P., & Bouzayen, M. (2015). Ethylene control of fruit ripening: Revisiting the complex network of transcriptional regulation. *Plant Physiol, 169*, 2380–2390. https://doi.org/10.1104/pp.15.01361.

Lombardo, V. A., Osorio, S., Borsani, J., Lauxmann, M. A., Bustamante, C. A., Budde, C. O., et al. (2011). Metabolic profiling during peach fruit development and ripening reveals the metabolic networks that underpin each developmental stage. *Plant Physiol, 157*(4), 1696–1710. https://doi.org/10.1104/pp.111.186064. Epub 2011 Oct 20. PMID: 22021422. PMCID: PMC3327199

Looney, N. E., & Killick, R. G. (1979). Apricot blossom thinning with dinitro-ortho-butylphenol. *Canadian Journal of Plant Science, 59*, 741–746. https://doi.org/10.4141/cjps79-115.

Maghuly, F., Fernandez, E. B., Laimer, M., Ruthner, S., Bisztray, G. D., & Pedryc, A. (2006). Microsatellite characterisation of apricot (Prunus armeniaca) cultivars grown in Central Europe. *Acta Horticulturae, 717*, 207–212. https://doi.org/10.17660/ActaHortic.2006.717.43.

Manganaris, A. G., Mainou, A., Goudaras, A., & Ledbetter, C. (1999). Identification of plum x apricot interspecific hybrids using isoenzyme polymorphism. *Acta Horticulturae, 488*, 361–368. https://doi.org/10.17660/ActaHortic.1999.488.57.

Manganaris, G. A., & Crisosto, C. H. (2020). Stone fruits: Peaches, nectarines, plums, apricots. In M. Gil, & R. Beaudry (Eds.), *Controlled and modified atmosphere for fresh and fresh-cut produce* (pp. 311–322). Elsevier.

Manganaris, G. A., Rasori, A., Bassi, D., Geuna, F., Ramina, A., Tonutti, P., & Bonghi, C. (2011). Comparative transcript profiling of apricot (*Prunus armeniaca* L.) fruit development and on-tree ripening. *Tree Genet. Gen, 7*, 609–616. https://doi.org/10.1007/s11295-010-0360-4.

Martínez-Gómez, P., Sozzi, G. O., Sánchez-Pérez, R., Rubio, M., & Gradziel, T. M. (2003). New approaches to Prunus tree crop breeding. *Journal of Food, Agriculture and Environment Issue, 1*, 52–63. https://doi.org/10.1234/4.2003.316.

Martínez-García, P. J., Hartung, J., Pérez de los Cobos, F., Martínez-García, P., Jalili, S., Sánchez-Roldán, J. M., et al. (2020). Temporal response to drought stress in several Prunus rootstocks and wild species. *Agronomy, 10*, 1383. https://doi.org/10.3390/agronomy10091383.

McLaren, G. F., Grandison, G., Wood, G. A., Tate, G., & Horner, I. (1999). *Summerfruit in New Zealand: Management of pests and diseases* (p. 14). University of Otago Press.

Mendelné Pászti, E., Bujdoso, G., Ercisli, S., Hrotkó, K., & Mendel, Á. (2023). Apricot rootstocks with potential in Hungary. *Horticulturae, 9(6), 720*(6), 1–15. https://doi.org/10.3390/horticulturae9060720.

Miller, N. F. (2014). Plants and humans in the Near East and the Caucasus: Ancient and traditional uses of plants as food and medicine, a diachronic ethnobotanical review. *Ethnobiology Letters, 5*(0), 22. https://doi.org/10.14237/ebl.5.2014.139.

Milošević, T., & Milošević, N. (2019a). Behavior of some cultivars of apricot (Prunus armeniaca L.) on different rootstocks. *Mitteilungen Klosterneuburg, 69*, 1–12.

Milošević, T., Milošević, N., Mladenović, J., & Jevremović, D. (2019b). Impact of Sharka disease on tree growth, productivity and fruit quality of apricot (*Prunus armeniaca* L.). *Scientia Horticulturae, 244*, 270–276. https://doi.org/10.1016/j.scienta.2018.09.055. ISSN 0304-4238.

Moale, C., Ghiurea, M., Sîrbu, C. E., Somoghi, R., Cioroianu, T. M., Faraon, V. A., et al. (2021). Effects of siliceous natural nanomaterials applied in combination with foliar fertilizers on physiology, yield and fruit quality of the apricot and peach trees. *Plants-Basel, 10*(11), 2359. https://doi.org/10.3390/plants10112395.

Mozaffari, M., Sadeghi, S., & Asefi, N. (2022). Prediction of the quality properties and maturity of apricot by laser light backscattering imaging. *Postharvest Biology and Technology, 186*, Article 111842. https://doi.org/10.1016/j.postharvbio.2022.111842.

Muñoz-Sanz, J. V., Zuriaga, E., Cruz-García, F., McClure, B., & Romero, C. (2020). Self-(In) compatibility systems: Target traits for crop-production, plant breeding, and biotechnology. *Front Plant Sci, 11*, 195. https://doi.org/10.3389/fpls.2020.00195.

Muhammad, M., Hussain, A., Ali, S., Akram, W., Roomi, I., Faiz, F., & Hussain, I. (2022). Geostatistical analysis of apricot shot hole disease and influence factors in district Nagar, Gilgit-Baltistan, Pakistan. *International Journal of Phytopathology, 11*(03), 227–238. https://doi.org/10.33687/phytopath.011.03.4241.

Musacchi, S., Abeti, C., Ancarani, V., Bucci, D., Gagliardi, F., & Serra, S. (2014). Summer pruning of apricot cvs. Bora® and carmen Top®. *Acta Hortic, 1058*, 85–93. https://doi.org/10.17660/ActaHortic.2014.1058.8.

Nas, Z., & Eşitken, A. (2023). Micropropagation of an apricot (*Prunus armeniaca* L.) rootstock candidate. *Erwerbs-Obstbau, 65*, 2357–2365. https://doi.org/10.1007/s10341-023-00967-9.

Nicolás-Almansa, M., Ruiz, D., Salazar, J. A., Guevara, A., Cos, J., Martínez-Gómez, P., et al. (2023). Phenotypic and molecular characterization of new interspecific Japanese plum × apricot hybrids (plumcots). *Scientia Horticulturae, 318*, Article 112131. https://doi.org/10.1016/j.scienta.2023.112131.

O'Connell, M. (2022). Crop load and canopy architecture affect yield and fruit quality of Golden May apricot. *Acta Horticulturae, 1346*, 287–294. https://doi.org/10.17660/ActaHortic.2022.1346.36.

Ortuño-Hernández, G., Fernández, M., Martínez-Gómez, P., Ruiz, D., & Salazar, J. A. (2024b). Ripening-related gene expression analysis revealed the molecular impact of 1-MCP application on apricot fruit softening, color, aroma, and antioxidant capacity. *Postharvest Biology and Technology, 216*, Article 113037. https://doi.org/10.1016/j.postharvbio.2024.113037.

Ortuño-Hernández, G., Sánchez, M., Ruiz, D., Martínez-Gómez, P., & Salazar, J. A. (2024a). Monitoring fruit growth and development in apricot (*Prunus armeniaca* L.) through gene expression analysis. *International Journal of Molecular Sciences, 25*, 9081. https://doi.org/10.3390/ijms25169081.

Pérez-Caselles, C., Burgos, L., Sánchez-Balibrea, I., Egea, J. A., Faize, L., Martín-Valmaseda, M., et al. (2023). The effect of silver nanoparticle addition on micropropagation of apricot cultivars (*Prunus armeniaca* L.) in semisolid and liquid media. *Plants, 12*, 1547. https://doi.org/10.3390/plants12071547.

Pantelidis, G., & Drogoudi, P. (2023). Exploitation of genotypic variation in chilling and heat requirements for flowering in Prunus armeniaca and Prunus persica (L.) Batsch cultivars. *Scientia Horticulturae, 321*, Article 112287. https://doi.org/10.1016/j.scienta.2023.112287.

Psota, V. (2012). Control of brown rot blossom blight (*Monilinia laxa*) on apricot in organic agriculture. *IOBC-WPRS Bulletin, 91*, 357–360.

Radi, M., Mahrouza, M., Jaouad, A., & Amiot, M. J. (2003). Influence of mineral fertilization (NPK) on the quality of apricot fruit (cv. Canino): The effect of the mode of nitrogen supply. *Agronomie, 23*(8), 737–745. https://doi.org/10.1051/agro:2003052.

Raina, J. N., Thakur, B. C., Suman, S., & Spehia, R. S. (2005). Effect of fertigation through drip system on nitrogen dynamics, growth, yield and quality of apricot. *Acta Horticulturae, 696*, 227–231. https://doi.org/10.17660/ActaHortic.2005.696.39.

Reig, G., Zarrouk, O., Font i Forcada, C., & Moreno, M.Á. (2018). Anatomical graft compatibility study between apricot cultivars and different plum based rootstocks. *Scientia Horticulturae, 237*, 67–73. https://doi.org/10.1016/j.scienta.2018.03.035.

Reighard, G. L., Ouellette, D., & Minas, I. S. (2022). Field performance of peach rootstock cultivars in South Carolina. *Acta Horticulturae, 1352*, 63–68. https://doi.org/10.17660/ActaHortic.2022.1352.8.

Remorini, D., Fei, C., Loreti, F., & Massai, R. (2015). Observations on nine peach rootstocks grown in a replant soil. *Acta Horticulturae, 1084*, 131–137. https://doi.org/10.17660/ActaHortic.2015.1084.16.

Rettke, M. A., & Dahlenburg, A. P. (1999). Effect of timing of hand thinning on productivity of Moorpark apricots destined for drying. *Aust J Exp Agric, 39*(7), 885–889. https://doi.org/10.1071/EA99018.

Rettke, M. A., Pitt, T. R., Maier, N. A., & Jones, J. A. (2006). Growth and yield responses of apricot (cv. Moorpark) to soil-applied nitrogen. *Australian Journal of Experimental Agriculture, 46*(1), 115–122. https://doi.org/10.1071/EA04207.

Rodrigo, J., Julian, C., & Herrero, M. (2006). Spring frost damage in buds, flowers and developing fruits in apricot. *Acta Hortic, 717*, 87–88. https://doi.org/10.17660/ActaHortic.2006.717.15.

Rodriguez, C. E., Bustamante, C. A., Budde, C. O., Müller, G. L., Drincovich, M. F., & Lara, M.v. (2019). Peach fruit development: A comparative proteomic study between endocarp and mesocarp at very early stages underpins the main differential biochemical processes between these tissues. *Frontiers in Plant Science, 10, 715*, 1–19. https://doi.org/10.3389/fpls.2019.00715.

Rubio, M., Martínez-Gómez, P., & Dicenta, F. (2023). Apricot breeding for multiple resistance to Plum pox virus and Apple chlorotic leaf spot virus. *Scientia Horticulturae, 309*, Article 111706. https://doi.org/10.1016/j.scienta.2022.111706.

Rubio, M., Ruiz, D., Martínez-Gómez, P., Egea, J., & Dicenta, F. (2018). Multi-resistance of apricot cultivars to the most important diseases in Spain. *Acta Hortic, 1214*, 231–234. https://doi.org/10.17660/ActaHortic.2018.1214.41.

Rubio, M., Salazar, J., Dicenta, F., Ruiz, D., Martinez-Gomez, P., & Martínez-García, P. J. (2019). Identification of quantitative trait loci (QTLs) linked to Apple chlorotic leaf spot virus (ACLSV) resistance in apricot. *Euphytica, 215, 170*, 1–8. https://doi.org/10.1007/s10681-019-2481-7.

Ruiz, D., Campoy, J. A., & Egea, J. (2007). Chilling and heat requirements of apricot cultivars for flowering. https://doi.org/10.1016/j.envexpbot.2007.06.008.

Ruiz, D., Egea, J., Gil, M. I., & Tomás-Barberán, F. A. (2005a). Characterization and quantitation of phenolic compounds in new apricot (Prunus armeniaca L.) varieties. *Journal of Agricultural and Food Chemistry, 53*, 9544–9552. https://doi.org/10.1021/jf051539p.

Ruiz, D., Egea, J., Tomás-Barberán, F. A., & Gil, M. I. (2005b). Carotenoids from new apricot (Prunus armeniaca L.) varieties and their relationship with flesh and skin color. *Journal of Agricultural and Food Chemistry, 53*, 6368–6374. https://doi.org/10.1021/jf0480703.

Ruiz-Sánchez, M. C., Domingo, R., Torrecillas, A., & Pérez-Pastor, A. (2000). Water stress preconditioning to improve drought resistance in young apricot plants. *Plant Science, 156*(2), 245–251. https://doi.org/10.1016/S0168-9452(00)00262-4.

Ruml, M., Vukovic, A., & Milatovic, D. (2010). Evaluation of different methods for determining growing degree-day thresholds in apricot cultivars. *International Journal of Biometeorology, 54*, 411–422. https://doi.org/10.1007/s00484-009-0292-6.

Salazar, J. A., Batnini, M. A., Trifi-Farah, N., et al. (2016). Quantitative trait loci (QTLs) identification and the transmission of resistance to powdery mildew in apricot. *Euphytica, 211*, 245–254. https://doi.org/10.1007/s10681-016-1734-y.

Salazar, J. A., Ruiz, D., Campoy, J. A., Tartarini, S., Dondini, L., & Martínez-Gómez, P. (2016). Inheritance of reproductive phenology traits and related QTL identification in apricot. *Tree Genetics and Genomes, 12*, 71. https://doi.org/10.1007/s11295-016-1027-6.

Salazar, J., Martinez-Gomez, P., & Ruiz, D. (2022). Varietal evaluation of postharvest behavior in apricot fruits. *European Journal of Horticultural Science, 87*, 1–14. https://doi.org/10.17660/eJHS.2022/010.

Scofield, C., Stanley, J., Schurmann, M., Marshall, R., Breen, K., Tustin, T., et al. (2020). Light interception and yield of sweet cherry and apricot trees grown as a planar cordon orchard system design. *Acta Horticulturae, 1281*, 213–222.

Seif El-Yazal, M., & El-Yazal, M. (2021). The control of apricot seed dormancy and germination by low temperature treatments. *Innovare Journal of Agricultural Sciences*, 11–15. https://doi.org/10.22159/ijags.2021.v9i1.40689.

Sharkawy Sahar, S. A., Asmaa, M., Alkolaly, & Kafsheer, D. A. (2023). "Use of biological and chemical compounds for the integrated management of apricot powdery mildew in Egypt.". *Asian Journal of Agricultural and Horticultural Research, 10*(4), 486–495. https://doi.org/10.9734/ajahr/2023/v10i4288.

Shi, S., Li, J., Sun, J., Yu, J., & Zhou, S. (2013). Phylogeny and classification of Prunus sensu lato (Rosaceae). *J Integr Plant Biol, 55*(11), 1069–1079. https://doi.org/10.1111/jipb.12095.

Solís-Solís, H. M., Calderón-Santoyo, M., Schorr-Galindo, S., Luna-Solano, G., & Ragazzo-Sánchez, J. A. (2007). Characterization of aroma potential of apricot varieties using different extraction techniques. *Food Chemistry, 105*(2), 829–837. https://doi.org/10.1016/j.foodchem.2007.01.061.

Southwick, S. M., Fritts, R. J., & Nito, N. (1995). Commercial thinning of stone fruit in California by gibberellins to reduce flowering. *Acta Hort, 394*, 135–147.

Southwick, S. M., & Yeager, J. T. (1999). Effect of rootstock, cultivar and orchard system on apricot production. *Acta Hortic, 488*, 483–488. https://doi.org/10.17660/ActaHortic.1999.488.77.

Stanley, J. (2016). Factors affecting fruit set and fruit quality along branch units of different apricot cultivars. *New Zealand Journal of Crop and Horticultural Science, 44*(3), 171–191. https://doi.org/10.1080/01140671.2016.1161652.

Stanley, J., Feng, J., & Olsson, S. (2015). Crop load and harvest maturity effects on consumer preferences for apricots. *J Sci Food Agric, 95*(4), 752–763. https://doi.org/10.1002/jsfa.6850.

Stanley, J., Marshall, R., Scofield, C., & Alspach, P. (2016). The effects of reflective cloth and adjusted fruit distribution on apricot fruit quality. *Acta Hort, 1130*, 241–248.

Stanley, J., Marshall, R., Tustin, S., & Woolf, A. (2014). Preharvest factors affect apricot fruit quality. *Acta Hort, 1058*, 269–276.

Stanley, J., Prakash, R., Marshall, R., & Schröder, R. (2013). Effect of harvest maturity and cold storage on correlations between fruit properties during ripening of apricot (Prunus armeniaca). *Postharvest Biology and Technology, 82*, 39–50. https://doi.org/10.1016/j.postharvbio.2013.02.020.

Stanley, J., Schurmann, M., Keene, L., Scofield, C., van Hooijdonk, B., Breen, K., et al. (2024). Canopy light irradiance, crop load and fruit quality in mature narrow-row planar cordon apricot canopies. *Acta Horticulturae* (in press).

Stanley, J., Scofield, C., Marshall, R., & Tustin, S. (2018). Early canopy development and production precocity in new orchard system designs for cherry. *Acta Horticulturae, 1228*, 45–50. https://doi.org/10.17660/ActaHortic.2018.1228.6.

Strickland, D., Carroll, J., & Cox, K. (2019). *Brown rot of stone and pome fruit*. https://hdl.handle.net/1813/43081.2. (Accessed 13 October 2024).

Temnani, A., Berríos, P., Zapata-García, S., Espinosa, P. J., & Pérez-Pastor, A. (2023). Threshold values of plant water status for scheduling deficit irrigation in early apricot trees. *Agronomy, 13*, 2344. https://doi.org/10.3390/agronomy13092344

Torrecillas, A., Corell, M., Galindo, A., Pérez-López, D., Memmi, H., Rodríguez, P., et al. (2018). Agronomical effects of deficit irrigation in apricot, peach, and plum trees. In I. Garcia-Tejero, & V. Duran (Eds.), *Water scarcity and sustainable agriculture in semi-arid environment: Tools, strategies, and challenges for woody crops* (pp. 87–109). London, UK: Elsevier.

Tursun, N., Isik, D., Demir, Z., & Jabran, K. (2018). Use of living, mowed, and soil-incorporated cover crops for weed control in apricot orchards. *Agronomy, 8*, 150.

Uğur, R. (2022). "Prunus microcarpa: A potential rootstock for apricots". *Turkish Journal of Agriculture and Forestry, 46*(1), 49–58. https://doi.org/10.3906/tar-2107-14. Available at https://journals.tubitak.gov.tr/agriculture/vol46/iss1/5. Available at.

Vávra, R., Kaplan, J., Schlesingerová, G., & Maryš, L. (2023). *Strategy to control plum pox virus damage*. https://orgprints.org/id/eprint/45928/1/67.PA_PPV%20control%20strategy_RV_5_IP_RV_LDi.pdf. (Accessed 6 October 2024).

Vavilov, N. I., & Dorofeev, V. F. (1992). *Origin and geography of cultivated plants*. Cambridge University Press.

Vavilov, N. I. (1951). The origin, variation, immunity and breeding of cultivated plants. *Soil Science, 72*(6), 482.

Velardo-Micharet, B., Agudo-Corbacho, F., Ayuso-Yuste, M. C., & Bernalte-García, M. J. (2021). Evolution of some fruit quality parameters during development and ripening of three apricot cultivars and effect of harvest maturity on postharvest maturation. *Agriculture, 11*, 639. https://doi.org/10.3390/agriculture11070639.

Wani, S. M., Masoodi, F. A., Ahmad, M., & Mir, S. A. (2018). Processing and storage of apricots: Effect on physicochemical and antioxidant properties. *J Food Sci Technol, 55*, 4505–4514. https://doi.org/10.1007/s13197-018-3381-x. PMID: 30333647. PMCID: PMC6170339.

Webster, A. D., & Spencer, J. E. (2000). Fruit thinning plums and apricots. *Plant Growth Regulation, 31*(1–2), 101–112. https://doi.org/10.1023/A:1006335620972.

Weinbaum, S. A., Uriu, K., & Muraoka, T. T. (1980). Relationship between K15NO3 application period and 15N enrichment of apricot blossoms and developing fruit. *Journal of Plant Nutrition, 2*, 699–706. https://doi.org/10.1080/01904168009362808.

Wimalajeewa, D. L. S. (1987). Seasonal variation in susceptibility of apricot to *Pseudomonas syringae* pv. *syringae* (bacterial canker), and site of infection in apricot and cherry. *Australian Journal of Experimental Agriculture, 27*, 475–479. https://doi.org/10.1071/EA9870475.

Xi, W., Feng, J., Liu, Y., Zhang, S., & Zhao, G. (2019). The R2R3-MYB transcription factor PaMYB10 is involved in anthocyanin biosynthesis in apricots and determines red blushed skin. *BMC Plant Biology, 19*, 287. https://doi.org/10.1186/s12870-019-1898-4.

Xiloyannis, C., Nuzzo, V., Dichio, B., Celano, G., & Montanaro, G. (2000). Characterization of training systems in relation to water use efficiency in apricot and kiwifruit plants. *Acta Horticulturae, 537*, 207–213. https://doi.org/10.17660/ActaHortic.2000.537.21.

Xu, M., Zhou, W., Geng, W., Zhao, S., Pan, Y., Fan, G., et al. (2021). Transcriptome analysis insight into ethylene metabolism and pectinase activity of apricot (*Prunus armeniaca* L.) development and ripening. *Scientific Reports, 11*(1), Article 13569. https://doi.org/10.1038/s41598-021-92832-6.

Yıldırım, H., Onay, A., Tilkat, E., & Akturk, Z. (2011). Micropropagation of the apricot (Prunus armeniaca L.) cv. Hacihaliloglu by means of single node culture. *Turkish Journal of Agriculture and Forestry, 35*, 55–64. https://doi.org/10.3906/tar-0904-2.

Yao, S., Guldan, S., & Heyduck, R. (2019). High tunnel apricot production in frost-prone northern New Mexico. *HortTechnology, 29*(4), 457–460. https://doi-org.ezpakl.plantandfood.co.nz/10.21273/HORTTECH04315-19.

Zhebentyayeva, T., Ledbetter, C., Burgos, L., & Llácer, G. (2012). Apricots. In *Fruit breeding* (pp. 415–458). https://doi.org/10.1007/978-1-4419-0763-9_12.

Zhou, W., Niu, Y., Ding, X., Zhao, S., Li, Y., Fan, G., et al. (2020). Analysis of carotenoid content and diversity in apricots (*Prunus armeniaca* L.) grown in China. *Food Chemistry, 330*, Article 127223. https://doi.org/10.1016/j.foodchem.2020.127223.

Zuriaga, E., Romero, C., Blanca, J. M., & Badenes, M. L. (2018). Resistance to Plum Pox Virus (PPV) in apricot (Prunus armeniaca L.) is associated with down-regulation of two MATHd genes. *BMC Plant Biology, 18*(1), 25. https://doi.org/10.1186/s12870-018-1237-1.

CHAPTER

7

Plum *(Prunus salicina)* and Prune *(Prunus domestica)*

Pedro Martínez-Gómez[1], Aleksandra Korićanac[2], María Nicolás-Almansa[1], David Ruiz[1], Macarena Farcuh[3], Florin Stănică[4], Chris Dardick[5], Theodore M. DeJong[6]

[1]*Department of Plant Breeding, CEBAS-CSIC, Murcia, Spain;* [2]*Fruit Research Institute, Čačak, Serbia;* [3]*Department of Plant Science and Landscape Architecture, University of Maryland, College Park, Maryland, U.S.;* [4]*Faculty of Horticulture, University of Agronomic Sciences and Veterinary Medicine, Bucharest, Romania;* [5]*USDA ARS Appalachian Fruit Research Station, Kearneysville, U.S.;* [6]*University of California, Department of Plant Sciences, Davis, California, U.S.*

Chapter outline

1. Origin and spread

Although the origin of plum is Central Europe, there is a great diversity of plum germplasm belonging to more than 25 species distributed in Asia, Europe, and America. Despite this wide diversity, only three species predominate in commercial production: the hexaploid ($2n = 6x = 48$) European plum or prune (*Prunus domestica* L.), the diploid ($2n = 2x = 16$) Japanese plum (*P. salicina* Lindell), and the diploid ($2n = 2x = 16$) sloe (*P. spinosa* L.). Although they belong to the same pomological classification, they are different crops in terms of uses, adaptation, origin, and domestication (Okie and Hancock, 2008; Potter, 2012). For thousands of years, people have been cultivating various fruit species, with plums and other *Prunus* species being particularly prominent (Al-Khayri et al., 2018). The earliest reference to plums can be found in classical texts, specifically in the writings of Archilochus Pollux in the 7th century BCE, where the term "prumnon" (plum) was mentioned. This species complex is believed to have

Temperate Tree Fruits and Nuts. https://doi.org/10.1016/B978-0-443-29137-1.00007-4

originated from five distinct centers: Western Asia, Europe, Western and Central Asia, North America, and China. It has been documented that various types of plums, such as *Prunus insititia* L. (Damson plum) from Western Asia, *P. domestica* L. (European plum) from Europe, *P. cerasifera* Ehrh. (cherry plum) from Western and Central Asia, *P. americana* Marsh. (American plum) from North America, and *P. salicina* Lindl. (Japanese plum) have distinct origins in different regions (Faust and Surányi, 1998).

The Caucasus Mountains near the Caspian Sea have been hypothesized as the center and place of origin of *P. domestica* and its ancestors by Faust and Surányi (1998). Around 6000 BC, plums migrated from this area to regions such as ancient Syria, Mesopotamia, Egypt, and Crete (Faust and Surányi, 1998). It is documented that natural hybrids between *P. cerasifera* Ehrh. and *P. spinosa* L. originated from Asia Minor and spread as seeds of these hybrids from Iran and Asia Minor, likely serving as the parent of *P. domestica* L. in Europe. In Central Asia, the oldest used term for plum is the Middle Persian word "ālūg," which was later expanded and associated with the oasis city of Bukhara, giving rise to the Pashto term "ālū bukhārā," meaning "plum" (Dal Martello et al., 2023).

Plum is one of the oldest fruit species, whose domestication proceeded parallel and independently on three continents: Europe, Asia, and North America (Milatović, 2019). Because two of the most important plum species hexaploidy *Prunus domestica* (2n = 48) and diploid *Prunus salicina* (2n = 16) have not been found in wild forms, it is considered that these species were selected and cultivated very early, probably a few thousand years ago. It is considered that growing of hexaploidy European plum (*Prunus domestica* L.) started in the area between the Caucasus Mountains and the Caspian Sea, from which it spread to Europe and areas with a moderate-continental climate, where it has been present for 6000 years (Neumüller, 2011). While autochthonous plum varieties were domesticated in North America, *Prunus salicina* was first cultivated in China, from where it was brought to Japan and then subsequently spread all around the world, called Japanese plum (Esmenjaud and Dirlewanger, 2007).

The origin of the European plum is not completely understood. According to Eryomine (1990), the European plum originated as a hybrid of the large-fruited form of diploid *Prunus cerasifera* (2n = 2x = 16) and tetraploid blackthorn (*Prunus spinosa*, 2n = 4x = 32), while Zohary (1992) suggests that *Prunus domestica* is an autopolyploid derived from *P. cerasifera*. Although neither *P. cerasifera* nor *P. spinosa* were identified in the northwest of China, some wild varieties of hexaploidy plum that resembled European plum were discovered there (Zhang et al., 1998), indicating that European plum may be the result of autopolyploidy of some primitive form of the genus *Prunus*. However, it was later found that these were likely feral trees due to the very narrow genetic diversity (Liu et al., 2007).

Archeological evidence indicates that plums were extensively used during the Neolithic period, although these were most likely small-fruited damson or cherry plums. European plum was cultivated in the Roman Empire, and the orchards were mainly established in the territories of today's Hungary, Romania, Serbia,

Bosnia, and Herzegovina. Monasteries and manor estates were the main locations for European plum cultivation throughout the Middle Ages. In the 19th century, planned hybridization began in England and became increasingly significant, leading to the identification of over 2000 cultivars by the beginning of the 20th century (Sottile et al., 2022).

Plums have a long history of domestication, with evidence of their cultivation in Asia, Europe, and North America (Okie and Hancock, 2008). In Switzerland, excavations of ancient lake dwellers' sites have uncovered stones from *Prunus spinosa* and *P. insititia*, suggesting their presence in the region (Al-Khayri et al., 2018). Early archaeological findings also suggested the existence of *P. insititia* and cherry plums (*P. cerasifera*) in Germany and Bulgaria (Watkins et al., 1976). While these plums are believed to have originated in Western or Central Asia (Faust and Surányi, 1998), the exact center of origin for each species remains unresolved (Dal Martello et al., 2023). This ambiguity is further compounded by the limited archaeological research conducted in Central Asia (Dal Martello et al., 2023). In 1870, Japanese plum was brought to the U.S. from Japan. New Japanese plum cultivars were bred using interspecies hybridization that included *P. simonii* as well as autochthonous American varieties. As a result, more than 100 cultivars were released, among them Santa Rosa, a leading plum cultivar in the U.S. until the end of the 1970s (Milatović, 2019).

2. Economic importance

Japanese plums and European plums/prunes are a very important temperate tree fruit, ranking second among the stone fruits, after peach. Japanese plum, European plum, and sloe are cultivated in more than 80 countries, with most production located in Central and Eastern Europe, the Mediterranean basin, California, and Chile (http://faostat.fao.org) (Figure 7.1). According to the Food and Agriculture Organization of the United Nations (FAO), in 2022 there were 2,575,703 ha dedicated to plum and sloe in the world with an annual production of 13,152,656 tons. China is first in terms of production volumes with 6,761,189 tons and 534,191 ha followed by Romania (665,730 tons), Serbia (488,593 tons), and Chile (424,886 tons) (http://faostat.fao.org). In general, production and cultivated areas increased from 2000 to 2018, with stabilization from 2018 onward (see Figure 7.1). Japanese plums and many European plums are primarily grown for fresh market consumption, while specific cultivars of *P. domestica* are primarily grown for producing dried prunes. Other uses of *P. domestica* and *P. spinosa* include processing into jams and beverages.

Plum can be considered a fruit of the Northern Hemisphere as it accounts for ca. 92% of the total production. Although Europe was the dominant plum production region during the 20th century, the leading producer among continents nowadays is Asia (63.2%) (see Figure 7.1).

Despite significant differences, particularly in terms of fruit characteristic, among diploid Japanese (*Prunus salicina*) and hexaploidy European plum (*Prunus*

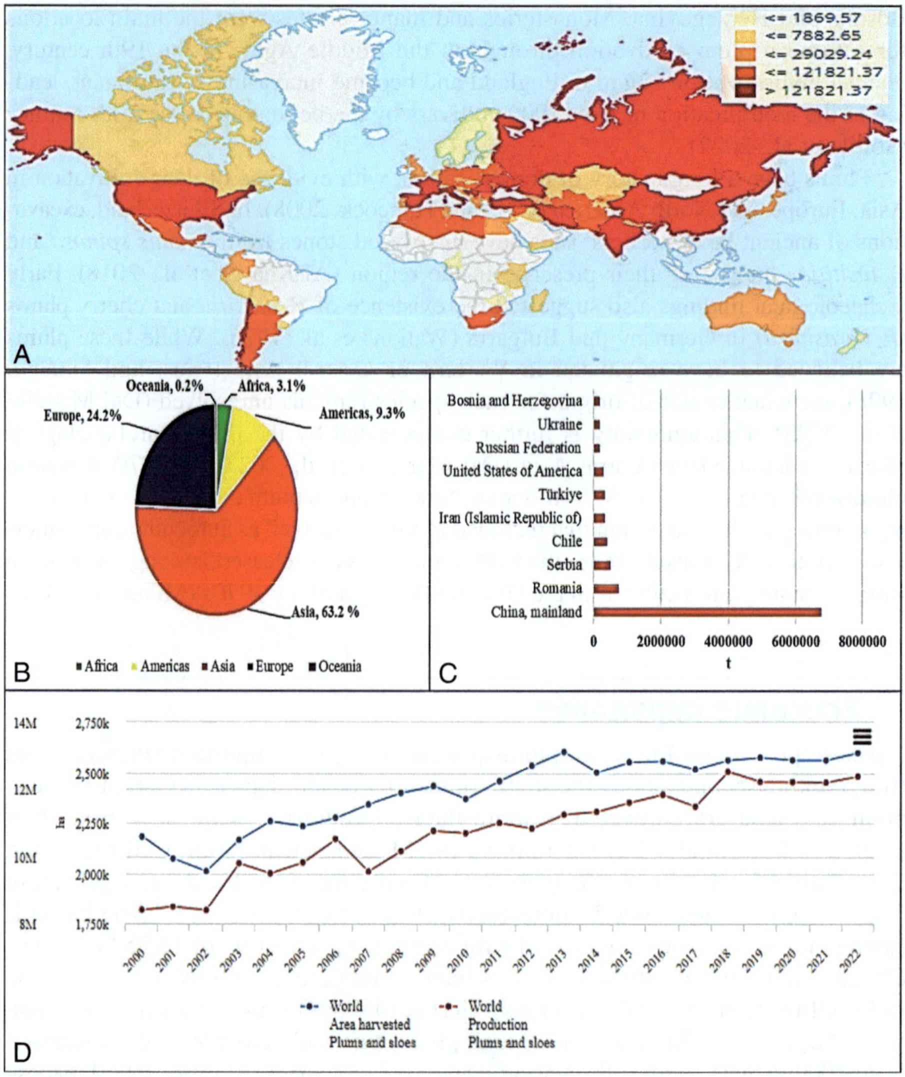

FIGURE 7.1

(A) Production quantities of plums and prunes by countries in 2022; (B) production share of plums and sloes by regions for the period 2000–2022; (C) production of plums: top 10 producers for the period 2018–2022; and (D) evolution of production/yield quantities during the period 2000–2022 (https://www.fao.org/faostat/en/#data/).

domestica), all available databases on production quantity, yield, and economic data refer to all plum species. However, it is estimated that two-thirds of all plum production is Japanese plum, which is dominant in Asia (82% of production), whereas 90% of plum production in Europe belongs to European plum (Milatović, 2019). China is the world's largest producer, accounting for 55% of global output with 6.7 million tons produced. It is followed by Romania, whose part is about 6%, and Serbia with a 4% share in total plum production (see Figure 7.1). Despite great plum production, mainly based on Japanese plum cultivars, China is not among the biggest exporters of plums because the fruit is primarily sold locally (Table 7.1). The European plum dominates fruit production in Romania in terms of both production volume and area. The leading cultivars are Stanley and Anna Späth, followed by Romanian cultivars (Centenar, Pescăruş, Agent, and Tuleu gras) (Butac et al., 2012). Similar to China, Romania is not an exporter of plums; on the contrary, it imports plums from neighboring countries such as Moldova. With an average annual production of 494,609 tons, Serbia is one of the biggest producers of European plum. However, production is characterized by great oscillations due to unfavorable weather conditions and alternate bearing. Dominant cultivars are Stanley, Čačanska Rodna, Čačanska Lepotica, and certain autochthonous cultivars (Milatović, 2019). Although a significant part of annual plum production is processed into plum brandy, fresh plums are also exported in considerable quantities.

Both plum varieties, diploid Japanese plum and hexaploid European plum, are significant for commercial trade. However, Japanese plum is dominant in long-distance fresh fruit trade since it withstands transport better due to more pronounced firmness. The export of Japanese plum cultivars to the European market during off-season months (from January to April) has provided Chile and South Africa a unique position in the fresh plum trade (Table 7.2). The plum production in these countries has been rising over the last years. Chile is currently the top exporter of both fresh and dried plums, and it has been positioned as the fourth largest world's producer in the period 2018 to 2022 (see Figure 7.1). In terms of export, Chile is followed by Spain, whose production is mainly based on *Prunus salicina*, located in the southern parts of the country. Export of plums grown in Spain begins quite early in April, and it is oriented toward Western European countries (the UK, the Netherlands, and Germany). Although producers are interested in more profitable stone fruits, plum is still the leading stone fruit imported in Europe in terms of volume (CBI, 2024).

Table 7.1 The leading plum exporters in the world, period 2020–2022 (FAOSTAT, 2024)

	Export quantity (t)			Export value (1000 USD)		
Country	**2020**	**2021**	**2022**	**2020**	**2021**	**2022**
Chile	125,718	121,177	158,113	191,201	205,462	236,867
Spain	90,016	111,292	94,274	122,063	158,700	133,300
South Africa	48,497	81,964	96,529	57,236	105,689	114,426
China, Hong Kong SAR	62,895	58,030	54,092	96,050	98,450	110,855
Turkey	47,756	38,622	47,435	31,660	26,896	24,112

Table 7.2 The leading plum importers in the world, period 2020–2022 (FAOSTAT, 2024)

	Import quantity (t)			Import value (1000 USD)		
Country	**2020**	**2021**	**2022**	**2020**	**2021**	**2022**
China, mainland	66,455	64,945	103,514	154,781	175,346	270,188
Russian Federation	74,937	63,193	60,611	60,746	54,921	52,832
China, Hong Kong SAR	63,934	61,818	60,148	121,403	117,921	121,996
Germany	52,413	50,215	48,463	79,887	78,472	67,087
Netherlands	40,987	54,006	55,212	70,215	94,900	85,104

The price of plum depends on several factors, and it also significantly differs among European and Japanese plum cultivars, which are regularly sold for higher prices. Even though plum is considered a mass-produced fruit, growers can achieve higher prices by offering attractive and tastier plums. Locally grown plums are also desirable because of freshness, lower carbon footprint, and increasingly important "farm to fork" strategy.

3. Botany and relevant species

The different species of plum trees are deciduous and considered medium in size, reaching between 8 to 10 meters in height. The flower buds give rise to cyme-like inflorescences on short branches (spurs), resulting in multiflowered shoots or mixed bouquets. These inflorescences usually appear on 1-year-old branches or previous-year growth on short, woody branches (spurs) that are several years old. The flowers are mainly white, although sometimes they can have a slight pinkish hue before petal fall. The flowers are composed of five sepals and petals, arranged alternately and solitary or in groups of two to five flowers with long pedicels. The fruit is a glabrous drupe usually covered with powdery wax, called bloom, which acts as a protective layer over the fruit, helping it retain moisture. Fruits can vary widely in size, shape, and color ranging from green and yellow to red, purple, or even black depending on the cultivar. Fruit size varies from 2 to 8 cm in length with a great diversity of shapes, from cordate (heart-shaped), elliptical (elongated), flattened, or round, generally presenting a shallow suture and a medium-sized peduncle. Fruits have a lignified endocarp of variable shapes, elongated or rounded and flattened, which contains a bitter seed with a brown testa and white cotyledons. Fruit have a fleshy mesocarp, generally adherent to the stone (lignified endocarp), which can also be greenish-white, yellowish-green, yellow, orange, red, to dark red (Okie & Hancock, 2008) (Figure 7.2).

FIGURE 7.2

The plum production cycle. (A) Flower buds; (B) flowering and pollination; (C) fruit development in the tree; (D) fruit maturation.

Courtesy Dr. Maria Nicolás.

The exact number of plum varieties has not been precisely determined yet, although according to the literature data, it varies between 19 and 40 (Butac, 2020). Plum production is mainly based on *Prunus domestica* and *Prunus salicina*. Furthermore, the largest number of plum cultivars originated from these two species. However, to preserve biodiversity and widen the genetic reservoir for breeding purposes, special attention should be given to plum germplasm accessions.

***Prunus domestica* L.** Although its origin has not been highlighted, the area between the Caucasus Mountains and Caspian Sea is assumed to be the area of origin. It can grow in the form of shrub or as a tree, whose height can vary from 4 to 8 meters. Fruit is a juicy monocarp drupe, usually of ovoid shape, and a weight ranging from 15 to 80 g. Fruit color is mostly dark blue or purple, although it can also be red, green, or yellow, and it is mostly covered with wax bloom. Due to pronounced variability, there are numerous forms that particularly differ in terms of fruit characteristics: prunes, Greengages, and Mirabelles. *Prunus domestica* is considered the most important plum species with over 3000 cultivars originating from it (Milatović, 2019). These fruits can be used for fresh consumption, drying, and processing. However, it is important to note that the vast majority of *P. domestica* is grown to produce dried prunes. These varieties tend to be somatic clones of the same cultivar (D'Agen), which is of ancient origin, suggesting global plum production has long been predominantly a monoculture. Likewise, genetic analyses have indicated that all *P. domestica* cultivars have relatively low diversity and likely originated from a very small number of founders (Zhebentyayeva et al., 2019).

***Prunus insititia* L.** The territory of today's Turkey is considered a place of origin of this plum species. Due to morphological similarities, it is also marked as a subspecies of European plum *(Prunus domestica* ssp. *insititia)* (Afanador-Barajas et al., 2022). This was further supported by genetic analyses where some germplasm selections named *P. insititia* grouped with *P. domestica* cultivars, while others formed distinct outgroups, suggesting the possibility that some represent either interspecific hybrids or distinct species (Zhebentyayeva et al., 2019). Nevertheless, it is more resistant to frost and drought than European plum and is commonly used as a rootstock. Its fruits are mainly used for processing.

***Prunus cerasifera* Ehrh.** This plum species is also known as cherry plum and Myrobalan plum. It is widely spread, although it is native to West Asia and the Balkan Peninsula (Afanador-Barajas et al., 2022). Due to its positive traits, such as cold hardiness, tolerance to drought and heat, resistance to bacterial spot, and good productivity, cherry plum represents valuable breeding material (Sottile et al., 2022). Its fruit is small though (5–20 g) and more suitable for processing into juices and jams than for fresh consumption. Myrobalan seedling is also extensively used as a rootstock for plums due to availability, hardiness, longevity, and adaptability to various soil types; it is the dominant rootstock in Central and East Europe (Cvetković and Glišić, 2020).

***Prunus spinosa* L.** Also known as blackthorn and sloe, *Prunus spinosa* can be found in a wild form, usually as a shrub, all across Europe, Asia, and North Africa

(Butac, 2020). Due to pronounced drought resistance and cold hardiness, it is used in breeding work (Sottile et al., 2022) and recently as a rootstock.

***Prunus salicina* Lindl.** Similar to *Prunus domestica*, *Prunus salicina* also does not have a known wild ancestor. However, it differs from European plum in several aspects: it is more vigorous; physiological dormancy period is shorter; it is more sensitive to winter frosts, early flowering, and higher sensitivity to spring frost; and mainly self-sterile (Milatović et al., 2017; Milošević and Milošević, 2022). Japanese plums have large, attractive, firm fruits that withstand cold storage and transport better than European plums and therefore are mainly used for fresh consumption.

***Prunus simonii*.** It has been grown in China, Japan, and Central Asia since ancient times. Due to its similarities to apricot, it is also known as Apricot plum (Butac, 2020). It has several positive fruit properties: small stone, high flesh firmness, and aromatic fruit, and for this reason it was utilized in a Californian breeding program of Japanese plum.

***Prunus ussuriensis*.** It is the most frost-resistant plum species, mostly known for its cold hardiness (Sottile et al., 2022).

***Prunus americana*.** It is the most important autochthonous plum in North America and can be found across the U.S. and in southern parts of Canada, commonly as roadside plum (Okie, 2001). Due to cold hardiness (up to -40°C), it is used in breeding work.

***Prunus angustifolia*.** According to Okie, *Prunus angustifolia* is the most common in southeastern parts of the U.S. Despite its resistance to disease, small fruit size and poor quality are the main limits for its use. Hybrids with Japanese plums have slightly bigger fruit, but the quality is not significantly improved (Okie, 2001).

4. Breeding and propagation

The objectives pursued in the improvement of the Japanese plum vary by country or region of production, but in general, they have focused on expanding the production calendar and availability of fruit in the market; obtaining optimal firmness to facilitate handling, storage, and transportation; increased organoleptic quality and appearance; increasing the size of the fruit; and improving postharvest fruit quality. In relation to the flowering date, the objectives may vary depending on the place of improvement. Recent emphasis has been on selection of extra-early cultivars adapted to growing in very warm areas and late-blooming cultivars to reduce the risk of spring frosts. Furthermore, characteristics such as pollen self-compatibility, climatic adaptability, productivity, and resistance to diseases and viruses, especially to Plum pox virus (PPV), are widely valued and desired in the genetic improvement of this species (Okie & Hancock, 2008). Interspecific hybrids using Japanese plum and apricot (plumcot) have been developed to include desirable apricot characteristics in plum, such as self-compatibility alleles and Sharka tolerance, together with new fruit typologies with attractive skin and/or flesh colors and a variety of valued organoleptic characteristics (Nicolás-Almansa et al., 2023).

In addition, these are the main objectives of European plum breeding programs: adaptability to different climatic conditions; self-fertility; early bearing; productivity; weak or moderate trunk vigor; resistance to disease; extension of the ripening period; and improvement of fruit quality, transportability, and storage potential (Butac et al., 2013; Butac, 2020; Milošević et al., 2021). As European plum cultivars can be used both for fresh consumption and various types of processing, the objectives of breeding differ significantly depending on the purpose of the fruit. The goals for cultivars aimed for fresh consumption are the attractive appearance of the fruit, uniform size, fruit weight above 35 g, regular shape (typically ovoid), and appropriate color (Milatović, 2019). The yellow color of the flesh is desirable, as well as the sweet-sour, harmonious flavor, pleasant aroma, and a high level of bioactive compound with health-beneficial effects. Cultivars aimed for processing should have fruit of uniform size, with a low stone share and high content of soluble solids, whilst elastic skin is of particular importance for plums intended for drying.

European plums are sold not only as fresh fruit but also as dried, particularly in the U.S. where the vast majority of product is sold as dried prunes. Breeding goals of fresh vs. dried types are significantly different. Fresh fruit types fall into a few different pomological categories that are genetically distinct including Mirabelles, Greengages, and European plums (Neumüller, 2011; Topp et al., 2012). These fruit types differ with respect to size, shape, and color and historically have different culinary uses. Due to high levels of PPV susceptibility, some heritage cultivars have become extremely difficult to produce, resulting in a shift toward less desired but PPV-tolerant cultivars. Unfortunately, there is no known strong genetic resistance to PPV in *Prunus domestica* germplasm, making it challenging to breed for new cultivars where PPV is widespread. One exception is the resistance found in a cultivar named Jojo that carries a classic "hypersensitive" resistance to PPV. European breeding programs have been using Jojo as a parent to develop new European plum varieties resistant to PPV (Zurawicz et al., 2013). Another plum cultivar resistant to PPV is the transgenic HoneySweet obtained in the U.S. (Ravelonandro et al., 2013); its resistance is based on gene silencing or RNA interference (RNAi). HoneySweet trees are grown in Czech Republic, Poland, Romania, and Spain under controlled conditions and high infection pression, and thus far no infections have been detected (Zagrai and Zagrai, 2020).

Both conventional methods (hybridization, clonal selection, induced mutations) and modern biotechnology methods can be used in plum breeding. Conventional methods, primarily hybridization, are still used in most breeding programs (Milošević and Milošević, 2018). Milatović (2019) states that the most widely used biotechnological methods for plums are various *in vitro* culture techniques, biochemical and molecular markers, and genetic transformations. The main disadvantages of conventional methods are a long reproductive cycle with a long juvenile period, complex reproductive biology, heterozygosity, a large area necessary for seedlings, as well as costs related to field work, which explains why biotechnological methods, including genetic engineering, are becoming more and more prevalent (Petri et al., 2018).

The largest number of breeding programs (21) was developed in 13 European countries: Serbia, Germany, Romania, Bulgaria, the United Kingdom, France, Italy, Sweden, Norway, Latvia, the Czech Republic, Belarus, and Moldova (Butac et al., 2013). Significant results in European plum breeding have been achieved in research centers in the U.S. and Canada.

On the other hand, genetic studies showed that the prune industry consists predominantly of a monoculture of a variety referred to as D'Agen or French prunes that likely date back to the 13th century (Zhebentyayeva et al., 2019). Numerous sports have been selected over the centuries resulting in a broad range of named French types. There is only one known breeding program that is solely focused on developing new prune cultivars for the dried prune market. This program is breeding *Prunus domestica* prune/plums to improve dried fruit characteristics that include dry-away ratio, flesh texture, skin thickness, size, and shape, as well as suitability with mechanical harvesting and processing including pitting. This program also focuses on horticultural characteristics such as pollen self-compatibility, time of bloom (chilling requirement), time of fruit maturity, fruiting precocity, and tree structural characteristics. Because *Prunus domestica* is a hexaploidy species, there is ample genetic diversity among seedling populations, and the major challenge is harnessing that diversity to meet breeding objectives. Recently, this program released a new yellow-fruited cultivar (UC-Yolo Gold) that has unique and exceptional dried fruit flavor. Currently, this program has several traditional blue/purple colored selections that are undergoing expanded industry testing for adaptability to current field horticultural practices and industry processing. Breeding efforts to improve the French prune have been challenging due to the very specific processing requirements of the industry.

5. Orchard management

Plum and prune cultivation presents a series of shortcomings and problems that need to be addressed. The list of traditional cultivars used is small, and many of the new cultivars come from international breeding programs, which can limit their adoption due to high costs and restricted access, lack of information about production requirements, and lack of adaptability to specific environments, often due to their need for high winter chilling (Ruiz et al., 2018). Likewise, the reliance on monocultures for prune places these industries in a precarious position with increasing temperatures and more erratic weather. The strict industry requirements for self-compatibility, size, skin thickness, and dry-away ratio make new cultivar development particularly difficult. A section that deserves to be highlighted for its importance and the difficulties it entails in the cultivation of this species is floral self-incompatibility. Growers need to take this into account in the design of plantations, with intercompatible and coincident pollinator cultivars (Guerra and Rodrigo, 2015). Another problem with Japanese plum is the mediocre commercial quality of some of the traditionally cultivated varieties. The growing competitiveness of the markets, together with the new demands of consumers, has led to the need to produce high-quality plums characterized by an attractive appearance, adequate texture and firmness, and increased organoleptic quality (Cuevas et al., 2015).

The plum and prune planting systems and canopies have had a slower evolution toward higher density orchards compared with apple or pear. This is partially a result of the lack of acceptable size-controlling rootstocks for plums as well as less research emphasis being placed on plums, because their production intensity tends to be overshadowed by other major tree crops such as apples, pears, and peaches. Planting distances have stabilized to around 4.0 to 4.5 m between rows and 1.5 to 2.5 m on the row. Vertical axis/Slender Spindle training systems are quite common, while Bi-axis (Bibaum) and Trident (Chandelier) are used to reduce tree vigor by dividing it among the axes (Iliescu et al., 2023; Stănică and Peticilă, 2011) (Figure 7.3). Most plum/prune orchards in the United States are planted at wider spacings than Europe and are trained to a traditional open vase structure. Many plum and prune orchards use interrow cover crops with sowed grass mixture, or natural resident vegetation that are regularly mowed. The tree rows are generally kept free of vegetation using herbicides or mechanical tools. Recently, several tests have been made to cover the rows with living mulch.

Most plum/prune trees require annual or semiannual pruning to maintain manageable tree structures and viable/productive fruit yields. Because most plums produce flowers on spurs (short lateral branches) that can remain productive for several years, trees may not require detailed pruning in every year. However, as the spurs age and canopies become crowded, it is advisable to do some pruning to stimulate new shoot growth and renew the producing spurs as well as remove

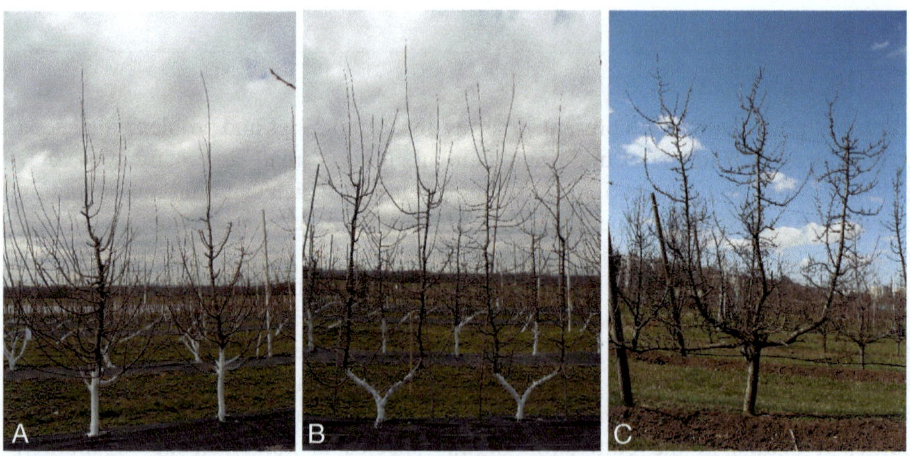

FIGURE 7.3

Plum and prune canopies: (A) Vertical axis/Slender Spindle, (B) Bi-axis (Bibaum), and (C) Trident (Chandelier).

Courtesy Prof. Florin Stănică.

some excess fruiting branches in cultivars that tend to set heavy crops (Yoshikawa and La Rue, 1989; Day and Buchner, 2012). Pruning requirements and practices vary by grower and plum cultivar. In recent years, some growers have been experimenting with mechanical topping and hedging with mixed results, because these types of blind pruning cuts tend to be heading cuts that can stimulate more regrowth than is desirable. In well-managed plum/prune orchards, pruning costs can account for up to one-third of a grower's annual production costs.

Plum trees tend to set heavy fruit loads under optimal growing conditions, and the fruit often needs to be thinned for it to attain marketable fruit size and sugar content. The same principles hold for fruit growth and thinning as with managing crop loads for peaches (Grossman and DeJong, 1995). Indeed, similar fruit growth and crop modeling techniques can be used to understand relationships between crop load, fruit size, and marketable yields for plums (Mariscal et al., 2002) as for peaches (Grossman and DeJong, 1994). Fruit thinning is done manually after fruit set with many fresh market plum cultivars and can represent up to one-third of the production costs for managing orchards in some years. In recent years with increased labor costs, some plum growers are trying mechanical thinning methods either during bloom or after fruit set to reduce costs. Prune growers often mechanically shake trees to reduce crop loads when fruit are about 1 cm in diameter (Buchner et al., 2012). Most attempts to use chemical thinners have had mixed and unreliable results.

Most fresh market plums are hand-harvested accounting for up to one-third of the annual production costs. Due to the large trees and more traditional tree spacings, much of this work is done on ladders, making harvests particularly expensive. In Europe, more orchards are set up to have more planar-oriented canopies to accommodate motorized platforms that increase worker efficiencies for all orchard operations, especially harvest, but these systems may have lower yields because more light is lost between row open space. Prunes that are used for drying are mechanically harvested using trunk shakers and large catch-frames to intercept the fruit before it contacts the ground. These systems are highly efficient, and the fruit are automatically sorted for small sizes in the field while larger fruit are deposited in large fruit bins and transported to drying facilities (Sibbett and Yeager, 2012; Day and Buchner, 2012).

Research on all these orchards' management practices has lagged behind what has been done in many of the more prominent temperate fruit tree crops for numerous reasons. One reason is that there is a very large variability among different plum types and cultivars with regard to tree structures and growth characteristics, so it is difficult to make uniform recommendations that fit all cultivars. Another reason is that, in many places where plums are grown, they occupy a secondary position and are less profitable for growers compared to other tree crops grown in the same region (such as apples, pears, peaches, and cherries), Thus, they have received less research attention and funding than competing crops. Clearly, there are needs and opportunities for more research on orchard management practices of plum/prune cultivation.

6. Postharvest physiology and technology

After harvest, plums and fresh prunes are transported to the packinghouse as soon as possible. Once at the packinghouse, they are dumped (mainly using dry bin dumps) and washed to remove any trash. Afterward, fruit sorting takes place with the objective of eliminating fruit with visual defects. Sorting is also conducted based on surface color targeting to divert fruit with high surface coloration to a high-maturity pack (Crisosto and Kader, 2015). Furthermore, fruit sizing, which predominantly segregates fruit based on weight or dimension, also takes place. Generally, plums and fresh prunes are volume-fill packed. Fruit are automatically filled by weight into shipping containers and packed into 12.6 kg (28 lb) volume-filled containers (Crisosto and Mitchell, 2002). The fruit surface of European plum is covered with wax bloom that prevents water loss and shriveling. Thus, all postharvest handling procedures should be carefully applied to minimize removing of wax bloom as a natural barrier. For this reason, European plums are typically not washed before cold storage.

More recently, the application of edible coatings of natural origin have gained importance in the postharvest plum industry. The use of coatings based on proteins or polysaccharides, as well as based on biodegradable materials such as starch, have shown potential for improving plum postharvest quality properties (Panahirad et al., 2019; Thakur et al., 2018). The big advantage is they have proven to extend storage life while posing no risk for human health or the environment. Prior to packing, plums and fresh prunes may undergo precooling in field bins via forced-air cooling, hydrocooling, or room cooling. On the other hand, already packed plums and fresh prunes are usually force-air cooled to near 0°C (32°F). Research has shown that forced-air cooling can enhance shelf-life and prevent mechanical damage in plums, particularly if this practice is conducted immediately after harvest (Martínez-Romero et al., 2003; Thompson et al., 2002). Modified atmosphere packaging can significantly reduce weight loss of plums during cold storage (Khan et al., 2018; Korićanac et al., 2023). Modified atmosphere in modified atmosphere packaging (MAP) is created passively by the synergistic action of the respiration of plums and permeability of the film to the gases, which enables formation of a micro-atmosphere around the fruit.

Plums for drying are mechanically harvested when they are very close to full maturity (3-4 lbf), and small fruit are typically sorted out by size in the field (Sibbett and Yeager, 2012) (Figure 7.4). Harvest timing is often influenced by ripening date and access to dehydrators, which is a major constraint to growers. Historically in the U.S. and currently to some extent in Chile and Argentina, prune plums were/are dried directly in the sun using ground cloths or wood trays. Today, virtually all fruit in California and Western Europe are collected on soft catch surfaces and arrayed on trays for drying in forced-air dehydrators at approximately 74°C (165°F) for 12 to 14 hours until the moisture level reaches around 20%. Additional drying is required to reduce moisture to a final storage content of 16% to 18%. Prunes can be bulk-stored in this condition for months until they are prepared for packaging by rehydrating

FIGURE 7.4

(A) UC-Yolo Gold fruit before drying and after drying in "natural condition" *(left center)* and (B) after processing (rehydrated, pitted, and treated with potassium sorbate).

Courtesy Dr. Theodore M. DeJong.

them with steam to ~30% moisture content followed by mechanical pitting and packaging (Sibbett and Yeager, 2012).

The optimum storage conditions for fresh plums are −1.1°C to 0°C (30°F to 32°F) with a relative humidity ranging between 90% to 95% to extend fruit postharvest life and maintain quality (Crisosto et al., 2004; Mitchell et al., 2008). Commercial storage and/or transportation conditions at temperatures between 0°C to 5°C (32°F to 41°F) have shown to be effective in delaying plum fruit softening and reducing weight loss as well as disease incidence. Nevertheless, storage at these temperatures can lead to the development of physiological disorders such as chilling injury (CI), rendering the fruit unmarketable (Crisosto et al., 2009; Palou et al., 2003). On the other hand, storage or transport of plum fruit at temperatures higher than 7.5°C (45.5°F) are effective for avoiding development of disorders such as CI but result in overripening, fruit softening, and enhanced senescence (Crisosto and Garner, 2008). Plum fruit freezing point has been reported to range between −2°C to −1°C (28°F to 30°F) depending on the soluble solids content. Regarding air circulation, it has been suggested that a velocity of approximately 15 m^3 min^{-1} should be used (Crisosto and Kader, 2015).

In general, plum fruit respiration and ethylene production rates vary widely depending on the genetic background and ripening behavior (Abdi et al., 1997; Farcuh et al., 2018, 2017; Kim et al., 2015) but also can be strongly influenced by the storage temperature (Crisosto and Kader, 2015). Plums that present a typical climacteric ripening behavior do not need exogenous ethylene application for proper and uniform ripening. However, slow ripening or suppressed climacteric ripening plum cultivars, i.e., Angeleno, Black Beauty, Casselman, Late Santa Rosa, Kelsey, Nubiana, Queen Anne, Red Rosa, and Roysum, can benefit from exogenous ethylene application of 100 μL L^{-1} for at least 24 h at 20°C (68°F) for a faster and more uniform ripening (Crisosto and Kader, 2015). On the other hand, plum cultivars with a typical climacteric ripening behavior can benefit from the use of 1-methylcyclopropene (1-MCP), which inhibits ethylene perception and thus prevents ethylene-dependent responses such as fruit softening and senescence (Farcuh et al., 2018; Sisler and Serek, 1997; Watkins, 2006). The capacity of 1-MCP to decrease plum ripening rate is well documented (Abdi et al., 1998; Candan et al., 2006; Farcuh et al., 2019, 2018; Martínez-Romero et al., 2003) helping to extend postharvest storage life and maintain quality. Generally, the recommended application for 1-MCP in plum corresponds to 0.5 $\mu L\ L^{-1}$ at 0°C (3 °F) for 24 h in a sealed room or tent.

Nevertheless, there has been work suggesting that applying 1-MCP (0.5 $\mu L\ L^{-1}$) during forced-air cooling reduces the application duration from 24 h to 6 h without affecting treatment performance, and this can be followed by storage at 10°C (50°F) avoiding chilling temperatures and saving considerable energy without affecting postharvest life of low-acid plum cultivars (Minas et al., 2013). It is important to consider that although the use of 1-MCP application in plums reduces softening during storage, transportation, and handling, it has also been shown to interfere with the normal ripening pattern of fruit during shelf-life as well as impact the development of physiological disorders during storage (Dong et al., 2002). These effects could be a result of altered hormonal balance as well as sugar metabolism that has been reported to occur as a result of 1-MCP applications in plum fruit with different ripening behaviors (Farcuh et al., 2019, 2018). Furthermore, it has been demonstrated that changes in ethylene and sugar metabolism can regulate flavonoid composition during storage in plums with different ripening behaviors (Farcuh et al., 2022).

Use of controlled atmosphere (CA) storage with fresh plums has been associated with benefits such as delaying the loss of flesh firmness and skin ground color evolution. Although CA conditions of 6% O_2 and 17% CO_2 have been suggested for reducing physiological disorders such as internal breakdown during storage, the effectiveness of this treatment varies according to the cultivar, preharvest management factors, maturity at harvest, as well as postharvest conditions (Crisosto and Kader, 2015). For European plums, a 0°C (32°F) temperature and a CA storage of 3 kPa O_2 and 7 kPa CO_2 has been recommended (Van De Geijn, 1993), while for Japanese plums firmer flesh has been recommended for Blackamber plums stored under CA of 1 to 2.5 kPa O_2 and 3 kPa CO_2 (Singh and Singh, 2012, 2013a). For Laetitia plums, the best CA condition for cold storage at 0.5°C (32.9°F) have been shown to be 2 kPa O_2 and 2 kPa CO_2. Furthermore, using CA conditions of low O_2 (3-5 kPa) and/or high CO_2 (10-15 kPa) to reduce softening and overripening in plums exposed to temperatures higher than 7.5°C (45.5°F) ended up causing softening and off-flavor development after extended storage (Crisosto and Garner, 2008). It has been reported that for some cultivars such as Angeleno, Casselman, Santa Rosa, Laroda, and Queen Anne, CA has a limited use for storing fruit for longer than 1 month (Crisosto and Kader, 2015).

Japanese and European plums (fresh prunes), as well as other stone fruits, are highly sensitive to low temperature, and the benefits of cold storage may be limited by the development of the physiological disorder of CI, which reduces their marketability (Farcuh and Hopfer, 2023; Wang et al., 2016). Chilling injury develops faster and more severely when fruit are stored between 2°C (35.6°F) and 8°C (46.4°F) than when stored at 0°C (32°F), and it typically expresses after cold storage, when the fruit is transferred to ambient temperature or shelf-life (20°C [68°F]) (Crisosto et al., 1999, 2004; Farcuh and Hopfer, 2023; Khan et al., 2011; Manganaris et al., 2008a). Chilling injury is manifested via several symptoms that include internal browning, flesh translucency (gel breakdown), reddening or bleeding, loss of flavor, and delayed softening or retarded ripening after prolonged cold storage (Crisosto et al., 2004; Minas et al., 2013; Pan et al., 2018; Taylor et al., 1995; Thakur et al., 2018). Internal browning (IB) is a CI symptom where the mesocarp of the fruit can change from a light to a darker brown in color, and the fruit loses juiciness. Over the course of development, browning will start just below the skin of the plum and spread inward toward the fruit's core. IB results from an oxidative process that causes the peroxidation of lipids and the reduction in the efficiency of the antioxidant systems, conferring damage to the cellular membranes (Singh and Singh, 2013a, 2013b). IB is associated with high temperatures during fruit ripening and delayed harvests (Crisosto and Kader, 2015). The symptom of flesh translucency is one of the most commonly observed CI symptoms, and it appears as a translucent gelatinous breakdown in the mesocarp tissue of the fruit, generally associated with the presence of water-soluble pectin (Candan et al., 2011; G A Manganaris et al., 2008b; Taylor et al., 1995).

Flesh reddening/bleeding corresponds to the intense red color development of the mesocarp tissue of the fruit and is usually a result of anthocyanin pigment accumulation (Farcuh and Hopfer, 2023; Wang et al., 2016). Several authors have reported that the presence of flesh translucency was concurrently observed with the development of flesh reddening in different Japanese plum cultivars (Abu-Kpawoh et al., 2002; Candan et al., 2011; Cantín et al., 2008; Crisosto et al., 1999; Manganaris et al., 2008a; Minas et al., 2013; Pan et al., 2016). The CI symptom of loss of flavor is mainly characterized by changes in fruit aroma volatiles and thus is generally considered as an invisible damage symptom (Farcuh and Hopfer, 2023). It is well known that the stage of maturity at harvest also plays a critical role in the development of CI, particularly in the susceptibility of plums to internal browning and flesh reddening where fruit with advanced maturity are more prone to developing CI (Figure 7.5) (Abdi et al., 1997; Singh and Singh, 2013b; Taylor et al., 1995). Nevertheless, some research has also suggested that late harvested Blackamber plums had a higher susceptibility to develop flesh translucency when stored at 5°C (41°F), while early harvested fruit displayed higher incidence of flesh bleeding/browning during storage at 0°C (32°F) or 5°C (41°F) (Crisosto et al., 2004). It has been reported that CI has been delayed in late season plums and fresh French prunes by storing CI-susceptible cultivars at −1.1°C (30°F). Nevertheless, to succeed in this approach and avoid freezing damage, high soluble solids contents of the fruit and an excellent thermostatic control are critical (Crisosto and Kader, 2015).

FIGURE 7.5

Fruit of European plum showing symptoms of physiological disorders (internal browning [IB] and gel breakdown).

Courtesy Ms. Aleksandra Koričanac.

7. Main diseases and pests and their management

Plum and prune production can be endangered by numerous diseases, and the causal agents could be fungi, bacteria, viruses, and phytoplasma. The most economically important fungi are *Monilinia* spp., *Polystigma rubrum*, *Tranzschelia pruni-spinosae*, *Taphrina pruni,* and *Thyrostroma carpophilum*. According to Holb (2003), pathogenic fungi belonging to the genus *Monilinia* are the biggest cause of losses in fruit production. In the absence of adequate protection measures, they can cause substantial losses, particularly in years when flowering and ripening seasons are rainy. In sensitive plum cultivars like Stanley, fruit losses caused by brown rot can exceed 50% (Milatović, 2023) (Figure 7.6). Among phytopathogenic bacteria *Pseudomonas syringae*, *Xanthomonas arboricola pv. pruni,* and *Xylella fastidiosa* represent the most dangerous causes of disease in plum. In addition, the most common causes of fruit diseases after harvest are fungi and bacteria, while other

FIGURE 7.6

Main fungi disease in plum and prune. (A) Pronounced symptoms of *Monilinia* spp. developed during cold storage. (B) The mycelium of *Monilinia* spp. developed on the surface of the plum fruit during cold storage. (C) Brown rot symptoms on (Stanley) fruit.
Courtesy Ms. Aleksandra Koričanac.

phytopathogenic organisms such as viruses can be the cause of the losses but are not the cause of fruit rot. After harvest, the most prevalent diseases of plum fruit are caused by fungus from the following genera: *Monilinia* (see Figure 7.6), *Botrytis, Rhizopus,* and *Penicillium*, and infection can occur in the orchard, during harvest, transport, postharvest handling, and cold storage as well (Milatović, 2019).

Although 27 plum viruses have been described in plum so far (Milatović, 2019), the most serious problem that affects plum, prune, and sloe cultivation around the world from a phytopathological point of view is the incidence of Sharka disease, caused by the *Plum pox virus* (PPV), which causes significant damage to this fruit species, affecting both the quality as well as production (see Figure 7.7). Sharka disease was discovered in Bulgaria in 1915, and now it is present in almost all continents. The symptoms are mainly pronounced on leaves and fruit, and rarely present on stone or branches. However, it is devastating for susceptible plum cultivars, affecting both yield and fruit quality (Dallot and Thébaud, 2021). PPV can be spread in two ways: by planting material or by vector—aphids. Unlike transmission of PPV by aphids, that is locally, usually near inoculum, planting material was key for spreading virus among long-distances including countries (Dallot and Thébaud, 2021). Thus, the primary preventive measure is the production of healthy, virus-free material, which can be successfully produced in traditional way by grafting, but only if all necessary precautions are taken (using resistant rootstocks, aphid control, isolated seedling production, and virus-free grafting twigs). Two rootstocks selected in Germany, Docera and Dospina, are resistant to PPV due to a mechanism of hypersensitivity (Neumüller et al., 2013). To potentially alleviate problems caused by PPV, which endangers plum production in many countries, various breeding strategies are used, including conventional methods based on resistance or hypersensitivity (Neumüller, 2011) and modern methods of genetic engineering and the creation of transgenic plum varieties (Scorza et al., 2013).

FIGURE 7.7

Progression of *Plum pox virus* (Sharka) symptoms in fruits and leaves of a plum tree during (A) spring and (B) summer.

Courtesy Dr. Manuel Rubio and Dr. P. Martínez-Gómez.

In addition, the plum tree is a source of infection for other species such as the peach and apricot trees, which can also be affected by this disease. This is particularly problematic since all cultivated Japanese plum cultivars are susceptible and the only known solution is tree removal (Rubio et al., 2011). Other diseases of plums include a handful of fungal and bacterial pathogens, the most common being bacterial canker, cytospora canker, brown rot, and black knot (Seethapathy et al., 2022).

In this context, production of virus-free planting material is a challenge, especially in Sharka-infested regions, such as Eastern Europe (Achim et al., 2017). One solution for European plums is the use of resistant rootstocks such as Myrobalan BN 4 Kr (Stănică et al., 2019) and resistant cultivars such as Jojo, Jofela, Jolina, Joganta, etc. or tolerant cultivars such as Roman and Alina (Butac et al., 2019). The propagation of clonal rootstocks is generalized for prune, using green and woody cuttings and *in vitro* micropropagation. Even so, there are still nurseries that are producing seedling rootstocks of Myrobalan or plum types for grafting (Stănică, 2007, Stănică et al., 2010). Recently, *ex vitro* micrografting was used to reduce the time and cost of grafted planting material.

Finally, the number of pests reported in plum and identified in different geographical regions is large. Severe economic loss can be caused by several of them: the plum fruit moth, plum sawfly, San Jose scale, aphids, twig borers, thrips, and mites (Seethapathy et al., 2022). Adequate identification of the pest and knowledge of its life cycle are crucial for effective control methods.

8. Current and future challenges in main production regions

Plum and prune tend to bear heavy crops that result in undersized fruits of low quality. Therefore, fruit thinning is necessary to increase fruit size and uniformity in color and to stimulate flower initiation to produce the subsequent year's crop. Various methods such as hand-thinning, mechanical thinning, and use of chemicals are used. Chemical thinning has superiority with respect to thinning cost but can be unpredictable and results can vary with environmental conditions. Early thinning of flowers with mechanical sting thinners can reduce the initial fruit set but can be risky in areas prone to spring frosts. Mechanical thinning by trunk shaking is practiced with prunes, and the results can be variable, but it is an economic option if the focus is primarily to reduce production of undersized fruit.

There is clearly a need for more and better cultivars for both Japanese and European plum production. In addition to dealing with Sharka resistance, many currently produced cultivars are not meeting consumer acceptance as expectations regarding flavor, aroma, and overall organoleptic quality increase. Some of the recently introduced plumcots have outstanding flavor characteristics and have increased consumer expectations for fresh plum quality. In addition, the dried prune industries around the world are dependent on a few derivative cultivars of French prune. This results in susceptibilities to crop loss due to synchrony of blooming among all trees in one region in the face of spring frosts or excessive heat during bloom. In addition, because there is only one cultivar grown in a region, the prune crops in that region are harvested at the same time, which leads to inefficient use of drying facilities.

Historically, plum production has taken a back seat to other more prominent fruit tree crops so research and experimentation on orchard management practices have received less attention than several other fruit tree crops such as apple, pear, peach, and cherry. More research is needed to identify rootstocks for plums that meet specific objectives. In addition, the growth and bearing habits of plum trees can vary greatly by cultivar and work to categorize specific plum growth and bearing habit types could help to improve recommendations for training systems and pruning of specific types.

References

Abdi, N., Holford, P., & McGlasson, W. B. (1997). Effects of harvest maturity on the storage life of Japanese type plums. *Australian Journal of Experimental Agriculture, 37*, 391–397.

Abdi, N., McGlasson, W. B., Holford, P., Williams, M., & Mizrahi, Y. (1998). Responses of climacteric and suppressed-climacteric plums to treatment with propylene and 1-methylcyclopropene. *Postharvest Biology and Technology, 14*, 29–39.

Abu-Kpawoh, J. C., Xi, Y. F., Zhang, Y. Z., & Jin, Y. F. (2002). Polyamine accumulation following hot-water dips influences chilling injury and decay in Friar plum fruit. *Journal of Food Science, 67*, 2649–2653.

Achim, G., Botu, I., Botu, M., Duţu, I., Stănică, F., & Mazilu, C. (2017). The rootstock's role in modernizing plum culture. *Acta Horticulture, 1175*, 49–54.

Afanador-Barajas, L. N., Wilches, A. V., Macana, Y. A. M., & Medina-Pérez, G. (2022). History, distribution, production and taxonomic classification of plum. In A. Gull, G. A. Nayik, S. Wani, & V. Nanda (Eds.), *Handbook of plum fruit production, postharvest science, and processing technology* (pp. 2–20).

Al-Khayri, J. M., Naik, P. M., Jain, S. M., & Johnson, D. V. (2018). Advances in date palm (*Phoenix dactylifera* L.) breeding. In J. M. Al-Khayri, S. M. Jain, & D. V. Johnson (Eds.), *Advances in plant breeding strategies: Fruits* (pp. 727–771). Springer.

Buchner, R. P., Southwick, S. M., Krueger, W. H., Yeager, J. T., & Gilles, C. K. (2012). Crop control. In R. P. Buchner (Ed.), *Prune production manual.* Publication 3507 (pp. 145–150). Richmond, CA. USA: Regents of the University of California Agriculture and Natural Resources.

Butac, M., Botu, M., Militaru, M., Mazilu, C. R., Duţu, I., & Nicolae, S. (2019). Plum germplasm and breeding in Romania. *Proceedings of the Latvian Academy of Sciences: Section B, 73*(3), 214–219. https://doi.org/10.2478/prolas-2019-0034, 720.

Butac, M., Bozhkova, V., Zhivondov, A., Milosevic, N., Bellini, E., Nencetti, V., et al. (2013). Overview of plum breeding in Europe. *Acta Horticulture, 981*, 91–98.

Butac, M., Ondrášek, I., & Krška, B. (2012). Behaviour of some plum cultivars in Lednice and Piteşti areas. *Acta Horticulture, 968*, 109–114.

Butac, M. (2020). Plum breeding. In A. Küden (Ed.), *Prunus* (pp. 1–24). Intech Open.

Candan, A. P., Graell i Sarle, J., & Larrigaudière, C. (2011). Postharvest quality and chilling injury of plums: Benefits of 1-methylcyclopropene. *Spanish Journal of Agricultural Research, 9*(2), 554–564.

Candan, A. P., Graell, J., Crisosto, C., & Larrigaudière, C. (2006). Improvement of storability and shelf-life of Blackamber plums treated with 1-methylcyclopropene. *Food Science and Technology International, 12*, 437–443.

Cantín, C. M., Crisosto, C. H., & Day, K. R. (2008). Evaluation of the effect of different modified atmosphere packaging box liners on the quality and shelf life of Friar plums. *Horttechnology, 18*, 261–265.

CBI, URL. https://www.cbi.eu/market-information/fresh-fruit-vegetables/plums/market-entry (accessed 16 11 2024).

Crisosto, C. H., Garner, D., Crisosto, G. M., & Bowerman, E. (2004). Increasing 'Blackamber plum (*Prunus salicina* Lindell) consumer acceptance. *Postharvest Biology and Technology, 34*, 237–244.

Crisosto, C. H., & Garner, D. (2008). Effects of controlled atmosphere on plums. *Cent Val Postharvest Newsl, 17*, 4–6.

Crisosto, C. H., & Kader, A. A. (2015). Plum and fresh prune. In K. C. Gross, C. Y. Wang, & M. Saltveit (Eds.), *The commercial storage of fruits, vegetables, and florist and nursery stocks. Agriculture handbook 66* (pp. 497–500). Washington, DC, USA: US Department of Agriculture, Agricultural Research Service.

Crisosto, C. H., Lurie, S., & Retamales, J. (2009). Stone fruits. In E. M. Yahia (Ed.), *Modified and controlled atmospheres for the storage, transportation, and packaging of horticultural commodities* (pp. 287–315). Boca Raton, Florida, USA: CRC Press/ Taylor & Francis Group.

Crisosto, C. H., Mitchell, F. G., & Ju, Z. (1999). Susceptibility to chilling injury of peach, nectarine, and plum cultivars grown in California. *Horticultural Science, 34*, 1116–1118.

Crisosto, C. H., & Mitchell, F. G. (2002). Postharvest handling systems: Stone fruits. In A. A. Kader (Ed.), *Postharvest technology of horticultural crops. University of California, agriculture and natural resources*, (pp. 345–352). Oakland, California, USA: Publication 3311.

Cuevas, F. J., Pradas, I., Ruiz-Moreno, M. J., Arroyo, F. T., Perez-Romero, L. F., Montenegro, J. C., et al. (2015). Effect of organic and conventional management on bio-functional quality of thirteen plum cultivars (*Prunus salicina* Lindl.). *PLoS One, 10*(8), Article e0136596.

Cvetković, M., & Glišić, I. (2020). *Plum – growing technology*, (p. 462). University of Banja Luka, Faculty of Agriculture (In Serbian). ISBN 978-99938-93-66-0.

Dal Martello, R., von Baeyer, M., Hudson, M., et al. (2023). The domestication and dispersal of large-fruiting *Prunus* spp.: A metadata analysis of archaeobotanical material. *Agronomy, 13*, 1027.

Dallot, S., & Thébaud, G. (2021). Improved knowledge on key epidemiological parameters to optimize Sharka management strategies. *Acta Horticulture, 1322*, 277–287.

Day, K. R., & Buchner, R. P. (2012). Growing pruned for fresh market. In R. P. Buchner (Ed.), *Prune production manual* (pp. 279–282). Richmond, CA. USA: Regents of the University of California Agriculture and Natural Resources. Publication 3507.

DeJong, T. M., & Castro, S. (2025). Forty years of prune breeding at UC Davis. *Acta Horticulture* (in Press).

Dinkova, H., Mihaylova, P., Dragoyski, K., & Vaneva, B. (2007). Living mulch: A promising system for maintenance of soil surface in young plum plantations established on sloping terrains. *Journal of Pomology, 159*, 101–106.

Dong, L., Lurie, S., & Zhou, H.-W. (2002). Effect of 1-methylcyclopropene on ripening of canino apricots and royal zee plums. *Postharvest Biology and Technology, 24*, 135–145.

Eryomine, G. V. (1990). New data on origin of *Prunus domestica* L. *Acta Horticulture, 283*, 27–30.

Esmenjaud, D., & Dirlewanger, E. (2007). Plum. In C. Kole (Ed.), *Fruits and Nuts: 4. Genome mapping and molecular breeding in plants* (pp. 119–135). Germany: Springer-Verlag Berlin Heidelberg.

FAOSTAT *Crops and Livestock Products*, URL. https://www.fao.org/faostat/en/#data/QCL (accessed 16 11 2024).

Farcuh, M., & Hopfer, H. (2023). Aroma volatiles as predictors of chilling injury development during peach (*Prunus persica* (L) Batsch) cold storage and subsequent shelf-life. *Postharvest Biology and Technology, 195*, Article 112137.

Farcuh, M., Li, B., Rivero, R. M., Shlizerman, L., Sadka, A., & Blumwald, E. (2017). Sugar metabolism reprogramming in a non-climacteric bud mutant of a climacteric plum fruit during development on the tree. *Journal of Experimental Botany, 68*, 5813–5828.

Farcuh, M., Rivero, R. M., Sadka, A., & Blumwald, E. (2018). Ethylene regulation of sugar metabolism in climacteric and non-climacteric plums. *Postharvest Biology and Technology, 139*, 20–30.

Farcuh, M., Tajima, H., Lerno, L. A., & Blumwald, E. (2022). Changes in ethylene and sugar metabolism regulate flavonoid composition in climacteric and non-climacteric plums during postharvest storage. *Food Chemistry Molecular sciences*, Article 100075.

Farcuh, M., Toubiana, D., Sade, N., Rivero, R. M., Doron-Faigenboim, A., Nambara, E., et al. (2019). Hormone balance in a climacteric plum fruit and its non-climacteric bud mutant during ripening. *Plant Science, 280*, 51–65.

Faust, M., & Surányi, D. (1998). Origin and dissemination of plums. In J. Janick (Ed.), *Horticultural reviews* (pp. 179–231). Wiley Online Library.

Grossman, Y. L., & DeJong, T. M. (1995). Maximum fruit growth potential following resource limitation during peach growth. *Annals of Botany, 75*, 561–567.

Grossman, Y. L., & DeJong, T. M. (1994). Peach: A simulation model of reproductive and vegetative growth in peach trees. *Tree Physiology, 14*, 329–345.

Guerra, M. E., & Rodrigo, J. (2015). Japanese plum pollination: A review. *Scientia Horticulture, 197*, 674–686.

Holb, I. J. (2003). The brown rot fungi of fruit crops (*Monilinia* spp.): I. Important features of their biology. *International Journal of Horticultural Science, 9*(3–4), 23–36.

Iliescu, L. M., Stan, E. G., Peticilă, G. A., Mihai, C. A., & Stănică, F. (2023). Performances of four Romanian prune cultivars grafted on two rootstocks under the Trident canopy. *Acta Horticulture, 1384*, 155–164.

Khan, A. S., Ahmed, M. J., & Singh, Z. (2011). Increased ethylene biosynthesis elevates incidence of chilling injury in cold-stored Amber Jewel Japanese plum (*Prunus salicina* Lindl.) during fruit ripening. *International Journal of Food Science and Technology, 46*, 642–650.

Khan, A. S., Singh, Z., & Ali, S. (2018). Postharvest biology and technology of plum. In S. A. Mir, M. A. Shah, & M. M. Mir (Eds.), *Postharvest biology and technology of temperate fruits* (pp. 101–147). Cham, Switzerland: Springer.

Kim, H.-Y., Farcuh, M., Cohen, Y., Crisosto, C., Sadka, A., & Blumwald, E. (2015). Non-climacteric ripening and sorbitol homeostasis in plum fruits. *Plant Science, 231*. https://doi.org/10.1016/j.plantsci.2014.11.002

Korićanac, A., Milatović, D., Popović, B., Mitrović, O., & Glišić, I. (2023). Modified atmosphere packaging of fruits of two European plum cultivars: Effect on physical, chemical and sensory properties. In *Book of proceedings of 'innovations in horticulture'* (pp. 93–104). Belgrade (Serbia): University of Belgrade, Faculty of Agriculture (In Serbian with abstract in English).

Liu, W., Li, D., Zhang, A., Feng, C., Yang, J., Yoon, J., et al. (2007). Genetic diversity and phylogenetic relationships among plum germplasm resources in China assessed with inter-simple sequence repeat markers. *Journal of the American Society for Horticultural Science, 132*, 619–628.

Manganaris, G. A., Vicente, A. R., Crisosto, C. H., & Labavitch, J. M. (2008b). Cell wall modifications in chilling-injured plum fruit (*Prunus salicina*). *Postharvest Biology and Technology, 48*, 77–83.

Manganaris, G. A., Vicente, A. R., Crisosto, C. H., & Labavitch, J. M. (2008a). Effect of delayed storage and continuous ethylene exposure on flesh reddening of Royal Diamond plums. *Journal of The Science of Food and Agriculture, 88*, 2180–2185.

Mariscal, M. J., Day, K. R., Basile, B., & DeJong, T. M. (2002). Modeling the vegetative and reproductive growth of plums. *Acta Horticulture, 584*, 35–42.

Martínez-Romero, D., Dupille, E., Guillén, F., Valverde, J. M., Serrano, M., & Valero, D. (2003). 1-Methylcyclopropene increases storability and shelf life in climacteric and non-climacteric plums. *Journal of Agricultural and Food Chemistry, 51*, 4680–4686.

Milatović, D., Đurović, D., & Zec, G. (2017). Growth and productivity characteristics of Japanese plum cultivars. *Journal of Agricultural Science, 62*, 133–141 (In Serbian with abstract in English).

Milatović, D. (2019). *Plum*, 978-86-913763-8-3 (p. 531). Čačak: Scientific Pomological Society of Serbia (In Serbian).

Milatović, D. (2023). *Stone fruit*, 978-86-903174-2-4 (p. 780). Čačak: Scientific Pomological Society of Serbia (In Serbian).

Milošević, N., Glišić, I., Đorđević, M., Radičević, S., & Jevremović, D. (2021). An overview of plum breeding at Fruit Research Institute, Čačak. *Acta Horticulture, 1322*, 7–11.

Milošević, T., & Milošević, N. (2022). Behaviour of Japanese plums in atypical environmental conditions of Serbia: Tree growth, productivity and fruit quality attributes. *Erwerbs-Obstbau, 64*, 513–522.

Milošević, T., & Milošević, N. (2018). Plum (Prunus spp.) breeding. In J. M. Al-Khayri, et al. (Eds.), *Advances in plant breeding strategies: Fruits* (pp. 165–215). Springer International Publishing AG.

Minas, I. S., Crisosto, G. M., Holcroft, D., Vasilakakis, M., & Crisosto, C. H. (2013). Postharvest handling of plums (*Prunus salicina* Lindl.) at 10 C to save energy and preserve fruit quality using an innovative application system of 1-MCP. *Postharvest Biology and Technology, 76*, 1–9.

Mitchell, F. G., Thompson, J. F., Crisosto, C. H., & Kasmire, R. F. (2008). The commodity. In J. F. Thompson, F. G. Mitchell, T. R. Rumsey, & C. H. Crisosto (Eds.), *Commercial cooling of fruits, vegetables, and flowers* (pp. 1–7). CA, USA: University of California Press.

Neumüller, M. (2011). Fundamental and applied aspects of plum (Prunus domestica) breeding. *Fruit, Vegetable and Cereal Science and Biotechnology, 5*, 139–156.

Neumüller, M., Mühlberger, L., Siegler, H., Hartmann, W., & Treutter, D. (2013). New rootstocks with resistance to plum pox virus for Prunus domestica and other stone fruit species: The Docera and Dospina rootstock series. *Acta Horticulture, 985*, 155–165.

Nicolás-Almansa, M., Ruiz, D., Salazar, J. A., Guevara, A., Cos, J., Martínez-Gómez, P., & Rubio, M. (2023). Phenotypic and molecular characterization of new interspecific Japanese plum × apricot hybrids (plumcots). *Scientia Horticulture, 318*, Article 112131.

Okie, W. R. (2001). Plum crazy: Rediscovering our lost Prunus resources. *Hort Science, 36*, 209–213.

Okie, W. R., & Hancock, J. F. (2008). Plums. In J. F. Hancock (Ed.), *Temperate fruit crop breeding* (pp. 337–357). New York: Springer.

Palou, L., Crisosto, C. H., Garner, D., & Basinal, L. M. (2003). Effect of continuous exposure to exogenous ethylene during cold storage on postharvest decay development and quality attributes of stone fruits and table grapes. *Postharvest Biology and Technology, 27*, 243–254.

Pan, H., Wang, L., Wang, R., Xie, F., & Cao, J. (2018). Modifications of cell wall pectin in chilling-injured Friar plum fruit subjected to intermediate storage temperatures. *Food Chemistry, 242*, 538–547.

Pan, H., Wang, R., Li, L., Wang, J., Cao, J., & Jiang, W. (2016). Manipulation of ripening progress of different plum cultivars during shelf life by post-storage treatments with ethylene and 1-methylcyclopropene. *Scientia Horticulture, 198*, 176–182.

Panahirad, S., Naghshiband-Hassani, R., Ghanbarzadeh, B., Zaare-Nahandi, F., & Mahna, N. (2019). Shelf life quality of plum fruits (*Prunus domestica* L.) improves with carboxymethylcellulose-based edible coating. *Hort Science, 54*, 505–510.

Petri, C., Alburquerque, N., Faize, M., Scorza, R., & Dardick, C. (2018). Current achievements and future directions in genetic engineering of European plum (*Prunus domestica* L.). *Transgenic Research, 27*, 225–240.

Potter, D. (2012). Basic information on the stone fruit crops. In C. Kole, & A. G. Abbott (Eds.), *Genetics, genomics and breeding of stone fruits* (pp. 1–21). New York: CRC Press. USA.

Ravelonandro, M., Scorza, R., Polak, J., Callahan, A., Krška, B., Kundu, J., et al. (2013). "HoneySweet" plum—a valuable genetically engineered fruit-tree cultivar. *Food and Nutrition Sciences, 4*, 45–49.

Rimbaud, L., Dallot, S., Gottwald, T., Decroocq, V., Jacquot, E., Soubeyrand, S., et al. (2015). Sharka epidemiology and worldwide management strategies: Learning lessons to optimize disease control in perennial plants. *Annual Review of Phytopathology, 53*, 357–378.

Rubio, M., García-Ibarra, A., Dicenta, F., & Martínez-Gómez, P. (2011). Plum pox virus (Sharka) sensitivity in *Prunus salicina* and *Prunus cerasifera* cultivars against a Dideron-type isolate. *Plant Breeding, 130*, 283–286.

Ruiz, D., Egea, J., Salazar, J. A., & Campoy, J. A. (2018). Chilling and heat requirements of Japanese plum cultivars for flowering. *Scientia Horticulture, 242*, 2018.

Scorza, R., Callahan, A. M., Dardick, C. D., Srinivasan, C., DeJong, T. M., Abbott, A. G., et al. (2013). Biotechnological advances in the genetic improvement of Prunus domestica. *Acta Horticulture, 985*, 111–118.

Seethapathy, P., Gothandaraman, R., Gurudevan, T., & Malik, I. A. (2022). Diseases, pests, and disorders in plum. Diagnosis and management. In A. Gull, G. A. Nayik, S. Wani, & V. Nanda (Eds.), *Handbook of plum fruit production, postharvest science, and processing technology* (pp. 133–176).

Sibbett, G. S., & Yeager, J. T. (2012). Fruit maturity and harvest management. In R. P. Buchner (Ed.), *Prune production manual. Publication 3507* (pp. 273–278). Richmond, CA. USA: Regents of the University of California Agriculture and Natural Resources.

Singh, S. P., & Singh, Z. (2013a). Controlled and modified atmospheres influence chilling injury, fruit quality and antioxidative system of Japanese plums (*Prunus salicina* Lindl.). *International Journal of Food Science and Technology, 48*, 363–374.

Singh, S. P., & Singh, Z. (2013b). Postharvest cold storage-induced oxidative stress in Japanese plums (*Prunus salicina* Lindl. cv. Amber Jewel) in relation to harvest maturity. *Australian Journal of Crop Science, 7*, 391–400.

Singh, S. P., & Singh, Z. (2012). Postharvest oxidative behaviour of 1-methylcyclopropene treated Japanese plums (*Prunus salicina* Lindell) during storage under controlled and modified atmospheres. *Postharvest Biology and Technology, 74*, 26–35.

Sisler, E. C., & Serek, M. (1997). Inhibitors of ethylene responses in plants at the receptor level: Recent developments. *Physiologia Plantarum, 100*, 577–582.

Sottile, F., Caltagirone, C., Giacalone, G., Peano, C., & Barone, E. (2022). Unlocking plum genetic potential: Where are we at? *Horticulture, 8*, 128.

Stănică, F., Dumitraşcu, M., Ancu, S. I., & Dutu, I. (2010). Propagation by green cuttings of some new promising Romanian Prunus rootstocks. *Acta Horticulture, 862*, 413–418.

Stănică, F., & Peticilă, G. A. (2011). *Înfiintarea plantatiilor pomicole (Orchards planting)* (p. 226). Editura Valahia University Press, 978-606-603-033-5.

Stănică, F. (2007). Propagation of some Prunus rootstocks by hardwood cutting on composed rooting substrates. *Acta Horticulture, 734*, 309–311.

Stănică, F., Zagrai, I., Feştilă, A., Zuccherelli, G., Zuccherelli, K., Neri, D., et al. (2019). BN 4 Kr - a promising Sharka resistant rootstock, XI ISHS int. Symp. On plum and prune genetics, breeding and pomology, freising-weihenstephan, Germany. *Acta Horticulture, 1260*, 203–208.

Taylor, M. A., Rabe, E., Jacobs, G., & Dodd, M. C. (1995). Effect of harvest maturity on pectic substances, internal conductivity, soluble solids and gel breakdown in cold stored Songold plums. *Postharvest Biology and Technology, 5*, 285–294.

Thakur, R., Pristijono, P., Golding, J. B., Stathopoulos, C. E., Scarlett, C. J., Bowyer, M., et al. (2018). Development and application of rice starch based edible coating to improve the postharvest storage potential and quality of plum fruit (*Prunus salicina*). *Scientia Horticulture, 237*, 59–66.

Thompson, J. F., Mitchell, F. G., & Kasmire, R. F. (2002). Cooling horticultural commodities. In A. A. Kader (Ed.), *Postharvest technology of horticultural crops* (pp. 97–112). CA, USA: University of California Press.

Topp, B. L., Russell, D. M., Neumüller, M., Dalbó, M. A., & Liu, W. (2012). In M. L. Badenes, & D. H. Byrne (Eds.), *Fruit breeding, vol. 8 of handbook of plant breeding* (pp. 571–621). New York, NY: Springer.

Van De Geijn, F. (1993). Meer duidelijkheid gewenst alvorens te investeren in CA-bevaring. *Fruitt, 25*, 20–21.

Wang, J., Pan, H., Wang, R., Hong, K., & Cao, J. (2016). Patterns of flesh reddening, translucency, ethylene production and storability of Friar plum fruit harvested at three maturity stages as affected by the storage temperature. *Postharvest Biology and Technology, 121*, 9–18.

Watkins, R. Cherry, Plum, Peach, Apricot and Almond: *Prunus* Spp. (Rosaceae).; 1976. Ed. Springer.

Watkins, C. B. (2006). The use of 1-methylcyclopropene (1-MCP) on fruits and vegetables. *Biotechnology Advances, 24*, 389–409.

Yoshikawa, F. T., & La Rue, J. H. (1989). Training and pruning. In J. H. La Rue, & R. S. Johnson (Eds.), *Peaches, plums, and nectarines, growing and handling for fresh market* (pp. 42–48). Oakland CA. USA: Regents of the University of California Division of Agriculture and Natural Resources. Publication 3331.

Zagrai, L. A., & Zagrai, I. (2020). Further evidence of durable resistance in HoneySweet transgenic plum under natural infection with D and Rec strains of Plum Pox Virus. *Scientific Papers Series B, Horticulture, LXIV*(1), 204–209. Print ISSN 2285-5653.

Zhang, J. Y., Liu, W. S., Lin, P. J., & Blanchet, P. (1998). Preliminary report about wild *Prunus domestica* L. In xinjiang (Western China). *Acta Horticulture, 478*, 347–350.

Zhebentyayeva, T., Shankar, V., Scorza, R., Callahan, A., Ravelonandro, M., Castro, S., et al. (2019). Genetic characterization of worldwide *Prunus domestica* (plum) germplasm using sequence-based genotyping. *Horticulture Research, 6*, 12.

Zohary, D. (1992). Is the European plum, *Prunus domestica* L., a *P. cerasifera* Ehrh. × *P. spinosa* L. allo-polyploid? *Euphytica, 60*, 75–77.

Zurawicz, E., Pruski, K., Szymajda, M., Lewandowski, M., Seliga, Ł., & Malinowski, T. (2013). *Prunus domestica* Jojo – good parent for breeding of new plum cultivars resistant to plum pox virus. *Journal of Agricultural Science, 5*(10), 5539.

CHAPTER

Kiwifruit†

8

Guglielmo Costa[1], George A. Manganaris[2], Evangelos Xylogiannis[3], Daniele Schiavi[4], Giorgio M. Balestra[4], Ross A. Ferguson[5]

[1]University of Bologna, Alma Mater Studiorum, Italy; [2]Cyprus University of Technology, Department of Agricultural Sciences, Biotechnology & Food Science, Lemesos, Cyprus; [3]RS Fruit S.A., Chariessa, Naoussa, Greece; [4]Department of Agriculture and Forest Sciences, University of Tuscia, Viterbo, Italy; [5]Formerly The New Zealand Institute for Plant and Food Research Ltd, Auckland, New Zealand

Chapter outline

† For Chapter Kiwifruit (Actinidia deliciosa, Actinidia chinensis), the publisher has provided a non-exclusive license to publish Photographs from The New Zealand Institute for Plant and Food Research.

Temperate Tree Fruits and Nuts. https://doi.org/10.1016/B978-0-443-29137-1.00008-6

1. Origin and spread

Commercial kiwifruit are large-fruited selections of *Actinidia chinensis* var. *chinensis* and *A. chinensis* var. *deliciosa* or their hybrids. Kiwifruit are the most recently domesticated of all fruit crops considered in this volume (Ferguson and Bollard, 1990; Huang and Ferguson, 2007). The name "kiwifruit" itself is likewise very recent, being first used in 1959 for the hairy, green-fleshed fruit of *A. chinensis* var. *deliciosa* (Ferguson and Bollard, 1990). Now the name is also used for the fruit of *A. chinensis* var. *chinensis,* which have less-hairy skins and flesh of different colors: yellow, green, or occasionally red. The name is sometimes wrongly shortened to "kiwi" or separated into two words "kiwi fruit."

The original home of kiwifruit is China where the most common name is mihoutao (monkey peach) (Figure 8.1). Although kiwifruit are widespread in Southern and Central China, it seems that they were not systematically cultivated until recently. Instead, fruit were simply collected from the wild (Ferguson, 1990).

Systematic cultivation began in New Zealand when seeds of the green-fleshed *A. chinensis* var. *deliciosa* were taken there in 1904. Cultivation was successful, and by 1910, fruit were being produced outside China. Plants were soon sold under the name "Chinese gooseberry," and a very small commercial orchard of only 14 vines was established in Whanganui, New Zealand, in the late 1920s, producing fruit in the early 1930s. This was the world's very first commercial plantation of kiwifruit. Subsequent orchards used plants derived from the one original introduction of seed—a classic example of domestication based on a single accession of wild germplasm, with almost every step on the process known and documented (Ferguson et al., 2024b).

The area planted with kiwifruit remained small for many years until the first commercial exports of the fruit from New Zealand in 1953. Even then the industry developed slowly, and by 1965 there were only about 100 ha planted in New Zealand. The fruit exported were well received, especially when it was realized that fruit of the cultivar Hayward responded better to the long sea voyage to overseas markets than fruit of other kiwifruit cultivars then available (Figure 8.2). The area

FIGURE 8.1

Wild *Actinidia chinensis* vine climbing up trees, Sichuan Province, China.

(Photo from The New Zealand Institute for Plant and Food Research Ltd.)

FIGURE 8.2

Hayward Wright *(left)*, selector of the *Actinidia chinensis* var. *deliciosa* Hayward cultivar *(right),* which was for many years the most important kiwifruit cultivar throughout the world.

(Photos from The New Zealand Institute for Plant and Food Research Ltd.)

planted with kiwifruit in New Zealand increased to about 1000 ha in 1975 and then expanded dramatically to 15,000 ha by 1985; fruit were produced primarily for export, and the quantities increased rapidly. From 1975, only fruit of the cultivar Hayward were accepted for export; the success of the New Zealand kiwifruit export industry can be largely ascribed to the selection of this remarkably good cultivar.

Successful promotion and excellent returns to New Zealand growers encouraged growers in other countries to grow kiwifruit. They were confident that kiwifruit could be managed to produce profitable crops and that the fruit produced would find a ready market. Commercial orchards were being established in Italy and France in the early 1970s and then in California (U.S.), Chile, Greece, Iran, Portugal, Spain, and Turkey. Growers in these countries initially adopted New Zealand practices and cultivars. Eventually, Hayward predominated in kiwifruit orchards throughout the world, and until 2000, it was unchallenged in international markets as *the* kiwifruit.

Successful exploitation of the green-fleshed kiwifruit, *A. chinensis* var. *deliciosa*, in New Zealand stimulated Chinese horticultural scientists to investigate the wild resources of kiwifruit in China. Starting in 1974, intensive surveys resulted in the selection of more than 1400 large-fruited plants of *A. chinensis* var. *chinensis* and *A. chinensis* var. *deliciosa* directly from the wild. These selections, as well as several cultivars developed earlier in New Zealand, were the starting material for the now very successful Chinese kiwifruit industry. It is a rare documented example of crop domestication by multiple selections from widely dispersed wild populations of germplasm (Ferguson et al., 2024b).

China was the first country to grow green-fleshed, yellow-fleshed, and red-fleshed kiwifruit. Subsequently, other countries already growing green-fleshed kiwifruit started breeding programs to develop yellow- and red-fleshed kiwifruit starting from limited accessions of germplasm. The first yellow-fleshed kiwifruit (Hort16A) entered international trade in 2000, and small quantities of red-fleshed kiwifruit are becoming available outside China. Kiwifruit domestication and the development of the cultivars grown have therefore followed several different pathways. Because the crop is so new, its origins are known in unusual detail.

2. Economic importance

The areas planted in the main producing countries are shown in Table 8.1. Statistics may vary according to source, and in any given year, production can be greatly affected by weather conditions, e.g., winter temperatures, frost, summer temperatures, and storms. Total world production of kiwifruit in 2022 was about 4,470,000 metric tons.

At least half the world's kiwifruit production is from China, and as orchards in that country reach full production, this proportion is likely to increase. The Food and Agriculture Organization's (FAO) statistics probably underestimate Chinese kiwifruit production; the Chinese Society for Horticultural Science estimated that

Table 8.1 Plantings and production of the main kiwifruit-producing countries (FAOSTAT, 2024)

Country	Plantings of kiwifruit (Ha producing area), 2022	Production (metric tons), 2022
China	199,080	2,380,000
Italy	24,040	523,000
New Zealand	14,690	604,000
Greece	13,560	320,000
Iran	9910	295,000
Chile	6640	115,000
Turkey	4220	101,000
France	3930	47,000
Portugal	1940	53,000

the kiwifruit production in China was almost a third higher at 3.22 million tons in 2020 (Zhong et al., 2022). A single province, Shaanxi, which accounts for about half the total Chinese production, produces more kiwifruit than any other country; Sichuan produces about one-sixth of the Chinese total.

The biggest recent changes in kiwifruit cultivation have been the rapid expansion of plantings and production in Greece, Iran, and Turkey and the decrease in production in Italy. Kiwifruit production represents about 0.5% of the total world fresh fruit production by weight. In most countries, kiwifruit are therefore a minor crop. In China, kiwifruit account for only about 1% of total fruit production by weight; in Italy, about 3% by weight; and in most other countries, even less. In New Zealand, however, kiwifruit are uniquely important, being the most widely planted horticultural crop and accounting for about 55% by value of all fruit and vegetable exports. The productivity of New Zealand kiwifruit orchards is much higher than that of other countries. This is probably due to the age of New Zealand orchards, the experience of orchardists, the exceptional soils, the good climatic conditions, and the particular cultivars grown.

About one-third of total kiwifruit production, very largely from green-fleshed and yellow-fleshed cultivars, is traded internationally. Such trade is particularly important for the Southern Hemisphere producers of kiwifruit. New Zealand essentially produces kiwifruit for export, over 90% of its production, about 39% of the total global exported volume; the other main exporters at present are Chile at 10% of total global exported volume, Italy 19%, and Greece 11% (Chamberlain, 2024). Exports from the Northern and Southern Hemispheres are currently approximately equal, leading to year-round supply in many markets. Exports from China (mainly to Russia) are still insignificant. The patterns of international trade will probably change with the rapidly increasing production in both Iran and Turkey.

3. Botany

3.1 Implications of kiwifruit botany for cultivation

The kiwifruit of commerce are selections of *Actinidia chinensis* Planch. or hybrids of this species (Figure 8.3). The genus *Actinidia* contains more than 50 species, mostly endemic to mainland China (Huang and Ferguson, 2007). Several *Actinidia* species have fruit with commercial potential, e.g., *A. arguta* (kiwiberries), *A. eriantha,* and *A. kolomikta*, but *A. chinensis* is by far the most important species commercially and is the most widely cultivated. It is currently divided botanically into three varieties: *A. chinensis* var. *chinensis*, *A. chinensis* var. *deliciosa,* and *A. chinensis* var. *setosa*. These taxa have sometimes been treated as separate species, but they are now considered as varieties of the same species, although the status of *A. chinensis* var. *setosa* (endemic to Taiwan) is still debated (Ferguson, 2024). Commercial kiwifruit are selections of *A. chinensis* var. *chinensis* and *A. chinensis* var. *deliciosa*. Wild populations of the two varieties are widespread through Southern and Central China, var. *chinensis* more to the east, at lower altitudes, and var. *deliciosa* more to the west, inland and at higher altitudes. Where the two varieties occur in the same region, var. *chinensis* is at lower, warmer altitudes, and var. *deliciosa* is at higher, colder altitudes. Differences between the two varieties are clear at the geographic extremes, but there is progression in morphology across the geographic range; where the varieties overlap, there is extensive gene flow despite the differences in ploidy (Ferguson and Huang, 2016).

Although cultivated kiwifruit can be considered as plants suitable for temperate regions, selections of the two botanical varieties can be expected to differ in their response to climatic factors, e.g., in their sensitivity to cold or their requirement for winter chilling. Overwintering buds of var. *chinensis* are exposed, covered only by bud scales, whereas those of var. *deliciosa* are deeply buried in the bark of the leaf axil. Most important for the marketing of kiwifruit and for consumers, fruit of var. *deliciosa* have green flesh, whereas those of var. *chinensis* usually have yellow flesh, sometimes with red.

FIGURE 8.3

The kiwifruit of commerce are large-fruited selections of *(left) Actinidia chinensis* var. *chinensis* and *(right) A. chinensis* var. *deliciosa*. The cultivars shown are *(left)* Zesy002 and *(right)* Hayward.

(Photo from The New Zealand Institute for Plant and Food Research Ltd.)

- Kiwifruit grow naturally in the moist, sheltered, relatively cool forests and scrub of the mountains of Central and Southern China (Huang and Ferguson, 2007). Kiwifruit vines have a high requirement for water as shown by their high transpiration rates, and they are very sensitive to even mild water deficits. Their large leaves and rapidly growing shoots in spring are easily damaged by wind. Hail nets, shade nets, and overhead shelter are increasingly used to improve vine growth and fruit quality.
- Kiwifruit are vigorous, twining climbers. In cultivation, they need strong and expensive support structures. This is a major cost several years before the vines start carrying fruit.
- If kiwifruit escape from cultivation, they can be formidable and invasive weeds.
- Vine vigor necessitates careful pruning to maintain a balance between strong vegetative growth and fruit production and to ensure that the canopy is not too dense.
- Kiwifruit are dioecious. Male plants have staminate flowers with stamens producing viable pollen but only rudimentary ovaries; female plants have pistillate flowers with fully developed syncarpous ovaries, but their pollen is nonviable, empty of cytoplasm. Dioecy is under the control of two genes: *SuF* (*Shy Girl* or *SyGI*), which suppresses ovary development, and *M* (*Friendly Boy* or *FrBy*), which ensures completion of pollen development (Akagi et al., 2019). Both genes are on the Y chromosome. Dioecy has practical consequences: pollen must be transferred from male vines to the flowers of female (fruit-bearing) vines. Other fruit crops that are self-incompatible also require pollenizers.
- Flower evocation occurs in spring/early summer, but flower differentiation occurs only in the following spring (Richardson and Varkonyi-Gasic, 2024). Weather conditions in the previous season as well as in winter and spring of the current season may affect the number of flowers.
- In cultivated kiwifruit, every flower that is set usually produces a fruit. In many other fruit crops, only a small proportion of flowers eventually result in fully developed fruit. Kiwifruit crop loads can therefore be initially determined by controlling the number of shoots that are expected to produce flowers.
- The fruit are berries with many hundreds of seeds embedded in the flesh (Ferguson, 1984). Fruit of the cultivar Hayward that are of commercially acceptable weight usually contain over 1000 seeds. This requires large amounts of pollen being transferred from the male vines to the female vines. In most commercial orchards, this is achieved by the introduction of honeybees so orchards must be designed such that the bees are effective. The alternative is artificial (mechanical) pollination.
- The fruit flesh of *A. chinensis* var. *deliciosa* cultivars is green when ripe owing to the retention of chlorophyll, whereas most other fruits lose their chlorophyll as they ripen (Nardozza and Burdon, 2024). Although the fruit of some selections of *A. chinensis* var. *chinensis* have green flesh, most selections have yellow flesh. This is due to the loss of chlorophyll as the fruit mature, unmasking the

yellow of carotenoids. Yellow-fleshed kiwifruit must be harvested at a stage that allows degreening. The consumer wants yellow-fleshed fruit, not yellow-green fruit.

3.2 Cultivars

There have been many recent molecular studies on kiwifruit, and some of these have direct implications for future breeding (Nazir et al., 2024). Genomic, transcriptomic, proteomic, and metabolomic studies have revealed even more clearly the genetic diversity within the genus *Actinidia*. Marker-assisted selection and mutation breeding have already proved their potential. Cisgenesis could be used to incorporate beneficial traits in new cultivars, as long as the methods used are accepted by consumers.

Female (fruiting) cultivars: These can be grouped according to the color of the fruit flesh: green, yellow, or red (Figure 8.4). Color is often associated with differences in flavor. The first kiwifruit cultivars selected were all green-fleshed (*A. chinensis* var. *deliciosa*), and in most countries growing kiwifruit, they still predominate, but in some countries, yellow-fleshed and red-fleshed cultivars (*A. chinensis* var. *chinensis*) are becoming more important (Ferguson et al., 2024a).

In 2020, 49.9% of Chinese kiwifruit orchards were of green-fleshed cultivars, 15.8% yellow-fleshed cultivars, and 32.6% red-fleshed cultivars (Zhong et al., 2022). Most Chinese selections are grown only in China. The red-fleshed cultivar Hongyang was the most widely planted cultivar, occupying 23.3% of the total area planted. Red-fleshed kiwifruit are uniquely important in China. In other countries, there is great interest but only limited plantings of red-fleshed cultivars. The same year, 2020, 82% of plantings in Italy were of green-fleshed cultivars (mainly Hayward) and 18% of yellow-fleshed cultivars (largely Zesy002), while plantings of red-fleshed cultivars were insignificant.

FIGURE 8.4

Slices of red-, green-, and yellow-fleshed kiwifruit cultivars.

(Photo from The New Zealand Institute for Plant and Food Research Ltd.)

For many years Hayward was the only female cultivar grown in New Zealand orchards, but by 2020, 53% of the producing area was green-fleshed cultivars (Hayward), 46% yellow-fleshed cultivars (Zesy002), and less than 0.1% red-fleshed cultivars. The decline in importance of New Zealand plantings of green-fleshed cultivars has continued. Two years later, in 2022, green-fleshed cultivars occupied 46% of the total producing area in New Zealand, and yellow-fleshed cultivars 54% of the area. However, the yellow-fleshed cultivar grown in New Zealand (Zesy002) is considerably more productive than Hayward, and that season only 38% of the fruit submitted for export was green-fleshed, while 62% was yellow-fleshed. Yellow-fleshed kiwifruit are therefore uniquely important in New Zealand, and as plantings mature, they will become even more important.

The selection criteria for female cultivars are much the same irrespective of flesh color. However, yellow-fleshed kiwifruit should be fully degreened when eating ripe. For red-fleshed kiwifruit, the intensity and distribution of red coloration are often variable and seem sensitive to environmental conditions. Possibly the most important selection criterion now is disease resistance, particularly to bacterial canker of kiwifruit (Psa) caused by *Pseudomonas syringae* pv. *actinidiae* (Scortichini et al., 2024). Psa eliminated Hort16A as a commercial cultivar, and Hongyang is also especially sensitive. Other criteria include vine attributes such as ploidy, crop load, winter chilling requirements, fruit size and shape, skin robustness, fruit flavor and texture, time of harvest, storage life, and health constituents. Storage life is particularly important for exporting countries. High vitamin C content is often used by marketers to promote kiwifruit consumption.

Increasingly, new kiwifruit cultivars are protected by plant variety rights or plant patents. Cultivars are owned by consortia, which limit plantings, provide technical advice, and control marketing. Good examples are Zespri Group Ltd, which largely controls the export of fruit grown in New Zealand, and Jingold SpA in Italy. Consortia often want year-round marketing and license production of their protected cultivars in both Northern and Southern Hemispheres.

Most female kiwifruit cultivars, especially those from China, are simply selections from the wild, e.g., Qinmei and Jintao. In a few cases, e.g., Hongyang, they are selected from seedling populations grown from seed collected in the wild.

The initial *A. chinensis* var. *deliciosa* cultivars developed in New Zealand, e.g., Hayward, were also seedling selections but at least one generation removed from the original introduction of seed to New Zealand. Although Hayward was selected almost a century ago, it is still an important cultivar wherever kiwifruit is grown; in many countries, it is by far the most important cultivar. Some cultivars have arisen as selections from open-pollinated seedlings of cultivars or named selections. An important example is Xuxiang, currently the second most widely planted kiwifruit cultivar in China. Numerous clonal selections have been made of Hayward, e.g., the Kramer strain, but the distinctiveness or the superiority of most of these clones is not obvious. Bud sports have also been identified. Only a few widely planted cultivars have so far resulted from controlled, planned crosses. The most successful to date are the New Zealand yellow-fleshed cultivars of *A. chinensis* var. *chinensis*,

Hort16A, and its replacement, Zesy002. The first commercially important interspecific hybrid is the Chinese Jinyan, a cross between *A. chinensis* var. *chinensis* and *A. eriantha*. However, the fruit dry matter (DM) of Jinyan fruit is usually less than 16%, with resultant poor taste and flavor (Liao et al., 2020).

Kiwifruit breeding for female cultivars by controlled crossing has serious limitations (Testolin and McNeilage, 2024):

- dioecy means that two good female parents cannot be directly crossed;
- the genetic contribution of the male parent to fruit characteristics is not known unless progeny testing is undertaken;
- differences in ploidy can make crossing difficult;
- crosses usually result in equal numbers of female and unwanted male seedlings;
- seedlings may take 4 or 5 years to reach sexual maturity, and females tend to take longer;
- young vines require expensive support structures and seedling populations occupy large areas of land; and
- most kiwifruit breeders outside China have access to only limited germplasm. However, owing to dioecy imposing obligatory outcrossing and the high chromosome number ($x = 29$), kiwifruit have high genetic heterozygosity.

Male (pollenizer) cultivars: These are just as important as female cultivars for fruit production (Seal et al., 2024). They must overlap in flowering with the chosen female, and orchards often contain several different males to ensure this as climatic conditions in spring can differentially affect the time of flowering of the various cultivars grown. Males and females should preferably be of the same ploidy. The male vines should carry large numbers of flowers that produce large quantities of viable, compatible pollen. They should be disease-resistant and shed the senescent flowers soon after flowering. If orchardists use artificial pollination, the criteria for selection of the males are different. For example, it would be an advantage if the male vines flowered before the female vines to allow time for collection and preparation of the pollen. Pollen yield, robustness, and viability are also important. Only pollen from hexaploid *A. chinensis* var. *deliciosa* vines is currently used in artificial pollination. Male cultivars have generally been selected from the wild, from seed collected in the wild, from seedlings on orchards, or as byproducts from breeding programs undertaken for new female cultivars.

3.3 Rootstocks

There has been only limited use of rootstocks intended to control vegetative growth, encourage flowering, or provide resistance to root diseases (Thorp and Barnett, 2024). Vines vegetatively propagated or propagated by tissue culture are on their own roots. Many orchards in New Zealand consist of seedlings of the *A. chinensis* var. *deliciosa* cultivar Bruno topworked. Grafting within *Actinidia* species is usually successful but often not between species. Somaclonal variation in kiwifruit plants propagated by tissue culture does not seem to be a problem. There is limited use of rootstocks. D1 Vitroplant is used in Italy for overcoming iron deficiency induced by high soil pH, and Miliang in China since it is tolerant of soilborne diseases. Bounty71 *(A. macrosperma)* appears to be tolerant of both drought and waterlogging and may be helpful in soils inducing Kiwifruit Vine Decline Syndrome (KVDS).

4. Orchard management

4.1 Introduction

A cultivated kiwifruit vine ideally consists of a single trunk, with leaders (cordons) trained horizontally in opposite directions along the support system. These form the permanent structure of the vine. Fruiting canes, usually removed after they have fruited, are trained at right angles to the leaders. The main aims of vine management are the control of excessive vegetative growth and obtaining a profitable compromise between fruit number, fruit size, and fruit quality (fruit DM content) as there is a market demand for larger, better-quality fruit (as indicated by higher DM content).

Management techniques have evolved since kiwifruit were first cultivated. Initially only the cultivar Hayward (*A. chinensis* var. *deliciosa*) was grown in New Zealand and management practices were developed accordingly, but conditions in other countries may vary considerably, calling for changes. The introduction at the beginning of this century of *A. chinensis* var. *chinensis* cultivars, generally characterized by higher fertility, required adaptation of the training methods and cultural management originally determined and refined for Hayward. The appearance of bacterial canker, Psa, (see Section 7.1, Section 9) meant that new methods to minimize infection and disease expression were required.

4.2 Support systems and orchard design

The support systems, T-bar and pergola, were originally developed in New Zealand but were adopted and still used in the main producing countries with some modifications. In the T-bar system, fruiting canes are allowed to arch toward the ground forming a tunnel of canopy; in the pergola system fruiting canes are held horizontal to form a canopy about 2 m above the ground. In the T-bar system, rows are generally 4.5–5.0 m apart with the vines separated by 2.0–2.5 m along the rows, and in the pergola system, rows are 4.50–5.0 m apart with the vines separated by 2.5–4.0 m along the rows (Figure 8.5A, B) (Costa et al., 2018). In Italy, the T-bar system is mainly used in northern areas, while the pergola system is the most common in central and southern areas (Costa and Testolin, 2022). Other training systems proposed are characterized by different canopy shapes and higher planting densities. The original planting density of 300–400 plants ha^{-1} increased up to 1000–1200 plants ha^{-1} for most cultivars, even in some circumstances in some countries (e.g., Italy) up to 2000 plants ha^{-1}. Other training systems have been also tested in Italy, such as the Spindle, V, or Tatura, nowadays essentially no longer used,

FIGURE 8.5

T-bar *(left)* and pergola *(center)* are the commonly used support systems. Geneva double curtain (GDC) is still present in some orchards with higher density planting *(right)*.

(Photo of T-bar system from The New Zealand Institute for Plant and Food Research Ltd.)

and the Geneva double curtain (GDC) still present in some orchards with higher density planting (4.75 m between rows and 1.0–1.25 m between vines) (Figure 8.5C). The main goal of many of these training systems is to increase yields in the initial years of orchard establishment (Testolin et al., 1990).

In New Zealand, the pergola system now predominates, especially for *A. chinensis* var. *chinensis* cultivars. A recent modification of this training system separates the replacement canes from the fruiting canopy (Figure 8.6); another is the "alternate year cropping system" that, in some circumstances, produces yields up to 60 t ha^{-1} (Thorp and Barnett, 2011) (see Figure 8.6).

4.3 Pruning

In general, winter pruning is based on replacing the canes that have already fruited with 1-year-old canes, which will produce the flowering shoots the following spring. These canes should be equidistant along the leader (c. 25–30 cm apart) to allow air circulation and light penetration into the canopy (Figure 8.7).

FIGURE 8.6

Replacement canes separated from the fruiting canopy of kiwifruit.

(Photo from The New Zealand Institute for Plant and Food Research Ltd.)

FIGURE 8.7

Kiwifruit productive canes should originate as close as possible to the permanent leader and be 25–30 cm apart.

The bud-load left on the vines is about 150,000–200,000 buds ha^{-1}, which, depending on the fertility of the buds, should ensure a yield of 25–30 t ha^{-1} for the green-fleshed cultivars (*A. chinensis* var. *deliciosa*) and up to 40–45 t ha^{-1} for yellow- and bicolor-fleshed cultivars (*A. chinensis* var. *chinensis*). Much higher yields are feasible, over 50 t ha^{-1} for Hayward and 90 t ha^{-1} for the yellow-fleshed Zesy002. The renewal canes selected should be 1 year old and originate as close as possible to the permanent leader, directly, or if necessary, from the base of 2-year-old canes close to the leader. Summer (green) pruning is also performed once or even twice during the season to eliminate vigorous shoots growing upright (water shoots, suckers) as these usually carry few floral shoots and can shade the lower canopy. Summer pruning can also correct some mistakes made during winter pruning. If too many buds have been left on the vines during winter pruning, summer pruning can help balance vegetative growth and fruit production, removing some of the canes and, at the same time, improving light penetration into the lower part of the vine. Pruning often stimulates shoot regrowth so stubbing (cutting back of a new shoot to a few buds) and crushing the tip of a growing shoot are effective ways of limiting excessive vegetative growth.

4.4 Fertilization

Nutrient balance: To ensure vine health and production, nutrient requirements must be met by maintaining the nutrient balance, the structure and fertility of the soil, and by replacing the main nutritional elements of vines, avoiding shortages or excesses. Even when certain mineral elements are present in the correct ranges in the soil, they may not be available to vines. This can occur if large amounts of macroelements, in many cases already present in excess, are applied once all together or repeatedly every season. This can increase competition between antagonistic elements and block their uptake. For example, potassium (K), magnesium (Mg), and calcium (Ca) are mutually antagonistic. This means that when an excess of one or more of these elements is present in the soil, the uptake of the others is blocked or severely limited. In addition, excess of certain macroelements can change the chemistry and structure of the soil, root function can deteriorate, and nutrient uptake can decrease. Clay with K can form highly dense microaggregates of small dimensions, which can move toward the deeper layers of the soil with the water movement and fill the soil pores, creating impermeable layers and causing waterlogging, while mechanically limiting kiwifruit root growth. On the other hand, clay together with Ca forms aggregates of larger dimensions, and this can maintain or even improve soil structure. Once nutrients are absorbed by the root system, they move in the xylem flow toward the upper organs of the vines through transpiration, but not all elements have the same phloem mobility. The most mobile are nitrogen (N), phosphorus (P), and K followed by Mg, while Ca, boron (B), manganese (Mn), zinc (Zn), and iron (Fe) are the least mobile, and their movement is highly dependent on transpiration rates.

Seasonal nutrient uptake and management: For correct nutrient management, the right conditions, nutrient balance, and porosity in the soil must be maintained, and the precise amounts of nutrients applied to satisfy the estimated annual crop requirements, avoiding antagonisms while considering the effects of changes in vine transpiration on the movement of least mobile elements. A fertilization plan must consider nutrient availability and the input of the various mineral elements in irrigation water in the volume of soil explored by the root system, the quantity and dynamics of absorption and partitioning of the various mineral elements in the different vine organs during the annual growth cycle, the minerals removed from the orchard, and for nitrogen, mineralization during the year. It is also important to evaluate all the nutrient dynamics in the orchard system. These include not only the soil and vine (permanent and seasonal biomass) but also any potential mowed cover crops or grass, the nutrients returned to the soil (leaves, thinned fruit, mulched pruning) being recycled by biomass mineralization and those removed in the harvested crop (Table 8.2).

In newly planted orchards, the nutrients absorbed are utilized for the establishment of vine structure. In mature orchards, nutrients are required for biomass development and the expected crop. Apart from N, which is highly mobile and must be replaced every season, nutritional plans should focus on the nutrients removed in the fruit from the orchard each season. These need to be replaced. K, Mg, and Ca are among the most important nutrients in kiwifruit. However, these elements have characteristics that influence their uptake by the fruit. Potassium is mobile and can move easily in the phloem, and potential shortages can be met by transport from the leaves to the fruit or through the phloem later in the season. On the other hand, Ca, the least phloem mobile, moves only during fruit transpiration for the first 7 weeks after fruit set (Quartieri et al., 2022). Thus, to maximize the uptake of replacement elements, they should be applied at the correct stage of the season based

Table 8.2 Mineral macronutrients removed annually by *Actinidia chinensis* var. *chinensis* vines, Zesy002 grafted onto Hayward in a double leader T-bar training system (889 plants ha^{-1}) with an average fruit production of 37.5 tons fresh weight ha^{-1}

	N	P	K	Ca	Mg
Plant part	**kg ha^{-1}**				
Fruit	76.8	10.0	74.1	10.8	4.6
Summer prunings	4.6	0.4	3.8	5.4	0.8
Winter prunings	27.2	2.8	13.2	17.1	5.1
Abscised leaves	32.9	2.1	12.8	168.0	27.1
Thinned fruits	1.5	0.2	1.3	0.3	0.1
Dead roots	23.6	2.6	41.1	26.2	6.7
Total	166.6	18.1	146.3	227.8	44.4

Adapted from Quartieri et al., 2022.

on the development of each organ and vine biomass at each phenological phase (Figure 8.8).

Under optimal conditions, vines have already met most of their nutrient needs by harvest. However, during the last stage of fruit growth and after harvest, when there is much less or no competition between shoots and fruit and before the leaves senesce, nutrients are redirected toward the permanent wood and organs. This process is crucial to avoid any potential deficiencies during budbreak and initial vegetative development the following spring.

4.5 Irrigation

Water transport and uptake: Water regulates several essential processes and has a variety of roles in plants, being essential to the biochemical and metabolic reactions that take place within the cell, controls the turgor pressure within it, and regulates stomata opening and closing, transpiration, and photosynthesis. Through transpiration, it regulates the temperature of the leaf and other tissues, and it solubilizes gases, mineral salts, and organic substances moving within the plant. Three stages make up the water flow through a plant: absorption by the roots, movement within the plant, and transpiration, by which water leaves the plant. Only a small quantity of water is lost via the cuticle during transpiration, most is through the stomata. The cellular potential decreases owing to water loss through transpiration, reaching progressively lower values throughout the day. The measure that is most frequently used to characterize the energy state of water in plants is the water potential (w) of a cell, tissue, or organ.

The water transport efficiency of a plant depends on the size and number of xylem vessels and the percentage of the effective conductive area. Small xylem vessels

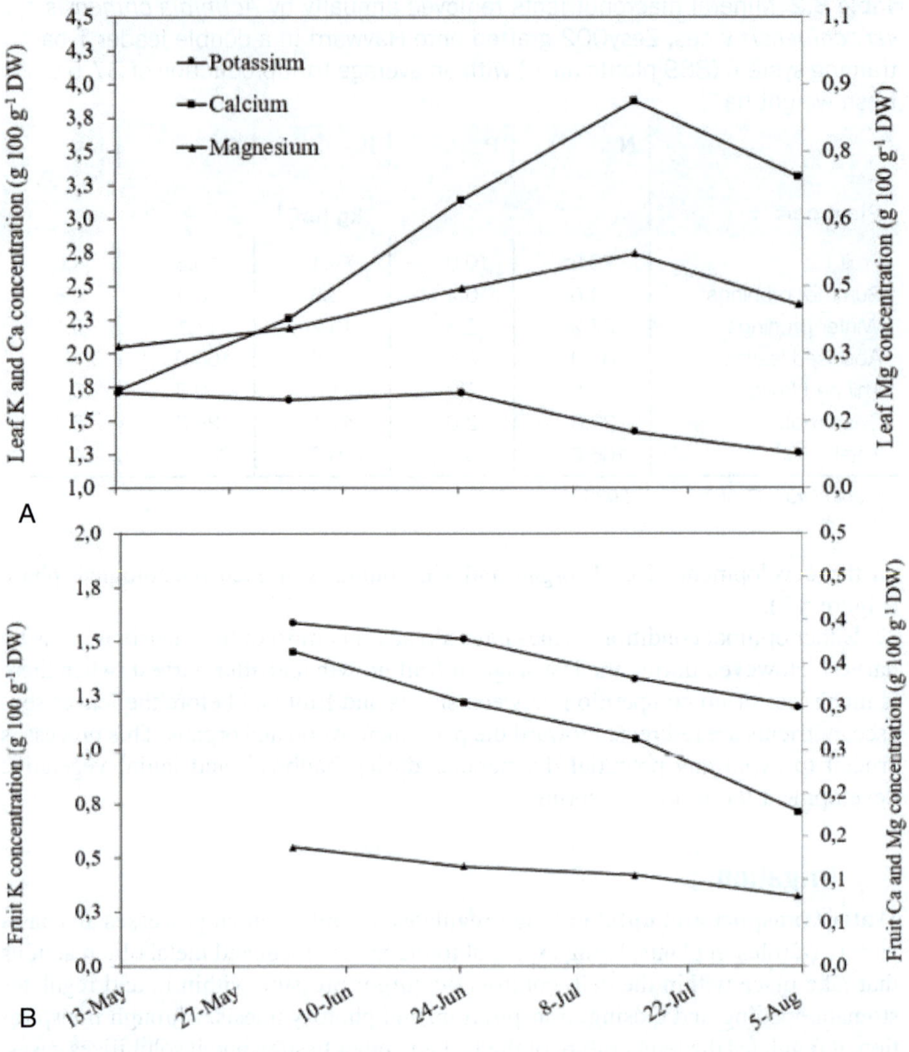

FIGURE 8.8

Seasonal concentration of potassium, magnesium, and calcium in leaves (A) and fruit (B) of *Actinidia chinensis* var. *chinensis* Zesy002 grafted onto Hayward, growing in Italy.

(Adapted from Quartieri et al., 2022.)

do not provide enough water to the leaves, leading to reduced leaf water potential. Large xylem vessels and simple perforated plates in kiwifruit improve transport efficiency, reducing oscillations in leaf water potential. The water in the vine moves mainly vertically, with resistance arising from the stomatal system, mesophyll, epicuticular layer, and boundary layer. The thickness, composition, and morphology of the cuticle depend on the cultivar, growing conditions, and leaf age.

The root system of kiwifruit vines is distinctive, featuring a large number of fibrous roots and a high root density largely restricted to the upper soil layers. In well-structured soils, they can grow root systems up to 20 t ha^{-1}. On the other hand, root development is impacted by soil management, drainage, and compaction, restricting growth and decreasing fruit quality. Kiwifruit root systems colonize the soil slowly but effectively, and waterlogging or reduced root turnover can be avoided by maximizing their development and density distribution (Hughes and Wilde, 1989). The relatively large xylem vessels of kiwifruit can transport large quantities of water from the soil to the leaves, but because of low leaf stomatal density, vines are very vulnerable to soil-water deficits (Dichio et al., 2013). Thus, kiwifruit vines are susceptible to waterlogging, which can result even in short-term water stress, hindering plant growth and reducing the ability of root development and generation of new white roots (D'Ippolito et al., 2022).

Irrigation systems and volumes: Soil properties, climate, water availability, cultivar characteristics, environmental impact, and investment costs must all be considered in designing and managing an irrigation system. The physical-mechanical properties, organic matter, and kind of management of the soil all affect its capacity to absorb and hold rain and irrigation water. Moreover, water quality and availability are two other factors that can affect the selection and proper operation of irrigation systems. Water must be available on demand to irrigate a kiwifruit orchard in accordance with its actual needs, which vary according to the soil, environment, irrigation season, system, and age. To maximize water use during fruit growth, the amount of soil that the roots are able to explore is crucial. Orchard age, planting density, rootstock, soil type, and management all affect how quickly the roots colonize the soil. The soil volume and the root hydrological properties allow calculation of the water retention capacity and the reserves that plants can access and are essential for managing and designing irrigation systems and for determining what should be the irrigated soil volume. Depending on fruit growth and kiwifruit root structure, the wetted volume should be 50–60 cm deep when irrigation is used.

Within kiwifruit orchards, drip, sprinkler, and a combination of the two are the most widely used irrigation techniques. Their combination is highly effective and can reduce water losses from evaporation and percolation, while meeting kiwifruit water requirements through the growing season (Figure 8.9). The difference between the amount of water in a soil at field water capacity (FWC) and the inducement point of water stress determines the reserve of easily available water (EAW) for plants. The flow rate, distance of emitters or drippers, and hydraulic conductivity are important factors to evaluate the lateral and vertical movement of water in the soil. Rainfall can be stored in the soil in substantial volumes, particularly outside the growing season when evaporation is low and water use is restricted.

The root system explores the entire volume of soil in two containers. Container 1, impacted by irrigation, is where most of the root system is concentrated. It is necessary to know the quantity of water that can be easily used to better manage the irrigation volume and frequency. It is often necessary, particularly in periods with high evapotranspiration demand in the environment, to intervene with irrigation several

FIGURE 8.9

Drip irrigation lines and microjet sprinklers in kiwifruit orchards.

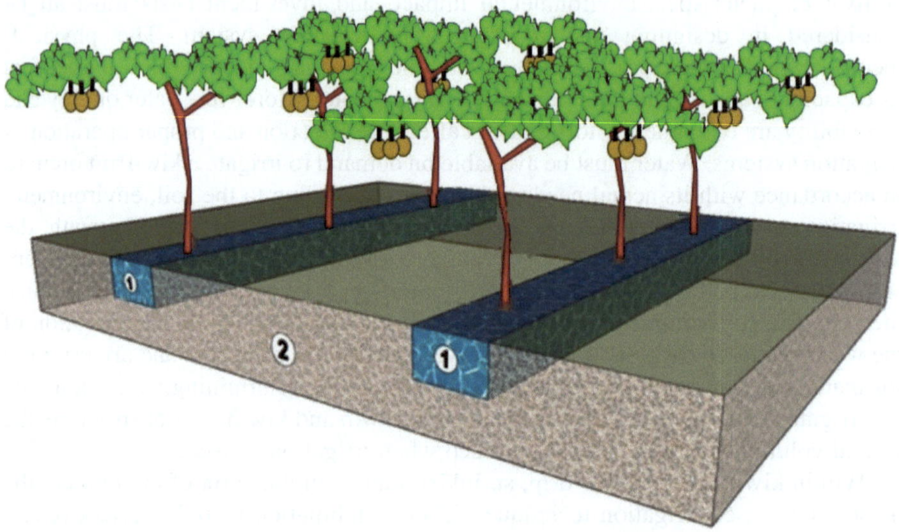

FIGURE 8.10

Representation of the two containers. Container 1 explores the root system and container 2 serves as water reserve.

(Adapted from Dichio et al., 2025.)

times a day. Container 2 serves as a water storage facility during periods of precipitation. Rainwater is a significant water reserve that needs to be handled via irrigation techniques (Figure 8.10).

4.6 Pollination

Kiwifruit are dioecious, and orchards must contain male (pollenizer) vines to ensure cross-pollination and fruit set of the selected fruiting cultivars (Costa et al., 1993, Figure 8.11).

FIGURE 8.11

Dioecy in *Actinidia chinensis* var. *deliciosa*. *(Left)* pistillate flower from female vine; *(right)* staminate flower with vestigial ovary and styles from male vine.

(Photo from The New Zealand Institute for Plant and Food Research Ltd.)

The distribution of male vines within the orchard (number/ha, proximity to the female vines, canopy space occupied) is important and has evolved over time. The initial ratio of 1 male:6 Hayward vines soon became 1:8 with male vines in every row. Other distribution patterns have been tested: e.g., the male plant planted between the two female plants and raised to a height of 2.0–2.5 m above the female plant canopy or vice versa, the male plants grown under the female plant fixed to a wire located 60 cm from the ground. In both cases, the shading caused by the upper vine created management problems and a reduction in yield.

In New Zealand in many pergola-supported orchards, male plants are planted in alternating rows with the females, the males being limited to a narrow strip and pruned heavily after flowering. Male and female plants are therefore close to each other, but the females occupy most of the orchard canopy, thereby increasing yields/ha (Figure 8.12) (Thorp et al., 2011).

Pollination is normally carried out by bees. A large number of active bees is required, and hives must be placed in optimal conditions since kiwifruit flowers do not produce nectar and are not particularly attractive to pollinating insects. The alternative is artificial pollination. Artificial pollination is now the main method used in kiwifruit orchards in Italy (Testolin and Ferguson, 2009), Southern China, and South Korea (Lee et al., 2019); elsewhere it is often used to supplement pollination by bees. It is also used when kiwifruit are grown under protection. Vines are often grown under hail nets or other overhead protection. Vines may also be grown in plastic tunnels or greenhouses to reduce Psa infection (Figure 8.13). The conditions in such protected plantings may be unsuitable for insect pollinators and often only fruiting cultivars are planted. Artificial pollination is then required.

Commercially available pollen is collected from flowers of male vines of *A. chinensis* var. *deliciosa*. (*A. chinensis* var. *chinensis* staminate flowers contain

FIGURE 8.12

Strip males: male vines are trained to form a "narrow strip" *(light green)* while female vines *(dark green)* are allowed to occupy the space between the two male rows.

(Photo from The New Zealand Institute for Plant and Food Research Ltd.)

FIGURE 8.13

Orchard grown under plastic tunnel or photovoltaic greenhouse.

insufficient pollen.) Pollen can transmit bacterial canker of kiwifruit, so the pollen used must be guaranteed Psa-free. Artificial pollination is carried out at maximum receptivity of the female flowers. Pollen (1–3 g pollen vine^{-1}) is applied using "dry" or "wet" pollen methods two or three times depending on the duration of flowering. Pollen applied dry is usually first diluted with *Lycopodium* spores, pine pollen, or charcoal. In wet pollination, the pollen is suspended in a liquid medium designed to protect pollen viability. Pollen is then applied at rates of 1–3 g L^{-1} with an application of 2 kg pollen ha^{-1} (Broussard and Pattemore, 2024).

4.7 Thinning

The inflorescences of female kiwifruit vines typically have three flowers—a terminal (king) flower and two lateral flowers—although there has been cultivar selection for primarily single-flowered inflorescences. Lateral flowers develop into fruit smaller than those from terminal flowers. Lateral fruit, malformed or poorly pollinated fruit are removed. As all kiwifruit flowers that are pollinated set fruit, thinning is usually required to ensure the development of high-quality, commercially marketable fruit to attain the desired commercial fruit size while maintaining a balanced vegetative-to-reproductive ratio. Many *A. chinensis* var. *chinensis* cultivars are very floriferous, and extensive fruit thinning is often necessary. This starts with the reduction of the total number of buds by removing canes that would produce floral shoots and can extend to the selective thinning of flowers and young fruitlets within inflorescences. Effective thinning can be labor-intensive, often requiring more than 500 h ha^{-1} to accomplish manually.

4.8 Girdling (cincturing)

Trunk girdling is routinely used in New Zealand and increasingly in Europe and Chile to increase flowering and improve fruit growth and development (Figure 8.14). Girdling prevents movement of carbohydrates resulting from photosynthesis in the canopy moving to the roots. Girdling early in the season can increase fruit size and later in the season can increase DM (carbohydrate) content of the fruit. Consumer enjoyment of kiwifruit increases as DM content increases.

4.9 Fruit bagging

Kiwifruit are easily damaged by strong winds or hail, and the skins of the fruit of many *A. chinensis* var. *chinensis* cultivars are particularly thin and susceptible. Fruit

FIGURE 8.14

Girdling performed on the trunk of a kiwifruit vines. The girdling can be performed every year.

bagging from soon after flowering until harvest is particularly common in China to increase fruit size, advance maturity, and reduce skin blemishes (Figure 8.15) (Patterson and Thorp, 2024). However, bagging of kiwifruit can reduce the concentrations of pigments in the fruit flesh, reduce the accumulation of DM, and reduce the ascorbic acid (vitamin C) content (Xu et al., 2022).

4.10 Plant bioregulators

Crop load in kiwifruit is initially determined during winter pruning by the number of buds laid down. Kiwifruit, however, have a winter chilling requirement, and in regions with milder winters, winter chilling may be inadequate with many buds not breaking and with many of the shoots that do break being vegetative. Hydrogen cyanamide-based products such as Dormex or Hi-Cane can increase budbreak, increase the proportion of flowering shoots, increase the number of flowers per shoot, and also reduce the number of lateral flowers. However, use of hydrogen cyanamide is banned in some countries, e.g., Italy. Alternative growth regulators and different management techniques are being tested.

Bioregulators, related to natural plant hormones, can be used to affect the growth of the fruit. The best known are based on cytokinin (N-(2-chloro-4-pyridyl)- N′-phenylurea, Forchlorfenuron, CPPU (commercial names Sitofex, Caplit), auxins, and a mixture of auxins and gibberellins, known as Spray Dünger Global. These were mainly used on Hayward but can also be used, mostly in China, on *A. chinensis* var. *chinensis* cultivars. Although their main benefit is increased fruit size, their use should not be a substitute for adequate pollination. They can cause undesirable side effects such as atypical fruit shape, reduced carbohydrate content resulting in poor taste, or premature softening of the distal end of the fruit. However, Sitofex can be used before flowering to suppress leaf spot symptoms caused by Psa.

FIGURE 8.15

Fruit bagging to increase fruit size and to reduce fruits skin blemishes.

5. Fruit growth, maturation, and harvest criteria

5.1 Fruit development

The fruit of kiwifruit are classified botanically as berries. They do not have a pit or stone, but instead the seeds are embedded in the fleshy pulp (the inner pericarp) (Figure 8.16). The fruit develops from the ovary of a single superior flower, the pericarp from the ovary wall, and the many seeds from the fertilized ovules. The ovary is syncarpous; that is, it consists of up to 40 fused carpels, as seen in a cross-section of a fruit. The weight of a fully grown kiwifruit depends on the cultivar, the crop load, and the number of seeds. The average size of Hayward kiwifruit can be about 100 g. Depending on the cultivar, kiwifruit take about 160–200 days to reach full size. Initial growth after pollination and seed set is by rapid cell division and enlargement followed by slower cell enlargement accompanied by starch accumulation. The increase in dry weight is essentially linear during development but there are several phases in the increase in fresh weight, largely due to differential rates of cell enlargement. The change in seed color from white to dark brown/black occurs up to several months before fruit growth stops.

The fruit flesh of *A. chinensis* var. *deliciosa* cultivars is green when ripe owing to the retention of chlorophyll, whereas most other fruits lose their chlorophyll as they ripen. Although the fruit of some genotypes of *A. chinensis* var. *chinensis* have green flesh, most cultivars have yellow flesh. This is due to the loss of chlorophyll late in fruit development unmasking the yellow of existing carotenoids, mainly β-carotene and xanthophylls, not to the synthesis of new carotenoids. The timing and rate of

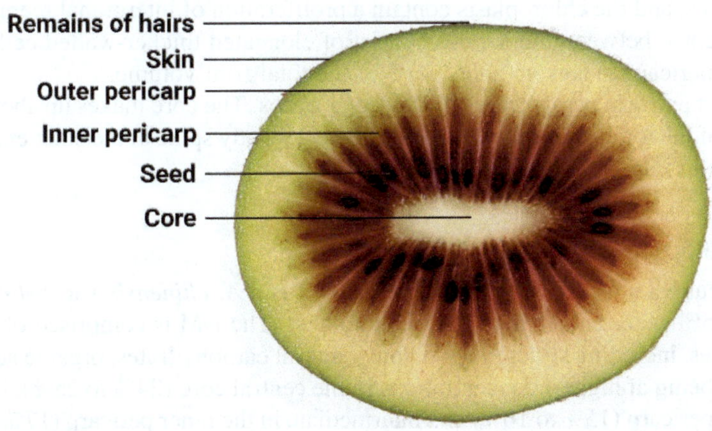

FIGURE 8.16

Cross-section of the fruit of an unnamed *A. chinensis* var. *chinensis* selection. As in the commercial cultivar Hongyang, anthocyanins are largely restricted to the locules, forming a red ring around the core.

(Photo from The New Zealand Institute for Plant and Food Research Ltd.)

chlorophyll loss (degreening) depends on the cultivar and climatic conditions. Yellow-fleshed kiwifruit must be harvested at a stage that allows full degreening. Even green-fleshed cultivars suffer some loss of chlorophyll during ripening and prolonged storage. Red-fleshed cultivars vary in the anthocyanin distribution: the pigments may be confined to the locules or spread more generally throughout the inner and outer pericarps. The amount of anthocyanins accumulated, and thus the intensity of red coloration at different stages of growth, can be variable even within a single cultivar.

5.2 Structure of the fruit

The fruit of a kiwifruit (Figure 8.16) can be divided into:

- Skin consisting of one or more layers of dead cells. Fruit of *A. chinensis* var. *chinensis* are usually covered by soft, fine hairs (trichomes), often lost at harvest. The skin can be rather delicate and easily damaged, e.g., by windrub. Fruit of *A. chinensis* var. *deliciosa* usually have more robust skins with much stiffer, coarser, multiseriate hairs that remain at harvest (although many are lost in commercial brushing and grading).
- Outer pericarp. Densely packed small parenchymatous cells (packed with starch granules when the fruit is still unripe) with occasional larger cells. There are also mucilaginous cells with raphides of calcium oxalate. The chloroplasts are similar to those in leaves. The outer pericarp occupies about 50% to 55% of the total volume of the mature fruit, depending on the cultivar.
- Inner pericarp. Seeds are suspended in two rows in the locules in a matrix of thin-walled parenchymatous cells. The locules contain many calcium oxalate raphides, and the chloroplasts contain a proliferation of intergranal membranes. The septae between the locules consist of elongated thicker-walled cells. The inner pericarp makes up about 40% of the total fruit volume.
- Core of uniform parenchyma and vascular traces. The core makes up about 5% to 10% of the total fruit volume. There can be a woody spike at the stalk end of the core; this can be a defect.

5.3 Composition of mature fruit

Fruit of cultivars of *A. chinensis* var. *chinensis* and *A. chinensis* var. *deliciosa* are 75% to 85% water and 15% to 25% DM content. The DM is comprised of primary metabolites, including structural and nonstructural carbohydrates, organic acids, and proteins, being at highest concentrations in the central core (24% to 25%), lowest in the outer pericarp (15% to 16%) and intermediate in the inner pericarp (17% to 18%) (MacRae et al., 1989). Starch is the major nonstructural carbohydrate stored during fruit development, up to 40% of the total DM (Nardozza and Burdon, 2024). As the fruit ripen, starch is converted into soluble sugars, mainly glucose and fructose, and these make up the bulk of the soluble solids content (SSC). The DM of fruit gives a reliable prediction of the SSC of the fruit when ripened (rSSC); DM is therefore used commercially as an indicator of fruit quality since the fruit sugar:acid ratio has a marked effect of perceived sweetness and consumer acceptability. DM is not in itself an indication of maturity, but the time at which DM stops increasing could be. The most striking feature of kiwifruit composition (apart from retention of most of the chlorophyll in green-fleshed cultivars) is the remarkably high vitamin C content, approximately 85 mg ascorbic acid 100 g^{-1} fresh weight in Hayward and nearly twice as much in some cultivars of *A. chinensis* var. *chinensis* such as Zesy002. High vitamin C content is the outstanding health benefit of kiwifruit with a single fruit able to supply most daily requirements. Potential disadvantages are the presence of allergens and calcium oxalate as raphides.

5.4 Fruit maturity

A mature fruit is defined as one that can ripen when harvested. Kiwifruit are harvested while still unripe and firm. There is no external sign when fruit are ready for harvest. Fruit sold immediately after harvest must ripen to good eating quality; fruit subjected to prolonged storage must remain firm, not develop disorders such as chilling injury, and still ripen to a state acceptable to the consumer. Mature fruit ripen gradually during cool storage or more rapidly at room temperature or with treatment with ethylene.

Harvest indices should be convenient and reproducible to ensure the quality of the fruit as far as possible (Burdon and Wang, 2024). The index based on SSC is still the most common harvesting criterion for many kiwifruit-producing areas; a minimum 6.2°Brix is required for harvesting Hayward. However, harvest indices have become more complex with the introduction of new cultivars with different fruit maturation properties, while advanced nondestructive tools are being used to define the optimum harvest period, usually matched with specific marketing periods (Burdon, 2019). Thus, there is no simple harvest index or absolute measure of when fruit can be harvested: this can depend on whether the fruit are to be sold to consumers immediately or subject to storage, possibly for many months, before being allowed to ripen.

The fruit maturation process and harvesting criteria are even more complex when referring to yellow-fleshed cultivars, where flesh color is a major component of a harvest index. The fruit of such cultivars are initially green and must undergo a degreening process (loss of chlorophyll) to receive the desired yellow color either before harvest or off vine, provided that they have degreened sufficiently on vine to allow completion of the process. Flesh color change is not necessarily linked with other aspects of fruit maturation and ripening, including SSC and firmness, and may be completed before ripening commences (Burdon and Wang, 2024).

Consumer acceptability is strongly influenced by the rSSC (Crisosto and Crisosto, 2001; Figure 8.17) that can be calculated based on the DM content at harvest. The SSC at harvest is indicative of the fruit development, while the DM content is indicative of the SSC at fully ripe stage and therefore of consumer enjoyment. Thus, a sufficient amount of carbohydrate accumulation in kiwifruit at preharvest level is a prerequisite to give an acceptably tasting fruit after harvest.

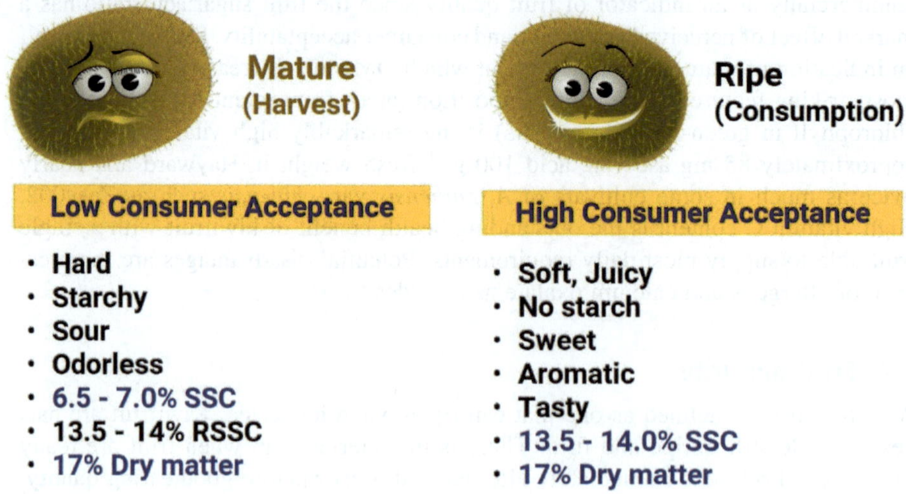

FIGURE 8.17

Qualitative attributes or mature and ripe kiwifruit

(Photo from Crisosto and Crisosto, 2001.)

5.5 Harvest indices

The usual parameters for determining when fruit can be harvested include flesh color (in yellow-fleshed cultivars), fruit firmness, and SSC. Date—the time after flowering—is not adequate. The various parameters are not always closely related, and there is no simple harvest index or absolute measure of when fruit can be harvested; this can depend on whether the fruit are to be sold to consumers immediately or are to be subject to storage, possibly for many months, before being allowed to ripen. The various parameters may have to be used sequentially, e.g., flesh color was the essential parameter in determining the time of harvest of the yellow-fleshed fruit of *A. chinensis* var. *chinensis* Hort16A and had to have reached 105°h–103°h (indicating the transition from green to yellow) (Minchin et al., 2003), but changes in SSC and firmness were then used to determine appropriate postharvest treatment (Burdon, 2019; Burdon and Wang, 2024). Even increases in SSC do not simply indicate a single physiological change, a cold-induced conversion of stored starch to soluble sugars, but could also partially result from a cessation of conversion to starch of the soluble sugars entering the fruit. Thus, there can be an interaction between the physiological state of the vine and its fruit and climate. Furthermore, the indices should be modified for individual cultivars and growing regions: what works for *A. chinensis* var. *deliciosa* Hayward in New Zealand will not necessarily work for Hayward in Italy or other countries, let alone different cultivars such as *A. chinensis* var. *chinensis* Zesy002. The ultimate fate of the fruit must also be considered.

The measurements used are usually destructive, e.g., SSC is measured on juice extracted from the fruit using a refractometer calibrated against sucrose or firmness using a penetrometer after a layer of fruit skin and flesh is removed. Fruit variability means that the particular fruit sampled may not be a good representative of all the fruit to be harvested. Furthermore, following changes in various parameters with time by consecutive measurements is likely to be more useful than single absolute measurements.

There are now nondestructive instruments that can be used to assess fruit maturity or ripening such as the Kiwi-Meter developed and patented by the University of Bologna (Kiwi-Meter) (Rocchi et al., 2016) and the F-751 Felix instrument device. Repeated repetitive measurements can be made of the same fruit as they are not removed from the vine or destroyed (Figure 8.18). The Kiwi-Meter allows the determination of an I^{AD} (index of absorbance difference) that correlates with the ripening evolution of the fruit, in particular with the flesh color (Huang et al., 2022), while the F-751 analyzers produced by Felix Instruments (U.S.) analyze the fruit SSC and DM directly in the field through near-infrared spectroscopy (NIR).

5.6 Harvest

All kiwifruit from a block are usually harvested together. The fruit are snapped from the fruit stalk and then transferred in bins to the packhouse. There, the fruit may be held at ambient temperatures for 2 or 3 days ("cured") partly to reduce the risk of stem end rot (SER) caused by *Botrytis cinerea.* Fruit may then be cool-stored in bins until packing is required or immediately graded and packed into trays for cool storage.

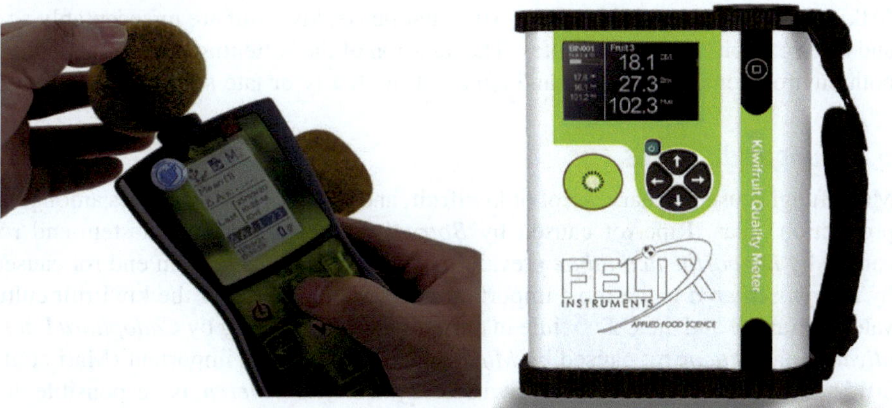

FIGURE 8.18

Nondestructive determination of kiwifruit maturity stage. Kiwi Meter - TR Turoni *(left)* and F751 Felix Instrument *(right).*

6. Postharvest physiology and technology

6.1 Introduction

Kiwifruit can be offered to consumers year-round since there is adequate production in both hemispheres, and some cultivars can be stored for up to 6 months. The aim of commercial postharvest storage is to slow the ripening of the harvested mature kiwifruit. Softening of the fruit occurs during ripening, and stored kiwifruit must be firm enough to withstand handling and transport, eventually to the consumer. Slowing ripening is achieved mainly by use of low temperatures or technologies such as controlled atmospheres (CA). However, there are certain limitations to quality after long-term storage. The main postharvest issues for kiwifruit are (i) the high sensitivity to exogenous ethylene, (ii) the susceptibility to *Botrytis cinerea* and other fungal rots, and (iii) the chilling-related disorders, with low temperature breakdown (LTB) being the most predominant, and (iv) the cost. The challenge for postharvest operators is to understand and manipulate those factors that influence the rate of deterioration and quality after storage, while additionally taking into consideration the preharvest factors.

Kiwifruit postharvest ripening is unusual for a climacteric-type fruit crop as it was initially categorized. In particular, kiwifruit do not produce endogenous ethylene at harvest until they become soft. However, they are highly sensitive to low concentrations of exogenous ethylene. For that reason, kiwifruit have been described as atypical with reference to a strict climacteric categorization (Antunes, 2007), making them an excellent model fruit crop to study ethylene-related ripening and softening syndrome.

After a period of rapid softening, the onset of endogenous "autocatalytic" ethylene production starts, which is considered to be the eating-ripe window for consumers, where the fruit are soft and produce characteristic ripe-fruit aroma volatiles (Atkinson et al., 2011) (Figure 19). After that phase, kiwifruit are unacceptably soft and often exhibit "off-flavor" notes. The duration of the softening phases depends on both environmental conditions and harvest time (early or late season).

6.2 Fungal diseases

Many fungi cause postharvest rot of kiwifruit, and their occurrence varies among the production areas. Ripe-rot caused by *Botryosphaeria dothidea* and stem-end rot caused by *Diaporthe actinidiae* prevail in New Zealand, while stem end rot caused by *Botrytis cinerea* is the most important postharvest disease of the kiwifruit cultivated in Europe and the U.S., while in these regions skin pitting by *Cadophora luteo-olivacea* and *Mucor* rot caused by *Mucor piriformis* are less important (Mari et al., 2015). Postharvest infection of kiwifruit by *Botrytis cinerea* is responsible for considerable losses whilst fruit are in cold rooms. Curing is a convenient biological treatment to prevent Botrytis rot, SER caused during harvest and the creation of sites where *Botrytis* can invade, and other disorders. Fruit are kept in a ventilated area for (a) 48 hours at 12°C–18°C; (b) 72–96 hours at 8°C–12°C; or (c) 96–120 hours at

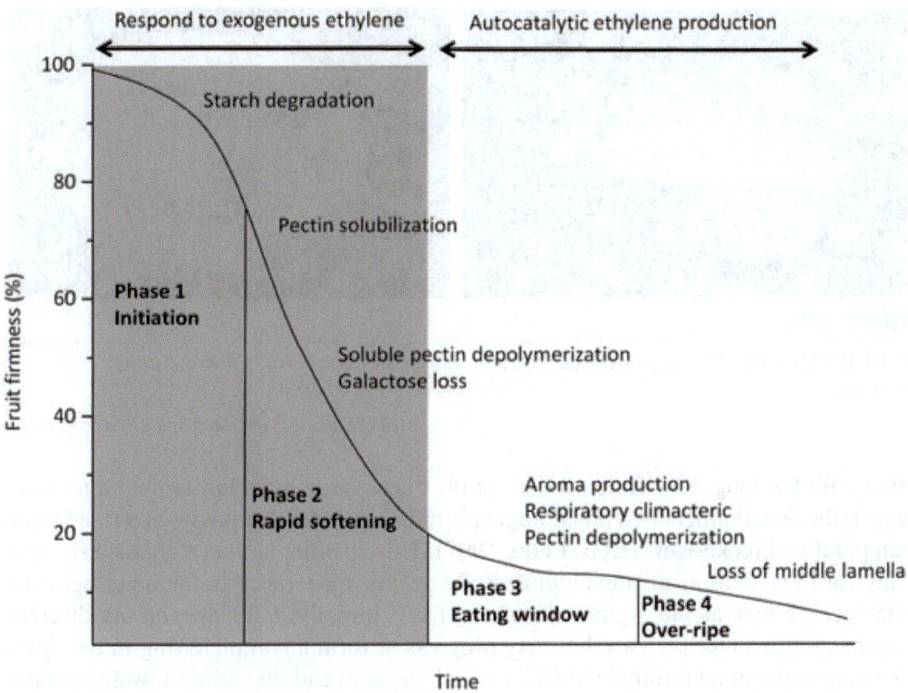

FIGURE 8.19

The distinct phases of kiwifruit softening. *Phase 1:* Fruit do not produce endogenous ethylene but are highly sensitive to exogenous ethylene; *Phase 2:* A period of rapid softening; *Phase 3:* starts with the onset of endogenous "autocatalytic" ethylene production (eating window); *Phase 4:* Fruit are unacceptably soft and often exhibit "off-flavor" notes.

(From Atkinson et al., 2011.)

5°C–8°C (Bautista Banos et al., 1997). Nowadays, curing tends to be applied almost by default as part of the logistics of handling harvested fruit irrespective of the risk of botrytis SER (Burdon and Wang, 2024). Temperature conditioning (holding at 7°C–10°C prior to storage at 0°C–1°C) is additionally applied to manage the off-vine degreening process of yellow-fleshed cultivars and to combat chilling-related disorders.

6.3 Chilling-related disorders

Besides fungal diseases, flesh softening and chilling injury related to storage breakdown disorder are the most important quality issues in the kiwifruit industry, being most evident in the yellow-fleshed cultivars (Figure 8.20). Kiwifruit are normally stored at ~0°C, and at this temperature, there is a risk of chilling damage, particularly for fruit harvested at a "low" maturity (early) when chilling-sensitive, and then

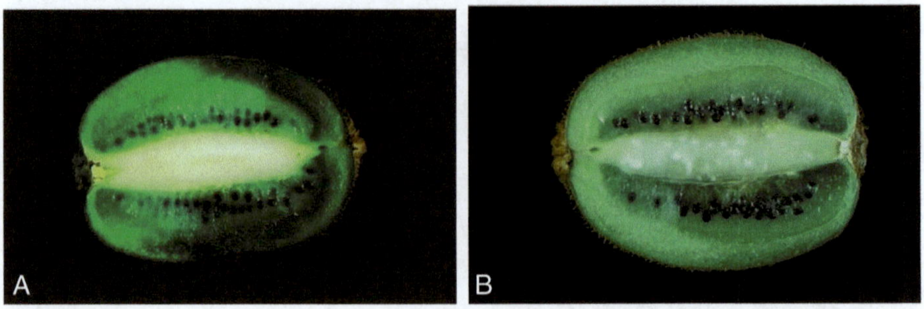

FIGURE 8.20

Chilling injury symptoms of kiwifruit, evident as translucency (A) and white core inclusions.

(UC Postharvest Research and Extension Center)

stored for too long. Kiwifruit are susceptible during prolonged low temperature storage to the development of a physiological storage disorder that has been termed "low temperature breakdown" (LTB, Lallu, 1997). This disorder has been characterized in Hayward kiwifruit as the development of a grainy appearance in the outer pericarp that occurs first at the stylar end of the fruit. Initially, LTB appears as discrete patches in the outer pericarp but may progress to form a complete ring in the outer pericarp and extend through the fruit toward the stem end. In addition, water-soaked areas may develop in either, or both, the inner and outer pericarp (Lallu et al., 1997; Burdon et al., 2007).

The susceptibility of kiwifruit to LTB is associated with preharvest temperatures and at-harvest SSC (Burdon et al., 2007). The incidence of LTB was lower in fruit from later harvests that had a higher SSC at harvest. It is concluded that acclimation by low preharvest temperatures increases the rate of SSC accumulation and reduces the susceptibility of Tomua kiwifruit to LTB (Burdon et al., 2007). A recent study was aimed at developing a mathematical model that could describe and predict evolution of storage breakdown disorder and firmness (Gwanpua et al., 2022). In this study, different concepts were used to model the three phases of softening often observed in kiwifruit. The initial rapid softening phase was modeled as a function of starch breakdown, the slow softening phase was modeled as a function of pectin degradation, while the final softening phase, often characterized by tissue collapse, was modeled as a function of chilling injury-related storage breakdown disorder.

6.4 Postharvest management protocols

There is an array of postharvest management protocols to extend storage potential. Ethylene catalytic oxidation and CA are the most common storage solutions; however, they are expensive in capital and energy costs. 1-Methylcyclopropene (1-MCP), an ethylene antagonist, is a commercial solution that requires a service fee but may sometimes result in nonuniform ripening. Ozone-enriched atmosphere is a relatively new solution offered in the portfolio of the kiwifruit postharvest management. The main principles of such postharvest management protocols are described as follows.

Ethylene contamination of storage atmospheres is a serious commercial risk. Consequently, ethylene monitoring and scrubbing systems are commonly used in kiwifruit coolstores. The most common postharvest management protocols deal with ethylene catalytic oxidation during cold storage (0°C −1°C). Ethylene can be removed from air by oxidizing it at the surface of an appropriate catalyst. Catalytic combustion of ethylene by the Swingtherm reversion method allows control of ethylene concentration at a desired level.

Ozone is a naturally occurring, highly reactive form of oxygen that can be generated economically for application to fruit susceptible to exogenous ethylene. Generally, ozone is reported to reduce rots caused by *Botrytis* and *Penicillium* and reduce mold build-up in packhouses and coolstores; both types of fungi are prevalent in kiwifruit (Burdon and Wang, 2024). Use of ozone was particularly beneficial to combat fungal diseases of kiwifruit (Minas et al., 2010) and extend storage potential (Minas et al., 2014). The challenge is that ozone is not stable. This limits application of ozone during transport since ozone had to be produced on demand.

1-MCP has been successfully applied in an array of commodities to combat softening of fruit, including kiwifruit. This treatment is being commercially applied, yet concerns have been raised about the fruit core not softening at the same rate as the pericarp.

Controlled atmosphere regimes (typically 4.5%−5.0% CO_2, 1.8%−2.0% O_2) are being applied for long-term storage. CA is usually used for storage to ensure that fruit are sufficiently firm to sustain subsequent grading and packaging (A. Woolf, personal communication). Ethylene scrubbers can be enclosed in packaging (Figure 8.21).

FIGURE 8.21

Potassium permanganate ($KMnO_4$) filters to control fruit softening in packed kiwifruit.

6.5 Behavior of different cultivars

Most early research on the storage conditions suitable for kiwifruit used the green-fleshed cultivar Hayward. It is now recognized that, in several ways, Hayward is exceptional, and this was probably critical for the early success of the kiwifruit export industry in New Zealand. The fruit mature particularly late in the season and have a particularly long storage life. The fruit seem to be not prone to chilling damage. Loss of chlorophyll during maturation and storage is relatively unimportant, whereas full degreening in yellow-fleshed cultivars is essential. Storage can spread the harvesting of the fruit and allows more orderly marketing. A good storage life is particularly important for those countries that produce kiwifruit for export. It is likely that new cultivars other than Hayward will be more demanding in the conditions required for storage of their fruit.

7. Main diseases, pests, and their management

Cultivation of kiwifruit demands care, since all *Actinidia* species and cultivars are susceptible to many different pests and diseases which can compromise the overall quality of production. This is particularly true in the modern context of climate change, in which traditional cropping areas for kiwifruit are being abandoned, owing to water shortages and increased temperatures, for new areas with more suitable environmental conditions. Furthermore, climate change greatly affects the biology and epidemiology of many plant pathogens and insects, which can exploit less severe winters and higher summer temperatures, and as consequence of those, allow weaker plants to proliferate and spread. The appearance of new diseases is one of the major threats to kiwifruit cultivation. For example, a new phytosanitary problem has recently been described in Italy: KVDS is a soilborne disease that damages the fine, young roots of *Actinidia* spp. (Figure 8.22). As a consequence, the whole plant seriously declines and eventually collapses and dies within a couple of years. Despite numerous efforts, no single pathogen responsible for this disease has been identified. Scientists investigated the role of asphyxiated soil conditions as drivers for the selection of Oomycetes, fungal-like populations (*Phytophthora* spp. and *Phytopythium* spp.), which could be related to the development of the symptoms (Donati et al., 2020). The crop losses due to this complex disease seriously threaten the future of kiwifruit cultivation in Italy and other Mediterranean countries, encouraging growers and researchers to work together to find alternative strategies to contain symptoms and avoid the spread of the disease. In discussions on control methods to protect kiwifruit from pests and diseases, it should be remembered that balanced management of the orchards in terms of rootstock selection, plant spacing, irrigation, nutrient supplements, use of pollen and hormones, coverage, and pruning and harvesting operations could make the difference between having healthy plants and more susceptible and diseased ones. Observing good agronomic practices and monitoring plants and environmental parameters are the winning keys

FIGURE 8.22

Kiwifruit Vine Decline Syndrome. Note the withering and the lack of canopy.

(Courtesy the authors.)

to prevent the spread of the diseases and maintain a high level of production (Di Francesco et al., 2021). The use of predictive models could also represent an effective approach to increasing the sustainability of kiwifruit protection strategies (Narouei-Khandan et al., 2022). The following is a list of the most common diseases of kiwifruit worldwide with brief descriptions of symptoms and useful control methods.

7.1 Diseases

Elephantiasis: This disease is provoked by several fungal microorganisms, mostly from the genera *Phialophora*, *Phomopsis*, *Cylindrocarpon,* and *Phaeoacremonium.* Recently, *Fusarium solani* was also linked to this disease. The main symptom is an abnormal swelling at the crown level (Figure 8.23). Trunk sections are crossed by browning. The whole plants have reduced aerial growth and production. Elephantiasis has been recorded in New Zealand and South American and European countries. No cure is currently available. Prevention-based strategies are strongly suggested, such as rotations before replacement plantings, as well as all the agronomic techniques that can sustain the diversity of the soil microbiome. Very stressed plants

FIGURE 8.23

Trunk affected by elephantiasis.

should be removed. Infected soil should be treated with lime before transplanting new plants (Prodi et al., 2008).

Wood decay: The main symptoms are related to deterioration of wood and cordons of the vines. Foliar symptoms appear as irregular necrotic spots on the upper surface of the leaf, which eventually prematurely falls (Figure 8.24). Severe attacks could lead to the wilt of branches and the whole plant. The disease is associated with different fungal microorganisms such as *Fomitiporia* spp., *Phaeoacremonium* spp., and *Phaeomoniella* spp., which have been reported in Italy, Greece, Turkey, and New Zealand. Cutting below the wilted branches and removing the decayed wood seems to be the only effective strategies so far (Di Marco et al., 2008).

Shoestring root rot: The fungus *Armillaria mellea* is an extremely polyphagous pathogen that attacks many tree species in all the continents. As a pathogen, the fungus provokes wood decay, growth reduction, and even death of the plant (Figure 8.25). Recently, more species belonging to the genus *Armillaria* have been associated with the disease. Symptoms are difficult to associate with the fungus until the appearance of characteristic mushrooms at the crown level. The fungus can survive as a saprophyte on dead woody remains. To avoid proliferation of this disease, soil drainage should be increased, infected plants should be removed with the roots, and the soil disinfected with lime.

FIGURE 8.24

Typical symptoms of wood decay on the canopy.

FIGURE 8.25

Shoestring root rot's typical symptoms of wood decay on a kiwifruit tree.

Bacterial canker: It represents one of the most threatening and pervasive disease of kiwifruit. It is caused by the Gram-negative bacterium *Pseudomonas syringae* pv. *actinidiae*, which has been recorded in almost all areas where kiwifruit are cultivated (Mazzaglia et al., 2012). Diseased plants present water-soaked polygonal necrosis on the leaves, and flowers can show browning and prematurely wither. However, the most significant symptoms are related to the presence of cankers on the lignified tissues of the plants, such as the branches and the trunk. From the cankers, a reddish gum can leak and solidify once outside, representing an inoculum for secondary infections (Figure 8.26). Diseased branches wither and eventually the whole plant will die, since the bacteria can move along the vascular system. The pathogen can overwinter in diseased trees and plant debris on the soil. The bacteria are easily spread on infectious plant material. They can enter the plants exploiting the picking scars made from fruit harvesting in autumn; even microlesions provoked by frosts can help the bacteria penetrate, as well as those made by pruning operations. The use of certified pollen is recommended to avoid spreading the disease, together with the practice of disinfecting pruning tools and using copper salts after pruning and harvesting operation. Localized irrigation should be preferred over above-canopy sprinklers. Cutting 100 cm below the cankers could be suitable to trying to save the vine; even if heavily infected, plants should be removed and destroyed, as well as other infectious debris. The use of antibiotics should be avoided because of the possibility of selecting and transferring antibiotic-resistance (Figure 8.27). The use of biocontrol agents, such as *Bacillus amyloliquefaciens*-based products, showed good reduction of the inoculum. Tolerant rootstock and varieties are available.

Bacterial blight: This disease is caused by the bacteria *Pseudomonas viridiflava* and *Pseudomonas syringae* pv. *syringae*, which have a biological cycle and are epidemiology similar to bacterial canker, even if the symptoms are limited to flowers and leaves (Figure 8.28). Interestingly, *P. viridiflava* exploits ice nucleation activity to spread and penetrate the host (Balestra et al., 2008).

Gray mold: As for many horticulture crops, gray mold (*Botryotinia fuckeliana* sexual form, *Botrytis cinerea* asexual form) represents a major threat to plants both in field and postharvest. The pathogen penetrates the host by the lesion of detached fruits. Even if rare, in this scenario, the smallest vines can wither and small black

FIGURE 8.26

Bacterial leak from cankers.

FIGURE 8.27

Plants removed to avoid bacterial canker from spreading.

FIGURE 8.28

Flowers buds infected by *Pseudomonas syringae* pv. *syringae (left)* and leaf symptoms by *P. viridiflava (right)*.

sclerotia appear. Infected fruits present soft rots and gray mold (Figure 8.29). To decrease the inoculum, some strategies can be adopted in the field, for example, reducing the shade amongst vines, decreasing the nitrogen supply. Iprodione and other hydantoin fungicides can be used after flowering and harvest. During postharvest storage, sanitation will play the major role in counteracting the infections. Fruit should be handled with care to avoid injuries and wounds. Reducing the humidity and the temperature is also a good strategy to rely on. In some cases, fruit coating with fungicides (e.g., fenhexamid) or other substances (e.g., chitosan) could improve the shelf-life of the products.

Skin pitting: This disease is currently one of the major threats in postharvest kiwifruit management. It has been detected in Italy as well as in South America.

FIGURE 8.29

Gray mold infection on fruit (outside – *left*; inside – *right*).

FIGURE 8.30

The fruit named kiwiberries are selections of *A. arguta* or, in some cases, hybrids of *A. arguta* and *A. melanandra* or, possibly, *A. kolomikta*.

(Photo from The New Zealand Institute for Plant and Food Research Ltd.)

The symptoms (oval dark lesions) are caused by the fungus *Cadophora luteo-olivacea*, which is particularly difficult to control owing to its latent behavior. Indeed, the pathogen can infect the fruits in the field, during the ripening, and remain inactive until the postharvest. Sanitation and dipping fruit with fungicides or other substances can be useful (Di Francesco et al., 2022).

7.2 Pests

Although slugs and snails, nematodes, and birds have all been reported as causing damage to kiwifruit, it is arthropods that are the most serious pests (Logan et al.,

2024). The pests on cultivated kiwifruit in China presumably came from wild *Actinidia* plants, but most of the pests in cultivated orchards in other countries are generalists or polyphagous, already on other crops. The most serious are listed here.

White peach scale *(Pseudaulacaspis pentagona)* is an Asian species that has spread to other countries. It can cause fruit blemishes, and serious infestations can reduce vine vigor. It can be controlled by insecticides or mineral oil. White peach scale is a serious pest in China, Italy, and other countries such as Greece, Turkey, and Iran.

Greedy scale *(Hemiberlesia rapax)* has a wide host range throughout the world but is serious on kiwifruit, particularly in New Zealand. On the fruit, it can cause export market restrictions. Control is by application of insecticides before flowering and mineral oil during a limited period after flowering.

Brown marmorated stink bug *(Halyomorpha halys)*. This dangerous pest is endemic to Eastern Asia and has spread to Europe and the Americas but not yet to New Zealand. In China, it damaged a third of the kiwifruit crop, and in Italy, it has recently been reported on the fruit of both *A. chinensis* var. *chinensis* (Zesy002) and *A. chinensis* var. *deliciosa* (Hayward). Partial control can be achieved by the application of pesticides.

Yellow-spotted stink bug *(Erthesina fullo)* is likewise a native of Eastern and Southern Asia. In China, it also can cause serious damage to fruit.

Summerfruit tortrix *(Adoxophyes orana)* is widespread in Asia and Europe as a pest of stone fruit and pome fruit, but the larvae can also attack kiwifruit, especially *A. chinensis* var. *chinensis* Hongyang in China.

Leafrollers endemic to New Zealand (brownheaded leafroller, *Ctenopseustis obliquana* and black-lyre leafroller, *Cnephasia jactatana*) can be very serious pests if left unmanaged. They can be effectively controlled by selective insecticides, including sprays containing *Bacillus thuringiensis*. Other leafrollers, such as *Proeulia auraria* and *P. chrysopteris,* are pests of kiwifruit and other fruit crops in South America.

Passionvine hopper *(Scolypopa australis)* has recently become a serious pest of kiwifruit in New Zealand. The nymphs and adults both secrete honeydew, which results in sooty mold contamination of the fruit.

Many other organisms have been reported as pests on kiwifruit, particularly in China. These are currently described as minor pests, but their geographic ranges may expand (as has that of the brown marmorated stink bug) or new kiwifruit cultivars may prove to be particularly susceptible. Some insects found on kiwifruit do not cause physical damage as such but are of concern as passenger pests, which may limit access to export markets because of quarantine restrictions. Kiwifruit are a relatively minor crop, and agrochemicals that might be effective against various pests might not be registered for use on kiwifruit. Residue limits imposed by importing countries also limit the use or the timing of application of some suitable agrochemicals. This was the stimulus for development of the KiwiGreen integrated control program in New Zealand in which sprays are applied during periods of high risk or only after monitoring of pests.

FIGURE 8.31

The kiwiberries Hortgem Tahi are small in size and have an edible skin. The vine in full production, the fruit "cluster," and the fruit details.

(Photo from The New Zealand Institute for Plant and Food Research Ltd.)

8. Cultivation of other *Actinidia* species

There is considerable genetic diversity within the genus *Actinidia* (Ferguson, 2024), but few species other than *A. chinensis* have been cultivated on a commercial scale. Crossing between *Actinidia* species is often possible, despite frequent differences in ploidy (Ferguson and Wu, 2016). Very few interspecific *Actinidia* hybrids have so far been commercialized. However, there is potential to introduce useful characteristics such as resistance to Psa or warm temperatures into commercial species.

A. arguta. Kiwiberries (Figures 8.30 and 8.31) are selections of *A. arguta* or, in some cases, hybrids of *A. arguta* and *A. melanandra* or, possibly, *A. kolomikta*. There are currently about 9000–10,000 ha of commercial kiwiberry orchards throughout the world: c. 7000 ha in China (Lu et al., 2024), the remainder in Chile, Japan, New Zealand, the United States, and various European countries (Debersaques et al., 2019). Most plantings are young, so production is still increasing.

Kiwiberries are vigorous climbers like kiwifruit and require much the same growing conditions but can tolerate much lower winter temperatures when dormant. Kiwiberries are therefore often grown in areas that are too cold for kiwifruit. Their fruit (6–20 g) are much smaller than those of kiwifruit cultivars. The fruit skins are hairless and edible but easily damaged when the fruit are young. There is considerable variation in fruit shape, flavor, and color—green, maroon, or red—of the skin and fruit flesh. No one cultivar predominates. At least one of the selections, but not all, with the name "Issai" sets fruit parthenocarpically (Mizugami et al., 2007). Most fruiting cultivars require pollenizers. Ripe kiwiberries can have an excellent flavor and they are ideal as a "snack" fruit. The fruit are consumed fresh, but a major constraint is the relatively short storage life.

A. eriantha. Superior-tasting genotypes have been selected from the wild (Liao et al., 2022), and limited commercial plantings have been established in Zhejiang Province, China. The most important attribute of *A. eriantha* is its very high fruit vitamin C content. Jinyan, a hybrid of *A. eriantha* x *A. chinensis* var. *chinensis*, is the first interspecific hybrid *Actinidia* cultivar to be grown extensively; however, its fruit vitamin C content is not exceptionally high (Zhong et al., 2012).

A. rufa. Cultivars, e.g., Kosui, that are interspecific hybrids, *A. rufa* x *A. chinensis* var. *chinensis*, have been commercialized in Japan (Kataoka et al., 2003).

9. Current and future challenges in main production regions

China, New Zealand, and Italy are currently the largest producers of kiwifruit. China is by far the largest producer, but its production is for domestic consumption. Kiwifruit exports are important for New Zealand, Italy, Greece, and Chile. Countries that rely on exports must meet the requirements of the importing countries, e.g., the requirements of GLOBALG.A.P.

The *A. chinensis* var. *deliciosa* cultivar Hayward is grown throughout much of the world and remains, except for China, a very important cultivar. Hayward, however, was originally selected for New Zealand conditions and requirements. It is possible that in the future new cultivars will have to be selected for specific countries or growing conditions or, possibly, different markets with different expectations. Marketers may increasingly specify the particular fruit sold and hence the cultivars grown. Most countries that grow kiwifruit rely on one or two cultivars (one for green-fleshed kiwifruit, one for yellow-fleshed) with only minor plantings of other cultivars. China is again an exception, but even there nearly a third of all kiwifruit plantings are of the *A. chinensis* var. *chinensis* cultivar Hongyang and very closely related selections. There are obvious advantages for growers to restrict their orchards to one or two cultivars as each cultivar tends to be unique in its management requirements. Likewise, there are advantages for marketers to promote and brand a limited number of products.

Reliance on a very limited number of kiwifruit cultivars does, however, have dangers, as does any monoclonal cultivation. The susceptibility of the *A. chinensis* var. *chinensis* cultivar Hort16A to bacterial canker caused by *Pseudomonas syringae* pv. *actinidiae* (Psa) resulted in devastating economic losses to New Zealand growers. The particularly rapid spread of the disease in New Zealand was largely due to kiwifruit being the dominant crop in a region where climatic conditions favored infection almost all year-round (Vanneste, 2017). Plant bacterial diseases are notoriously difficult to control, and the Psa pandemic with loss of Hort16A as a commercial crop probably cost the New Zealand industry at least $NZ1billion (c. $US600 million) in lost export earnings. Hort16A was rapidly replaced by a new cultivar, ('Zesy002'), which had been selected from a breeding program for other attributes but was, fortuitously, less susceptible to Psa. The New Zealand kiwifruit industry recovered but is now again almost entirely dependent on two clones, green-fleshed Hayward, as before, and with the yellow-fleshed Zesy002 replacing Hort16A. Despite enhanced biosecurity precautions, the lack of genetic diversity means a very real risk of another disease or pest pandemic. The same is largely true for the kiwifruit industries of other countries. The cultivar Hongyang (*A. chinensis* var. *chinensis*) is likewise very susceptible to Psa, but the very high prices currently received for its fruit may compensate Chinese growers for the risk.

Dioecy imposes serious restraints on kiwifruit cultivation: at least 10% of the orchard must be dedicated to male plants that flower at the same time as the female (fruiting) vines, large quantities of pollen must be transferred from the male vines to the female vines, the pollen must be effective in setting seed, orchards must be saturated with honeybees as the flowers, lacking nectar, are not particularly attractive to them. If the weather is cold, windy, or wet, the bees may not work but stay in their hives. The alternative, artificial pollination, is expensive with separate orchards for male vines providing the pollen for application. Dioecy in kiwifruit is not absolute: inconstant males ("fruiting males") of *A. chinensis* var. *deliciosa* have some bisexual flowers producing small fruit. By using such an inconstant male in a crossing program, McNeilage and Steinhagen (1998) produced a true, stable hermaphrodite with large fruit, albeit not of adequate commercial quality. In this case, hermaphroditism was inheritable, so a hermaphrodite cultivar with good quality fruit should be achievable using traditional breeding methods. Eventually hermaphroditism could be introduced into *A. chinensis* var. *chinensis* and other *Actinidia* species, such as kiwiberries. Hermaphrodite kiwifruit can also be obtained using a genetic engineering procedure, CRISPR-cas9-mediated mutagenesis, which has resulted in self-fertility in the male *A. chinensis* var. *chinensis* cultivar Bruce and in the female cultivar Hort16A (Akagi et al., 2019; Richardson and Varokonyi-Gasic, 2024). Consumers would have to accept genetically engineered kiwifruit for such approaches to be commercially useful. A fruiting cultivar that is self-fertile would be a great advantage to growers, although the costs involved and the loss of production for several years might not justify replacing existing cultivars in orchards. If the fruit were of acceptable quality, most consumers would be unaware of the change. However other changes could result from marketers responding to what they perceive to be consumer expectations: changes in flesh color, changes in flavor, increase in health constituents such as vitamin C, or decreases in the content of raphides of calcium oxalate or allergens. Consumers can sometimes be more adaptable than might be expected as shown by the rapid acceptance of Zesy002 fruit replacing those of Hort16A after the Psa epidemic forced the removal of Hort16A as a commercial cultivar. Perhaps differences between the fruit of two cultivars are not that important if both fruit are generally acceptable.

Kiwifruit are demanding in the climatic conditions required for growth and good crop production (Richardson et al., 2024), and growers are particularly vulnerable to climatic stress (Cradock-Henry, 2017). Kiwifruit need a long growing season without frosts, summers that are not too hot and winters that are not too cold but cold enough to provide the winter chilling to overcome bud endodormancy. They need a regular supply of water. They are easily damaged by high winds or hail. The fruit of *A. chinensis* var. *chinensis* cultivars with their thinner, more delicate skins without stiff hairs are more susceptible to damage by wind or sunburn than those of *A. chinensis* var. *deliciosa*. If the climate continues to warm as expected, then the warmer winters would make commercial production of *A. chinensis* var. *deliciosa* Hayward only marginally profitable in many parts of New Zealand where it has long been grown (Vetharaniam et al., 2022). The same is likely to be true in other countries growing Hayward kiwifruit. In New Zealand, a lack of winter chilling can be counteracted by application of hydrogen cyanamide, but this is not permitted in many other countries. A more acceptable solution might be introgression of a low winter-chilling requirement from species such as *A. rufa* (Kataoka et al., 2014). Frosts can be devastating resulting in the total loss of the crop of the current year as well as requiring costly management of vines to repair the fruiting canopy. Warmer climatic conditions might in some areas reduce the risks of frosts; however, warmer winters might also lead to earlier budbreak resulting to greater damage from early spring frosts. Adverse weather conditions are expected to become more frequent because of climate change. The solution for kiwifruit might be more overhead shelter or cultivation in protected environments, e.g., in Korea kiwifruit are grown under cover to reduce the risk of Psa. More intense or more frequent rain may affect soil conditions to favor the spread of diseases such as KVDS.

Pruning is a major cost to growers. Much effort is spent in attempting to get the right balance between vegetative growth and production of fruit, and the most profitable compromise between fruit number, fruit size, and fruit quality, as predicted by DM content. Higher temperatures during the growing season encourage vegetative growth at the expense of fruit quality with carryover effects the following season (Richardson et al., 2004). Increased vegetative growth means greater costs in summer pruning. Rootstocks show promise for control of plant vigor and vegetative growth and for promoting more consistent budbreak without the use of hydrogen cyanamide. Artificial intelligence and automation and mechanization might improve efficiency of cultural management and reduce costs of operations carried out with traditional methods, which require the use of numerous workers.

References

Akagi, T., Pilkington, S. M., Varkonyi-Gasic, E., Henry, I. M., Sugano, S. S., et al. (2019). Two Y-chromosome-encoded genes determine sex in kiwifruit. *Nature Plants, 5*, 801e809. https://doi.org/10.1038/s41477-019-0489-6

Antunes, M. D. C. (2007). The role of ethylene in kiwifruit ripening and senescence. *Stewart Postharvest Review, 3*(2), 9. https://doi.org/10.2212/spr.2007.2.9

Atkinson, R. G., Gunaseelan, K., Wang, M. Y., Luo, L. K., Wang, T. C., Norling, C. L., Johnston, S. L., Maddumage, R., Schroder, R., & Schaffer, R. J. (2011). Dissecting the role of climacteric ethylene in kiwifruit (*Actinidia chinensis*) ripening using a 1-aminocyclopropane-1-carboxylic acid oxidase knockdown line. *J. Exp. Bot, 62*, 3821–3835. https://doi.org/10.1093/jxb/err063

Balestra, G. M., Mazzaglia, A., & Rossetti, A. (2008). Outbreak of bacterial blossom blight caused by Pseudomonas viridiflava on *Actinidia chinensis* kiwifruit plants in Italy. *Plant Disease, 92*, 1707. https://doi.org/10.1094/PDIS-92-12-1707A

Bautista, B. S., Long, P. G., & Ganesh, S. (1997). Curing of kiwifruit for control of postharvest infection by *Botrytis cinerea*. *Postharvest Biol Tec, 12*(2), 137–145. https://doi.org/10.1016/S0925-5214(97)00043-4

Broussard, M. A., & Pattemore, D. E. (2024). Kiwifruit pollination. In A. C. Richardson, J. N. Burdon, & A. R. Ferguson (Eds.), *Kiwifruit: Botany, production and uses* (pp. 207–227). Wallingford, U.K: CABI. https://doi.org/10.1079/9781800620933.0013

Burdon, J. (2019). Kiwifruit biology: The commercial implications of fruit maturation. *Horticultural Reviews, 46*, 385–421. https://doi.org/10.1002/9781119521082.ch8

Burdon, J., Lallu, N., Francis, K., & Boldingh, H. (2007). The susceptibility of kiwifruit to low temperature breakdown is associated with pre-harvest temperatures and at-harvest soluble solids content. *Postharvest Biology and Technology, 43*, 283–290. https://doi.org/10.1016/j.postharvbio.2006.09.011

Burdon, J. N., & Wang, R. (2024). Postharvest: Fresh fruit harvest, storage and supply. In A. C. Richardson, J. N. Burdon, & A. R. Ferguson (Eds.), *Kiwifruit: Botany, production and uses* (pp. 353–374). Wallingford, U.K: CABI. https://doi.org/10.1079/9781800620933.0020

Chamberlain, J. E. (2024). Global industry and markets. In A. C. Richardson, J. N. Burdon, & A. R. Ferguson (Eds.), *Kiwifruit: Botany, production and uses* (pp. 16–28). Wallingford, U.K: CABI. https://doi.org/10.1079/9781800620933.0002

Costa, G., Ferguson, R., Huang, H., & Testolin, R. (2018). Main changes in the kiwifruit industry since its introduction: Present situation and future. *Acta Horticulturae, 1218*, 1–16. https://doi.org/10.17660/ActaHortic.2018.1218.1

Costa, G., Testolin, R., & Vizzotto, G. (1993). Kiwifruit pollination: An unbiased estimate of wind and bee contribution. *New Zealand Journal of Crop and Horticultural Science, 21*, 189–195. https://doi.org/10.1080/01140671.1993.9513767

Costa, G., & Testolin, R. (2022). Recent changes in the kiwifruit (*Actinidia* spp.) industry since its introduction: Italy as a case study. *Acta Horticulturae, 1332*, 1–10. https://doi.org/10.17660/ActaHortic.2022.1332.1

Cradock-Henry, N. A. (2017). New Zealand kiwifruit growers' vulnerability to climate and other stresors. *Regional Environmental Change, 17*, 245–259. https://doi.org/10.1007/s10113-016-1000-9

Crisosto, C. H., & Crisosto, G. M. (2001). Understanding consumer acceptance of early harvested Hayward kiwifruit. *Postharvest Biology and Technology, 22*, 205–213. https://doi.org/10.1016/S0925-5214(01)00097-7

D'Ippolito, I., Mang, S. M., Elshafie, H. S., Camele, I., Scillitani, G., Mastrodonato, M., Sofo, A., Mininni, A. N., & Xylogiannis, E. (2022). Morpho-anatomical and microbiological analysis of kiwifruit roots with KVDS symptoms. *Acta Horticulturae, 1332*, 131–136.

Debersaques, F., Latocha, P., & Decorte, J. (2019). Kiwiberry. *Actinidia arguta.* Grower's manual. Szczecin, Poland: Polish Dendrology Society.

Di Francesco, A., Di Foggia, M., Baldo, D., Ratti, C., & Baraldi, E. (2022). Preliminary results on *Cadophora luteo-olivacea* pathogenicity aspects on kiwifruit. *European Journal of Plant Pathology, 163*, 997–1001. https://doi.org/10.1007/s10658-022-02518-6

Di Francesco, A., Di Foggia, M., Vittoria, A., & Baraldi, E. (2021). Post-harvest nonconventional and traditional methods to control *Cadophora luteo-olivacea*: Skin pitting agent *of Actinidia chinensis* var. *deliciosa* (A. Chev.). *Horticulturae, 7*, 169. https://doi.org/10.3390/horticulturae7070169

Di Marco, S., & Osti, F. (2008). Foliar symptom expression of wood decay in *Actinidia deliciosa* in relation to environmental factors. *Plant Disease, 92*, 1150–1157. https://doi.org/10.1094/PDIS-92-8-1150

Dichio, B., Di Biase, R., Calabritto, M., Green, S., Mastroleo, M., Tuzio, A. C., Laterza, D., Xylogiannis, E., & Mininni, A. N. (2025). Precision irrigation strategy based on a feedback adjustment mechanism by soil moisture monitoring in kiwifruit orchard. *Acta Hortic*. In press.

Dichio, B., Montanaro, G., Sofo, A., & Xiloyannis, C. (2013). Stem and whole-plant hydraulics in olive (*Olea europaea*) and kiwifruit (*Actinidia deliciosa*). *Trees, 27*(1), 183–191. https://doi.org/10.1007/s00468-012-0787-3

Donati, I., Cellini, A., Sangiorgio, D., Caldera, E., Sorrenti, G., & Spinelli, F. (2020). Pathogens associated to kiwifruit vine decline in Italy. *Agriculture, 10*(4), 119. https://doi.org/10.3390/agriculture10040119

FAOSTAT. (2024). *Kiwifruit statistics*. Available at https://www.fao.org/faostat/en/?&data.)#search/kiwi%20fruit. (Accessed 28 March 2024).

Ferguson, A. R. (1984). Kiwifruit: A botanical review. *Horticultural Reviews, 6*, 1–64. https://doi.org/10.1002/9781118060797.ch1

Ferguson, A. R. (1990). The kiwifruit in China. In I. J. Warrington, & G. C. Weston (Eds.), *Kiwifruit: Science and management*. (pp. 155–164). Auckland: Ray Richards Publisher in association with the New Zealand Society for Horticultural Science. + 3 plates.

Ferguson, A. R. (2024). Biology of the genus *Actinidia*. In A. C. Richardson, J. N. Burdon, & A. R. Ferguson (Eds.), *Kiwifruit: Botany, production and uses* (pp. 29–51). Wallingford, U.K: CABI. https://doi.org/10.1079/9781800620933.0003

Ferguson, A. R., & Bollard, E. G. (1990). Domestication of the kiwifruit. In I. J. Warrington, & G. C. Weston (Eds.), *Kiwifruit: Science and management*. (pp. 165–246). Auckland: Ray Richards Publisher in association with the New Zealand Society for Horticultural Science. + 3 plates.

Ferguson, A. R., Fang, J., & Costa, G. (2024a). Fruiting cultivars. In A. C. Richardson, J. N. Burdon, & A. R. Ferguson (Eds.), *Kiwifruit: Botany, production and uses* (pp. 52–71). Wallingford, U.K: CABI. https://doi.org/10.1079/9781800620933.0004

Ferguson, A. R., & Huang, H.-W. (2016). Cytology, ploidy and ploidy manipulation. In R. Testolin, H.-W. Huang, & A. R. Ferguson (Eds.), *The kiwifruit genome. Compendium of plant genomes* (pp. 55–63). Cham, Switzerland: Springer International Publishing. https://doi.org/10.1007/978-3-319-32274-2_5

Ferguson, A. R., Huang, H., & Costa, G. (2024b). History of kiwifruit: Evolution of a global crop. In A. C. Richardson, J. N. Burdon, & A. R. Ferguson (Eds.), *Kiwifruit: Botany, production and uses* (pp. 1–15). Wallingford, U.K: CABI. https://doi.org/10.1079/9781800620933.0001

Ferguson, A. R., & Wu, J.-H. (2016). Interploid and interspecific hybridization for kiwifruit improvement. In A. S. Mason (Ed.), *Polyploidy and hybridization for crop improvement* (pp. 339–363). Boca Raton, Fl: CRC Press.

Gwanpua, S. G., Zhao, M., Jabbar, A., Bronlund, J. E., & East, A. R. (2022). A model for firmness and low temperature-induced storage breakdown disorder of Hayward kiwifruit in supply chain. *Postharvest Biol Tec185*. https://doi.org/10.1016/j.postharvbio.2021.111789

Huang, H.-W., & Ferguson, A. R. (2007). Genetic resources of kiwifruit: Domestication and breeding. *Horticultural Reviews, 33*, 1–121. https://doi.org/10.1002/9780470168011.ch1

Huang, W., Wang, Z., Zhang, Q., Feng, S., Burdon, J., & Zhong, C. (2022). Maturity, ripening and quality of donghong kiwifruit evaluated by the kiwi-meter™. *Horticulturae, 8*, 852. https://doi.org/10.3390/horticulturae8090852

Hughes, K. A., & Wilde, R. J. (1989). The effect of poor drainage on the root distribution of kiwifruit vines. *New Zealand Journal of Crop and Horticultural Science, 17*, 239–244. https://doi.org/10.1080/01140671.1989.10428038

Kataoka, I., Kokudo, K., Beppu, K., Fukuda, T., Mabuchi, S., & Suezawa, K. (2003). Evaluation of characteristics of *Actinidia* interspecific hybrid Kosui. *Acta Horticulturae, 610*, 103–108. https://doi.org/10.17660/ActaHortic.2003.610.11

Kataoka, I., Matsumoto, H., Kawano, A., Beppu, K., Ohtani, M., et al. (2014). Selection of low-chill kiwifruit adapting to warm climate by utilizing *Actinidia rufa* native to southwestern part of Japan. *Acta Horticulturae, 1059*, 85–88. https://doi.org/10.17660/ActaHort.2014.1059.9

Lallu, N. (1997). Low temperature breakdown in kiwifruit. *Acta Hort, 444*, 579–585. https://doi.org/10.17660/ActaHortic.1997.444.89

Lee, K. Y., Cho, H., Yoon, H. J., Lee, S., & Ko, H. (2019). Comparison of the foraging activity and pollination effects of honeybees (Apis mellifera L.) and bumblebees (*Bombus terrestris* L.) in gold kiwi (*Actinidia chinensis* P. Haegeum). *Journal of Apiculture, 34*, 295–304. https://doi.org/10.17519/apiculture.2019.11.34.4.295

Liao, G., Jiang, Z., He, Y., Zhong, M., Huang, C., et al. (2022). The comprehensive evaluation analysis of the fruit quality in *Actinidia eriantha* pollinated with different pollen donors based on the membership function method. *Erwerbs-Obstbau, 64*(1), 91–96.

Liao, G.-L., Xu, X.-B., Liu, Q., et al. (2020). A special summer pruning method significantly increases fruit weight, ascorbic acid, and dry matter of kiwifruit (Jinyan, *Actinidia eriantha* x *A. chinensis*). *HortScience, 55*(10), 1698–1702. https://doi.org/10.21273/HORTSCI15158-20

Liao, G., Xu, X., Huang, C., Zhong, M., & Jiao, D. (2021). Resource evaluation and novel germplasm mining of *Actinidia eriantha*. *Scientia Horticulturae, 282*, Article 110037. https://doi.org/10.1016/j.scienta.2021.110037

Logan, D. P., Zhuang, Q., & Pollini, A. (2024). Pests and their management. In A. C. Richardson, J. N. Burdon, & A. R. Ferguson (Eds.), *Kiwifruit: Botany, production and uses* (pp. 332–352). Wallingford, U.K: CABI. https://doi.org/10.1079/9781800620933.0019

Lu, X.-M., Wang, J., Li, Z.-Z., & Wang, Y.-C. (2024). Scientific and breeding value of wild *Actinidia* Sect. *Leiocarpae, Acta Horticulturae* (in press).

MacRae, E. A., Bowen, J. H., & Stec, M. G. H. (1989). Maturation of kiwifruit (*Actinidia deliciosa* cv. Hayward) from two orchards: Differences in composition of the tissue zones. *Journal of the Science of Food and Agriculture, 47*, 401–416. https://doi.org/10.1002/jsfa.2740470403

Mari, M., Spadoni, A., & Ceredi, G. (2015). Alternative technologies to control postharvest diseases of kiwifruit. *Stewart Postharvest Review, 11*(4). https://doi.org/10.2212/spr.2015.4.1

Mazzaglia, A., Studholme, D. J., Taratufolo, M. C., Cai, R., Almeida, N. F., & Goodman, T. (2012). *Pseudomonas syringae* pv. *actinidiae* (PSA) isolates from recent bacterial canker of kiwifruit outbreaks belong to the same genetic lineage. *PLoS ONE, 7*, Article e36518. https://doi.org/10.1371/journal.pone.0036518

McNeilage, M. A., & Steinhagen, S. (1998). Flower and fruit characteristics in a kiwifruit hermaphrodite. *Euphytica, 101*, 69–72. https://doi.org/10.1023/A:1018398924819

Minchin, P. E. H., Richardson, A. C., Patterson, K. J., & Martin, P. J. (2003). Prediction of final weight for *Actinidia chinensis* 'Hort16A' fruit. *New Zealand Journal of Crop and Horticultural Science, 31*, 147–157. https://doi.org/10.1080/01140671.2003.9514247

Minas, I. S., Karaoglanidis, G. S., Manganaris, G. A., & Vasilakakis, M. (2010). Effect of ozone application during cold storage of kiwifruit on the development of stem-end rot caused by *Botrytis cinerea*. *Postharvest Biology & Technology, 58*, 203–210. https://doi.org/10.1016/j.postharvbio.2010.07.002

Minas, I. S., Vicente, A. R., Dhanapal, A. P., Manganaris, G. A., Goulas, V., Vasilakakis, M., Crisosto, C. H., & Molassiotis, A. (2014). Ozone-induced kiwifruit ripening delay is mediated by ethylene biosynthesis inhibition and cell wall dismantling regulation. *Plant Science, 229*, 76–85. https://doi.org/10.1016/j.plantsci.2014.08.016

Mizugami, T., Kim, J. G., Beppu, K., Fukuda, T., & Kataoka, I. (2007). Observation of parthenocarpy in *Actinidia arguta* selection Issai. *Acta Horticulturae, 753*, 199–203. https://doi.org/10.17660/ActaHortic.2007.753.22

Nardozza, S., & Burdon, J. N. (2024). Fruit development: Growth, maturation and ripening. In A. C. Richardson, J. N. Burdon, & A. R. Ferguson (Eds.), *Kiwifruit: Botany, production and uses* (pp. 273–294). Wallingford, U.K: CABI. https://doi.org/10.1079/9781800620933.00016

Narouei-Khandan, H. A., Worner, S. P., Viljanen, S. L. H., van Bruggen, A. H. C., Balestra, G. M., & Jones, E. (2022). The potential global climate suitability of kiwifruit bacterial canker disease (*Pseudomonas syringae* pv. *actinidiae* (Psa)) using three modelling approaches: CLIMEX, maxent and multimodel framework. *Climate, 10*, 14. https://doi.org/10.3390/cli10020014

Nazir, M. F., Lou, J., Wang, Y., Zou, S., & Huang, H. (2024). Kiwifruit in the omics age: Advances in genomics, breeding, and beyond. *Plants, 13*, 2156. https://doi.org/10.3390/plants123152156

Patterson, K. J., & Thorp, G. (2024). Vine management systems. In A. C. Richardson, J. N. Burdon, & A. R. Ferguson (Eds.), *Kiwifruit: Botany, production and uses* (pp. 184–206). Wallingford, U.K: CABI. https://doi.org/10.1079/9781800620933.0012

Prodi, A., Sandalo, S., Tonti, S., Nipoti, P., & Pisi, A. (2008). Phialophora-like fungi associated with kiwifruit elephantiasis. *Journal of Plant Pathology, 90*(3), 487–494. http://www.jstor.org/stable/41998543.

Quartieri, M., Toselli, M., Sorrenti, G., Baldi, E., Germani, M. A., & Xilogiannis, E. (2022). Dynamic of nutrient uptake and partitioning within yellow-fleshed (*Actinidia chinenss* var. *chinensis*) organs. *Acta Horticulturae, 1333*, 203–208. https://doi.org/10.17660/ActaHortic.2022.1333.26

Richardson, A. C., Marsh, K. B., Boldingh, H. L., Pickering, A. H., Bulley, S. M., Frearson, N. J., et al. (2004). High growing temperatures reduce fruit carbohydrate and vitamn C in kiwifruit. *Plant, Cell & Environment, 27*, 423–433. https://doi.org/10.1111/j.1365-3040.2003.01161.x

Richardson, A. C., & Varkonyi-Gasic, E. (2024). Axillary bud, shoot and flower development. In A. C. Richardson, J. N. Burdon, & A. R. Ferguson (Eds.), *Kiwifruit: Botany, production and uses* (pp. 228–254). Wallingford, U.K: CABI. https://doi.org/10.1079/9781800620933.0014

Rocchi, L., Vidoni, S., Ceccarelli, A., Fiori, G., & Costa, G. (2016). Use of the DAindexTM for monitoring fruit ripening evolution in *A. chinensis* to precisely assess harvesting time "in planta". *Journal of Berry Research, 6*, 373–379. https://doi.org/10.3233/JBR-160116

Scortichini, M., Spinelli, F., & Templeton, M. D. (2024). The bacterial canker of kiwifruit caused by *Pseudomonas syringae* pv. *actinidiae*. In A. C. Richardson, J. N. Burdon, & A. R. Ferguson (Eds.), *Kiwifruit: Botany, production and uses* (pp. 317–331). Wallingford, U.K: CABI. https://doi.org/10.1079/9781800620933.0018

Seal, A. G., Testolin, R., & Fang, J. (2024). Male cultivars. In A. C. Richardson, J. N. Burdon, & A. R. Ferguson (Eds.), *Kiwifruit: Botany, production and uses* (pp. 72–84). Wallingford, U.K: CABI. https://doi.org/10.1079/9781800620933.0005

Testolin, R., & Ferguson, A. R. (2009). Kiwifruit (*Actinidia* spp.) production and marketing in Italy. *New Zealand Journal of Crop and Horticultural Science, 37*, 1–32. https://doi.org/10.1080/01140670909510246

Testolin, R., & McNeilage, M. A. (2024). Genetic improvement, kiwifruit genome and dioecy. In A. C. Richardson, J. N. Burdon, & A. R. Ferguson (Eds.), *Kiwifruit: Botany, production and uses* (pp. 127–149). Wallingford, U.K: CABI. https://doi.org/10.1079/9781800620933.0009

Testolin, R., Messina, R., & Peterlunger, E. (1990). Kiwifruit growth and yield as affected by in-row spacing. *Acta Horticulturae, 282*, 151–158. https://doi.org/10.17660/ActaHortic.1990.282.19

Thorp, G., & Barnett, A. M. (2024). Kiwifruit rootstocks. In A. C. Richardson, J. N. Burdon, & A. R. Ferguson (Eds.), *Kiwifruit: Botany, production and uses* (pp. 85–93). Wallingford, U.K: CABI. https://doi.org/10.1079/9781800620933.0006

Thorp, T. G., & Barnett, A. M. (2011). Increasing productivity of alternate year cropping systems for Zespri® Gold kiwifruit (Hort16A). *Acta Hortic, 913*, 321–326. https://doi.org/10.17660/ActaHortic.2011.913.42

Vanneste, J. L. (2017). The scientific, economic, and social impacts of the New Zealand outbreak of bacterial canker of kiwifruit (*Pseudomonas syringae* pv. *actinidiae*). *Annual Review of Phytopathology, 55*, 377–399. https://doi.org/10.1146/annurev-phyto-080516-035530

Vetharaniam, I., van den Dijssel, C., Stanley, J., Richardson, A., & Müller, K. (2022). Projecting suitability of kiwifruit by chill requirements for different locations in New Zealand under climate change. *Acta Horticulturae, 1332*, 455–462. https://doi.org/10.17660/ActaHortic.2022.1332.60

Xu, Y., Liu, Y., Li, W., Yang, C., Lin, Y., Wang, Y., Chen, C., Wan, C., Chen, J., & Gan, Z. (2022). The effects of bagging on color change and chemical composition in Jinyan kiwifruit (*Actinidia chinensis*). *Horticulturae, 8*(6), 478. https://doi.org/10.3390/horticulturae8060478

Zhong, C., Huang, W., Wang, Z., Li, L., Li, D., Zhang, Q., Zhao, T., & Zhang, P. (2022). The breeding progress and development status of the kiwifruit industry in China. *Acta Horticulturae, 1332*, 445–454. https://doi.org/10.17660/ActaHortic.2022.1332.59

Zhong, C., Wang, S., Jiang, Z., & Huang, H. (2012). Jinyan, an interspecific hybrid kiwifruit with brilliant yellow flesh and good storage quality. *HortScience, 47*, 1187–1190. https://doi.org/10.21273/HORTSCI.47.8.1187

CHAPTER

Fig

9

Uygun Aksoy[1], **Moshe Flaishman**[2]

[1]*Faculty of Agriculture, Department of Horticulture, Ege University, İzmir, Turkey;*
[2]*Institute of Plant Sciences, Volcani Research Center, Rishon Lezion, Israel*

Chapter outline

1. Origin and spread

The genus *Ficus* belongs to the Moraceae family with more than 700 species, mostly found in the subtropic and tropic areas (Berg, 2003). The common fig (*Ficus carica* L.) originated in the region stretching from Western Asia toward the Mediterranean region. This region, from Turkey extending to Northern India and embracing the mountains of Asia Minor (Turkey) and Persia (Iran), is a crucial part of fig history (Condit, 1947). The common fig is one of the five founder fruit trees or woody vine species that established horticulture in the late prehistoric Near East (Levant), along

Temperate Tree Fruits and Nuts. https://doi.org/10.1016/B978-0-443-29137-1.00009-8

with olive, grapevine, date palm, and pomegranate (Zohary et al., 2012). Early archaeological evidence exploitation of the wild common fig is around Israel and Syria, with the earliest evidence from the 780,000-year-old Middle Paleolithic, and of cultivated fig that dates back to 10,000 BCE (Kislev et al., 2006). Common fig was important throughout history as confirmed by the archaeobotanical fruit and seed remains from the Levant area dating back 7000 years to the Chalcolithic period (Weiss, 2015).

From its origins, the fig spread through human activity and trade to the Mediterranean regions with suitable climate conditions. It has been an important food crop for thousands of years and is thought to be highly beneficial in the diet. Many cultivars, mostly unnamed, have been developed or existed as human migration brought the fig to many places beyond its natural range (Flaishman et al., 2008). Figs were introduced to Italy before recorded history. Pliny gives details of 29 types of fig (Condit, 1947). Further, figs were introduced into England sometime between 1525 and 1548. Later, different fig types were taken to China, Japan, India, South Africa, and Australia. In 1550, it was reliably reported to be grown in Chinese gardens. The first figs in the New World were planted in Mexico in 1560. Figs reached Virginia in the Eastern United States by 1669 and were introduced into California when the San Diego mission was established in 1769. Subsequently, many distinctive varieties were received from Europe. The Smyrna fig was brought to California in 1881—1882, but it was not fruitful until 1900 when the pollinating wasp was introduced to make commercial production possible. Fig was introduced to South Africa and Australia after colonization during the 18th century and became a familiar dooryard plant in the West Indies, and at medium and low altitudes in Central America and northern South America (Condit, 1947).

The Mediterranean region remains ideal for fig production, but today, the fig tree can be found in many climates worldwide, including warm, tropical, and colder climates. In Florida, South America, China, Japan, and India, only the common fig type that does not need pollination is grown. Chile and Argentina grow the types suited to cooler zones, while Brazil and China grow fig types suited to tropical areas.

2. Economic importance

Fig tree is a species commonly found around the Mediterranean basin, either wild or cultivated, and available as fresh during summer months but then sun-dried to extend its shelf-life or processed for further consumption and trade. Its ability to be dried provided ease in distribution and storage throughout history (Aksoy, 1998). Starting from the end of the 19th century, fig fruit gained wider recognition via introduction to the Americas and Eastern Asia. Its movement toward the east was rather slow due to the rainy and humid climate of Southern and Southeast Asia. The rich fig genetic resources allowed wider adaptation, giving birth to multiple uses. Recent developments in management techniques and soilless culture of common type varieties availed new fig producer regions for the local and/or international fresh fruit market.

In countries outside its traditional territory such as in the Netherlands or the United Kingdom, fig is valued as an exotic species with an ornamental value.

Various parts of the fig tree such as the fruit, the leaf, and the latex can be utilized; however, its main economic importance lies in its fresh or dried fruit trade. Fresh figs are traded and consumed more locally, whereas dried figs have been traded internationally for centuries. The fig fruit trade statistics including the United Nations Food and Agriculture Organization (UN/FAO) and the International Trade Center (ITC) trade databases have limitations or discrepancies, either due to joint reporting of fresh and dried figs or lack of or inaccurate reporting of national statistics. Thus, export-import figures of the same year quoted from different sources may vary, thus an explanatory note must be given when necessary. Observatory of Economic Complexity (https://oec.world/en/profile/hs/figs-freshdried) reported fresh and dried fig trade together totaling to $630 million USD in 2022. Top exporters were Turkey with a value of $320 million USD followed by Afghanistan ($70.6 million), Austria ($28.8 million), Spain ($1.7 million), Germany ($21.1 million), and Iran ($20.5 million). Austria and Germany are not producers, but they rank among the top exporting countries due to their significant re-exports. Top importers in 2022 were listed in decreasing order as Germany ($89 million), India ($65.3 million), France ($53.4 million), United States ($48.7 million), and Canada ($26 million). Germany, being the top importer, does not consume totally but is a transit country and distributes fresh and dried figs mainly to other European countries. On the other hand, the U.S., even though a fig producer, imports additional figs for its own consumption. The Trade Map of the International Trade Center (www.trademap.org) displays the export and import figures of fresh and dried figs together, and classified under the same product code. When trade of fresh and dried figs are added together, there are greater variations regarding the value versus quantity (Tables 9.1 and 9.2).

Almost all dried figs traded are sun-dried, meaning that circa 4 kg of fresh yields to 1 kg of sun-dried. Quality also plays a crucial role, especially in the value of unit dried fig. Globally traded quantity of fresh fig is much less compared to sun-dried fig volumes. This limitation is due to the short harvest, storage and shelf-life, and fragile nature of the fresh fruit. In this respect, those that reach international markets are sold at rather high unit prices. Except for a few varieties, such as the Black Mission, which is marketed as fresh and dried, varieties for sun-drying are not mostly suitable for the fresh market; similarly, major varieties destined for fresh consumption are not suitable for sun-drying because of lower sugar and dry matter contents.

To appropriately evaluate the figures for the quantity and value of fig exports and imports, it is necessary to understand basically whether the producer country exports only fresh, only dried, or both. There are countries that solely export fresh figs, as in the case of Brazil or Peru, or only dried figs, as in Iran and Afghanistan. On the other hand, Turkey and Spain export both fresh and dried figs. To be more explanatory and give an example of the quantity versus value, in 2022 statistics, Uzbekistan exported 7177 tons of figs in total, worth a value of $2.02 million, while Spain exported 6569 tons for a nearly 9-fold higher total value ($19.67 million). Uzbekistan exported only

Table 9.1 Global exporters of fig, fresh or dried, as quantity (tons) and value ($1000) (Source: https://www.trademap.org/Index.aspx, Product 080420)

Country	2019		2020		2021		2022	
	Quantity	Value	Quantity	Value	Quantity	Value	Quantity	Value
Turkey	84,991	286,516	88,584	293,555	91,224	330,514	95,411	306,964
Afghanistan	12,093	85,737	19,179	137,919	27,151	16,760	16,990	70,969
Iran	14,038	48,697	14,415	42,580	13,609	40,436	14,055	42,009
Uzbekistan	212	263	403	660	457	343	7177	2017
Spain	7568	20,613	6771	20,131	8069	25,123	6569	19,672
Syria	4461	10,575	2808	6585	3985	10,833	4904	16,376
Austria	3907	15,569	5656	22,970	5598	24,018	4741	17,948
Germany	3872	19,230	4784	23,156	5126	25,859	4446	21,197
Netherlands	3774	18,768	3795	18,715	4292	24,867	3487	20,575
Greece	3610	16,474	3303	14,092	3185	14,881	3384	16,421
Saudi Arabia	7145	6964	11,043	9898	3990	3842	3238	2264
United Arab Emirates	3551	4046	4647	4955	2376	5021	2682	8231
Italy	2081	6593	2881	9215	3203	11,505	2192	7313
U.S.	2304	13,491	2030	11,091	2193	11,105	1836	9770
Brazil	1377	6671	1184	4839	1569	6956	1568	6342
World	**171,125**	**61,507**	**88,438**	**673,329**	**192,540**	**766,884**	**186,111**	**624685**

Table 9.2 Global importers of fig, as quantity (tons) and value ($1,000) of fresh or dried figs (Source: https://www.trademap.org/Index.aspx, Product 080420)

	2019		2020		2021		2022	
Importing countries	**Quantity**	**Value**	**Quantity**	**Value**	**Quantity**	**Value**	**Quantity**	**Value**
Germany	17,988	63,176	21,175	76,380	20,345	83,293	20,484	68,905
France	17,574	60,929	21,092	69,375	19,701	71,759	16,619	57,960
U.S.	12,304	44,551	14,916	46,414	14,925	51,426	15,982	54,954
China	2768	4275	3745	6005	11,313	16,321	14,691	21,207
India	11,940	96,147	18,103	135,355	23,286	163,802	14,326	65,299
Kazakhstan	256	149	417	254	658	367	7826	2920
United Kingdom	7496	28,447	6606	24,415	7075	29,777	7195	28,087
Austria	7234	27,993	8404	30,048	7323	31,284	7036	24,243
Russian Federation	5795	9721	6596	11,930	5630	12,098	6903	14,388
Canada	4121	23,387	4616	24,563	4870	27,316	5271	29,966
Italy	5814	23,366	4722	20,047	5795	26,791	5102	23,123
Netherlands	4168	18,481	5378	22,109	6537	303,364	4949	22,930
United Arab Emirates	3876	10,170	3738	9085	4189	10,403	4411	11,225
Switzerland	3590	18,083	4142	21,512	4035	22,622	4057	20,233
Saudi Arabia	3600	7702	2342	4977	2979	7221	3426	115,999
World	**156,842**	**606,323**	**180,404**	**672,159**	**193,570**	**776,292**	**192,211**	**651,499**

sun-dried figs of medium quality, whereas Spain exported sun-dried (mostly as processed) and fresh figs. Saudi Arabia, United Arab Emirates, South Africa, Peru, Mexico, and lately India are listed as exporters of fresh figs. The European countries and the U.S. are the leading importers, and China and India are countries where dried fig imports are increasing rapidly.

To get a better picture of the fig trade, the International Nut and Dried Fruit Council (INC, 2023) figures that are confined to dried figs only are given. INC reports that 178,974 metric tons of dried figs were traded in 2021, and Turkey supplied 51% of global exports. Following Turkey, Afghanistan (15%), Austria (5%), Spain (5%), Greece (4%), and others (20%) were the main dried fig exporters. The major destinations are the European Union led by Germany. The leading countries importing dried figs from the global market (as metric tons) in 2021 are Germany (28,132), India (22,559), France (16,202), U.S. (15,046), the Netherlands (7947), UK (6749), Russian Federation (6688), Italy (6440), Canada (5165), China (4065), Switzerland (3860), and Austria (3422). As a weak point of this dataset, Iran is not included as a dried fig exporter, whereas Iran exports over $40 million worth of dried figs especially to Asian markets (see Table 9.1). As previously mentioned, countries like Germany, Austria, the Netherlands, and Singapore are regional distributors of fig and are therefore included in trade statistics as exporters due to re-exportation.

According to the database of the Aegean Exporters' Associations (AEA), which is the body responsible for dried and fresh fruit exports in Turkey, dried fig exports from Turkey surpassed 70,000 tons during 2022 and 2023. The U.S. is the leading importer sourcing from Turkey, followed by Germany and France. Turkish dried fig export is widespread and reaches 120 countries globally.

The global fig trade is growing relatively slow with the emergence of new producer countries. Afghanistan, Uzbekistan, Morocco, and Syria are traditionally fig-growing countries but entered the global trade scene as new producers of dried figs as prices went up due to demand; they were able to enhance their market quality and fulfill requirements of the international trade standards. New consumer groups, such as in China, India, and South American countries, started to demand dried and fresh figs. The quantities of fresh figs traded internationally are relatively small due to the highly perishable nature of the fig fruit. Most of the fig trade routes are destined to regional markets within the continent, e.g., Iran to Asian countries, and Peru or Mexico to the U.S. There is an exception for Brazilian or South African fresh figs exported to Europe as an out-of-season delicacy or organically certified Turkish dried figs exported to the U.S. to fulfill the demand of rapidly growing organic markets. The amount of fresh and dried figs destined to global markets are closely related to their total production amount and share of marketable quality. Figs produced are consumed within the same year, and there are negligible amounts of stock for dried figs.

The average unit prices vary according to the fresh and dried fig quality, namely fruit size and number of defective fruits, e.g. crack, sunburn, and mechanical damage. The total amount of fruit production is affected largely by climatic conditions and demand versus amount to be traded determines the price. In 2022, the overall average fresh fig Free on Board (FOB) price was $2.50 per kg for Bursa Black figs, whereas $3.60 in 2023. The prices also vary according to the destination due to higher losses during long transfers and to the harvest time. Varietal peculiarities play a crucial role in the market demand; however, quality standards on fresh or dried figs define the marketable quality of a specific market. As an example, severe fruit cracking is a defect in the UN/ECE standards on figs, so a severely cracked dried fig of cv. Sarılop is considered as defective fruit in the European market. On the other hand, Sabz, the Iranian dried fig variety, has small size and is always cracked from the ostiole end; however, it can be successfully marketed in the South or Southeast Asian markets.

Even if the volumes for the fresh fig trade are much smaller than dried fig, Turkey is the leading country in the fresh fig trade that solely relies on cv. Bursa Black. Indicatively, during 2022 and 2023, 24 and 19 tons were exported at a value of $58.7 and $68.3 million, respectively. Major importers of Turkish fresh figs of Bursa Black variety are Germany, Russian Federation, Austria, UK, France, Switzerland, Poland, and Hong Kong as of 2023 (AEA, 2024).

3. Botany and relevant species

3.1 Botany

The fig tree is a deciduous tree that, depending on the variety, can have a shape ranging from open to compact and from pendulous to upright to spreading. Tree size and density also depend on the variety. Exceptional specimens exist that have attained 9 to 12 m in height with a 10-m spread, although commercially grown trees are much smaller.

The vegetative parts of the fig tree include roots, stems, and leaves. The fig tree has a system of fibrous roots that spread out to up to three times the canopy diameter. This system is typically very shallow; although, in some soils, it can reach great depths. Fig stems can grow into a small tree or large shrubs with light-color softwood. The bark is smooth, but very few varieties may have fissures. The trunks and larger branches may have bark tubers. The terminal bud, or those just below, produces new growth that bear fruits yearly and are 20 to 200 cm, depending on the variety. Fig leaves are typically large (10-25 cm long), deeply lobed with three to seven sinuses, rough and shiny on the upper surface, and soft and hairy underneath. The lamina is obovate to nearly orbicular, either undivided or with highly pronounced lobation. Fig varieties exhibit highly pronounced heterophylly, with the leaf lamina differing in size and shape. The alternate leaves are borne on long petioles. Kadota fig cultivar leaves have no lobes, whereas others may show shallow lobing or deep dissected leaves (Flaishman, 2022).

3.2 The fig reproductive morphology

Like all *Ficus* species, the fig bears a unique closed inflorescence structure, the syconium, which is the edible fruit. The syconium is an enlarged, fleshy, involute, hollow, and almost closed receptacle with multiple ovaries lined on its inner surface. Small flowers and when fruit ripens numerous one-seeded fruits (drupelets) line the interior cavity. It commonly has a bulbous shape with a small opening, called the ostiole or the eye, at the distal end leading to a bract-lined tunnel eventually opening to the syconium cavity that allows access to pollinators. *F. carica* is a gynodioecious species but functionally dioecious, bearing "male" and "female" figs on separate plants. Female trees have long-styled female flowers that do not host the fig wasp (*Blastophaga psenes* L.) larvae, and their figs produce seeds but no pollen. Male trees have short-styled female flowers that can host the pollinator wasp larvae, and their fig fruits produce pollen but no fertile seeds (Figure 9.1).

Female fig trees can produce two crops: breba (spring) and the main (summer) crop. Breba figs are the early crop produced in the spring on previous year's wood. Breba figs are generally juicy yet have lower sugar content and are thus accepted as inferior quality compared to the main crop. However, breba fruit are demanded by the markets with high prices because they are precocious. The main crop develops later in the season on the current year's growth. In some varieties, the main crop fruit set at the very end of the season, thus in some sources, female figs are reported to have three crops.

The fig varieties are classified into three horticultural types based on cropping and pollination needs, namely Smyrna, San Pedro, and Common (Flaishman et al., 2008) (Figure 9.2). Varieties of the Smyrna type, like Sarilop (Calimyrna), Marabout, Kalamata, Bursa Black, and Zidi, develop summer crops upon pollination.

The syconia shrivel and fall at the end of the anthesis period unless the flowers are pollinated. Usually, the Smyrna-type varieties do not produce a breba crop. Most of the fig varieties classified as the Common type, known also as having persistent syconia, like Brown Turkey, Kadota, Mission, Summer Figaro, Autumn-Honey (late Figaro), and Sierra, produce fair to large main crops without pollination. Caprification of common-type figs sometimes leads to an increase of fig size, changes on both

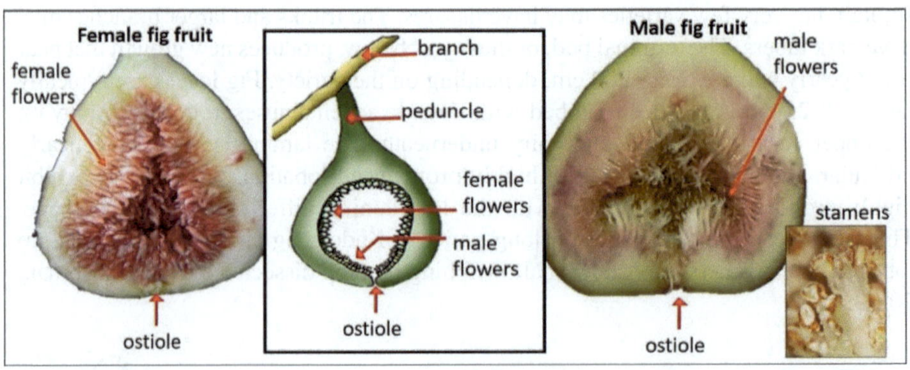

FIGURE 9.1

The fig syconium. Female *(left)* and male *(right)* fig fruit syconia with cross-sectional diagram. *(Modified from Moshe Flaishman)*

FIGURE 9.2

Diversity of fig varieties, varieties exemplifying the three horticultural types. Among these, Autumn Honey (= Late Figaro) is a cultivar released from a breeding program at Volcani ARO, Israel, whereas others spread from unknown sources as farmers' selections. *(Modified from Moshe Flaishman)*

skin and pulp color, which enhances fruit taste but also increases tendency to split (Rosianski et al., 2016). Common-type fig trees may initiate breba crops. Fig trees of the San Pedro type, like Nazareth, Dauphine, Desert King, and early Figaro varieties initiate fair to large breba crops. These persist to maturity without any known stimulus and fully develop and ripen into tasty fruits. An adequate second crop of syconia is initiated, but usually, only those in which the flowers have been fertilized develop to maturity stage.

Fig fruit exhibits significant differences regarding its appearance, taste, and flavor. The phenotypic characterization of fig fruit involves assessing various morphological and pomological traits to identify and distinguish different cultivars or genotypes, revealing significant diversity among fig cultivars. The key points regarding the phenotypic characterization of fig fruit are fruit size, shape, apex shape, skin and pulp color, fruit firmness, and ostiole size. This information is crucial for germplasm conservation, selection, and breeding programs to improve fig cultivation. Fig descriptors prepared by IPGRI and CIHEAM (2003) or UPOV (2010) for international guidance as well as those based on national descriptors on figs or studies on relevant topics could be used as a reference to describe the fig germplasm (Wang et al., 2022).

3.3 Taxonomy and related species

Ficus carica is a type of *Ficus* species that belongs to the subgenus Ficus and is classified as a member of the mulberry family Moraceae, order Rosales. Other well-known edible Ficus species include the Sycomorus fig (*Ficus sycomorus*), Creeping or Climbing Fig (*Ficus pumila*), and Bedu or Punjab fig (*Ficus palmata*). (Barolo et al., 2014).

The *Ficus* species is characterized by two distinguishing features: (i) a unique inflorescence structure, commonly called syconium (Condit, 1947); and (ii) unique specialized mutualism and coevolution carried by the fig pollinator wasps of the family Agaonidae (superfamily Chalcidoidea). The association between fig and fig wasp dates back approximately 60–100 million years (Rønsted et al., 2005; Cruaud et al., 2012).

The common fig is the only deciduous *Ficus* tree, and it is highly adaptable to different growing conditions. Fig has been cultivated since ancient times for its nutritious fruit and is now widely grown worldwide in different climate zones. The general biology of the root, stem, leaf, and fruit illustrates the complex adaptation of fig trees to different climates. Its botanical characteristics and diversity within the large Ficus genus renders it an economically important species. The unique syconium structure of the fig tree and its development in male and female trees are well adapted to the specific mutualism between the fig fruit and the fig wasp.

4. Propagation and breeding

4.1 Propagation

Seeds, cuttings, or suckers can be used to propagate fig trees. However, seedlings from seeds will most probably not inherit the desired traits due to the mixed genetics. On the other hand, fig fruit species, even if there could be differences in rooting rates among varieties are easily propagated vegetatively. Thus, the propagation method by cuttings is still the most used with the hardwood cuttings performed as the best vegetative propagation method for fig genotypes. For multiplication, pieces of hardwood (1-3 years old) are taken from true-to-type and healthy parent material during the dormant season and kept under appropriate conditions until planted. Most of the commercial fig orchards are established with rooted cuttings. Additional methods for propagating fig trees include branch layering, grafting, or utilizing laboratory techniques for plant tissue culture to produce commercially a high number of disease-free saplings (Flaishman et al., 2008; Özer, 2022a).

Fig propagation using rootstocks via grafting is an additional costly step unless the aim is to change an existing variety or for other specific purposes. Rootstocks can be used for disease, pest or cold hardiness resistance, to increase vigor, or to develop multivariety fig trees. Disease-resistant rootstocks are an important technique in fig propagation to combat soilborne diseases and pathogens affecting tree health and productivity. Recently, new rootstocks resistant to Ceratocystis canker and the fig tree borer *Batocera rufomaculata* were developed by using interspecific hybrids with wild *Ficus erecta* and *F. palmata,* respectively (Kamimori et al., 2022; Peer et al., 2022).

Rootstock choice may also consider cold hardiness, as some tropical fig species used as rootstocks may be less hardy in colder climates. Hardy rootstocks ensure that the grafted tree can withstand certain lower temperatures. In addition, some rootstock cultivars, such as Desert King, Emalyn's Purple, and wild seedlings, are chosen for their exceptional vigor and growth rates. Grafting fig varieties onto rootstocks is commonly conducted through cleft grafting, side grafting, or budding. Fig trees grown on interspecific hybrid rootstock that requires grafting will be more resistant to diseases and more tolerant of drought and poor soil conditions. This can help confirm that fig growers have healthy and productive trees. Utilizing proper grafting techniques is important to ensure a successful graft union and avoid transmission of viruses.

4.2 Breeding

Understanding the unique biology of *F. carica* is essential for successful breeding programs. Fig has a unique floral arrangement and a complex breeding system. Fig trees have two morphs: caprifigs (male) and female edible figs (see Figure 9.1). Caprifigs have male flowers and short-styled female flowers in the syconium. Edible figs have syconia containing only long-styled female flowers. Figs rely on a specific pollinator, the fig wasp, *Blastophaga psenes* L., for pollination. The wasp lays eggs in the ovaries of the short-styled flowers and dies inside the caprifig. Its newborn wasp emerges from the spring caprifig fruit, the profichi, covered with pollen and pollinates the long-styled flowers in edible figs. This process, called caprification, is crucial for fruit set in the second and main crop of Smyrna and San Pedro types with caducous (nonpersistent) syconia (Figure 9.3). Only pollinated long-styled flowers in edible figs develop into mature fruits with seeds having an endocarp. Seeds from such fruits can produce trees of both caprifig and edible fig morphs (Galil, and Neeman, 1977).

Conventional breeding utilizing diverse genetic resources focuses on crossing specific fig varieties to achieve desired characteristics like fruit quality, disease resistance, or extended shelf-life. The selection of parent trees with desired characteristics is crucial. Fig breeding combines ancient practices and modern techniques (Condit 1947; Peer et al., 2022). Nowadays, there are limited breeding programs for figs worldwide. Breeding programs were initiated in California at the beginning of the 20th century, producing mainly Smyrna-type dried fruit (Condit, 1947). Breeding continued by the California fig industry and resulted in the release of common-type Sierra and Sequoia cultivars (Stover, et al., 2007).

In the Israeli fig breeding program, initiated in 2005, around 8000 hybrid seedlings were assessed yearly to identify the best parents and cross combinations. This led to the selection of 20 cultivars, named Figaro, with one early- and one late-

FIGURE 9.3

Caprifigs prepared for caprification *(left)* and hung on the female fig tree *(right)*. *(From Uygun Aksoy)*

ripening cultivar now being commercially exploited. The midsummer cultivars have high sugar contents, close ostiole, and no cracks, ensuring pest resistance, high-quality fruits, and extended shelf-life. Most of the new cultivars in commercial orchards are common types with parthenocarpy, suitable to produce fruit without pollination. This simplifies cultivation and rapid extension of fig production worldwide in areas that do not have the male caprifig, and/or the fig wasp, like the bred cultivar Sweet City Red, grown in many locations today across China.

As another breeding strategy, interspecific hybrids between *F. carica* × *F. pumila*, *F. carica* × *F. palmata*, and *F. carica* × *F. auriculata* were obtained. Leaves, shoots, and fruits of the hybrids were then characterized. While some of the interspecific hybrid fruits showed fig characteristics, others exhibited new kinds of fruit species. Currently, the hybrid progeny of *Ficus carica* × *Ficus palmata* showing resistance to the fig tree borer, *Batocera rufomaculata,* is used as a rootstock (Peer et al., 2022). Molecular tools like marker-assisted selection (MAS) use DNA markers to identify plants with desired traits. Such DNA markers that can distinguish male and female fig seedlings are being explored and are being used to accelerate fig breeding efforts (Mori et al., 2017; Wang et al., 2021). While conventional breeding has progressed, challenges such as limited shelf-life, cold and drought hardiness, and better fruit-eating quality, such as improved flavor, size, and color, are still waiting for a solution or improvement.

Fig trees are heterozygous diploid ($2n = 26$) plants with low DNA 2C values and with an estimated 36,000 genes (Mori et al., 2017). Modern breeding techniques with direct gene manipulation like genetic modification (GM) and genome editing are already established in fig (Peer et al., 2022) and may have the potential to create new cultivars that are resistant to diseases, alter their flavor or nutritional profile, or have a longer shelf-life.

5. Orchard management

Climate change predicts that drought and high temperatures will become the main stress factors affecting fruit tree production, including fig. Fig bearing diversity can provide a wider adaptation to climate change and currently practiced in different climatic zones worldwide (Flaishman, 2022). Proper site selection, soil preparation, and irrigation system installation are crucial for successful fig orchard establishment. Usually, fig orchards are established in mild temperate and subtropical regions with moderate relative humidity, mild and rainy winters (dormant period), and dry summers (fruit growing and maturation period). Fig requires some winter chilling and can be grown at sea level to 1700 m in areas with low rainfall if the short chilling requirement is fulfilled. Figs are not frost-tolerant, and growing tips of the yearly growth, overwintering male fruits, and the fig wasp are more sensitive to temperatures below -10°C.

Figs can adapt to various soil types and prefer sandy-clay loam soil with a pH between 6.0 and 8.0, but growth may be slower in poor or rocky soil. Well-drained soil is crucial. The average planting density may vary and expressed as between and on-row distances between 3 to 6 m and 4 to 6 meters are applied, though some recent experiments in Brazil and Israel have used a higher density of 3 x 2 m as between and on-row distances. Figs respond well to organic matter. A soil depth of 1.0 to 1.5 m is sufficient for fig growth. Appropriate growing conditions will increase revenue from growing fresh figs. Therefore, fruit production under a net or by using hydroponics in greenhouses may increase fruit quality and allow better control over the time of fruit yield (Flaishman et al., 2008).

Numerous fig varieties exist, varying in fruit characteristics, productivity, and adaptation to local conditions. Key selection criteria include fruit size, color, flavor, texture, ripening season, productivity, and resistance to biotic and abiotic stresses. Cold areas with short summers will limit many varieties due to insufficient heat and light. Orchards with Smyrna-type fig varieties require caprification. The male fig trees are grown separately, with one male tree required for every 50 female trees. The general practice of caprification is to distribute the profichis (male syconia with ripen pollens in early summer) with ripen pollen at intervals of 4 or 5 days over a period of about 3 weeks, corresponding to the receptivity of female figs. Common-type fig cultivars do not require pollination, and fruit usually ripen from August to October. With the appropriate climate conditions, they can produce additional late yield. On the other hand, other common-type fig cultivars, like Autumn-Honey (late-Figaro) and Sierra, produce a single, main, late crop between September and December. Some cultivars of the San Pedro type, like Nazareth, early-Figaro and

Desert King, produce high-quality breba crops at the beginning of summer in addition to the main summer-fall crop.

Fig orchards are rapidly growing under rainfed irrigation wherever the climate allows. The equivalent of 650 mm of annual rainfall is sufficient to produce a good crop. When growing in hot climate conditions, irrigation is commonly provided via sprinklers or drip systems. Irregular or excessive irrigation can contribute to ostiole-end splitting, especially toward ripening in some varieties and favor fungal growth and internal or surface rots.

Fertilization requirements depend on soil fertility, tree age, and yield potential. Nitrogen, phosphorus, and potassium are the routinely applied primary nutrients. Fig trees may benefit from water-soluble fertilizers to provide a steady supply of macro- and micronutrients. Young fig trees (1-2 years old) should receive nutrients balanced with their growth, starting when new growth appears and stopping before the end of summer. Excessive nitrogen fertilization can lead to unnecessary vegetative growth and reduced fruit quality. Late application of nitrogenous fertilizers and/or irrigation may trigger frost damage if growth does not terminate prior to early frosts. Nutrient demands and thus the application of fertilizers may vary according to the fig variety, e.g., for common-type varieties, mature fig trees should receive fertilizer applied three times yearly in late winter, mid-spring, and mid-summer after fruit harvesting to assist long shoots and a high number of fruits (circa 20-25) on the yearly growth. On the contrary, for Smyrna-type varieties where the shoot length is short, shoot growth terminates early summer and number of fruits per shoot are fewer (circa 5-7), winter application of required nutrients will be adequate. Nutrient requirements differ according to various balanced fertilizations based on leaf analysis.

Fig trees can be trained in various systems, such as an open-center vase, central leader, modified central leader, or trellising (Figure 9.4). In its traditional agroecosystems, fig trees are commonly trained as an open-center vase system, where the sapling is topped from circa 75 cm, and short laterals are removed, and only three

FIGURE 9.4

(A) Fig orchards: Traditional open vase training in a lowland orchard (Aydın-Turkey) (*From Uygun Aksoy*). (B) Fig trees on terraces in Weiyuan-China (*From Uygun Aksoy*). (C) Intensive orchard training in a greenhouse (Gilgal, Israel). *(From Moshe Flaishman)*

or four branches are left to develop the primary scaffold. During the first 5 years, the tree should be trained by selecting three or four main branches distributed evenly around the trunk and removing all other branches to develop a robust tree architecture. The fig tree, mostly those of common-type varieties, can be trained on a trellis or wire system. When trellis training is employed, a basic scaffold of horizontal branches is first established. During the late winter or early spring, vertical shoots from the previous year's growth are shortened to stimulate vigorous new shoots that will bear the current year's fruit crop. Around 30 new shoots may come up from the horizontal branches in the spring. Most of the new growth is thinned down to leave no more than 8 to 10 shoots, 25 cm apart from each other. Pruning includes the typical winter-spring pruning and a "nipping" process applied several times during the growth season. Nipping removes the apical meristem from green, rapidly growing shoots preventing vigorous vegetative growth, and thus, nutrients and plant hormones are diverted into the developing fruit. New shoots are trained onto a trellis of 2 meters in height which allows easier picking. Proper crop load management is vital for ensuring fruit quality and sustainable yields. To control the crop load of figs, summer pruning or shoot pinching is used to eliminate excess fruit production in common-type varieties where the apical dominancy is weak.

Pruning is essential for maintaining tree structure, promoting light and air circulation, and regulating crop load. The common fig requires specific pruning, training, and cropping system to optimize its growth and fruit production. In fig, pruning is typically performed during dormancy or after harvest in cases where there is no frost risk to remove dead, diseased, or overcrowded branches. Mature trees should be pruned by thinning branches and heading back long shoots to maintain tree shape and balance.

The fig fruiting cycle is 120 to 150 days. Some varieties produce one crop per year, others two. Cropping systems like the breba and main crop cycles are managed differently. With summer (second crop) fruit production of Smyrna- and San Pedro-type varieties, the fruits are produced on the current season's shoots. Usually, the canopy is pruned down during the winter. In breba (first crop) cropping of common-and San Pedro-type varieties, leaving full-length branches during the winter pruning can help retain the breba crop (first crop) figs produced at the previous year's shoot growth. Fruit thinning is generally unnecessary for improving fruit size or quality in figs and is typically not practiced in figs. However, deformed and infected fruit are wilted or removed in some orchards during fruit growth and development. In addition, after harvest, any fruits that fail to ripen are removed (Can, 2022; Jafari et al., 2022).

6. Fruit maturation and harvest criteria

Development of female fig fruit follows a typical double-sigmoid growth curve based on fruit diameter, including two rapid growth phases (I and III), which are separated by the slower growth phase II (Flaishman et al., 2008). Notably, the ripening process (phase III) of summer crop fig is extremely rapid, occurring within a week, and in as little as 3 days in the peak summer season. Accordingly, ripening-related parameters of size, flavor, and texture undergo major changes within a very limited time. Compared to the seemingly uneventful phase II, fruit size may increase twice to three-fold and softness increases drastically during phase III (Rosianski et al., 2016). With respect to fruit sweetness, more than 70% of the total dry weight and 90% of the fruit's total sugar content accumulate during this short ripening phase (Flaishman et al., 2008). Detailed developmental analysis of water and dry matter accumulation during these final ripening days indicated a temporal separation between the increase in volume due to increasing water content and a subsequent sharp increase in sugar content. Fig fruit represents an extreme example of sugar import and accumulation, with calculated import rates that are one order of magnitude higher than those of all other sugar-accumulating fruit species (Lama et al., 2022). To shed light on the metabolic changes occurring during this period, tissue-specific analysis of gene-expression patterns indicated that sugar metabolism and sugar transporter gene expression regulation undergo a global transition (Lama et al., 2022). The major soluble sugars in fig fruit are glucose and fructose, the products of sucrose hydrolysis, whereas sucrose is present but at low concentrations (Trad et al., 2012).

The ripening process is typically variety dependent and occurs in the main fig crop from June to December, depending additionally on the prevailing climatic conditions. Knowledge of the optimum maturity stage is important to ensure the best quality and flavor. One of the most reliable indicators of fig maturity is skin color. The exact color varies depending on the cultivar, but most figs will change from green to a deeper color as they ripen. For shipping, figs are collected daily just before they reach the fully ripe stage, but yield to a soft pressure, usually indicated by small cracks in the skin in some varieties. Indicatively, Black Mission figs should be light to dark purple rather than black for shipping. While picking, fresh figs should be firm to the touch but not hard. If the fig is too soft, it is likely overripe, highly perishable, and may have begun to ferment. To assess fig firmness, one can gently press on the fig fruit. In some cultivars, the fig's stem (neck) begins to drop slightly as the fruit ripens. This is a visual indicator that the fig is approaching full maturation. In some varieties, the ostiole may begin to split slightly as the fruit ripens and excessive splitting can indicate overripeness that renders the fig more susceptible to spoilage and could be classified as a defective quality. In summary, the key maturity for harvesting indices for fresh figs are skin color and flesh firmness, whereas fruits destined for drying should be allowed to partially dry on the tree before harvest and further sun-dried in drying yards (Figures 9.5 and 9.6).

7. Postharvest physiology and technology

The fig fruit is one of the most delicate and complex fruit because of its morphology, an aggregate fruit composed of two botanically different tissues, the receptacle and the inflorescence. Both abscisic acid (ABA) and ethylene-dependent processes take

FIGURE 9.5

Sarılop (= Calimyrna) fig fruit at different stages of ripening on the tree, shriveled fruit is ready to fall onto the ground for further sun-drying on trays. *(From Uygun Aksoy)*

place during fruit ripening. The high sugar accumulation just prior to the harvest and a fleshy succulent peel that also possess sugars, the short storage and shelf-life period, the different types of varieties with persistent (parthenocarpic) or caducous (non-parthenocarpic) syconia formation and consecutive ripening of fruits on the shoot make the harvest period quite short (e.g., 2 weeks) or long (e.g., 4-5 months) according to the intervals between each ripening fruit. These properties, when coupled with the taste, flavor, and health-promoting properties render the fig, a rare and valuable fruit crop species, appreciated by consumers. Its shelf-life is quite short, even if the currently available knowledge and technology is utilized. Thus, the postharvest stage and technology aim to preserve quality during storage and shelf-life to supply fruit throughout a long marketing season without reducing fruit appreciable qualitative attributes upon harvest.

Proper postharvest handling of fresh fig fruit starts by picking at the right stage of maturity for harvest as previously described. The first step is to provide precooling and reduce fruit temperatures. This is very important considering the high temperatures prevailing during the fig harvest period. Cold conditions must be maintained

FIGURE 9.6

Open field sun-drying process: (A) Under plastic tunnel, Aydın-Turkey *(left)*; (B) under tunnel, Calabria-Italy on traditional wooden trays *(center)*; and (C) sun-drying at open field conditions (Aydın, Turkey) *(right)*. *(From Uygun Aksoy)*

throughout the value chain, together with high humidity levels to preserve quality and prevent moisture loss from the fruit. The temperature requirements of fresh fig may vary between -1°C to 6°C according to the cultivar and maturity stage at harvest, their morphology and constituents, and relative humidity (RH) above 90%. Even under these recommended conditions, the reported storage periods range between 7 to 30 days, and an additional shelf-life of 1 to 3 days is foreseen when kept at 20°C. At postharvest stage, major quality losses occur in fresh fig fruit, such as shriveling due to moisture loss (more apparent in varieties with longer neck fruits) and decay caused by yeasts and molds.

Initial sorting and grading are mostly conducted at the orchard to remove substandard or defective fruit that cannot be sold in the destined markets for both fresh and sun-dried fig production. In fig production, physiological disorders such as ostiole-end cracks (or split) and sunburn are highly widespread. The nutritional status and varietal characteristics of the fig tend to affect the frequency and severity (Aksoy and Anaç, 1994; İrget et al., 2008). Severe cracking (more than one-third of the fruit length) or sunburn (more than one-third of the surface area) are considered as defects in quality standards and cannot be marketed (Figure 9.7).

For fresh figs, it is vital to provide shade and cool conditions during this initial stage if fruit cannot be readily transferred to the cold conditions of the packing house. Precooling through forced air reduces the initial fruit temperature to storage temperatures rapidly enhancing storage and shelf-life time (Dussan-Sarria et al., 2010). Since fresh figs are highly perishable and can be easily damaged, the packing house operations are mostly done manually. However, in regions where volumes are bigger, grading and sizing are performed mechanically in packing houses. Packaging is important, especially if the destination is a distant market (Figure 9.8).

Various research work is being carried out to determine the ideal controlled atmosphere conditions for figs that will reduce respiration rate, ethylene and production of aroma volatiles, delay discoloration and senescence and preserve fruit firmness. Optimal controlled atmosphere conditions generally presented for figs are: 0°C, 90% to 95% RH, 15% to 20% CO_2 and 5% to 10% O_2. However, these

FIGURE 9.7

(A) Ostiole-end crack or split fig fruit cv. Sarılop *(From Mehmet Özkul)*; (B) sunburn or sun-scalded fig fruit cv. Sarılop. *(From Uygun Aksoy)*

FIGURE 9.8

Preparing fresh Bursa Black figs for the export market. *(Courtesy Alara AgriBusiness, Bursa-Turkey)*

optimal values may differ significantly according to the varieties and/or stages of maturity, and thus, designing specific research work is always recommended if the conditions are not set for a specific variety. In practice, controlled atmospheric storage is not widely applied due to the short storage period when cost-benefit analysis is made. Active or passive modified atmosphere packaging (MAP) and edible coatings proved successful in fresh fig.

Dried figs are much more durable commodities however, cold storage conditions are recommended to extend their marketing periods with well-kept quality. Recommended cold storage conditions for sun-dried figs (with moisture contents up to ca. 25%) are 2°C-4°C and 60% to 65% RH. Under high humidity conditions, dried fruit absorb moisture that may lead to fermentation due to yeast and mold growth and elevate secondary mycotoxin formation. On the contrary, if relative humidity levels are low and/or temperatures are high in the storage room, dried fig fruit loses water, and white sugar crystals form over the fruit surface. In processing plants, dried figs undergo different steps prior to packaging as follows: defective or substandard fruit are removed, may (circa 4% salt solution) or may not be washed, packed naturally or shaped into special forms such as "lerida, layer, garland, macaroni, pulled" that the global trade still demands. These stages require intensive hand labor. As an alternative to sun-drying, recently fresh figs are dried artificially and processed to deliver low-moisture dried snacks (Şen, 2022).

8. Main diseases and pests and their management

8.1 Diseases

Fig is one of the fruit trees that is quite widespread in natural flora, especially around the Mediterranean basin, thus as a misconception, it is generally thought to be resistant to diseases and pests. However, fig diseases have been reported two millennia ago, back in the 3rd century BCE during the time of Theophrastus and even in the Old Testament of the Bible. Today, except for a few cases that are found detrimental, diseases mainly result in yield and quality losses. Despite some of the very early studies and reports released during the 1920-1930s, the major focus on fig diseases and pests appeared after significant cases were reported in commercial fig production, especially during the last decades.

8.1.1 Fungal diseases

Fig diseases are mostly of fungal origin and can be divided as those attacking **roots, stem, and branches,** such as honey fungus *(Armillaria mellea)*, white root rot *(Rosellinia [Dematophora] necatrix)* (Figure 9.9), root and butt rot *(Ganoderma sessile)*, Texas root rot *(Phymatotrichum omnivorum)*, mushroom root rot *(Armillaria tabescens)*, fig foot rot *(Ceratocystis fimbriata* [anamorph: *Chalara* sp.]), pink limb blight *(Erythricium salmonicolor)*, Phomopsis dieback *(Phomopsis*

FIGURE 9.9

Complete and sudden death of a fig tree in summer affected by White root rot (*Rosellinia [Dematophora] necatrix*). *(From Uygun Aksoy) cinerescens),* Nectriella stem gall *(Nectriella pironii/Kutilakesa pironii)*, damping off (rotting of the stem and roots in nursery greenhouses) *(F. solani, A. alternata* and *P. ultimum)*; **leaves,** such as leaf rust *(Physopella fici)*, leaf spot *(Cylindrocladium scoparium* or *Cercospora fici)*, web blight *(Thanatephorus cucumeris/Rhizoctonia solani)*, or **fruit,** by growing inside the fruit cavity resulting in internal rots, such as endosepsis (*Fusarium* sp.) or smut *(Aspergillus niger)*, or those growing on the fleshy skin resulting in surface rots, such as Anthracnose *(Glomerella cingulata/Colletotrichum gloeosporoides)* or *Botrytis* sp. Among these pathogens, as in *Alternaria* sp., it may attack fruit and cause surface (e.g., skin lesions) and/or internal rots and/or damage the leaf resulting in Alternaria leaf spot.

8.1.2 Bacterial diseases

Regarding bacterial agents that are less common compared to fungal pathogens, bacterial diseases of figs reported are Bacterial canker *(Phytomonas fici)* and Crown gall *(Agrobacterium tumefaciens)* (Elbeaino, 2022; Flaishman et al., 2022; Michailides, 2003; Mostowfizadeh-Ghalamfarsa et al., 2022). Root rot agents, which may cause decline of fig trees, may vary according to the fig production region. In California, *Armillaria* sp., and in Brazil, *Ceratocystis fimbriata,* are reported to cause fig tree loss, whereas in Turkey, *Rosellinia (Dematophora) necatrix* is the prime pathogen causing root rot in the fig-producing Aegean region. Fruit rots are more widespread and may create economic losses in both fresh and dried fruit production by lowering the quality and marketable yields. Since they appear mostly at the time of fruit maturation and their incidence depends on the seasonal climatic conditions, preventive nonchemical or chemical measures are not highly effective. Additionally, some of these fungal agents, either saprophytic or pathogenic, developing inside the fruit cavity or on the skin may be responsible for toxin formation as aflatoxins, ochratoxins, and other toxins, even in rare cases (Michailides, 2003; Özer, 2022b).

8.1.3 Virus diseases

Among the virus diseases, fig mosaic disease (FMD) is one of the most widespread and has been known for decades, with early reports in 1928 and 1933. FMD may first affect the leaves and then later unripe fruit, and may retard growth, reducing yields and quality. Under stress conditions and high temperatures, it becomes vigorous and may cause a complete die out of young nursery trees. Until 2006 when molecular studies started to pave the way, all major fig diseases focused on bacterial and fungal pathogens; the information on viruses, viroids, or phytoplasmas was rather rare and controversial. During recent years, much research work has been done to identify and classify virus species through molecular techniques.

The virus species identified in figs are mostly cosmopolitan and found in almost all producing regions. Even if widespread globally, the economic damage caused can vary based on the tree's health and vigor. In this respect, the parent material used for propagation must be virus-free since figs generally multiply by rooting cutting or rarely by grafting that helps transmission. However, studies have put forth that many virus species are transmitted by aerial vectors like aphids, mealybugs, and leafhoppers. *Aceria ficus* is the most known vector for virus species and, more specifically, a soilborne vector as the nematode *Xiphinema diversicaudatum,* is a transmitter of the Strawberry latent ringspot virus in nature. However, these vectors can be selective in transmission of target virus species (Elbeiano, 2022).

For the management of fig diseases, if cuttings are going to be used for establishing the orchard, healthy propagation material must be provided from known and approved origins. If saplings from a commercial nursery are to be used, then they must be certified healthy nursery trees. Further management in a fruiting orchard should aim for sanitation, including proper training and pruning to allow adequate aeration to reduce accumulation of moisture that may favor fungal growth; proper nutrient and irrigation management and removal of infected plant parts should be performed wherever possible. Keeping the soil surface covered or avoiding tillage, especially prior to fruit maturation, reduces dust and thus the spread of fungal agents embedded in top-soil layers, especially surface rots in fig fruit. Vectors carrying fungus into the fruit cavity, like mites, dried fruit beetles, vinegar fruit flies, or the fig wasp, need to be closely monitored for their role in carrying infections. In the case of fig wasp, its presence is inevitable in fig varieties requiring pollination for fruit set, so the best solution is to use clean and healthy male figs to prevent carrying over, especially of *Fusarium* sp., which generally originates from male fruits.

8.2 Pests

The degree of economic loss that occur because of fig pests may vary according to the variety, prevailing climatic conditions, and operations performed at the orchard. Major fig pests can be classified under four groups based upon the plant part they infest and the damage they create: (i) **trunk and branch:** stem-borer *(Bactrocera rufomaculata)* (Figure 9.10A) and fig bark beetle *(Hypoborus ficus)* attack mostly weakened or old tissues; (ii) **leaf and shoot:** those feeding on new shoots, buds, and leaves as fig wax scale *(Ceroplastes rusci)*, fig psylla *(Homotoma ficus)*, and fig weevil *(Otiorhynchus davricus)*; (iii) **immature fruit:** Black fig fly *(Silba*

FIGURE 9.10

Major pests of fig: (A) Stem borer *(Batocera rufomaculata) (left) (From Wikipedia),* (B) Black fig fly *(Silba adipata) (center),* and (C) dried fruit beetle (*Carpophilus sp.*) on a rotten fig fruit *(right). (From B. Eşref Tutmuş and C. Uygun Aksoy) adipata)* (Figure 9.10B); and (iv) **storage pests:** that attack fig fruit at full maturity or overripe stage while on the tree or at the drying yard and that continue their activity in storage: Fig moth (or Tropical warehouse moth) *(Cadra cautella)*, Indian meal moth *(Plodia interpunctella)*, saw-toothed grain beetle *(Oryzaephilus surinamensis)*, dried fruit beetle (sap beetle, *Carpophilus hemipterus* and other *sp.)*, and dried fruit mite *(Carpoglyphus lactis)* (Figure 9.10C).

Additionally, soilborne nematodes root-knot nematodes *(Meloidogyne* spp.*)* and lesion nematodes *(Pratylenchus musicola)* may create problems by damaging the roots and limiting the water and nutrient uptake. They could be carried on with the nursery trees or present at the orchard soils (Ben-Yakir and Costa, 2022; Elbeaino, 2022; Turanlı, 2022).

Among the storage pests, Lepidopterous moths especially fig moth creates problems in dried fig sectors in Turkey and Greece if not well managed. On the other hand, dried fruit beetles are the major pest in California dried fig production. Until methyl bromide (MeB) was banned globally due to its ozone depletion property, it was widely and effectively used to control storage pests at a low cost. After the ban, special production protocols are developed to reduce the infestation of the storage pests at the preharvest stage in the orchard and further within the storage or processing facilities. In marketing standards of dried figs, there are maximum allowed limits for pest damaged dried figs and/or parts (UN/ECE, 2016). Lepidopterous moths are attracted to light, so sanitation as the removal of infested fruit and debris prior to new harvest or whitewashing storage rooms, keeping the top of the drying trays covered to prevent moths from laying eggs, and avoiding lights at the drying yard or storage rooms at nights could help reduction of initial infestation (Turanlı, 2022). On the other hand, fermentation of fig fruit due to yeasts or molds attract Nitidulid (*Carpophilus* sp.) beetles and vinegar flies, so measures taken to minimize fermentation or rotting, e.g., avoiding excess nitrogen fertilization or late irrigation in commercial sun-drying of figs are recommended. Bait traps, if managed timely and effectively, can help to reduce *Carpophylus* sp. and vinegar fly populations. To prevent further distribution, dried figs are treated in processing plants either by chemicals, e.g., phosphine gas or by alternative treatments allowed in organic production, e.g., fumigation through short-term modified atmosphere (as carbon dioxide) or deep freezing (Turanlı, 2022).

In commercial fig production, storage pests are the major challenge for the dried fig sector, and other pests seldom cause significant economic losses in fresh fig production. The impact of fig pests may alter in time or according to the regions since new varieties, production methods, or ecosystems are introduced. For example, Black fig fly was a significant pest in the Middle Eastern and North African fig-growing regions; however, it became more widespread globally, damaging fruit at unripe stage and causing yield losses in fresh fig production. An alert was raised in the U.S. in 2022 as an invasive species. Similarly, there are cases where fig wax scale or fig psylla were reported as becoming a significant pest locally.

9. Current and future challenges in main production regions

Fig production is concentrated more around the Mediterranean basin or in regions with Mediterranean climates such as in the U.S., India, or Iran. The bulk of the global fig production comes from rain-fed orchards managed as low-input where fruit are marketed as sun-dried. A major challenge in such orchards is climate change, especially shortage and irregularity of rainfall and extreme temperatures during fruit development and ripening. Currently, fig varieties Sarılop (syn. Calimyrna) in Turkey and Sabz in Iran destined for sun-dried market are adapted to arid or semi-arid conditions and are grown traditionally as rain-fed for decades. Any significant shortages in total precipitation, below 500 mm and 200 mm in Turkey and Iran, respectively, including those in winter, may be a threat for dried fig production by limiting new growth and fruit development. Shifting of rainy periods toward fruit ripening and sun-drying may similarly cause quality and quantity losses in global dried fruit supplies. These two varieties destined for sun-drying are Smyrna-type fig varieties requiring pollination for fruit set in which the fig wasp (*Blastophaga psenes* L.) carries pollens from the male fruit to the female fruit. Adverse weather conditions such as harsh winds, rain, or extreme temperatures during the caprification period (June-July in the Northern Hemisphere) will hinder fruit set and create yield losses in Smyrna and a second crop of San Pedro-type fig varieties. In this regard, male fig varieties are susceptible to low winter temperatures e.g., below -10°C, since the fig wasp or the fruit may freeze, creating a lack of male caprifig fruit required for pollination. In common-type fig varieties, these are not seen as major challenges. Another threat for producing regions in the future could be newly emerging pests, diseases, and weeds because of climate change. Currently, Black fig fly *(Silba adipata)* and spotted wing drosophila *(Drosophila suzukii)* started to cause harm in commercial fig orchards and the first fly already raised an alert in Southern California as an invasive pest (Britt et al., 2022; Wilson et al. 2022). Additionally, in fig-producing regions where temperature and humidity levels increase, fungal diseases may become a significant threat for fresh fruit production and trade. Additionally, mycotoxins are already causing safety concerns and could become a more serious problem. Changes in climate may cause shifts in regions where mycotoxins are currently found or may change the prevailing mycoflora or prevalence of toxigenic species (Zingales et al., 2022).

In global agriculture, hand labor is becoming scarce or expensive due to decreasing and aging of the rural population. Fig is a rather delicate fruit requiring hand labor, especially during harvest for the fresh market, caprification in Smyrna and second crop San-Pedro-type varieties, and at postharvest stages. In sun-drying, the current varieties dry partially on the tree and then drop onto the ground, making harvest easier by collecting from the surface. When dried fig trade flows of the late 19th and early 20th centuries are evaluated, Mediterranean countries like Spain, Portugal, and Italy were top ranking in dried fig exportation, including destinations such as the U.S. However, because of increased labor costs, these countries lost their competitiveness, and Turkey, Iran, and lately Afghanistan became the major producers and exporters of dried figs. In these countries, fig production is mainly realized as family farming on small plots. Therefore, if new producer countries emerge, or the current leaders lose the "family farming" typology because of an aging population, and if some other regions offer lower prices, another significant shift may become inevitable. In this respect, studies focusing on lowering production costs, e.g., breeding for higher yields and quality, tree training methods, and harvesting aids, are crucial to make harvest more compact, easier, and less labor-demanding.

The common fig has rich genetic resources, which gives an opportunity to benefit from this diversity. This diversity may limit the production of a single variety in larger quantities as it is currently demanded by enlarging markets. A challenge in the coming years could be to select durable and high-yielding fig varieties among common types for fresh consumption. As consumers with high purchasing power demand fresh figs, one of the main challenges will be to offer fresh figs all year-round either through new varieties, new locations, production techniques (e.g., open and protected cultivation, management of tree architecture for longer fruit bearing), or extended storage and shelf-life. Another appealing aspect for the producer and consumer, but perhaps a challenge for the packer, will be to combine fresh fig fruit (or apply the same for dried figs) and pack different tastes, skin and internal colors, or shapes together in the same pack. The diversity in some products, as in tomatoes, is already available as mixtures to the consumer.

Fig fruit is recognized as a functional food and is consumed for its health benefits. In this respect, there are maximum allowed limits for pesticides or contaminants set at national or international levels, e.g., Codex Alimentarius as the competent UN authority set limits to guarantee that the products on the global market are safe and thus satisfy consumers' expectations. There is no consensus on many of these active substances or limits, and both the content and set limits vary among producer and/or importer countries, creating discrepancies in the market. The market standards that determine the fruit quality also differ. For example, sugaring, which is a natural phenomenon in dried fruit, is accepted as a defect in the U.S. Standard on figs (USDA, 2001), whereas it is not a defect in the UN/ECE standard (UN/ECE, 2016). Likewise, national and international standards on organic produce may allow labeling of soilless grown figs as organic in U.S., but it is not allowed in the European Union (EU). Geographic indication is another quality system that is favored in fig trade and contributes to rural development by valorizing local fig varieties and traditional processing methods. Existence of too many standards or labeling schemes may on the one hand increase cost, especially for small farmers, and create confusion among rare customers but help to differentiate products.

References

Aksoy, U., & Anaç, D. (1994). The effect of calcium chloride application on fruit quality and mineral content of fig. *Acta Horticulturae, 368*, 754–762. doi.org/10.17660/ActaHortic.1994.368.89.

Aksoy, U. (1998). Why figs? An old taste and a new perspective. *Acta Horticulturae, 480*, 25–26. https://doi.org/10.17660/ActaHortic.1998.480.1

AEA. (2024). Aegean Exporters' Associations' database, Dried Fruit and Products Board, Türkiye. https://en.eib.org.tr

Barolo, M. I., Mostacero, N. R., & López, S. N. (2014). *Ficus carica* L. (Moraceae): An ancient source of food and health. *Food Chemistry, 164*, 119–127.

Ben-Yakir, D., & Costa, A. (2022). Pests of fig trees in *Advances in fig Research and sustainable production Flaishman and Aksoy*, 259–278. https://doi.org/10.1079/9781789242492.0015a. CAB International.

Berg, C. C. (2003). *Flora malesiana* precursor for the treatment of *Moraceae* 1: The main subdivision of *Ficus*: The subgenera. *Blumea, 48*, 167–7.8.

Britt, K. E., Gordon, P. E., Faber, B. A., & Rios, S. I. (2022). Houston Wilson, first report of black fig fly, *Silba adipata* (Diptera: Lonchaeidae), in the United States. *Journal of Integrated Pest Management, 13*(1), 12. https://doi.org/10.1093/jipm/pmac009

Can, H. Z. (2022). Fig orchard establishment. In *Advances in fig Research and sustainable production*. In *Flaishman and Aksoy* (pp. 83–97). CAB International. https://doi.org/10.1079/9781789242492.0007.

Condit, I. J. (1947). The fig. *Massachusetts: Chronic botanic* (p. 222p). Mass, USA: Waltham.

Cruaud, A., Rønsted, N., Chantarasuwan, B., Chou, L. S., Clement, W. L., Couloux, A., et al. (2012). An extreme case of plant–insect codiversification: Figs and fig-pollinating wasps. *Systematic Biology, 61*, 1029–1047.

Dussan Sarria, S., Honorio, S. L., & Teruel, B. J. (2010). Precooling characteristics of the fig fruits 'Roxo de Valinhos' packed in wooden box. *Acta Horticulturae, 864*, 283–288. https://doi.org/10.177660/ActaHortic.2010.864.37

Elbeiano, T. (2022). Fig pathogens: Viruses, viroids, and phytoplasmas in *Advances in fig Research and sustainable production*. In *Flaishman and Aksoy* (pp. 279–292). CAB International doi.org/10.1079/9781789242492.0016.

Flaishman, M. A. (2022). Fig morphology and development in *Advances in fig Research and sustainable production Flaishman and Aksoy*, 47–58. https://doi.org/10.1079/9781789242492.0004a. CAB International.

Flaishman, M. A., Peers, R., Dvora, N., & Aksoy, U. (2022). Fig diseases in *Advances in fig Research and sustainable production*. In *Flaishman and Aksoy* (pp. 293–307). https://doi.org/10.1079/9781789242492.0017. CAB International.

Flaishman, M. A., Rodov, V., & Stover, E. (2008). The fig: Botany, horticulture, and breeding. *Horticultural Reviews, 34*, 113–196. https://doi.org/10.1002/9780470380147.ch2

Galil, J., & Neeman, G. (1977). Pollen transfer and pollination in the common fig (*Ficus carica* L.). *New Phytologist, 79*, 163–171.

İrget, E. M., Aksoy, U., Okur, B., Ongun, R., & Tepecik, M. (2008). Effect of calcium-based fertilization on dried fig (*Ficus carica* L. cv. Sarılop) yield and quality. *Scientia Horticulturae, 118*, 308–313.

INC. (2023). *Dried figs, statistical yearbook*. https://inc.nutfruit.org_wp-content_uploads_2023_05_Statistical-Yearbook.

International Union for the Protection of New Varieties of Plants of Figs (UPOV Code: *Ficus carica* L.). (2009). Guidelines for the conduct of tests for distinctness, uniformity, and stability. Available online: https://www.upov.int/edocs/mdocs/upov/es/tc/45.

IPGRI and CIHEAM. (2003). *Descriptors for fig. International plant genetic resources institute*. Rome, Italy Paris, France: International Centre for Advanced Mediterranean Agronomic Studies.

Jafari, M., López-Corrales, M., Galán, A. J., Galván, A. I., Hosomi, A., Nogata, et al. (2022). Orchard establishment and management. In *The fig: Botany, production and uses* (pp. 184–230). CABI. GB.

Kamimori, M., Isobe, T., & Yakushiji, H. (2022). Evaluation of ceratocystis canker resistance, vegetative growth, and fruit production of 'Masui Dauphine' fig (*Ficus carica*) grafted on 'Reikodai 1 Go'BC1 of an interspecific hybridization of *F. carica* and. *F. Erecta. The Horticulture Journal, 91*, 337–344.

Karami, H., Rezaei, M., Sarkhosh, A., Rahemi, M., & Jafari, M. J. G. P. (2018). Cold hardiness assessment in seven commercial fig cultivars (*Ficus carica* L.). *Gesunde Pflanzen, 70*, 195–203.

Kislev, M. E., Hartmann, A., & Bar-Yosef, O. (2006). Early domesticated fig in the Jordan Valley. *Science, 312*, 1372–1374.

Lama, K., Chai, L. J., Peer, R., Ma, H., Yeselson, Y., Schaffer, A. A., et al. (2022). Extreme sugar accumulation in late fig ripening is accompanied by global changes in sugar metabolism and transporter gene expression. *Physiologia Plantarum, 174*, Article e13648.

Michailides, T. J. (2003). In R. C. Ploetz (Ed.), *Diseases of fig in* Diseases of tropical fruit crops. CABI Digital Library https://www.cabidigitallibrary.org/doi/pdf/10.5555/20033099835.

Mori, K., Shirasawa, K., Nogata, H., Hirata, C., Tashiro, K., Habu, T., et al. (2017). Identification of RAN1 orthologue associated with sex determination through whole genome sequencing analysis in fig (*Ficus carica* L.). *Sci. Rep., 7*, Article 41124.

Mostowfizadeh-Ghalamfarsa, R., Bolboli, Z., Salmaninezhad, F., & Michailides, T. J. (2022). Fungal, bacterial and nematode diseases in figs in *the Fig: Botany, Production and Uses*. In *Sarhosh, yavari and ferguson* (pp. 292–323). https://doi.org/10.1079/9781789242881.0012. CABI.

Özer, K. B. (2022a). Fig propagation. In *Advances in fig research and sustainable production* (pp. 70–82). https://doi.org/10.1079/9781789242492.0006. CAB International.

Özer, K. B. (2022b). Mycotoxins in figs. In , *Flaishman and AksoyAdvances in fig research and sustainable production* (pp. 318–340). https://doi.org/10.1079/9781789242492.0019. CAB International.

Peer, R., Namdar, D., Modi, A., & Flaishman, M. A. (2022). Conventional and molecular breeding in fig. *Advances in fig research and sustainable production* (pp. 416–438). CAB International. https://doi.org/10.1079/9781789242492.002.

Rønsted, N., Weiblen, G. D., Cook, J. M., Salamin, N., Machado, C. A., & Savolainen, V. (2005). 60 million years of co-divergence in the fig–wasp symbiosis. *Proceedings of the Royal Society B: Biological Sciences, 272*, 2593–2599.

Ramazanian, A., Amiri, A., & Crisosto, C. H. (2022). Postharvest physiology and physiological disorders of fresh fig fruits in *the Fig: Botany, Production and Uses*. In , *Vol. 12. Sarhosh, yavari and ferguson* (pp. 384–397). https://doi.org/10.1079/9781789242881.0016. CAB International.

Rosianski, Y., Freiman, Z. E., Cochavi, S. M., Yablovitz, Z., Kerem, Z., & Flaishman, M. A. (2016). Advanced analysis of developmental and ripening characteristics of pollinated common-type fig (*Ficus carica* L.). *Scientia Horticulturae, 198*, 98–106.

Şen, F. (2022). Postharvest handling of dried figs in *Advances in fig Research and sustainable production*. In *Flaishman and Aksoy* (pp. 232–256). https://doi.org/10.1079/9781789242492.0014. CAB International.

Stover, E., Aradhya, M., Ferguson, L., & Crisosto, C. H. (2007). The fig: Overview of an ancient fruit. *Hort Science, 42*, 1083–1087.

Tırkaz, E., Balcı, B., & Aksoy, U. (2022). World fig production and trade, *Advances in fig Research and sustainable production*. In *Flaishman and Aksoy* (pp. 472–494). https://doi.org/10.1079/9781789242942.0027. CAB International.

Turanlı, F. (2022). Postharvest pests of dried figs. In Flaishman & Aksoy (Eds.), *Advances in Fig Research and Sustainable Production*. Chapter 18. (pp. 308–316). CAB International. http://doi.org/10.1079/97817892492.0018

Trad, M., Ginies, C., Gaaliche, B., Renard, C. M., & Mars, M. (2012). Does pollination affect aroma development in ripened fig [*Ficus carica* L.] fruit? *Scientia Horticulturae, 134*, 93–99.

UN/ECE. (2016). *UNECE Standard DDP-14 concerning the marketing and commercial quality control of dried figs* (p. 7). https://unece.org/fileadmin/DAM/trade/agr/standard. http://dry/dry%20e/14DriedFigs_E2016.pdf.

UPOV. (2010). International Union for the Protection of New Varieties of Plants Guidelines for the Conduct of Tests for Distinctness, Uniformity and Stability Fig, UPOV Code: Ficus_Car *Ficus carica* L. https://www.upov.int/meetings/en/doc_details.jsp?meeting_id=17482&doc_id=177669

USDA. (2001). *United States United States standards for grades of dried figs*. Department of Agriculture. https://www.ams.usda.gov/sites/default/files/media/Dried_Figs_Standard.pdf.

Wang, X., Song, M., Flaishman, M. A., Chen, S., & Ma, H. (2021). AGAMOUS gene as a new sex-identification marker in fig (*Ficus carica* L.) is more efficient than RAN1. *Frontiers in Plant Science, 12*, Article 755358.

Wang, Y., Gu, X., Wu, P., Zhang, L., Deng, C., Tao, T., Zheng, C., & Ma, H. (2022). Comparative analysis of test guidelines for distinctness, uniformity and stability of Fig (*Ficus carica* L.) formulated by China and UPOV. *Journal of Fruit Science, 39*, 2172–2180. https://doi.org/10.13925/j.cnki.gsxb.20220057

Weiss, E. (2015). "Beginnings of fruit growing in the old world"—two generations later. *Israel Journal of Plant Sciences, 62*, 75–85.

Wilson, H., Gordon, P., Faber, B., & Rios, S. (2022). Pest alert-black fig fly, univ. Of calif. Agriculture and natural resources. https://www.cdfa.ca.gov/countyag/postings/files/Wilson_et_al_-_P.pdf.

Zingales, V., Taroncher, M., Martino, P. A., Ruiz, M. J., & Caloni, F. (2022). Climate change and effects on molds and mycotoxins. *Toxins, 14*, 445. https://doi.org/10.3390/toxins14070445

Zohary, D., Hopf, M., & Weiss, E. (2012). *Current state of the art in* Domestication of Plant in the old World (4th ed., pp. 1–8). Oxford: Clarendon. https://doi.org/10.1093/acprof:osobl/9780199549061.003.0001

CHAPTER

Persimmon 10

Cristina Besada[1], Akihiko Sato[2], Takuya Tetsumura[3], Alejandra Salvador[1], Zhengrong Luo[4], Alejandro Tena[1]

[1]*Valencian Institute of Agricultural Research (IVIA), Valencia, Spain;* [2]*Experimental Farm, Kindai University, Wakayama, Japan;* [3]*Faculty of Agriculture, University of Miyazaki, 1-1 Gakuen Kinbanadai-nishi, Miyazaki, Japan;* [4]*Key Laboratory of Horticultural Plant Biology, Huazhong Agricultural University, Wuhan, China*

Chapter outline

1. Origin and spread

Persimmon (*Diospyros kaki* Thunb., 2n = 6x = 90; 9x = 135) is a fruit tree crop, native to China (Yonemori et al., 2000). Although it is unclear from which species persimmon originated, *D. roxburghii* (synonym of *D. glandulosa,* 2n = 2x = 30), which is widespread in Southern China and Southeast Asia, has been reported as the presumed ancestor. *D. glandulosa* has low cold tolerance and the fruit surface is densely hairy, whereas persimmon is found in the temperate zone and has little or no hair on the fruit surface. In Eastern Asian countries, persimmon has been grown as an important fruit crop. From the descriptions in old Chinese documents, it is estimated that the persimmon has been cultivated for at least 2000 years. Currently, persimmon is widely distributed in China, except for the northern provinces. In Japan, there are records of the sale and use of persimmons dated back into the 8th century, while in Korea, it is thought to have been introduced around the 14th century.

Persimmon seems to have diffused into Western areas from East Asia through caravans along the Silk Road in ancient times, but large scale of the diffusion in

Temperate Tree Fruits and Nuts. https://doi.org/10.1016/B978-0-443-29137-1.00010-4

the Mediterranean basin has been relevant only in the last two centuries. Spanish and Portuguese colonial trades and immigrants from oriental countries contributed to the expansion of persimmon to Europe and Oceania (Giordani, 2022). In the 19th century, persimmon was introduced to France and were cultivated in Provence and other Mediterranean basin regions (Bellini & Giordani, 2005). It was also introduced to Italy in the 19th century and cultivated in the Florence region. Initially, persimmon was introduced not for fruit use but as an ornamental tree in Europe (Badenes et al., 2013). After that, in Spain, it was cultivated on a small scale as a home fruit tree, and then several local cultivars were developed through selection from chance seedlings, bud mutation, and artificial selection. Persimmon is thought to have been introduced to the Balkan Peninsula in the middle of the 20th century.

Persimmon was introduced into the United States in 1856, and several cultivars were imported through the U.S. Department of Agriculture (USDA) in the 1870s. Japanese and Chinese immigrants also brought many cultivars into California. In Brazil, persimmon was introduced from France at the end of the 19th century. At the beginning of the 20th century, Japanese people began to immigrate to Brazil. Several Japanese cultivars, including Fuyu, Hiratanenashi, and Hachiya were introduced directly by Japanese immigrants. Thus, Japanese and Chinese immigrants played a key role in the development of persimmon cultivation in the United States and Brazil.

2. Economic importance

Persimmon cultivation has been prevalent in Asian countries, mainly in China, Republic of Korea, and Japan (Table 10.1). According to the latest Food and Agriculture Organization (FAO) statistics (FAOSTAT Database, 2022), the world persimmon production was estimated at 4,436,474 tons in 2022. China is the largest producer, accounting for more than 70% of world persimmon production (3,470,450 tons) yet mainly destined for the domestic market.

Production has also spread to temperate and subtropical regions of the world. The Mediterranean basin has a suitable climate for the crop, and this area has experienced the greatest expansion of persimmon production in recent decades. The most significant expansion took place in Spain, which has recently reached a production of more than 430,000 tons, becoming the second largest producer country after China (MAPA, 2022). Italy and Israel are also notable for their significant production of persimmons in the Mediterranean region. In the Americas, Brazil boasts the longest tradition of persimmon cultivation, currently ranking sixth in terms of production volume (164,439 tons). In recent years, South American countries such as Peru and Chile have also started growing persimmons, recognizing the opportunity to supply fruit to the Northern Hemisphere during the off-season. The production of persimmons in the U.S. remains modest, with no official data available. In Oceania, Australia and New Zealand have a long experience of growing persimmons,

Table 10.1 Evolution of persimmon production (in tons) since 2000 in the main producing countries. List of countries in order of highest to lowest production in 2022

	Year					
Country	**2000**	**2005**	**2010**	**2015**	**2020**	**2022**
China	1,615,791	2,212,111	2,647,501	3,010,529	3,344,505	3,470,425
Spain	33,000	51,950	125,280	268,764	–	**436,015***
Rep. of Korea	287,847	363,822	390,630	303,526	198,817	231,523
Japan	–	285,900	189,400	242,000	193,200	216,100
Azerbaijan	70,266	108,965	142,188	146,566	185,247	184,291
Brazil	63,300	164,849	167,215	192,327	158,762	164,439
Uzbekistan	16,000	21,000	38,000	70,000	85,100	80,724
Italy	42,450	51,332	48,165	47,322	**48,114***	**54,021***
Irán	4347	10,851	10,199	34,405	30,683	30,472
Israel	14,206	48,000	28,201	25,000	21,908	30,000
Pakistan	–	–	–	–	–	21,799
Nepal	360	443	690	2401	2574	2919
Oceania	3473	3681	3289	3189	122,767	2807
New Zealand	2821	3000	2594	2474	1850	2060
Australia	652	681	696	715	740	747
Chile	356	450	437	451	614	647
Mexico	247	369	363	141	198	234
Bhutan	–	618	129	135	207	95
Slovenia	–	499	805	2138	–	–

Due to the absence of data in the FAO statistics, production data from the official agriculture institutions of the country are presented. (MAPA, 2022; ISTAT, 2022) although production volumes are not very high. On the African continent, persimmon production is still low.

In China, the world's largest producer, persimmons are primarily grown for the domestic market and have traditionally been eaten as soft fruit. Nowadays, postharvest treatments that allow the marketing of firm fruit are being implemented in China and other Asian countries. In Europe, the primary commercialization of persimmons is as firm fruit, with exportation playing a key role. In Spain, the leading European producer, over 80% of the production is being exported. Indeed, much of the expansion of the persimmon industry in recent decades has been driven by export market opportunities, with Spain being the major supplier of persimmon to Europe. The introduction of postharvest technology at the end of the last century, which allows the removal of astringency while preserving firmness, and other technologies that help delay chilling damage during cold storage, were the main triggers for the expansion of persimmon exports.

The growing awareness among consumers regarding the impact of their food choices on health has also contributed to the steady demand for persimmons in markets. Its nutritional profile, rich in vitamin C, antioxidants, and fiber, makes it a healthy choice for health-conscious consumers. In Asian countries, where persimmon consumption has a long tradition, the commercialization of persimmon-derived products, such as dried persimmons or persimmon vinegar, is significant. In countries where the crop has been introduced more recently, consumption is almost exclusively based on the fresh product. However, recent studies in Spain have shown high consumer acceptance of dried persimmon slices and persimmon chips. Therefore, this type of product could find a place in the European market in the near future and be a good alternative to add value to production surpluses or fruit discarded for aesthetic reasons (Castillo et al., 2023; Llorca et al., 2023; Cervera-Chiner et al., 2024).

3. Botany and relevant species

The genus *Diospyros* consists of ca. 500 species (Zhang & Luo, 2022). This genus is distributed mainly in the tropical and subtropical areas, including tropical Asia, Central to South America, Africa, and the island of Madagascar, except for some species that are distributed in temperate zones. This genus is the largest genus in the *Ebenaceae* family. Some tropical species accumulate tannins not only in their fruits but also inside the wood so their wood material turns black, such as ebony. Therefore, these species are used as high-quality wood and craft material. Some species such as persimmon *(D. kaki)*, date-plum *(D. lotus)*, and American persimmon *(D. virginiana)* are distributed in temperate zones. Among these three species, persimmon *(D. kaki)* is the most economically important, as most of the commercialized cultivars belong to it. *D. lotus* and *D. virginiana* are mainly used as rootstocks.

Persimmon (*Diospyros kaki* Thunb.): Persimmon is a diclinous deciduous tree. Many commercially important cultivars such as Fuyu, Hiratanenashi, and Jiro in Japan; Rojo Brillante in Spain; Triumph (Sharon fruit) in Israel; and Tamopan in

China bear only female flowers. Cultivars such as Zenjimaru and Akagaki bear both female and male flowers and are used as pollinizers in Japan.

A particularity of persimmon compared to other fruits is that the fruit accumulates tannins in the flesh during fruit growth (Yonemori et al., 2000). Some persimmon cultivars remain astringent at commercial maturity while others undergo a natural loss of astringency. Thus, persimmon cultivars are classified into two types: the astringent type and the nonastringent type. Hume (1914) classified persimmon cultivars into two groups based on differences in flesh coloration linked to pollination (seed presence): (1) **pollination-constant** (PC), when the fruit flesh does not change color by seed formation, and (2) **pollination-variant** (PV), when the flesh darkens by seed formation. Actually, this classification is not only related to a change in flesh color but also to the loss of astringency. Thus, the browning or darkening of the flesh has been associated with a process of tannin insolubilization and the subsequent loss of astringency, which is linked to the production of volatile compounds in the seeds. Combining these two classification methods, persimmon cultivars have been commonly classified into four types (Table 10.2) as follows:

- **Pollination constant and nonastringent (PCNA):** A cultivar whose astringency disappears on the tree regardless of number of seeds, and the color of the flesh does not change.
- **Pollination-variant and nonastringent (PVNA):** A cultivar whose astringency disappears on the tree when fruit contains seeds. When the fruit is seeded, large amounts of brown specks will appear in the flesh, turning the flesh brown and leading to the disappearance of astringency. If the fruit contains few seeds, brown specks occur only around the seeds so that flesh remains astringent.
- **Pollination-variant and astringent (PVA):** A cultivar whose flesh produces small number of brown specks only around the seeds so that flesh remains astringent regardless of seed number.
- **Pollination constant and astringent (PCA):** A cultivar whose flesh remain astringent. The flesh does not contain brown specks even if seed are produced. Consequently, flesh color does not change.

Table 10.2 Classification of different cultivars based on their pollination and astringency patterns

Cultivar type			
	Non-PCNA		
PCNA	**PVNA**	**PVA**	**PCA**
Fuyu Jiro Gosho Luotian Tianshi	Zenjimaru Nishimurawase Kaki Tipo	Hiratanenashi Rojo Brillante Triumph	Saijo Tamopan Ichidagaki

Out of the four types of persimmon cultivars, only the PCNA type undergoes a natural loss of astringency at commercial maturity, regardless of the number of seeds. Thus, it is common to distinguish between PCNA and non-PCNA cultivars. The non-PCNA (PCA, PVA, and seedless fruit of PVNA) cultivars must be allowed to overripe on the tree or undergo postharvest treatment to ensure the loss of astringency. For this reason, PCNA types are the most important at the commercial level in several countries because the natural loss of astringency facilitates handling.

PCNA cultivars are divided into two types based on their origin: Japanese PCNA cultivar (J-PCNA) and Chinese PCNA cultivars (C-PCNA). J-PCNA and C-PCNA have different origins, and the mechanisms of natural astringency removal on trees are also different between them. Until the 16th century, old documents described only non-PCNA cultivars. In the 17th century, J-PCNA cv. Gosho was first documented in the literature. In the early 20th century, the Agricultural Research Station collected and investigated persimmon cultivars across Japan. Among them, Fuyu and Jiro were recommended for cultivation as superior cultivars. These J-PCNA cultivars then spread around the world. Fruit from all J-PCNA cultivars are fat in shape, possess a late ripening time, and are prone to cracking. Therefore, it is assumed that J-PCNA cultivars are related to each other. C-PCNA cultivars were initially distributed near the Dabie Mountains, which straddles Luotian and Macheng counties in the Hubei Province in China (Wang et al., 2005; Yonemori et al., 2005). The cultivar Luotian Tianshi is estimated to have been cultivated for over 900 years. In addition to Luotian Tianshi, several C-PCNA cultivars have been found in this area.

***Date-plum (Diospyros lotus* L.):** Date-plum ($2n = 2x = 30$) is a deciduous and dioecious tree that is widely distributed from West to East Asia, including Japan, China, and Iran (Yonemori et al., 2000). The fruits are small and round to oval in shape. Young fruits are astringent, but when overripe, the astringency disappears and they become edible. Rootstocks of this species are cold tolerant and for that reason is used in cold regions of Japan and China. This species, however, is graft-incompatible with some cultivars such as Fuyu.

***American persimmon (Diospyros virginiana* L.):** American persimmon (common persimmon, $2n = 4x = 60$ or $6x = 90$) is a cold-resistant species native to the eastern United States (Yonemori et al., 2000). The trees are dioecious, but recently monoecious type was produced through breeding programs. The fruit is very similar to date-plum. Similar to other species, the fruits are astringent, but the astringency disappears when the fruits are overripe. In the United States, the fruits are eaten as overripe or dried fruit and are also used in jams and pies. Because American persimmon has tolerance to soil salinity, they are used as rootstocks for persimmon cultivation in semi-arid areas such as Spain (Gil-Muñoz et al., 2018).

Other temperate species include *Diospyros oleifera* ($2n = 2x = 30$) and *Diospyros rhombifolia* ($2n = 4x = 60$ or $6x = 90$). *D. oleifera* is native to China and is mainly grown for purposes of extracting persimmon oil, but in northern Vietnam, the overripe fruit is eaten. *D. rhombifolia* is also native to China, and the tree is dwarf with very small leaves and fruits. It is used for ornamental purposes (Yonemori et al., 2000).

4. Breeding and propagation

Breeding: A common objective of all persimmon breeding programs is to obtain nonastringent fruit (PCNA type) of superior quality. This means good size and appearance, high organoleptic quality, resistance to fruit cracking, early or late ripening, long shelf-life, high productivity, high parthenocarpy ability, and resistance to various pests and diseases. In the case of Mediterranean-producing countries where production is concentrated on a single astringent cultivar, consideration is also being given to broadening the varietal range with astringent cultivars of higher quality that can extend the commercial season.

PCNA cultivars show qualitative inheritance while the differences within non-PCNA (PVNA, PVA, and PCA) cultivars show quantitative inheritance. Since most of the existing breeding programs are focusing on enhancing the genetics of PCNA-type cultivars, including Japanese PCNA (J-PCNA) and Chinese PCNA (C-PCNA), this section will focus particularly on the breeding of this type of cultivar. Recent advancements in persimmon genetic improvement worldwide are described in detail elsewhere (Sato, 2022).

For many years, the introduction of new cultivars came only from the selection of seedlings resulting from natural hybridization and bud mutation. Cultivars as important as Fuyu, Hiratanenashi, Rojo Brillante, or Triumph have such origins. Japan was the first country where a breeding program of persimmon was established. Thus, persimmon breeding in Japan was initiated in the 1930s. Later, programs were established in other producer countries like China, Korea, Italy, Spain, and Brazil. Since the Japanese breeding program is the oldest one, it is interesting to comment on how its approach has evolved, as it gives us an idea of the evolution of genetic improvement in persimmon.

J-PCNA cultivars are the predominant PCNA type worldwide, and Japanese breeding programs have focused on releasing superior PCNA cultivars. Given that the PCNA trait in J-PCNA cultivars is recessive to non-PCNA genotypes (Ikeda et al., 1985; Yamada & Sato, 2002), crosses among J-PCNA genotypes have become the primary method for producing PCNA cultivars. However, the number of superior PCNA cross-parents has been limited, leading to inbreeding depression in the late 1980s, particularly in fruit size, tree vigor, and productivity (Sato & Yamada, 2016). To address this, a backcross program using PCNA ([non-PCNA × PCNA] × PCNA) was initiated in 1990. This process was inefficient, with only 15% of the offspring being PCNA (Ikeda et al., 1985), and all offspring had to be grown to the fruiting stage. To mitigate this inefficiency, molecular markers linked to the PCNA/non-PCNA locus were developed in Japan for discriminating PCNA and have been applied to practical breeding (Kanzaki et al., 2010).

Breeding new cultivars through hybridization is a slow process. In theory, it takes 12 to 17 years (Figure 10.1), but in practice, it can take up to 25 years. To date, the release of new persimmon cultivars developed through crossbreeding has happened only in Japan and Korea (Sato & Yamada, 2016). To date, the National Agriculture and Food Research Organization (NARO) in Japan has released 12 PCNAs,

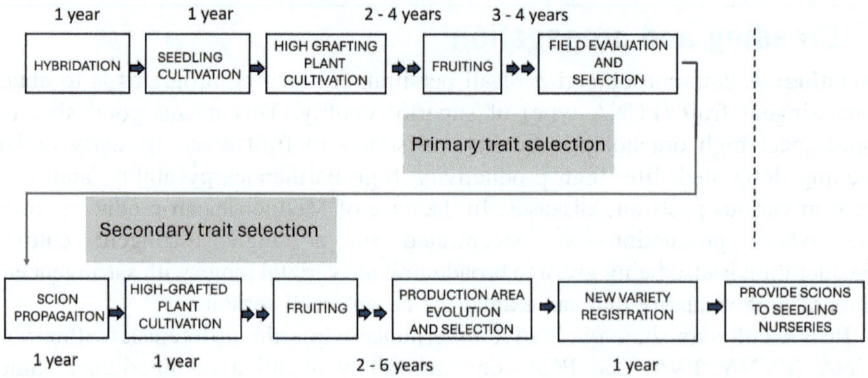

FIGURE 10.1

A protocol for genetic improvement of J-PCNA persimmon. (Yamada, 2011)

including high-quality J-PCNA cultivars like Taishu and two pollination variant and astringent (PVA) cultivars for table use. Korea has released six PCNA cultivars obtained by J-PCNA × J-PCNA crosses (Table 10.3).

The C-PCNA cultivars differ in origin and astringency removal mechanism from J-PCNA cultivars and dominates the non-PCNA trait (Ikegami et al., 2006). Therefore, crossing C-PCNA × non-PCNA or J-PCNA is highly interesting, and it has been the major breeding strategy in the Chinese breeding program established at Huazhong Agricultural University (HZAU) in China. In this program, a reliable protocol involving traditional cross-embryo rescue, astringency, and sexuality identification by marker-assisted selection (MAS), *in vitro* shoot top-grafting, and simultaneous evaluation in multiple nursery lands and production areas was established for rapid genetic improvement of C-PCNA (Xu et al., 2017).

These techniques will make it possible to obtain high-quality C-PCNA cultivars in the not too distant future. However, until now, the obtention of C-PCNA cultivars in China primarily result from seedling variation. Examples include the PCNA type Eshi 1 for fresh-eating and the androecious-type Male 8 as pollen donor. HZAU, Northwest A&F University, and the Non-timber Forestry Research Institute of the Chinese Academy of Forestry Sciences have also developed new cultivars through hybrid breeding (Table 10.4).

Over the years, approaches other than hybridization have been followed. Thus, for example, gamma irradiation of the astringent cultivar Rojo Brillante was assayed in the Spanish breeding program.

In addition to breeding cultivars, persimmon breeders have also focused on rootstocks. Grafting persimmons onto rootstocks is a common practice aimed at improving orchard management and increasing yield. Traditionally, rootstocks from *Diospyros* species such as *D. kaki*, *D. lotus*, and *D. virginiana* have been used, typically sourced from seedlings, leading to genetic variability. Efforts have been directed toward developing clonal rootstocks through breeding programs.

Table 10.3 The persimmon scion cultivars released by NARO*, Japan, and Korea

Cultivar	Type	Parents	Fruit weight (g)	Sugars (%)	Maturation period	Origin	Reference*
Suruga	PCNA	Hanagosho × Okugosho	250	17	Nov. 21-Dec. 10	Japan	Iikubo et al. (1961)
Izu	PCNA	Fuyu × Okitsu-1	240	15	Late Oct.	Japan	Hirose et al. (1971)
Shinsyu	PCNA	Okitsu-20 × Okitsu-1	240	17~18	Late Oct.	Japan	Yamane et al. (1991a)
Yoho	PCNA	Fuyu × Jiro	280	16~17	Beginning Nov.	Japan	Yamane et al. (1991b)
Taishu	PCNA	Fuyu × II iG-16	400	17	Beginning Nov.	Japan	Yamane et al. (2001)
Yubeni	PCNA	Matsumoto-wase Fuyu × F-2	280	17.5	Late Nov.	Japan	Yamada et al. (2003)
Soshu	PCNA	Izu × 109-27	250	14~15	Sep. 21-Oct. 10	Japan	Yamada et al. (2004)
Kanshu	PCNA	Shinsyu × 18-4	248	18	Late Oct.	Japan	Yamada et al. (2006)
Kishu	PCNA	Izu × Akitsu-5	350	16	Late Oct.	Japan	Yamada et al. (2009)
Taiho	PCNA	Okitsu-20 × Taishu	336	16	Mid to Late Nov.	Japan	Sato et al. (2018b)
Taiga	PCNA	Kanshu × Akitsu-19	324	16-17	Mid Oct.	Japan	Sato et al. (2019a)
Rigyoku	PCNA	Kanshu × Akitsu-19	278	18	Mid Oct.	Japan	Sato et al. (2019b)
Taigetsu	PVA	Kurokuma × Taishu	447	14.6	Beginning Nov.	Japan	Yamada et al. (2012b)
Taiten	PVA	Kurokuma × Taishu	450	16	Mid Nov.	Japan	Yamada et al. (2012c)
Jowan	PCNA	Shinshu × Taishu	200	16.4	Late Sep.	Korea	Ma et al. (2018)
Romang	PCNA	Ro-19 × Oku-gosho	185	18.6	Mid Sep.	Korea	Ma et al. (2018)
Wonmi	PCNA	Fuyu × Taishu	220	15.1	Mid Sep.	Korea	Ma et al. (2020)
Gampung	PCNA	Daeandangam × Taishu	413	14.7	Late Oct.	Korea	Ma et al. (2020)
Dannuri	PCNA	Danyeon 104 × Taishu	320	18.4	Mid Oct.	Korea	Kim et al. (2021)
Wonchu	PCNA	Shinshu × Taishu	324	15.1	Beginning Oct.	Korea	Ma et al. (2021)

The references in the table can be found in Sato (2022).

Table 10.4 Preliminary progress in persimmon genetic improvement in China

Cultivar	Source or main characteristics	References
Eshi 1	Natural hybrids, PCNA	Zhang et al. (2013)
Chuantianshi 1	Jiro bud mutation, PCNA	Chen et al. (2006b)
Chuantianshi 2	Jiro bud mutation, PCNA	Chen et al. (2006a)
Gold 1	Yoho bud mutation, PCNA	Chen et al. (2021)
Huangjin-fangshi	Niuxinshi × Zenjimaru, non-PCNA	Zheng (2012)
H1-36-1	Huashi 1 × Luotian Tianshi, PCNA	Unpublished data
H9-1	Eshi 1 × Male 1, PCNA	Unpublished data
Male 8 (Pollen donor)	Natural hybrids, androecious	Zhang et al. (2018)

**The references in the table can be found in Luo et al. (2022).*

Japan's NARO focuses on breeding dwarfing rootstocks. Desired traits for persimmon dwarfing rootstocks include graft compatibility, high yield efficiency, reduced biennial bearing, precocity, ease of propagation, fruit quality maintenance or enhancement, adaptability, disease and pest resistance, and longevity. In China, rootstock compatibility with PCNA cultivars is a priority, reflecting regional preferences. Significant advancements in persimmon rootstock improvement have been made in Japan and China (Table 10.5). Other programs aim to enhance adaptability to climate change. Salinity tolerance is increasingly important in certain regions,

Table 10.5 Genetic improvement for rootstock of persimmon in Japan and China

Cultivar	Source or main characteristics	Origin	References
MKR1	Dwarf	Japan	Tetsumura et al. (2013)
Shizukadai-1	Dwarf	Japan	Hattori et al. (2015)
Shizukadai-2	Dwarf	Japan	Hattori et al. (2015)
Hourakudai	Dwarf	Japan	Yakushiji et al. (2021)
Yalin 6	Compatible with Fuyu	China	Wu et al. (2021)
Xiaoguo-tianshi	Compatible with Fuyu scions	China	Hu et al. (2017)
L824	*D. lotus* seedling, compatible with Fuyu	China	Shi et al. (2020)
L848	*D. lotus* seedling, compatible with Fuyu	China	Shi et al. (2020)
Taifu	Compatible with Fuyu scions, cold resistance	China	Lü et al. (2021)
Nantong-xiaofangshi	Dwarf	China	Dong et al. (2019)

driving demand for tolerant rootstocks. This is the case of Spain, where a rootstock breeding program emphasizing salinity resistance has been established (Badenes et al., 2022).

Propagation: Persimmon propagation of cultivars by cuttings or layering has proven to be very difficult, so they are usually raised by grafting or budding on seedlings. There are three main rootstock species. Seedlings of *D. kaki* have taproots, few fine roots, and are deep-rooted and have high grafting affinity for all cultivars, although they are highly sensitive to lime-filled soils. For this reason, *D. kaki* seedling rootstock are used in regions where sweet persimmons are grown, namely in Japan, South Korea, New Zealand, and Australia. Seedlings of *D. lotus* grow better than those of *D. kaki*; are more cold-, drought-, and alkali soil-tolerant; and are being used all around the world, including East Asia, the Middle East, the Mediterranean region, and Eastern Europe. However, they have low grafting compatibility with some persimmon cultivars, low salt tolerance, and susceptibility to crown gall disease. Seedlings of *D. virginiana*, a native of the United States, are used in Israel because of their strong salt tolerance and are well adapted to damp soil. However, they produce many root suckers and are not always uniform in size or vigor.

In terms of rootstock propagation, historically, the primary method has been seedlings produced by cross-pollination, leading to genetic heterogeneity among orchard trees. To address this issue, clonal rootstocks from tissue culture and stem or root cuttings were intensively developed at the end of the 20th century (Tetsumura et al., 1991; Tao & Sugiura, 1992). Advances in cutting propagation techniques have improved success rates, facilitating rapid multiplication of rootstocks. Additionally, tissue culture methods such as shoot tip culture and micropropagation provide precise control over plant growth and development. Shoot tip culture initiates cultures from actively growing shoots or dormant winter buds, while micropropagation multiplies shoots on nutrient media supplemented with cytokinins. Techniques like green branch mist propagation and late June planting of leaf-bud cuttings from root suckers have been recommended for specific rootstock varieties like Miyazaki kaki rootstock 1 (MKR1) (Hejazi et al., 2018). Moreover, new rootstock types such as Xiaoguo-tianshi from Hubei's Dabieshan Mountain area are gaining popularity for their grafting compatibility with popular cultivars like Taishu and Fuyu. Research from institutions like HZAU has demonstrated rapid propagation of Xiaoguo-tianshi through *in vitro* culture (Liu et al., 2023). These methods enable the mass production of genetically identical rootstocks, ensuring uniformity in orchard establishment and management.

5. Orchard management

Persimmon tree is adaptable to a wide range of climatic conditions. Temperate and subtropical regions are suitable for growing persimmons. However, because of its subtropical origin, its cold tolerance is weaker than that of apples, and trees are often damaged by winter temperatures below -10°C. Moreover, high temperatures and long hours of sunlight are required to completely lose astringency in fruits of non-astringent cultivars (PCNA), and average annual temperatures of 14°C-15°C or higher are desirable for their cultivation. Persimmons have such low chilling requirement that they can be grown even in subtropical regions, so lack of winter cooling is rarely a problem, and delayed defoliation in winter is not an issue. Although snow covering the ground in winter is no problem, the young shoots in spring are susceptible to frost, so growing persimmons, especially cultivars that sprout earlier, should be avoided in areas and sites that are prone to late spring frosts. Early fall frosts may cause skin blemishes on the fruit.

Persimmon trees are very susceptible to wind, so they need to be protected in areas where strong winds blow during the growing season. The young leaves are tender in spring, easily damaged by wind, infected with *Botrytis cinerea*, which causes defoliation in summer. Strong winds in late summer and fall cause defoliation, which reduces sugar content and other fruit qualities.

Persimmon trees grow well in a range of soil. Although it tolerates clay better than other fruit trees, it grows best in deep, fertile, well-drained soils with a pH of 5.5-6.5. When growing persimmons in saline soils, special care must be taken when selecting rootstocks as they are susceptible to chlorine damage. It should be noted that persimmon trees, unlike citrus, are sensitive to soil moisture deficiency (Ballester et al., 2022), and fruit yield and size is affected by the amount of water applied (Kanety et al., 2014).

The most widely used method for calculating irrigation requirements is that proposed by the Food and Agriculture Organization of the United Nations (FAO) in its FAO Irrigation and Drainage Paper no. 56 (Crop Evapotranspiration: Guidelines for Computing Crop Water Requirements), which estimates water requirements using a procedure that considers two factors: (1) climatic variables (air temperature and humidity, solar radiation, and wind speed) that affect evaporative demand or reference evapotranspiration (ETo) and (2) a factor related to the crop, called crop coefficient (Kc). In this model, the crop water requirement or evapotranspiration (ETc) is calculated as ETc = ETo * Kc. At present, no scientific studies have been carried out to quantify the Kc of persimmon. However, a number of empirical experiments have been carried out, which have made it possible to define preliminary (experimental) crop coefficients and therefore indicative water requirements.

The crop coefficient obtained (Figure 10.2) varies over the course of the season and the age of the orchard, because the older the trees, the greater the water consumption. The highest Kc values occur in the months of August and September (Northern Hemisphere), during a period of rapid linear fruit growth and when the tree has completed its vegetative development; during these late summer months, Kc can reach values close to unity. Thus, it is possible to estimate the irrigation needs of the persimmon in a fairly approximate way. Bearing in mind the parameters involved in the calculation, the methodology proposes a weekly calculation of irrigation needs to consider both weather variations and those of the crop itself. Supplying the tree's water needs through drip irrigation (Figure 10.3) enables a more controlled and efficient water usage compared to traditional flood irrigation.

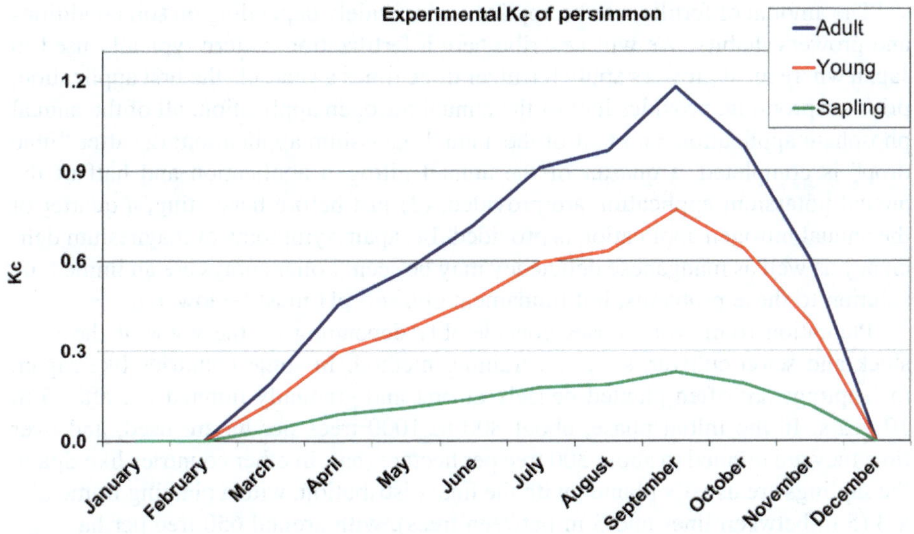

FIGURE 10.2

Temporal evolution of the experimental crop coefficient (Kc) of persimmon as a function of the age of the plantation and the months of the year. Experiment performed in Spain with Rojo Brillante persimmon. *Figure courtesy Luis Bonet (IVIA).*

FIGURE 10.3

Drip irrigation at Rojo Brillante orchard in Valencia, Spain. *Photo courtesy Luis Bonet (IVIA).*

The amount of fertilizer to be applied varies widely depending on soil conditions and growers' habits. We will describe here a fertilization pattern typically used in Japan where most growers apply fertilizer three times a year: (1) the first application, prior to sprouting, provides half of the annual nitrogen application, all of the annual phosphate application, and half of the annual potassium application; (2) after "June drop" is completed, a quarter of the annual nitrogen application and half of the annual potassium application are provided; (3) just before harvesting, a quarter of the annual nitrogen application is provided. In Japan, symptoms of magnesium deficiency as well as manganese deficiency may be seen. Foliar sprays are an immediate solution to these problems, but fundamentally, soil pH must be lowered.

Plantation framework varies considerably depending on the vigor of the rootstock and scion cultivar, soil, and training method. In some countries like Japan, the saplings are often planted densely at first and gradually thinned out after 5 to 10 years. In the initial phase, about 800 to 1000 trees per ha are used, and over time they are thinned to about 300 tree per hectare (ha). In other countries like Spain, the saplings are usually planted with the final distribution, with a planting frame of 5 x 3 (5 m between lines and 3 m between trees), with around 650 tree per ha.

When allowed to grow freely, persimmon trees are more or less spherical in shape. Therefore, they are often trained in a modified center leader system with well-shaped lateral branches; they are also trained in an open-center system. Since dwarf rootstocks are not yet available, the use of pergolas to maintain low tree height is an alternative, and this practice has been adopted in several orchards in Japan (Figure 10.4).

FIGURE 10.4

Fuyu trees trained to the pergolas in Gifu, Japan. *Photo by Takuya Tetsumura.*

In Italy, most fruit trees, including persimmons, are grown on the oblique-branch palmette system. This tree form is superficially flattened, making it easier to access equipment, especially during harvesting (Figure 10.5). Persimmon trees appear to be easy to train to the palmette system, and high yields can be expected by using highly productive cultivars.

The most important point in pruning persimmon trees is that normally only the top two or three buds at the end of the season's growth contain flower buds. If the shoot is heavily cut back, it will not flower, and severe pruning will force excessive vegetative growth and reduce the yield. As for cultivars that are not planted as pollen donors, such as the sweet persimmon Taishu, shoots bearing male flowers should be removed completely, as most of them produce shoots with male flowers, reducing the yield (Yonemori et al., 1992).

In most countries, statistical data on yields show 10 t/ha for sweet persimmons and 20 t/ha for astringent persimmons, but the actual yields vary greatly depending on the cultivar. Some nonastringent cultivars yield more than 20 t/ha, while astringent cultivars such as Kaki Tipo yield 30 to 40 t/ha and Triumph 40 to 50 t/ha or more. The yield varies greatly depending on cultivation management, soil, climate, and other factors.

In some countries, thinning is a habitual practice to improve fruit quality, mainly size and sugar content, and to reduce biennial bearing, which is a major problem in certain persimmon cultivars. Usually, 20 or more leaves per fruit are recommended to obtain good fruit size, but large fruit cultivars such as Taishu require 30 leaves per fruit to produce a good-quality fruit crop. Thinning has the desirable effect of

FIGURE 10.5

Harvesting Kaki Tipo trained to the palmette system in Faenza, Italy. *Photo by Takuya Tetsumura.*

eliminating pest prone areas between adjacent fruit. Therefore, in some countries like Japan, the fruit is usually thinned by hand so that there is only one fruit per bearing shoot. The best time to thin persimmons is before flowering rather than at the fruitlet stage. However, because the early physiological drop of persimmon fruitlet is sometimes severe, the number of flowers to be left is the double or triple the number of fruits to be harvested. The end of "June drop" is the start of the final thinning. In other countries like Spain, thinning is not a habitual practice.

6. Fruit maturation and harvest criteria

Fruit growth in weight and diameter follows a typical double-sigmoid curve, consisting of two rapid growth stages, stage I and stage III, separated by a period of slow growth (stage II) (Zheng et al., 1990). Persimmon maturation has been well studied. Although initially there was some doubt as to whether or not they were climacteric, they are now undoubtedly classified as climacteric on the basis of their respiratory activity and ethylene emission pattern. They produce a relatively small (below 5 nL $g^{-1}h^{-1}$) but significant amount of ethylene during ripening. They are also very sensitive to exogenous ethylene, which even at low levels triggers fruit ripening, accelerating softening and in certain scenarios aggravates postharvest disorders reducing shelf-life and commercial value (Besada et al., 2010a).

Similar to numerous other commodities, the most representative event observed during persimmon ripening is the change in external coloration. In parallel with chlorophyll degradation and carotenoid synthesis, persimmon skin transitions from a uniform green color to yellow-orange or orange-reddish tones upon reaching commercial maturity. The intensity of the orange/red tone depends on the variety (Woolf & Ben-Arie, 2011). Concurrent with the process of getting red coloration during ripening, the flesh also shifts from white to orange, with certain cultivars displaying a more pronounced orange tone than others. Cryptoxanthin predominates among the carotenoids found in various persimmon cultivars, alongside lutein, violaxanthin, zeaxanthin, and β-carotene (Del Bubba et al., 2009).

Flesh firmness declines in parallel to fruit coloration. The gradual softening of the fruit is attributed to microstructural changes of the flesh, involving the progressive degradation of parenchyma cells, resulting in diminished cell volume and increased cell deformation. With advancing maturity, there is a notable degradation of the middle lamella, accompanied by the loss of cell wall constituents and the leakage of solutes into intercellular spaces, culminating in a general loss of intercellular adhesion in later stages of maturity (Salvador et al., 2007; Tessmer et al., 2016).

Both astringent and nonastringent persimmon cultivars exhibit high levels of astringency when immature. During maturation, both types experience a reduction in soluble tannins, but the extent of this reduction varies. Nonastringent cultivars gradually decrease their soluble tannin content until it becomes sensorially imperceptible in the early stages of orange coloration. In contrast, astringent cultivars exhibit a slower polymerization of tannins, maintaining a soluble tannin content of ca. 1% at commercial maturity stage (homogeneous orange coloration), which corresponds to high sensory astringency levels (Tessmer et al., 2014). Most astringent cultivars undergo a total loss of astringency when overripe, as the fruit becomes very soft. However, some cultivars retain perceptible astringency even when overripe. For example, in the Rojo Brillante cultivar, astringency becomes undetectable as firmness declines to values close to zero, with tannin content dropping to approximately 0.04%. In contrast, the Giombo cultivar, when soft, retains about 0.5% soluble tannins, and its astringency remains perceptible (Tessmer et al., 2016). Persimmon maturation also entails chemical changes that contribute to its unique flavor, including sugar accumulation and accumulation of volatile compounds. Glucose, fructose, and sucrose are the predominant sugars in persimmons, increasing steadily throughout fruit development until reaching a plateau before harvest maturity (Besada & Salvador, 2018).

Due to the close correlation between changes in skin coloration and the physicochemical changes occurring during fruit ripening (Salvador et al., 2007), in practice, external fruit coloration is the most used parameter as a nondestructive harvesting index in persimmons. The commercial quality standards established by the Working Party on Agricultural Quality Standards of the United Nations Economic Commission for Europe (UNECE, 2020) aim to facilitate international trade, encourage high-quality production, and protect consumer interests. These standards specify that the maturity and development of persimmons at harvest must allow the fruit to continue ripening and reach a satisfactory degree of ripeness. The minimum requirement for commercialization established by this commission is that at least the lower one-third of the fruit should be yellow, or the fruit should be in the process of turning color. Although this is the basic requirement, in practice, most persimmon cultivars are suitable for harvesting once they have attained a uniform yellow/pale orange coloration. Utilizing a color chart can be useful in determining the optimal harvest time (Figure 10.6).

Sugar content is also a parameter used in the harvesting of certain persimmon cultivars, especially nonastringent ones, although it should be noted that the same variety may behave differently depending on the climatic conditions where it is grown. For example, in some areas of Australia, the Fuyu cultivar is recommended to be harvested when the fruit has reached a soluble solid content (SSC) of 15°Brix, whereas in Japan, this cultivar is typically harvested with a higher SSC, closer to 18°Brix.

The use of SSC as a maturity index is discouraged for astringent cultivars because this measure includes soluble tannins, making it a poor indicator of sugar content in this type of cultivar. Although it is not usually used as a harvest index, fruit firmness at the time of harvest plays a key role in postharvest behavior and is a parameter that must be considered when deciding on the correct handling of the fruit (Vilhena et al., 2022a).

To prolong the harvesting period and improve commercialization, various preharvest treatments have been evaluated. For instance, field applications of gibberellic acid (GA_3) at the onset of color change have proven highly effective in delaying harvest in certain persimmon varieties (Ben-Arie et al., 1996; Besada et al., 2008).

FIGURE 10.6

Color chart for persimmon fruit used in certain Asian countries. *Photo by Alejandra Salvador and Cristina Besada.*

Ethephon can be used to accelerate fruit ripening on the tree but can be risky as it shortens postharvest life (Vilhena et al., 2022b).

7. Postharvest physiology and technology

Once harvested, persimmon fruit are transported to packing houses where they are subjected to handling operations prior to be commercialized. In these packing houses, deastringency treatment will be carried out in the case of astringent cultivars and the fruit will be cold stored if required. In addition, the packaging and preparation of the fruit for marketing will be performed in the packing line.

Packing lines are used to wash, treat, grade, and pack fruit. Special care must be taken in the design of packaging lines since persimmon fruit are very sensitive to mechanical damage. Impacts of fruit on the line should be minimized by carefully tipping the boxes and padding the contact areas. It is also possible to dump the fruit into a water tank, where the fruit are moved by rollers or water currents.

FIGURE 10.7

Darkening of the skin in the equatorial area of Rojo Brillante persimmon caused by mechanical damage in roller brushes. *Photo by Alejandra Salvador and Cristina Besada.*

Persimmon fruit are susceptible to external damage from rubbing on the packing line as they pass through the brushing rollers, thus these must be very soft and kept in good condition (Figure 10.7). Mechanical damage on the packaging line can cause internal disorders. For astringent cultivars, two types of flesh disorders have been described, depending on the level of astringency of the fruit at the time of mechanical impact: flesh browning and pinkish bruising (Figure 10.8). Flesh browning occurs in fruit that have been mechanically damaged after removal of astringency. This browning begins in the superficial zone of the fruit and then spreads to the interior regions and may affect the overall flesh. However, pinkish bruising manifests as isolated areas of the flesh (near the skin). It does not progress inward and is detected in fruit with high astringency that have been mechanically impacted (Novillo et al., 2014). From a commercial point of view, the manifestation of flesh browning implies a greater loss of quality than pinkish bruising. Therefore, it is recommended to apply the deastringency treatment after handling and packing operations in the packing line.

Deastringency treatment: For fresh marketing, astringent persimmon cultivars must undergo postharvest treatments to eliminate astringency. The traditional method for removing astringency consist of enhancing the ripening process by treating the fruit with ethylene (10 ppm at 20°C) or immersion in ethephon (50-500 ppm) (Park & Lee, 2005). However, fruit treated by these methods become very soft, which limits their handling and shortens their postharvest life. For this reason, these treatments have become a minority procedure. Current deastringency treatments are based on exposing the fruit to anaerobic conditions by exposing them to carbon dioxide (CO_2) or ethanol vapor (Vilhena et al., 2022c). These treatments trigger the accumulation of acetaldehyde, a volatile compound that acts as a binding bridge

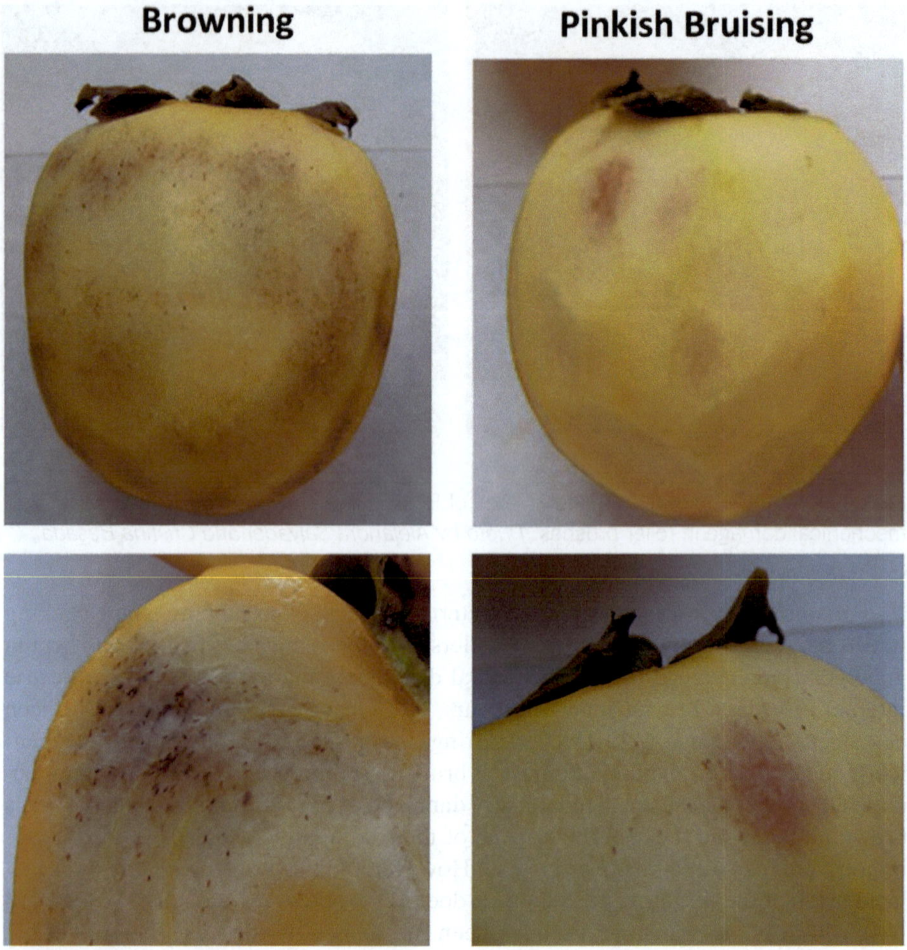

FIGURE 10.8

Internal browning and pinkish bruising associated with mechanical damage in Rojo Brillante persimmon. *Photo by Alejandra Salvador and Cristina Besada.*

for proanthocyanidins, causing their insolubilization and, consequently, loss of astringency, without affecting fruit firmness (Figure 10.9).

The deastringency treatment based on subjecting the fruit to a high concentration of CO_2 is currently the most widely used by the industry in developed countries. Its efficacy depends on different factors, such as CO_2 concentration, temperature, fruit maturity stage, and treatment duration (Besada et al., 2010b). Currently, the most widely used commercial method consists of treating fruit in chambers at 95%-100% CO_2 at 20°C and 90% relative humidity for 24-36 hours (Salvador et al., 2007; Ancillotti et al., 2019). Concentrations of 95%-100% CO_2 ensure efficiency and are important to guarantee a homogeneous distribution of the gas inside

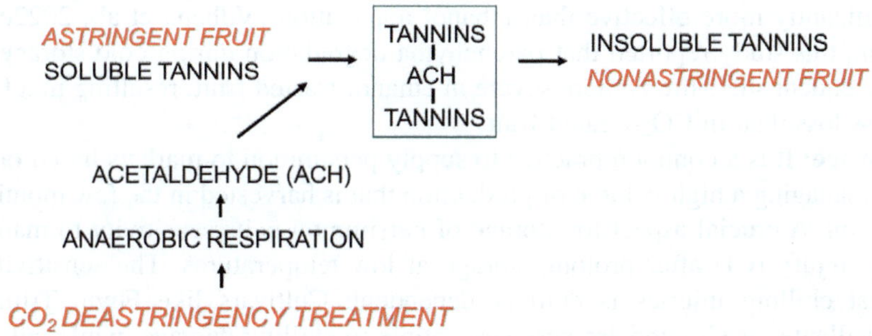

FIGURE 10.9

Outline of the process of astringency removal by CO_2 treatment.

the chambers. Treatments at lower concentrations need longer exposure times and/or can result in fruit with residual astringency. Moreover, low temperatures during treatment delay the deastringency process, and it has been shown that astringency removal is more effective at higher temperatures.

For years, to verify the effectiveness of the treatment, a sample from the treated batch was taken and destructive methods based on the reaction of soluble tannins with ferric chloride were used. Recent research suggests the use of hyperspectral imaging as a nondestructive method for evaluating astringency (Munera et al., 2019).

Exposure to high concentrations of CO_2 implies a stress for the fruits, affecting the parenchyma integrity and changing the redox state (Novillo et al., 2014). Because of that, this treatment may result in internal disorders manifested during postharvest life, especially if it is excessively prolonged and the fruit is cold-stored after astringency removal. Symptoms of flesh browning and fruit softening during cold storage of cultivars such as Niuxin, Sangjudungsi, and Rojo Brillante have been associated with CO_2 treatment (Min et al., 2018; Win et al., 2019; Fathi-Najafabadi et al., 2020). To minimize these alterations, it is recommended to apply the deastringency treatment after cold storage rather than before it (Win et al., 2019). However, in some marketing scenarios, such as shipments to distant overseas countries at low temperatures, the deastringency treatment is necessarily applied before the fruit is subjected to cold storage. In these cases, an acclimatization period at 20°C in air for 24 hours after the deastringency treatment facilitates the diffusion of CO_2 out of the fruit before cold storage and helps to prevent flesh browning (Fathi-Najafabadi et al., 2020).

In addition to treatment with high concentrations of CO_2, the elimination of astringency by the application of ethanol vapor is also widely used commercially in some countries such as Brazil, where it is the main treatment used because it is a less expensive alternative with good results. This treatment consists of steam application of ethanol at a concentration of about 1.7 mL of ethanol per kg of fruit for 24 hours. Nevertheless, most research studies comparing the efficacy of ethanol and CO_2 methods for removing astringency from persimmon show that CO_2 treatment is significantly more effective than ethanol application (Vilhena et al., 2022c). In addition, this study reported that parenchyma degradation during cold storage and the subsequent shelf-life is more severe in ethanol-treated fruit, resulting in a faster firmness loss than in CO_2-treated fruit.

Storage: It is a common practice to supply persimmon to markets based on demand, managing a high volume of production that is harvested in the few months of the season. A crucial aspect for storage of persimmon is its sensitivity to manifest chilling injury (CI) after prolong storage at low temperatures. The sensitivity to manifest chilling injuries is cultivar dependent. Cultivars like Fuyu, Triumph, Rojo Brillante, or Giombo are very susceptible to chilling damage, while cultivars like Hachiya have been reported to have low susceptibility (Salvador et al., 2005; Besada et al., 2014; Tessmer et al., 2019).

CI symptoms in persimmon are generally related to changes in flesh texture. Flesh gelling is the main symptom in cultivars such as Fuyu and Yohou (Besada et al., 2010a; Li et al., 2018), and consist in the development of a gel-like consistency in the flesh parallel to an increase in skin transparency that allows one to see the gel through it. Flesh gelling is associated with a complete disruption of cell walls and membranes, which led to the total loss of the initial parenchyma structure (Besada et al., 2015). Another of the most common symptoms of CI in persimmon is a drastic firmness loss, which has been described for Rojo Brillante and Triumph cultivars, in which flesh gelling development has been also observed under certain storage conditions (Tsviling et al., 2003; Salvador et al., 2005) (Figure 10.10). In addition, after a large storage period at low temperatures, a structural change in flesh can occur without firmness loss but adopting a hard and rubbery texture. Other chilling-related disorders less frequently described are related with loss of flavor, sweetness, juiciness, and titratable acidity, as well as flesh browning.

FIGURE 10.10

Flesh gelling as one of the main chilling injury symptoms of Rojo Brillante persimmon. *Photo by Alejandra Salvador and Cristina Besada.*

The CI symptoms can manifest during cold storage, but their incidence and severity become more marked after transferring the fruit from low to ambient temperatures. In persimmon, damages are induced faster at 5°C-7°C than at 0°C-1°C, which is the recommended temperature for cold storage (Tessmer et al., 2019). Due to the susceptibility of persimmons to CI, many efforts have been made to find strategies that allow prolonging fruit storage. The postharvest application of 1-methylcyclopropene (1-MCP) prior to storage has been reported as the most efficient treatment for reducing CI symptoms in persimmon. 1-MCP is a nontoxic compound that acts as a competitive inhibitor of ethylene perception by preventing ethylene binding and eliciting subsequent signal transduction and translation (Valero et al., 2016). Some commercial names of 1-MCP are EthylBloc, SmartFresh, SmartTabs, and EthylBloc Sachet, which contain different 1-MCP concentrations. A pretreatment with 1-MCP has been demonstrated to reduce flesh gelling and softening during cold storage in many persimmon cultivars sensitive to low temperatures (Harima et al., 2003; Salvador et al., 2004; Besada et al., 2008).

Currently, and for the most cultivars, 1-MCP is routinely applied to cold-stored fruit by the industry in developed countries. The commercial combination of 1-MCP with other storage technologies like modified atmosphere packaging or controlled atmosphere is also typical (Zhang et al., 2018; Zhao et al., 2020).

Although 1-MCP is usually applied at postharvest, recent studies have reported the effectiveness of applying 1-MCP at preharvest as a liquid spray to trees (Harvista, Philadelphia, PA, U.S.). Vilhena et al. (2022b) have reported that the application of Harvista 3 days before harvesting maintained the fruit firmness to the same extent as the postharvest application after cold storage.

8. Main diseases and pests and their management

The main diseases of persimmon are the circular leaf spot, caused by the fungus *Plurivorosphaerella nawae* (Hiura & Ikata), O. Hassan & T. Chang (synonym *Mycosphaerella nawae* Hiura & Ikata), and the anthracnose caused by the fungus *Colletotrichum horii* Weir & Johnston.

The circular leaf spot was present mostly in humid areas from Eastern Asia until it was detected in the Mediterranean basin in 2008, where it spread rapidly (Berbegal et al., 2010). This disease can cause severe yield losses if the fungus is not managed in spring, under Mediterranean conditions, or in summer, under humid conditions of Eastern Asia. This fungal species reproduces by sexual spores (ascospores) that develop in leaf litter on the ground of persimmon orchards after leaf fall. The mature ascospores formed on the leaf litter are the primary inoculum source of *P. nawae* in spring and summer (Kwon & Park, 2004; Vicent et al., 2015). *Plurivorosphaerella nawae* ascospores are released from the leaf litter in spring or summer, but disease symptoms are visible only in September to October (Berbegal et al., 2010; Vicent et al., 2012). During this period, the fungus causes necrotic spots on persimmon leaves that induce a defoliation followed by the premature fruit abscission. The necrotic areas are light brown and darken in the perimeter area. The affected leaves acquire a yellowish color.

Colletotrichum horii causes anthracnose in the field, and it is one of the main problems in Eastern Asia (Kwon et al., 2013). It affects young shoots and fruit and can cause up to 50% of premature fruit fall in heavily affected orchards. The infestation occurs from the first week of May until the end of September in Korea (Kwon et al., 2013). There are two main methods to manage both fungal diseases. The preventive use of fungicides during the infestation period and the removal of leaf litter from the ground just after leaf fall. It is important to highlight that application of fungicides once disease symptoms have developed are not effective against these diseases (Kwon & Park, 2004; Vicent et al., 2015).

There are numerous arthropod pests that can affect persimmon. While mealybugs, whiteflies, and fruit flies are the main arthropod pests of persimmon under Mediterranean conditions, stink bugs, thrips, and scale insects affect this crop in Eastern Asia. The relative importance of these groups and the main species are variable among regions and years. Although several mealybug species have been identified in persimmon in the world, the mealybug *Pseudococcus longispinus* (Targioni Tozzetti) (Hemiptera: Pseudococcidae) has become the main arthropod pest in Spain (Figure 10.11). This mealybug increases its density from June until harvest, when up to 80% of fruit can be infested (Plata et al., 2023a). The mealybug can be controlled by the parasitoid *Anagyrus fusciventris* (Girault) (Hymenoptera: Encyrtidae), but insecticides may be necessary during the summer (Plata et al., 2023b).

FIGURE 10.11

An adult female of the mealybug *Pseudococcus longispinus* infests a persimmon fruit. *Photo by Alejandro Tena (IVIA).*

FIGURE 10.12

Adults of the whitefly *Dialeurodes citri* lay eggs in persimmon leaves in spring. *Photo by María Antonia Gómez-Martínez.*

The whiteflies *Dialeurodes citri* (Ashmead) (Figure 10.12) and *Paraleyrodes minei* Iaccarino (Hemiptera: Aleyrodidae) infest the underside of the leaves from spring to fall. Both species excrete a high amount of honeydew, and sooty mold can cover leaves and fruit. *Dialeurodes citri* is more abundant from April to August and *P. minei* from July to harvest under Mediterranean conditions. The Mediterranean fruit fly *Ceratitis capitata* (Wiedemann) (Diptera: Tephritidae) has been documented causing damage in most of the Mediterranean countries, yet of limited extent (Tena et al., 2015).

In Eastern Asia, stink bugs, such as *Plautia stali* Scott and *Halyomorpha halys* (Stal) (Hemiptera: Pentatomidae) among other species, are the main persimmon arthropod pest (Son et al., 2009; Lee et al., 2009; Yamada, 2011). Stink bugs move from the forest to the persimmon orchards where they feed on fruit in summer and fall, making them unmarketable. Damage severity varies among stink bug species and the attack period, but it can also produce premature fruit fall (Lee et al.,

2009). The intensive application of broad insecticides to manage stink bugs can cause the resurgence of other pest species (Morishita, 2005).

There are many species of thrips that have been identified in persimmon orchards (Alim et al., 2018). Adults and immatures feed on the surface of developing fruit, usually under the calyces. This feeding habit causes patches or rings of scar tissue around the apex as fruit mature. Thrips can attack persimmon fruit from fruit setting until harvest depending on the species and tend to develop resistance to insecticides (Morishita, 2008).

9. Current and future challenges

A common challenge for persimmon-growing countries around the world is the impact of climate change. The increase in global temperatures can adversely affect fruit development and quality, especially during critical periods such as flowering and fruit set. Additionally, changes in precipitation patterns increase the risk of droughts or floods, impacting production and tree health. A clear example of this is the exacerbation of salinity problems due to absence of rainfall in Spain, where salinity is already a problem in some regions. Thus, a breeding program has been initiated to develop salt-tolerant rootstocks in anticipation of potentially severe scenarios in future years.

Furthermore, climate change is intricately linked to occurrences of extreme weather events, such as heatwaves, untimely frosts, and storms, which can inflict direct harm upon crops and diminish yields. Moreover, ecosystem modifications and the proliferation of climate-induced diseases and pests can precipitate a decline in productivity and fruit quality, necessitating escalated interventions and field labor, thereby incurring significant economic burdens. In countries where production is centered on a single cultivar, varietal diversification is key, as it would minimize the risk of a total loss of production in the event of a new pest.

Another common challenge for most producing countries is the lack of generational succession, with the average age of farmers gradually increasing. Achieving higher profits could motivate young people to take over, and there are several key aspects that need to be addressed to facilitate this transition. The development of nonastringent varieties that are highly productive and of high quality would offer a significant advantage for several reasons. In countries where production is already based on nonastringent varieties, it would allow for higher yields, while in countries where production is based on astringent varieties, it would facilitate and reduce the cost of postharvest handling by eliminating the need for deastringency treatments.

Additionally, labor-saving in production could also play a key role. To achieve this, it is necessary to develop dwarfing rootstocks for persimmon cultivation, and breeding programs are already working on this in countries like Japan. In the case of Asian countries such as China and Japan, increasing export volumes also poses a significant challenge, as consumption is currently primarily domestic. Thus, for example, in Japan the export volume does not reach 1% of total production. To promote exportation from Asian countries, it is necessary to implement storage technologies and breeding new cultivars with enhanced postharvest performance.

Another key aspect being addressed by research is the development of products derived from surpluses or damaged fruit, thus adding value to them. It is worth noting that losses during the pre- and postharvest stages can exceed 30% of production in some cases. While dried persimmon is widely known and appreciated in Asian countries, it is not yet produced in Europe, and it represents a good possibility to expand the supply of persimmon in European markets with a product with a very long shelf-life.

References

Alim, M. A., Song, J., Seo, H. J., & Choi, J. J. (2018). Monitoring thrips species with yellow sticky traps in astringent persimmon orchards in Korea. *Applied Entomology and Zoology, 53*(1), 75−84.

Ancillotti, C., Caprini, C., Scordo, C., Renai, L., Giordani, E., Orlandini, S., Furlanetto, S., & Del Bubba, M. (2019). Phenolic compounds in Rojo Brillante and Kaki Tipo persimmons at commercial harvest and in response to CO_2 and ethylene treatments for astringency removal. *LWT, 100*, 99−105.

Badenes, M. L., Gil-Muñoz, F., & Tetsumura, T. (2022). Rootstock breeding and propagation. In R. Tao, & Z. Luo (Eds.), *The persimmon genome* (pp. 47−64). Switzerland: Springer Nature.

Badenes, M. L., Martinez-Calvo, J., & Naval, M. M. (2013). The persimmon breeding program at IVIA: Alternatives to conventional breeding of persimmon. *Acta Horticulturae, 996*, 71−75.

Ballester, C., Badal, E., Bonet, L., Testi, L., & Intrigliolo, D. S. (2022). Determining transpiration coefficients of 'Rojo Brillante' persimmon trees under Mediterranean climatic conditions. *Agricultural Water Management, 271*, Article 107804.

Bellini, E., & Giordani, E. (2005). Germplasm and breeding of persimmon in Europe. *Acta Horticulturae, 685*, 65−75.

Ben-Arie, R., Saks, Y., Sonego, L., & Frank, A. (1996). Cell wall metabolism in gibberellin-treated persimmon fruits. *Plant Growth Regulation, 19*, 25−33.

Berbegal, M., Pérez-Sierra, A., Armengol, J., Park, C. S., & García-Jiménez, J. (2010). First report of circular leaf spot of persimmon caused by *Mycosphaerella nawae* in Spain. *Plant Disease, 94*(3), 374-374.

Besada, C., Arnal, L., & Salvador, A. (2008). Improving storability of persimmon cv. Rojo Brillante by combined use of preharvest and postharvest treatments. *Postharvest Biology and Technology, 50*(2−3), 169−175.

Besada, C., Jackman, R. C., Olsson, S., & Woolf, A. B. (2010a). Response of 'Fuyu' persimmons to ethylene exposure before and during storage. *Postharvest Biology and Technology, 57*(2), 124−131.

Besada, C., Llorca, E., Novillo, P., Hernando, I., & Salvador, A. (2015). Short-term high CO_2 treatment alleviates chilling injury of persimmon cv. Fuyu by preserving the parenchyma structure. *Food Control, 51*, 163−170.

Besada, C., Novillo, P., Navarro, P., & Salvador, A. (2014). Effect of a low oxygen atmosphere combined with 1-MCP pretreatment on preserving the quality of 'Rojo Brillante' and 'Triumph' persimmon during cold storage. *Scientia Horticulturae, 179*, 51−58.

Besada, C., Salvador, A., Arnal, L., & Martínez-Jávega, J. M. (2010b). Optimization of the duration of deastringency treatment depending on persimmon maturity. *Acta Horticulturae, 858*, 69–74.

Besada, C., & Salvador, A. (2018). Postharvest biology and technology of persimmon. In E. S. A. Mir, & M. A. S.y M. M. Mir (Eds.), *Postharvest biology and technology of temperate fruits* (pp. 1–415). Cham: Springer International Publishing.

Castillo, M., Pons-Gómez, A., Albert-Sidro, C., Delpozo, B., & Besada, C. (2023). Acceptance, sensory characterization and consumption contexts for dehydrated persimmon slices, chips, leathers and powder: A consumer study. *Foods, 12*(10), 1966.

Cervera-Chiner, L., Vilhena, N. Q., Larrea, V., Moraga, G., & Salvador, A. (2024). Influence of temperature on 'Rojo Brillante' persimmon drying. Quality characteristics and drying kinetics. *LWT*, Article 115902.

Del Bubba, M., Giordani, E., Pippucci, L., Cincinelli, A., Checchini, L., & Galvan, P. (2009). Changes in tannins, ascorbic acid and sugar content in astringent persimmons during on-tree growth and ripening and in response to different postharvest treatments. *Journal of Food Composition and Analysis, 22*, 668–677.

Dong, Y., Wang, P., Jiang, M., & Qu, S. (2019). Antioxidant and the dwarfing candidate gene of "Nantongxiaofangshi" (*Diospyros kaki* Thunb.). *Oxidative Medicine and Cellular Longevity*, Article 1629845.

FAOSTAT Database. (2022). Retrieved from http://www.fao.org/faostat.

Fathi-Najafabadi, A., Salvador, A., Navarro, P., Gil, R., & Besada, C. (2020). Effect of temperature during and immediately after CO_2-deastringency treatment on internal flesh browning after cold storage of persimmon fruit. *Scientia Horticulturae, 268*, Article 109363.

Gil-Muñoz, F., Peche, P. M., Climent, J., Forner, M. A., Naval, M. M., & Badenes, M. L. (2018). Breeding and screening persimmon rootstocks for saline stress tolerance. *Acta Horticulturae, 1195*, 105–110.

Giordani, E. (2022). History and current status of worldwide production. In R. Tao, & Z. Luo (Eds.), *The persimmon genome* (pp. 1–10). Switzerland: Springer Nature.

Harima, S., Nakano, R., Yamauchi, S., Kitano, Y., Yamamoto, Y., Inaba, A., & Kubo, Y. (2003). Extending shelf-life of astringent persimmon (*Diospyros kaki* Thunb.) fruit by 1-MCP. *Postharvest Biology and Technology, 29*(3), 319–324.

Hattori, N., Kamada, N., Isobe, T., Anma, S., Kagami, H., Araki, Y., & Taneishi, M. (2015). Development of new dwarfing persimmon (Diospyros kaki Thunb.) rootstock cultivars 'Shizukadai 1' and 'Shizukadai 2. *Bulletin of Shizuoka Research Institute of Agriculture and Forestry*, 61–67 (In Japanese with English summary).

Hejazi, Z., Ishimura, S., Honsho, C., & Tetsumura, T. (2018). Effects of planting time, irrigation system, rooting medium, and IBA concentration on cutting propagation of the persimmon dwarfing rootstock 'MKR1. *The Horticulture Journal, 87*, 184–192.

Hu, M. J., Chen, L., Liu, Y. F., Zhang, Q. L., & Luo, Z. R. (2017). Potential of Xiaoguo tianshi and niuyanshi (Diospyros kaki Thunb.) as novel rootstocks for PCNA persimmon. *Journal of Fruit Science, 34*(1), 50–58.

Hume, H. H. (1914). A kaki classification. *Journal of Heredity, 5*, 400–406.

Ikeda, I., Yamada, M., Kurihara, A., & Nishida, T. (1985). Inheritance of astringency in Japanese persimmon. *Journal of the Japanese Society for Horticultural Science, 54*, 39–45.

Ikegami, A., Eguchi, S., Yonemori, K., Yamada, M., Sato, A., Mitani, N., & Kitajima, A. (2006). Segregations of astringent progenies in the F1 populations derived from crosses between a Chinese pollination-constant nonastringent (PCNA) 'Luo Tian Tian Shi' and

Japanese PCNA pollination-constant astringent (PCA) cultivars of Japanese origin. *HortScience, 41*, 561–563.

ISTAT. (2022). Retrieved from https://www.istat.it/it/agricoltura.

Kanety, T., Naor, A., Gips, A., Dicken, U., Lemcoff, J. H., & Cohen, S. (2014). Irrigation influences on growth, yield, and water use of persimmon trees. *Irrigation Science, 32*, 1–13.

Kanzaki, S., Akagi, T., Masuko, T., Kimura, M., Yamada, M., Sato, A., Mitani, N., Utsunomiya, N., & Yonemori, K. (2010). SCAR markers for practical application of marker-assisted selection in persimmon (*Diospyros kaki* Thunb.) breeding. *Journal of the Japanese Society for Horticultural Science, 79*, 150–155.

Kwon, J. H., Kim, J., Choi, O., Gang, G. H., Han, S., & Kwak, Y. S. (2013). Anthracnose caused by Colletotrichum horii on sweet persimmon in Korea: Dissemination of conidia and disease development. *Journal of Phytopathology, 161*(7–8), 497–502.

Kwon, J. H., & Park, C. S. (2004). Ecology of disease outbreak of circular leaf spot of persimmon and inoculum dynamics of *Mycosphaerella nawae*. *Research in Plant Disease, 10*(4), 209–216.

LÜ, X. R., Ai, C. X., Zhu, C. Y., Wang, S., Geng, J. J., & Wang, W. J. (2021). Effect of plant growth regulators and citric acid on the rooting of softwood cuttings of persimmon rootstocks. *J-GLOBAL ID:202202245897299925 Reference number:22A0814183*.

Lee, H. S., Chung, B. K., Kim, T. S., Kwon, J. H., Song, W. D., & Rho, C. W. (2009). Damage of sweet persimmon fruit by the inoculation date and number of stink bugs, *Riptortus clavatus, Halyomorpha halys* and *Plautia stali*. *Korean Journal of Applied Entomology, 48*(4), 485–491.

Li, J., Han, Y., Hu, M., Jin, M., & Rao, J. (2018). Oxalic acid and 1-methylcyclopropene alleviate chilling injury of 'Youhou' sweet persimmon during cold storage. *Postharvest Biology and Technology, 137*, 134–141.

Liu, B., Du, X. Y., Chen, W. X., Guo, D. Y., Xu, L. Q., Zhang, Q. L., & Luo, Z. R. (2023). Establishment and optimization of in vitro rapid propagation technology of 'Xiaoguo Tianshi' as compatible rootstock for persimmon. *Journal of Fruit Science, 40*, 1980–1991.

Llorca, E., Pons-Gómez, A., & Besada, C. (2023). Physico-chemical and microstructural changes during the drying of persimmons with different disorders. Consumer acceptance of dried slices as a criterion to valorize discards. *LWT, 182*, Article 114882.

Luo, Z. R., Zhang, Q. L., Chen, W. X., Xu, L. Q., & Guo, D. Y. (2022). Persimmon industry in China. *Acta Horticulturae, 1338*, 25–30.

MAPA. (2022). Retrieved from https://www.mapa.gob.es/es/agricultura/estadisticas/.

Min, D., Dong, L., Shu, P., Cui, X., Zhang, X., & Li, F. (2018). The application of carbon dioxide and 1-methylcyclopropene to maintain fruit quality of 'Niuxin' persimmon during storage. *Scientia Horticulturae, 229*, 201–206.

Morishita, M. (2008). Pyrethroid-resistant onion thrips, Thrips tabaci Lindeman (Thysanoptera: Thripidae), infesting persimmon fruit. *Applied Entomology and Zoology, 43*(1), 25–31.

Morishita, M. (2005). Resurgence of Japanese mealybug, *Planococcus kraunhiae* (Kuwana), in persimmon induced by a synthetic pyrethroid cypermethrin. *Annual Report of Kansai Plant Protection, 47*, 125–126.

Munera, S., Aleixos, N., Besada, C., Gómez-Sanchis, J., Salvador, A., Cubero, S., Talens, P., & Blasco, J. (2019). Discrimination of astringent and deastringed hard 'Rojo Brillante' persimmon fruit using a sensory threshold by means of hyperspectral imaging. *Journal of Food Engineering, 263*, 173–180.

Novillo, P., Salvador, A., Llorca, E., Hernando, I., & Besada, C. (2014). Effect of CO2 deastringency treatment on flesh disorders induced by mechanical damage in persimmon. Biochemical and microstructural studies. *Food Chemistry, 145*, 454–463.

Park, Y. M., & Lee, Y. J. (2005). Ripening responses of 'Fuyu' persimmon fruit to exogenous ethylene and subsequent shelf temperature. *Acta Horticulturae, 685*, 151–156.

Plata, Á., Gómez-Martínez, M. A., Beitia, F. J., & Tena, A. (2023b). Do hyperparasitoids disrupt the biological control of *Pseudococcus longispinus* in persimmon? *Biological Control, 185*, Article 105310.

Plata, Á., Gómez-Martínez, M. A., Beitia, F. J., & Tena, A. (2023a). Mealybugs in mediterranean persimmon: Fruit infestation, seasonal trend and effect of climate change. *Entomologia Generalis, 43*(6), 1161–1169.

Salvador, A., Arnal, L., Besada, C., Larrea, V., Quiles, A., & Pérez-Munuera, I. (2007). Physiological and structural changes during ripening and deastringency treatment of persimmon fruit cv. 'Rojo Brillante. *Postharvest Biology and Technology, 46*, 181–188.

Salvador, A., Arnal, L., Monterde, A., & Cuquerella, J. (2004). Reduction of chilling injury symptoms in persimmon fruit cv. 'Rojo Brillante' by 1-MCP. *Postharvest Biology and Technology, 33*(3), 285–291.

Salvador, A., Arnal, L., Monterde, A., & Martínez-Jávega, J. M. (2005). Influence of ripening stage at harvest on chilling injury symptoms of persimmon cv. Rojo Brillante stored at different temperatures. *Food Science and Technology International, 11*(5), 359–365.

Sato, A. (2022). In S. Cultivar Breeding, E. R. Tao, & Z. R. Luo (Eds.), *The persimmon genome* (pp. 29–45). Nature Switzerland AG: Springer.

Sato, A., & Yamada, M. (2016). Persimmon breeding in Japan for pollination-constant non-astringent (PCNA) type with marker-assisted selection. *Breeding Science, 66*, 60–68.

Shi, J. C., Wei, P., Wang, M. K., Nie, M. T., Wu, J. Y., Wang, R. Z., Yang, Y., & Guan, C. F. (2020). Investigation on the germplasm of 'Fuyu' sweet persimmon affinity rootstock in early. *China Fruits*, (2), 53–57.

Son, J. K., Ji Eun Yun, & Park, C. G. (2009). Insect pest problems of sweet persimmon in Korea. *Acta Horticulturae,* 833, 325–330.

Tao, R., & Sugiura, A. (1992). Micropropagation of Japanese persimmon (Diospyros kaki L.). In E. Y. P. S. Bajaj (Ed.), *Biotechnology in agriculture and forestry 18: High-tech and micropropagation II* (pp. 424–440). Springer.

Tena, A., Pérez-Hedo, M., Catalán, J., Juan-Blasco, M., & Urbaneja, A. (2015). Fitófagos plaga asociados al cultivo del caqui. In *Badenes et al. "El cultivo del caqui.* Spain: Generalitat Valenciana.

Tessmer, M. A., Appezzato-da-Glória, B., & Kluge, R. A. (2019). Evaluation of storage temperatures to astringency 'Giombo' persimmon: Storage at 1C combined with 1-MCP is recommended to alleviate chilling injury. *Scientia Horticulturae, 257*, Article 108675.

Tessmer, M. A., Besada, C., Hernando, I., Appezzato-da-Glória, B., Quiles, A., & Salvador, A. (2016). Microstructural changes while persimmon fruits mature and ripen. Comparison between astringent and non-astringent cultivars. *Postharvest Biology and Technology, 120*, 52–60.

Tessmer, M. A., Kluge, R. A., & Appezzato-da-Glória, B. (2014). The accumulation of tannins during the development of 'Giombo' and 'Fuyu' persimmon fruits. *Scientia Horticulturae, 172*, 292–299.

Tetsumura, T., Ishimura, S., Hidaka, T., Hirano, E., Kuroki, S., Uchida, Y., & Honsho, C. (2013). MKR1, a dwarfing rootstock for kaki, decreases early fruit drop. *Acta Horticulturae, 996*, 257–263.

Tetsumura, T., Tao, R., & Sugiura, A. (1991). Effect of cytokinin types on the *in vitro* propagation of Japanese persimmon (*Diospyros kaki* Thunb.). *Plant Tissue Culture Letters, 8*, 209−211.

Tsviling, A., Nerya, O., Gizis, A., Sharabi-Nov, A., & Ben-Arie, R. (2003). Extending the shelf-life of 'Triumph' persimmons after storage with 1-MCP. *Acta Hort, 599*, 53−58.

UNECE. (2020). *UNECE STANDARD FFV-63 concerning the marketing and commercial quality control of persimmons*. New York and Geneva: United Nations.

Valero, D., Guillén, F., Valverde, J. M., Castillo, S., & Serrano, M. (2016). Recent developments of 1-methylcyclopropene (1-MCP) treatments on fruit quality attributes. In M. S. Siddiqui (Ed.), *Eco-friendly technology for postharvest produce quality* (pp. 185−201). Academic Press.

Vicent, A., Bassimba, D. D., Hinarejos, C., & Mira, J. L. (2012). Inoculum and disease dynamics of circular leaf spot of persimmon caused by *Mycosphaerella nawae* under semi-arid conditions. *European Journal of Plant Pathology, 134*, 289−299.

Vicent, A., Mira, J. L., Taberner, V., & Palou, L. (2015). Enfermedades fúngicas de importancia en campo y almacén. In *Badenes et al. "El cultivo del caqui*. Spain: Generalitat Valenciana.

Vilhena, N. Q., Gil, R., Vendrell, M., & Salvador, A. (2022b). Effect of preharvest 1-MCP treatment on the flesh firmness of 'Rojo Brillante' persimmon. *Horticulturae, 8*, 350.

Vilhena, N. Q., Quiles, A., Gil, R., Llorca, E., Fernández-Serrano, P., Vendrell, M., & Salvador, A. (2022a). Slight changes in fruit firmness at harvest determine the storage potential of the 'Rojo Brillante' persimmon treated with gibberellic acid. *Horticulturae, 8*, 140.

Vilhena, N. Q., Tessmer, M. A., Hernando, I., Kluge, R. A., Quiles, A., & Salvador, A. (2022c). Structural changes caused by CO2 or ethanol deastringency treatments in cold-stored 'Giombo' persimmon. *Agronomy, 12*, 2464.

Wang, R., Yang, Y., Ruan, X., & Li, G. (2005). Native non-astringent persimmon in China. *Acta Horticulturae, 685*, 99−102.

Win, N. M., Yoo, J., Lee, J., Jung, H. Y., & Kang, I. K. (2019). Effect of different CO2 deastringency application timing on fruit quality attributes and physiological disorders in cold-stored 'Sangjudungsi' persimmon fruit. *Horticultural Science and Technology, 37*(3), 395−403.

Woolf, A. B., & Ben-Arie, R. (2011). *Persimmon (Diospyros kaki L.). En Postharvest biology and technology of tropical and subtropical fruits* (pp. 166−194e). Woodhead Publishing.

Wu, K. Y., Gong, B. C., Xu, Y., & Zhang, P. S. (2021). Effects of different plant growth regulators on hardwood cuttage of persimmon rootstock Yalin 6. *Acta Agriculturae Zhejiangensis, 33*(7), 1256−1263.

Xu, J. C., Gan, J. P., Xiang, J., Zhang, Q. L., Xu, L. Q., Guo, D. Y., & Luo, Z. R. (2017). Advance of genetic improvement technology for Chinese PCNA persimmon. *Acta Horticulturae Sinica, 44*, 987−998.

Yakushiji, H., Sugiura, H., Yamasaki, A., Azuma, A., & Koshita, Y. (2021). Tree growth, productivity, and fruit quality of 'Fuyu' persimmon trees onto different dwarfing rootstocks. *Scientia Horticulturae, 278*, Article 109869.

Yamada, M. (2011). *Cross breeding in woody fruit crops. Tokyo: yokendo (in Japanese).*

Yamada, M., & Sato, A. (2002). Segregation for fruit astringency type in progenies derived from crosses of 'Nishimurawase' pollination constant non-astringent genotypes in oriental persimmon (Diospyros kaki Thunb.). *Scientia Horticulturae, 92*, 107−111.

Yonemori, K., Ikegami, A., Kitajima, A., Luo, Z., Kanzaki, S., Sato, A., Yamada, M., Yang, Y., & Wang, R. (2005). Existence of several pollination constant non-astringent type persimmons in China. *Acta Horticulturae, 685*, 77–83.

Yonemori, K., Kameda, K., & Sugiura, S. (1992). Characteristics of sex expression in monoecious persimmon. *Journal of the Japanese Society for Horticultural Science, 61*, 303–310.

Yonemori, K., Sugiura, A., & Yamada, M. (2000). Persimmon genomics and breeding. *Plant Breeding Reviews, 19*, 191–225.

Zhang, J., Lu, J., Mantri, N., Jiang, L., Ying, S., Chen, S., Feng, X., Cao, Y., Chen, Z., Ren, L., & Lu, H. (2018). An effective combination storage technology to prolong storability, preserve high nutrients and antioxidant ability of astringent persimmon. *Scientia Horticulturae, 241*, 304–312.

Zhang, Q., & Luo, Z. (2022). Origin, evolution, taxonomy and germplasm. In R. Tao, & Z. Luo (Eds.), *The persimmon genome* (pp. 11–27). Switzerland: Springer Nature.

Zhao, Q., Jin, M., Guo, L., Pei, H., Nan, Y., & Rao, J. (2020). Modified atmosphere packaging and 1-methylcyclopropene alleviate chilling injury of 'Youhou' sweet persimmon during cold storage. *Food Packaging and Shelf Life, 24*, Article 100479.

Zheng, G. H., Taira, S., Yonemori, K., & Sugiura, A. (1990). Fruit growth and ripening of Japanese persimmons (Diospyros kaki Thunb.) grown at two locations with different temperature conditions. *Journal of the Japanese Society for Horticultural Science, 59*, 471–477.

CHAPTER

Pomegranate *(Punica granatum)*

11

Olaniyi A. Fawole[1,2], Yardjouma Silue[1,2], Pavlina Drogoudi[3], Giuseppe Ferrara[4], María Serrano[5]

[1]*South African Research Chairs Initiative in Sustainable Preservation and Agroprocessing Research, Faculty of Science, University of Johannesburg, South Africa;* [2]*Postharvest and Agroprocessing Research Centre, Department of Botany and Plant Biotechnology, University of Johannesburg, South Africa;* [3]*Hellenic Agricultural Organization (ELGO)—"DIMITRA," Institute of Plant Breeding and Genetic Resources, Naoussa, Greece;* [4]*Department of Soil, Plant and Food Sciences, University of Bari "Aldo Moro," Italy;* [5]*Department of Applied Biology, Escuela Politécnica Superiorde Orihuela—Centro de Investigación e Innovación Agroalimentario y Agroambiental, University Miguel Hernández, Orihuela, Spain*

Chapter outline

Temperate Tree Fruits and Nuts. https://doi.org/10.1016/B978-0-443-29137-1.00011-6

1. Origin and spread

Pomegranate (*Punica granatum* L.) derives its name from the Latin words "pomum" and "granatus," meaning "apple" and "seeded," respectively. Initially referred to as "Punic Apple" or *"Malum granatum,"* meaning "Seeded Apple," the name evolved to *Punica granatum* following the Punic Wars, reflecting both the fruit's scarlet color *(puniceus)* and its high seed count *(granatum)*. The pomegranate holds significant cultural and religious importance, being mentioned in both the Bible and the Quran, and is often seen as a symbol of fertility, abundance, perfection, and sanctity (Nigro and Spagnoli, 2018). It is among the oldest cultivated crops, alongside apples, grapes, figs, dates, and olives (Jithender et al., 2017; Vandana et al., 2017). While the exact origin of the pomegranate is debated, the majority of sources point to the Transcaucasia-Caspian region, particularly within modern-day Iran (Sharma et al., 2020; Verma et al., 2010). This region includes the northern Iranian Plateau, Afghanistan, Central Asia, northern India, and Anatolia, considered the homeland of wild pomegranate ancestors (Figure 11.1) (Morton, 1987). The domestication of the pomegranate began in the Transcaucasia-Caspian region and northern Turkey during the Neolithic era, subsequently spreading to neighboring regions such as India and China (Still, 2006). The fruit then spread farther across Asia to the Mediterranean basin, reaching North Africa and Europe, notably Egypt, Morocco, Spain, and Italy. It was likely introduced to Spain around 600 BCE by the Phoenicians, who were prominent Mediterranean traders and probably introduced this species in Italy around the same period. A chamber tomb with frescoed red pomegranates in the upper part of the walls of the tomb dating to the 3rd or 4th century BCE is present at Egnazia, in Puglia region, Italy (Giancaspro et al., 2017). In the 1500s, Spanish sailors carried the pomegranate to the Americas, introducing it to Mexico and California (Holland and Bar-Ya'akov, 2018).

2. Economic value, worldwide distribution, and global production

Pomegranate used to be an underutilized crop cultivated for local markets, and only in the recent decades, due to the growing demand for fruit rich in nutritional and healthy active compounds, there was a substantial worldwide expansion in its cultivation. Pomegranate can be consumed as fresh or frozen juice, arils, dried arils, and processed as sauce. Moreover, seeds and peels, waste from the juice industry, and leaves are also

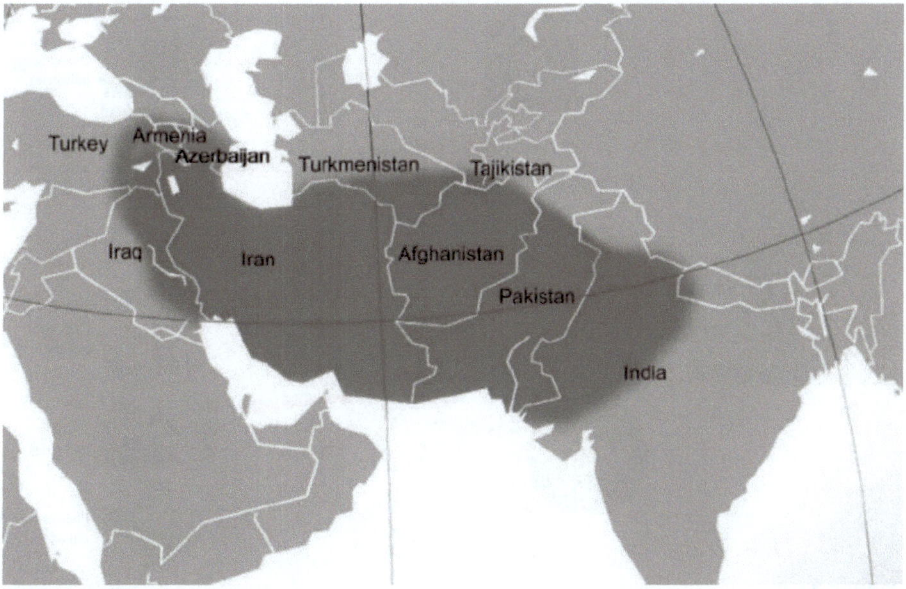

FIGURE 11.1

Center of origin of wild ancestors of pomegranate.

(From Morton, 1987)

valuable as a crop of significant economic importance due to its high nutritional value, offering high returns on investment that can transform subsistence farming into sustainable farming and combat poverty levels (Figure 11.2). The pomegranate is grown mainly in the arid and semi-arid areas of tropical and subtropical regions. Major pomegranate-producing countries include Iran, the United States, India, China, Turkey, Spain, Morocco, Italy, Egypt, and South Africa (Holland et al., 2009).

Tracing global production and trade patterns is not possible due to the lack of a unique harmonized system code. Consequently, the Food and Agriculture Organization (FAO) does not collect specific data on pomegranate production, and organizations such as the USDA and the European Union have ceased collecting such data. Therefore, pomegranate production and trade statistics are analyzed country by country using various sources, including data agencies, scientific papers, reports, and producer/exporter associations. Table 11.1 presents a collation of pomegranate production by country in different years, compiled by Pienaar (2021), while Table 11.2 shows the availability of pomegranates in the international market.

3. Botany and relevant species

Punica granatum L. is a vigorous perennial plant that traditionally belongs to the monogenic *Punicaceae* family, which contains only one genus, *Punica.* This genus includes two species: *Punica protopunica* Balf. (wild pomegranate) and *Punica*

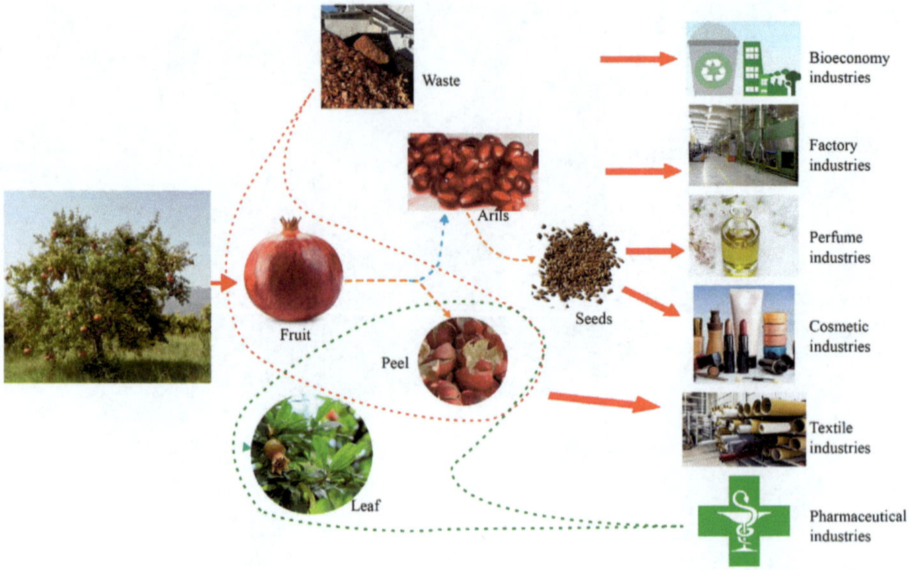

FIGURE 11.2

Use of pomegranate in different industries.

(Adapted from Kalantari Khalil Abad et al., 2023, Abad et al., 2023)

granatum L. (cultivated pomegranate). However, according to Berger et al. (2016), pomegranate, a diploid fruit species (2n = 16), is now considered a member of the *Lythraceae* family.

Additionally, molecular and morphological studies have reported a close relationship between the genus *Punica* and the *Lythraceae* family, particularly regarding leaf characteristics (Graham and Graham, 2014; Huang and Shi, 2002). The currently cultivated pomegranates, exhibiting considerable morphological and genetic diversity (Ferrara et al., 2023; Giancaspro et al., 2023), resulted from the artificial selection and domestication from growers and breeding efforts carried out by humans over centuries. Consequently, following a natural classification system and assigning a specific botanical variety to the diverse pomegranate variants is difficult.

The pomegranate tree has hermaphrodite and functionally male flowers on the same tree, a condition referred to as andromonoecy. The hermaphrodite flowers have well-formed female (stigma, style, ovary) and male (filaments and anthers) parts and have been referred to as "fertile," "vase-shaped," and "bisexual" flowers (Wetzstein et al., 2011). Because the bisexual flowers are the type that set fruit, they are commonly referred to as "female" flowers. The abundance of bisexual flowers has a profound impact on yield, since this will affect the number of fruits produced (Drogoudi et al., 2012). The pomegranate has been described variously as self-pollinated, self- and cross-pollinated, highly cross-pollinated, or often cross-pollinated. Recent investigations demonstrated that it is capable of both open and self-pollination.

Table 11.1 Global production of pomegranate

Country	Year	Hectares	Total production (Tons)	Export (Tons)
India	2019	262,000	3,034,000	67,891
China	2017	120,000	1,697,100	12,602
Iran	2016	94,000	915,000	11,498
Turkey	2019	28,587	559,171	155,189
Egypt	2019	33,096	380,000	127,447
U.S.-California	2018	8606	218,052	28,091
Afghanistan	2019	18,018	194,386	39,056
Azerbaijan	2019	22,948	181,060	22,267
Tunisia	2019	13,257	98,045	10,416
Algeria	2017	91,900	84,870	-
Iraq	2017	5970	82,370	-
Tunisia	2017	88,600	78,000	10,416
Spain	2019	5645	72,115	53,803
Syria	2017	5450	69,820	-
Saudi Arabia	2017	1230	65,310	3874
Italy	2017	3000	30,000	-
Peru	2019	2327	48,320	40,191
Greece	2018	4189	41,888	16,624
Pakistan	2018	7300	37,300	-
Israel	2013	2500	31,250	9528
South Africa	2019	1024	12,984	6369
Chile	2015	705	12,000	5325
Argentina	2014	800	10,000	2207
Mexico	2019	1068	7144	3016
Australia	2019	400	4000	526
Total		**823,620**	**7,994,095**	**626,336**

(From Pienaar (2021) POMASA, South Africa)

4. Breeding and propagation

Pomegranate research is primarily driven by the fresh market and industrial processing sectors, which have evolved significantly over time. Consequently, breeding efforts have recently gained interest in aligning with market requirements, resulting in many cultivars worldwide. These requirements include the physical, nutritional, and sensory properties of both fruits and arils. The main objectives of pomegranate breeding related to its tree are high productivity, disease and pest resistance, frost resistance, and dwarfing habit. Resistance to cracking is also an important trait affecting commercial yield (Drogoudi et al., 2005; SaeiAhagh et al., 2015).

Regarding fruit, important attributes encompass various physical and chemical traits, including fruit weight, size, rind thickness, aril weight, compositional quality,

Table 11.2 Availability of pomegranate in the world

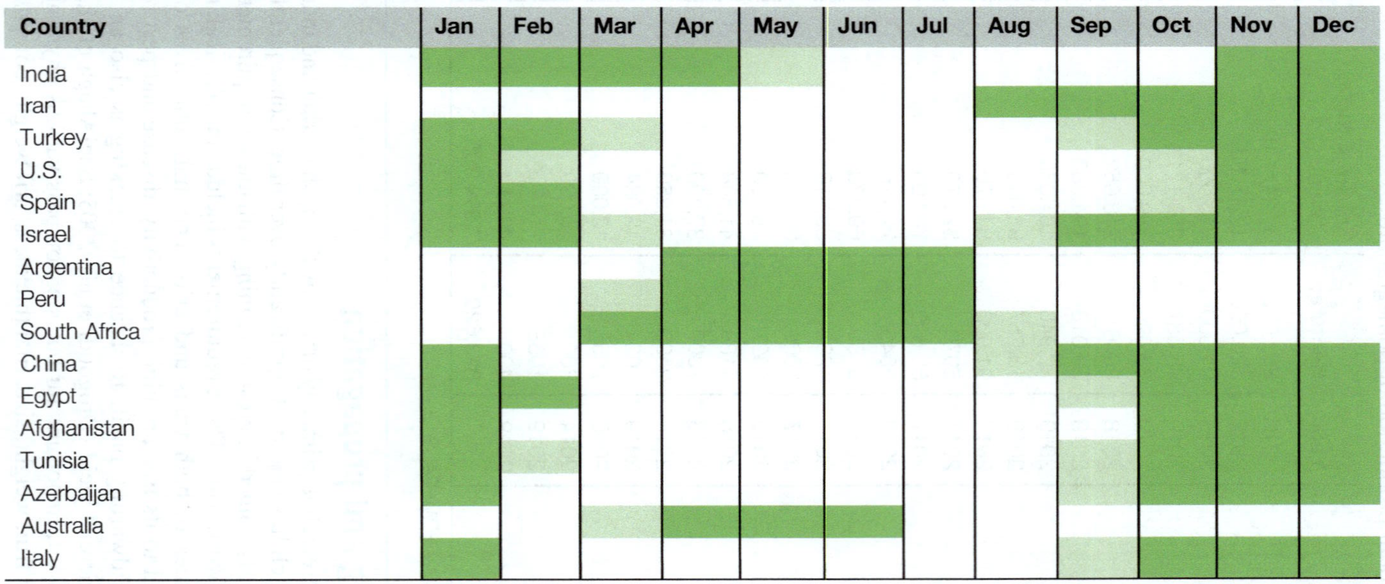

From Kahramanoğlu (2019)

fruit and aril color, sweetness, acidity, and juice content (Alcaraz-Mármol et al., 2015; Okatan et al., 2015). There is growing interest in cultivars with soft seeds (arils), as most commercial pomegranate cultivars have hard or semi-hard seeds (Alcaraz-Mármol et al., 2015; Niu et al., 2018). Standard techniques used for pomegranate breeding include marker-assisted selection, genomic engineering, and conventional breeding methods such as seedling selection and hybridization followed by selection (Chang et al., 2019; Martínez et al., 2006; Yazici and Çevi̇;K, 2022). Good storability, development of new export-suitable cultivars, a wider harvesting and ripening period, and identification of genetic markers for important traits such as healthful compounds and fruit quality remain important goals (Holland and Bar-Ya'akov, 2018; Jacobo, 2012).

Pomegranate propagation methods can be classified into two categories: sexual propagation (by seed) and asexual propagation (cutting, layering, grafting). The most common methods for pomegranate propagation are seedling, grafting, and cutting, with a preference for cutting due to its quick, easy, economical, and convenient nature for obtaining true-to-type trees in a short time. The maturity of the wood used for cuttings, the media, and the timing of cutting are crucial factors that significantly influence the rooting and growth of cuttings (Deshmukh et al., 2019; Leila et al., 2011). It has been reported that semi-hardwood stem cuttings in vermiculite medium under plastic tunnels are optimal for rooting success and a good sprouting percentage (Deshmukh et al., 2019; Hussain et al., 2012). Other factors influencing cutting success include the condition of the parent plant, planting time, rootstocks, rain, temperature fluctuations, and postcutting treatments, including the application of hormones (Rajkumar et al., 2017).

Furthermore, pomegranate plants can be propagated by layering (air-layering and mound/stooling) and grafting (Karimi, 2011; Kaur and Kaur, 2021). Like cutting, grafting is also significantly impacted by factors such as the type and age of rootstocks and scions, as well as environmental conditions (Karimi and Nowrozy, 2017).

5. Agroclimatic requirements and orchard management

The pomegranate grows in various agroclimatic conditions and up to an altitude of 500 m above mean sea level, yet for successful commercial production, it requires a hot and dry climate providing that irrigation facilities are available. Close to the fruit ripening stage (autumn months), rain and high relative humidity are damaging, inducing cracking and russeting (Drogoudi et al., 2021). Differences in day/night temperature are needed close to ripening for achieving best rind and juice coloration (Ghasemi-Soloklui et al., 2023). Winds may result in fruit peel scratches and damage from thorns. Four key practices are commonly reported for appropriately managing pomegranate orchards: soil preparation, planting and management, irrigation, and fertilization and canopy management (pruning and training).

5.1 Soil preparation, planting, and management

Soil preparation represents the most important operation to allow optimal development of the root systems and thus the whole plant, as it increases the porosity and permeability even in the deepest soil layers. In the case of rocky soils, machines for grinding the rocks can be used in some cultivation areas (Ferrara et al., 2012). During soil preparation, organic and mineral fertilizers (N, P, K) are often applied to improve soil fertility. Sometimes, as in the case of heavy soils or frequently occurring heavy rains (as in the late years), raised beds are prepared in correspondence with the rows to avoid excessive water presence close to the plant collar, which can increase the risk of diseases (Ferrara et al., 2020). Weed control is a fundamental practice for correct pomegranate orchard management because of the competition of the weeds for water and nutrients, particularly in hot and/or arid regions. Weed control is exerted through mulching, mowing, tilling, or applying herbicides and is essential in the first years after planting. Once the soil is prepared, planting is done in late winter or early spring, with a density of 400-700 plants per hectare or up to 1200-1600 in high-density systems. In the Northern Hemisphere, plantation commonly occurs from December to February, whereas in the Southern Hemisphere, from August to September. The young trees can be damaged and even broken by the wind, and poles can be used to sustain the first steps of the establishment, particularly in windy areas. The row-orientation needs to be in a north-south direction in northern areas of cultivation to ensure maximum light exposure, since the lack of lighting was related to lower yield and higher damages from cracking and russeting (Drogoudi et al., 2021).

5.2 Irrigation

Irrigation is crucial in pomegranate cultivation due to its impact on productivity and fruit quality, especially in arid and semi-arid regions (Bhagat et al., 2022). Irrigation strategies such as surface, drip, deficit, and microirrigation optimize water use while ensuring plant growth, development, yield, and fruit quality. Proper irrigation can save 32% of water, reduce fruit cracking by 10.9%, increase yield, and produce larger arils with higher juice amounts (Kumar et al., 2012). However, the irrigation effect on pomegranate is cultivar-dependent (Adiba et al., 2021). Drip irrigation is the most widely applied among these strategies in commercial orchards (Holland et al., 2009), and emitters should not be too close to the collar of the tree either to stimulate a wider and longer development of the root absorbing system or to avoid excessive water presence, which can cause fungal diseases (Ferrara et al., 2020). Moreover, pomegranate can be considered a moderately sensitive crop to saline conditions, and irrigation with moderate saline waters can be adopted (Intrigliolo et al., 2020). However, although pomegranate can stand arid and salty soil conditions, pomegranate water requirements can be high if relevant yield needs to be achieved in areas with hot and dry summers. The use of plant and soil sensors, such as precision agriculture devices, can be useful tools for better management of the water requirements of a pomegranate orchard, especially to save this important resource in semi-arid and arid areas. Techniques such as regulated deficit irrigation and partial root-zone drying (PRD) can also be applied to reduce irrigation water use and increase water use efficiency with positive effects on water savings (20%), flowering and fruit set, and with no reduction of the yield. Appropriate management of the irrigation can be used to advance fruit ripening and thus harvesting because of both a faster peel coloration and accumulation of sugars (Intrigliolo et al., 2020).

5.3 Fertilization

Fertilization schedules should be based on the requirements of the crop (age, phenological stages, cultivar, etc.) and the pedoclimatic conditions, in particular the soil's chemical and physical properties. Soil management practices such as mulching, cover crops, limited tillage, etc. can deeply affect the pomegranate nutrient requirements. A high level of nitrogen reserves in the pomegranate trees stimulates the development of flowers and shoots at bud break and a higher fruit set. Nitrogen (N) should be available, particularly in the first phenological stages, to have a better flower formation and development of shoots and leaves. Small split application of N can make this nutrient available when is needed, thus avoiding waste. It has been reported that an increase in N levels from 350 to 500 g per plant improves fruit growth and size, but higher levels (750 g) decrease fruit weight (Maity et al., 2020). Foliar calcium fertilization has been shown to reduce fruit cracking in the cultivar Ardestani while improving fruit biochemical quality (Davarpanah et al., 2018). Additionally, iron and zinc fertilization have been found to increase anthocyanin concentrations in pomegranate arils, alleviating colorless arils disorder (Asadi et al., 2019). Spraying nanomicronutrients (1000 and/or 1500 μg/mL) as foliar fertilization can increase yield, physical and biochemical properties, nutritional status, and polyphenol levels in the fruits (Hussein et al., 2023). Moreover, the application of a combination of organic and silicon, calcium, and potassium-containing fertilizers has proven effective in managing carob moth disease (*Ectomyelois ceratoniae* Zell) in Iranian pomegranate orchards (Torshiz et al., 2020). Organic fertilizers, either in combination or alone, such as vermicompost and chicken and cattle manures, are gaining importance in pomegranate cultivation for enhancing vegetative growth, fruit quality, and yield (Lazare et al., 2020; Silva and Silva, 2020). Application of compost (25 kg/tree) or humic acid (25 g/tree) together with N, P, K improved fruit yield and quality in a pomegranate orchard. Use of inorganic fertilizers with organic ones both stimulated tree development and yield (Maity et al., 2020).

5.4 Canopy management

Pomegranate trees can be grown from a vase-shaped (open center) up to a multitrunk system (bush), depending on tradition, country, type of mechanization, labor intensity, etc. In many commercial orchards, the open center training system is commonly adopted in Mediterranean countries, but in later years, other training systems have been adopted such as a Y-shaped system, central leader, pergola system, etc. (Figure 11.3; Ferrara et al., 2020). Canopy management aims to improve the light needed for flower differentiation and yield, and aeration that is important to reduce disease incidence. It also aims to provide a robust structure to sustain the heavy load of fruits during ripening. Canopy management is made during summer and winter pruning. There is not much information available on pomegranate winter and summer pruning compared with other fruit crops such as grapes, peaches, or apples. Heavy or light/absent pruning can have different effects: in the absence of pruning, there is an increase of small shoots, whereas intense pruning stimulates the strong growth of the shoots, in particular, water sprouts (Ferrara et al., 2020). Summer pruning in pomegranate is used to manage the trellis and the size of the tree (trimming/heading of shoots and water sprouts); regulate the canopy microclimate (water sprouts and shoots removal); increase carbohydrates and nitrogen content in the fruiting shoots (desuckering); and improve the fruit quality and size (fruit thinning).

Pruning and training are essential practices in horticulture to regulate tree and branch shape and canopy, reduce shading effects on flower bud induction and fruit development, and improve air circulation and light penetration. These practices enhance fruit quality and reduce pest and disease incidence (El-Bolok et al., 2023; Hiremath et al., 2018; Nagar et al., 2018). In pomegranate research, pruning and training have been widely reported to affect fruit size (volume and weight), marketable yield, fruit quality, the number of fruits per tree, peel thickness, and to reduce fruit cracking, incidence and severity of sunburn, pests, and diseases (El-Bolok et al., 2023; Niazi et al., 2020; Sharma and Singh, 2018). However, the influence of these practices on pomegranate growth and quality depends on the intensity of application and the cultivar (Hiremath et al., 2018; Mustafa et al., 2023).

FIGURE 11.3

Training in Y-shape system that allows the fruit to develop under the canopy providing protection from sun scalds.

6. Fruit maturation and harvest criteria

Pomegranates, like all nonclimacteric fruits, do not continue to ripen after harvesting and must be picked only when fully ripe (Kader et al., 1984). Early harvesting results in fruit with poorly colored arils and unacceptable flavor (soluble solids/acid ratio), while late harvesting increases susceptibility to spoilage and reduces storage time. Many preharvest and postharvest factors, including maturity at harvest, significantly influence the composition of pomegranates, particularly the quality and quantity of bioactive compounds (Fawole and Opara, 2013a, 2013b; Gözlekçi et al., 2011; Mphahlele et al., 2014; Weerakkody et al., 2010).

During fruit maturation, titratable acidity (TA), total phenolic content, and antioxidant activity decrease, while total soluble solids (TSS) and the color of the fruit skin and juice increase until harvest. Fruit size, TA, soluble solids, and skin color are commonly reported as harvest criteria to ensure good-quality pomegranates. At harvest, the TA of pomegranates can be between 0.13% and 4.98%, while the soluble solids content (SSC) varies between 8.3% and 20.5%, depending on the cultivar/variety, time of harvesting, season, and country of origin (Ferrara et al., 2011, 2014). Although the TSS/TA ratio is considered the most reliable indicator for meeting market requirements (Fawole and Opara, 2013a), skin color is mostly used for grading and has recently been promoted as a maturity index due to its nondestructive nature. Furthermore, the optimal harvesting time for pomegranates depends on the cultivar and growing conditions. Some studies have found that 143 days after full bloom (DAFB) is the optimum time for harvesting many pomegranate cultivars (Fawole and Opara, 2013c; Khodabakhshian et al., 2021).

7. Physiological disorders

One of the most important causes limiting commercial yield worldwide is cracking. It is induced by differences in the growth rate between peel and flesh of the fruit and the pressure imposed by the quickly expanding arils on the stretched peel (Singh et al., 2020). Cultivars differ in their susceptibility to cracking, and damages were related to increasing trends in fruit volume, lower calcium (Ca) ion and higher pectin content in peel, resulting in decreased elasticity (SaeiAhagh et al., 2015). Rainfall in previously water-stressed pomegranate trees induced cracking due to an asymmetric increase in fruit turgor pressure; aril turgor increased to a much greater extent than peel turgor (Galindo et al., 2014).

Russeting is also a physiological disorder in pomegranate appearing as tiny cracks in the cuticle and a cork-like surface in the peel, being restricted in only a portion of fruit located mainly in the sides or lower part of the fruit and in some cases covering its surface. Russeting damages may be substantial in cv. Wonderful, affecting up to 70% of fruit in some orchards and years in areas with relative humidity in Northern Greece. Spraying during the fruit developmental period with acetyl salicylic acid was found to reduce russeting by up to 57% (Drogoudi et al., 2021).

Fruit from the northern side of the canopy had greater cracking and russeting symptoms, while temperature fluctuation during ripening was related to higher symptoms. The presence of russeting damage may also be related to increased relative humidity and water movement as symptoms were higher in years with higher values, in the north side of the canopy, and often occurred in the exposed stylar end of the fruit.

Sunburn is a major physiological disorder in pomegranates that appears because they are terminal-bearing plants, with generally thin branches that bend with elevated fruit weight as the season progresses (Melgarejo et al., 2004). This exposes fruit parts that had previously developed in the shade and are extremely sensitive to sunburn. Spraying with kaolin may reduce damage from sunscald (Melgarejo et al., 2004); however, for commercial production directed to the fresh market, an early application in mid-summer can minimize the dust remaining on the rind. Training in a Y-shape or pergola system may provide protection from sunburn in pomegranate (Figure 11.3).

8. Postharvest physiology

8.1 Respiration and ethylene

The respiration of pomegranate refers to the production of carbon dioxide (CO_2) and the consumption of oxygen (O_2) in the fruit and arils. As a nonclimacteric fruit, pomegranate has a relatively low respiration rate after harvesting. For example, the respiration rate of minimally processed fresh arils of the pomegranate cultivar Mollar de Elche stored at 5°C was found to be 1.15 mL $CO_2 \cdot kg^{-1} \cdot h^{-1}$ (López-Rubira et al., 2005). Ethylene plays a crucial role in regulating the quality of both climacteric and nonclimacteric fruits during storage. It has been reported that pomegranates produce only trace amounts of ethylene, with less than 0.2 $\mu L \cdot kg^{-1} \cdot h^{-1}$ for the Wonderful cultivar after 84 days of storage (Sayyari et al., 2011). Another study reported that ethylene production in whole fruits of three pomegranate cultivars ranges between 0.04 $\mu L \cdot kg^{-1} \cdot h^{-1}$ and 0.76 $\mu L \cdot kg^{-1} \cdot h^{-1}$ after 90 days of storage (Wan et al., 2023). Both respiration rate and ethylene production are significantly influenced by storage conditions, including temperature, relative humidity, the gas composition of the storage atmosphere, storage time, and postharvest treatments. These factors must be carefully managed to maintain fruit quality during storage.

8.2 Water loss

One of the main challenges in preserving the quality of pomegranate fruit during storage is water loss (weight loss), which leads to shriveling of the fruit surface and browning of the skin and arils. These changes reduce the fruit's freshness and market value. The high susceptibility of pomegranates to water loss can be attributed to several factors. Previous studies have shown that the water loss in pomegranates is due to the porous nature of their skin (Fawole et al., 2013; Lufu et al., 2023). Respiration and transpiration are significant causes of water loss in pomegranates. During storage, the transpiration rate increases due to the fragmentation of the waxy cuticle and the widening of microcracks on the fruit surface (Lufu et al., 2021). Additionally, the susceptibility to water loss can be influenced by factors such as rind thickness, the impact level of bruise damage, storage duration, water stress, water vapor pressure deficit, packaging type, and storage temperature (Maghoumi et al., 2023b; Pathare et al., 2023).

9. Postharvest technologies for preserving the pomegranate fruit

Apart from weight loss, other main challenges in the postharvest storage of pomegranates (both skin and arils) include browning, chilling injury, decay, husk scald, and aril discoloration (Çandir et al., 2019; Porat et al., 2016). To address these issues, researchers have developed several postharvest treatments such as low-temperature storage, modified atmosphere packaging (MAP), controlled atmosphere storage (CAS), as well as heat treatments, polyamines, salicylic acid, coatings and waxing, and fungicide treatments, among others.

9.1 Cold storage

Temperature is a major factor influencing the storage of fruits and vegetables, including pomegranates, due to its ability to reduce various physiological, biochemical, and enzymatic activities such as water loss and respiration (Kader, 2013; Pathare et al., 2023; Tano, 1997). Low temperatures decrease disease incidence and rotting of pomegranates, thereby extending their shelf-life. Cold storage is the most used technology for preserving pomegranates and can be combined with other technologies or treatments (Artés et al., 2000; Molaei et al., 2021).

Arendse et al. (2014) reported that storing the Wonderful pomegranate cultivar at 5°C with 92% relative humidity for 2 to 3 months allows for good physicochemical and mechanical properties preservation. Similarly, Fawole and Opara (2013b) found that the cultivars Bhagwa and Ruby could be preserved for 16 weeks under the same conditions. Several authors have indicated that whole pomegranate fruit can be stored at 10°C for 3 to 4 months (Nanda et al., 2001). Although low temperatures can prolong the shelf-life of whole pomegranates to several months, the preservation of pomegranate arils is limited to a few days, regardless of the cultivar. Arils are more sensitive to textural, microbial, and nutritional deterioration than whole fruits. Therefore, optimal storage of pomegranate arils requires the lowest possible temperature. The arils of different pomegranate cultivars (Arakta, Bhagwa, and Ruby) have been successfully stored at 1°C and 4°C for 14 days while maintaining their nutritional properties (O'Grady et al., 2014). Studies have reported that the shelf-life of pomegranate arils ranges from 7 to 18 days at temperatures between 0°C and 5°C, depending on packaging conditions, particularly the material used (Ayhan and Eştürk, 2009; López-Rubira et al., 2005). However, for both whole fruits and arils, excessively low temperatures can lead to various undesirable symptoms known as chilling injury (Figure 11.4), which include skin browning, surface pitting, and desiccation (Valero et al., 2015). Prolonged storage of pomegranates under 5°C can induce chilling injury (Maghoumi et al., 2023b). Several postharvest treatments have been developed to address these chilling injury issues.

9.2 Control of relative humidity

Relative humidity is another important factor in managing pomegranate storage. It is defined as the amount of moisture in the atmosphere and significantly influences the transpiration rate (Kader, 2013). Low relative humidity induces high water loss, leading to skin desiccation, which becomes dark and hard. Consequently, prolonged storage under low humidity conditions results in fruits with poor attractiveness, reducing their market value. On the other hand, a water-saturated atmosphere (high relative humidity) can induce condensation on the fruit surface and facilitate fungal development (Tano, 1997). The optimal relative humidity for pomegranate storage ranges from 90% to 95% (Belay et al., 2018; Sidhu et al., 2019).

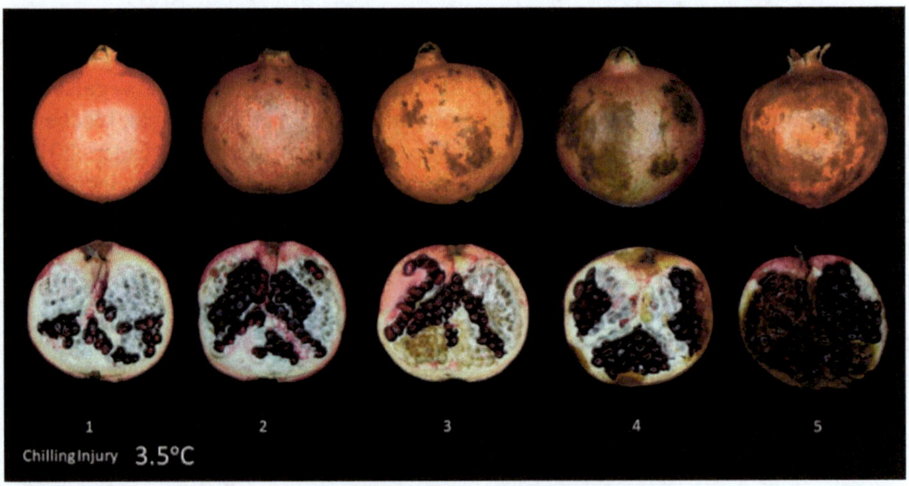

FIGURE 11.4

Chilling injury hedonic scale: expressed as surface (browning, pitting, dehydration), albedo (browning), and arils (browning) disorders. 1 = No symptom, 2 = surface (<25%), albedo (none), arils (none); 3 = surface (25%-50%), albedo (<25%), arils (none); 4 = surface (50%-70%), albedo (25%-50%), arils (some); 5 = surface (70%-100%), albedo (>50%), arils (>50%).

From Maghoumi et al., (2023a)

9.3 Atmospheric control methods

Modified and controlled atmosphere applications are commonly used for postharvest maintenance of pomegranates. The gas composition in fresh produce storage is a crucial factor that significantly influences postharvest quality and shelf-life. The main gases affecting fruit storage are carbon dioxide (CO_2), oxygen (O_2), and ethylene (C_2H_4). Even though pomegranates are a nonclimacteric fruit, effective gas management is essential to avoid physiological alterations and symptoms such as browning of the skin and arils, skin shriveling, and microbial development. Two methods are used to create an optimal atmosphere for fruit storage: controlled atmosphere storage (CAS) and modified atmosphere packaging (MAP).

9.3.1 MAP storage

MAP is a passive or active dynamic process of altering the gaseous composition within a package, relying on the interaction between fruit respiration rate (RR) and the permeability of the packaging material without further control exerted over the initial gas composition (Caleb et al., 2012; Mahajan et al., 2007). Hence, MAP involves adjusting the permeability of the packaging material to match the fruit's respiration rate, continually maintaining the desired gas composition during storage.

The major challenge of MAP is to reduce the respiration rate of fruit. For pomegranates, being nonclimacteric, MAP aims to reduce water loss (weight loss), shrinkage, scald development, aerobic enzyme activities, microbial growth, and maintain postharvest fruit quality, particularly freshness (Ayhan and Eştürk, 2009; Sahel et al., 2018). MAP bags reduce weight loss and decay incidence, improve fruit quality, and extend the shelf-life of pomegranates. Moreover, MAP has been used to reduce internal and external husk chilling injury, weight and firmness loss, and RR of the pomegranate cultivar Mollar de Elche stored at low temperatures, even below -2°C (Çandir et al., 2019; Lorente-Mento et al., 2024). Other researchers have confirmed the effectiveness of MAP in prolonging the shelf-life of minimally processed pomegranate arils (Ayhan and Eştürk, 2009; Moradinezhad and Ranjbar, 2023).

9.3.2 Controlled atmosphere storage

CAS involves maintaining a constant and optimal level of CO_2 and O_2 in a hermetic enclosure (cold chamber) containing the fruit, using a mechanical or electronic system or a CO_2 absorption system (Tano, 1997). In CAS, the gas composition of the enclosure is known and strictly maintained constantly over the storage period. Like MAP, CAS primarily aims to delay respiration rate and ethylene production and to control biochemical activities involving oxygen. Maghoumi et al. (2022) reported that pomegranate fruit storage in natural air varies from 3 to 4 months, while it ranges from 4 to 6 months in a controlled atmosphere with 3%-5% O_2 and 10%-15% CO_2. Furthermore, CAS has been effective in reducing peel browning (Ben-Arie and Or, 1986). According to Pierce and Kader (2003), an atmosphere of 5% O_2 and 15% CO_2 is most suitable for pomegranate storage at 7.5°C, extending shelf-life to 5 months while avoiding fungal infections and husk scald. It is well known that decreasing oxygen concentrations and increasing carbon dioxide concentrations can inhibit microbial development. However, high levels of CO_2 can produce undesirable effects, including internal and external browning and membrane disintegration.

9.4 Coatings and waxing

In postharvest management, "coatings" refers to the application of a protective layer on the surface of a product to delay the ripening process, reduce water loss, and prevent fruit decay and deterioration. Like MAP, coatings and waxing create a microatmosphere around the fruit, reducing gas exchange and allowing low oxygen levels for the fruit (Figure 11.5). Coatings and waxing reduce water loss, respiration rate, and ethylene biosynthesis, delay spoilage, maintain or enhance fruit quality, and prolong shelf-life. Unlike MAP, coatings are composed strictly of edible materials such as chitosan, alginate, gums, pectin, starch, wheat gluten, gelatin, beeswax, and carnauba wax. Hence, they offer a sustainable method for preserving fruit quality after harvesting. Edible coatings are mostly applied for pomegranate aril storage, with chitosan being the most used material (Table 11.3).

9.5 Heat treatments

Also known as thermal treatments, heat treatments include hot water and hot air treatments. Extensive studies have reported on the effects of heat treatments on pomegranate fruit, arils, and juice (Colantuono et al., 2018). For example, the application of hot air treatments (35°C for 60/120 min or 45°C for 30 min) on Rabab-e-

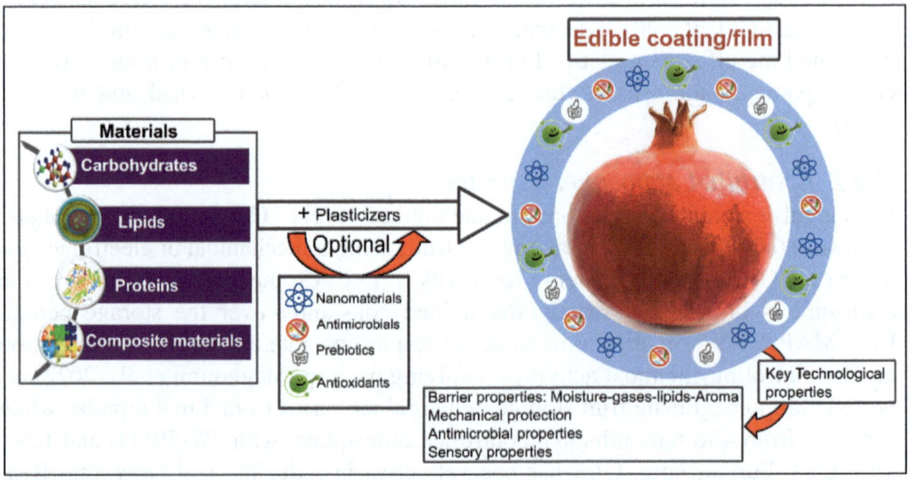

FIGURE 11.5

Preparation process of edible coatings and their properties.

Table 11.3 Some coatings and waxing for extending the shelf-life of both pomegranate whole fruit and arils

Fruit part and cultivars	Coating materials	Storage conditions	Key findings	Reference
Pomegranate arils cv. Malase Saveh	Chitosan-based coatings, either with or without clove essential oil or nanoparticles	5°C and 90% RH for 60 days	Enriched coatings were the most effective coating for extending shelf-life and controlling undesirable microbial, physicochemical, and sensory alterations of pomegranate arils.	Hasheminejad and Khodaiyan, 2020
Pomegranate arils cv. Mollar de Elche	Aloe vera-based coatings alone or in combination with acids (ascorbic and citric acids)	3°C and 90% RH for 12 days	It retains arils firmness, reduces the number of both mesophilic aerobics and yeast and molds and increases total anthocyanins and phenolics levels and sensory scores.	Martínez-Romero et al., 2013
Pomegranate arils cv. Tarom	Chitosan-based coatings	4°C and 95% RH for 12 days	Chitosan coating inhibited bacterial and fungal growth on the surface of arils and suppressed the activity of polyphenol oxidase (PPO).	Ghasemnezhad et al., 2013
Pomegranate fruit cv. Rabbab-e-Neyriz	Chitosan-based coatings	2°C and 5°C and 90% RH for 135 days	Chitosan slowed down anthocyanin degradation in pomegranate arils and prevented color deterioration.	Varasteh et al., 2012
Pomegranate fruit cv. Malase Torshe Saveh	Resin wax, carnauba wax, and chitosan	4.5°C and 90% RH for 120 days	Resin and carnauba waxes reduce respiration rates and weight loss more than chitosan, while carnauba wax maintains significantly higher fruit quality and bioactive compounds than other materials.	Meighani et al., 2015
Ready-to-eat pomegranate arils	Flaxseed gum-based coating with essential oil of lemongrass	5°C for 12 days	Reducing weight loss, delaying the ripening process, and controlling biochemical properties and color.	Yousuf and Srivastava, 2017

Continued

Table 11.3 Some coatings and waxing for extending the shelf-life of both pomegranate whole fruit and arils—*cont'd*

Fruit part and cultivars	Coating materials	Storage conditions	Key findings	Reference
Minimally processed pomegranate arils cv. Bhagwa	Guar gum based with grape pomace extract	10°C for 8 days	The coating reduced mesophilic growth and maintained the sensory properties, phenolic compounds, and antioxidants. It also reduces weight loss.	Saurabh et al., 2022
Pomegranate arils cv. Zagh	Guar gum-based coating incorporated with Mentha piperita L. essential oil	4°C and 85% RH for 28 days	Guar gum (alone or combined with EO) reduced spoilage and exhibited the highest total phenol and antioxidant content.	Izadi et al., 2023
Pomegranate fruit cv. Wonderful	Gum arabic coatings with or without lemongrass essential oil and pomegranate peel extract	5°C and 90% RH for 6 weeks	Arabic gum coatings with/without LM and PP reduced weight loss and fruit decay.	Kawhena et al., 2022

RH, Relative humidity; EO, Essential oils; LM, Lemongrass oil; PP, Pomegranate peel extract.

Neyriz pomegranate arils reduced decay and increased the TSS/TA ratio of the juice after storage for 15 days at 5°C and 70%-80% relative humidity (Sedaghat et al., 2022). Hot water treatment (50°C for 3 min) combined with MAP has been shown to reduce fungal decay, husk scald, and weight loss of pomegranate fruit cv. Hicaznar stored at 6°C for 6 months (Ozdemir and Candir, 2021).

9.6 Polyamines

Polyamines are low-molecular-weight compounds with at least two amino groups produced by living organisms during cellular metabolism, playing an important role in plant growth (Chen et al., 2019). Common polyamines used in pomegranate storage include putrescine, spermidine, and spermine. Studies have demonstrated the capacity of polyamines to improve fruit storage, including pomegranates. For instance, putrescine and spermidine have been shown to reduce respiration rate, aril discoloration, and weight loss during the storage of pomegranate cv. 'Bhagwa' at 5°C and 8°C for 75 days (Waskar et al., 2015). Additionally, putrescine increased chilling injury tolerance and TSS content of the pomegranates. Similar observations were reported by Fawole et al. (2020), who treated pomegranate fruit (cv. Wonderful) with putrescine (2 mmol/L) before storing them at 5°C and 95% relative humidity for 4 months.

10. Postharvest technologies of minimally processed pomegranate arils

Although the storage of fresh produce aims to extend the production season, minimally processing allows for the convenient availability and diversity of pomegranate products, such as minimally processed arils (or ready-to-eat arils) and other products (Dokhanieh et al., 2016; El-Beltagi et al., 2023), even with increased antioxidant activity (Pontonio et al., 2019). The high number and diversity of products derived from pomegranates, along with their benefits for human health, explain the increasing economic value of this fruit. As indicated earlier, pomegranates are susceptible to external damage, including cuts, sunburn, bruising, and cracking (Fawole et al., 2022). As a result, minimal processing has emerged as a viable strategy to utilize externally damaged but internally sound fruits, thereby adding economic value (López-Rubira et al., 2005). However, the postharvest management of pomegranate arils presents several challenges, and the shelf-life of pomegranate arils can be limited to 5 to 10 days (Caleb et al., 2013). Surface damage to the arils can accelerate respiration rates and trigger adverse biochemical changes, leading to off-flavors, textural degradation, and enzymatic browning (Dokhanieh et al., 2016; El-Beltagi et al., 2023). Physical postharvest treatments for the preservation of pomegranate arils include: (1) MAP, (2) ultraviolet-C treatment, (3) organic acid treatment, (4) ozone treatment, and (5) edible coatings, which are briefly described next.

Modified atmosphere packaging (MAP) effectively manages the physiological responses of pomegranate arils, reducing weight loss and maintaining physiological stability (Caleb et al., 2013). Optimized films balance O_2 consumption and CO_2 production, crucial for maintaining aril quality (Belay et al., 2018). MAP stabilizes physicochemical properties such as TSS, TA, and pH under various storage temperatures (Caleb et al., 2013; Hussein et al., 2015). It also affects phytochemical content, with varying effects on anthocyanins and total phenolic content depending on storage conditions (Belay et al., 2018; Ghasemnezhad et al., 2013). Additionally, MAP significantly reduces microbial growth, keeping counts within permissible limits and extending shelf-life (Banda et al., 2015; Caleb et al., 2013; Hussein et al., 2015).

Ultraviolet-C treatment (200–280 nm) is used for surface disinfection and maintaining the quality of fresh-cut produce. It reduces microbial loads and can stabilize TA and SSC, although its effects on anthocyanins and phenolics vary (López-Rubira et al., 2005).

Organic acid treatment includes acids such as salicylic acid (SA) and methyl jasmonate (MeJA), which are effective postharvest treatments for pomegranate arils. They reduce weight loss, maintain firmness, and preserve phytochemical content. Combined SA and MeJA treatments have been reported to enhance aril quality during storage (El-Beltagi et al., 2023).

Ozone treatment, known for its antibacterial properties, reduces microbial populations and improves the storage life of minimally processed crops. Buluc and Koyuncu (2022) showed that ozone treatment extends the shelf-life of pomegranate arils by reducing decay and microbial growth, though its impact on phytochemical content can be variable.

Edible coatings, such as *Aloe vera* gel, chitosan, and flaxseed gum, modify the microenvironment around the arils, acting as barriers against moisture loss and delaying respiration and senescence. Studies have shown that these coatings can maintain firmness, reduce color changes, and preserve phytochemical content, effectively extending the shelf-life of pomegranate arils (Amiri et al., 2021; Martínez-Romero et al., 2013).

11. Main diseases

11.1 Diseases and their symptoms

Like many other fruits, pomegranates are susceptible to various diseases before and after harvesting, mostly caused by fungi and bacteria (Petersen et al., 2010). Common causal pathogens include *Botrytis cinerea, Alternaria alternata, Penicillium implicatum, Colletotrichum* spp., *Coniella granati, Aspergillus niger,* and *Xanthomonas axonopodis* pv. punicae (Jain and Desai, 2018; Mincuzzi et al., 2022; Teksür, 2015). These pathogens can significantly affect fruit quality (Figure 11.5) and economic viability of pomegranate production, leading to enormous losses. These losses include: (i) qualitative losses: shriveling, nutritional value loss, poor appearance due to the development of lesions, rot, and discoloration; (ii) quantitative losses: reduced physical volume, weight loss, and shortened shelf-life; and (iii) economic losses: reduced potential income, market value loss, and lower prices.

According to Ammar and El-Naggar (2014), pomegranate disease symptoms can be summarized as follows: fruit surface rots, internal fruit infections, and stem-end rots (Figure 11.6). Pomegranate infections can occur through various pathways, including field contamination during flowering or harvesting and injuries or wounds

FIGURE 11.6

Pomegranate disease symptoms. (A) Bacterial blight *(Xanthomonas axanopodis)*; (B) Gray mold *(Botrytis* spp.*)*; (C) Black spot *(Alternaria alternata)*; (D) Anthracnose *(Colletotrichum* spp.*)*; (E) *Coniella granati* rot; (F) Black Heart rot *(Aspergillus* spp., *Alternaria* spp.*)*; (G) Fruit rot *(Fusarium oxysporum)*; (H) Blue mold *(Penicillium* spp.*)*; (I) Wilt *(Ceratocystis* spp.*)*.

on the fruit (cracking) due to poor irrigation, improper handling after harvesting, and transportation. Infections can remain latent until fruit maturation during storage. Common sources of pomegranate pathogens include infected orchards (due to infected plant debris, stems, weeds, or fruits left on the tree or soil), wounds from mechanical damage or insect attacks, sunburn, and fruit cracking (Holland et al., 2009). The major pomegranate diseases are presented in Table 11.4.

11.2 Disease treatments

Efficient disease management is essential for ensuring pomegranate production. Pomegranate disease control often encompasses an integrated approach combining chemicals, biological control, and cultural and handling practices. Nonetheless, fungicides are particularly used for managing fungal infections because fungi are the highest causal agents of pomegranate diseases. The effectiveness of fungicides in managing pomegranate pathogens both in the field and during postharvest storage has been widely demonstrated. Fungicides, composed of natural substances and synthetic chemicals, are commonly used for controlling pomegranate rots (Munhuweyi et al., 2016). Among these, the most-studied fungicides include Merivon, Luna Experience, Copper Oxide, Azoxystrobin, Fludioxonil, and Prochloraz (Xavier et al., 2020). Fludioxonil has been found to reduce fruit decay caused by *Penicillium* spp significantly (Opara et al., 2015) and *Botrytis cinerea* (Kahramanoğlu et al., 2018). In postharvest applications, fungicides are applied alone or in combination with other postharvest treatments or technologies. Sustainable methods such as biopesticides and biocontrol are also used to manage pomegranate soilborne, leaf, fruit, and aril diseases (Chavan et al., 2016; Elango et al., 2019; Raut et al., 2019).

12. Current situation and future perspectives

Pomegranate (*Punica granatum* L.), revered in many cultures and religions as a symbol of abundance, fertility, and prosperity, has become one of the most popular fruits globally due to its palatable taste, health benefits, extensive distribution, and significant economic impact. These attributes have driven extensive research on pomegranate cultivation and postharvest management.

This review has highlighted diverse research efforts, including variety selection, genomic exploration, orchard management, and postharvest technologies. Genomics-assisted breeding programs have been implemented, while propagation is primarily achieved through cutting and grafting. To preserve fruit quality, various strategies such as cold storage, modified atmosphere packaging, chemical treatments, and nondestructive techniques (e.g., spectroscopy-based methods and computer vision) have been widely investigated, particularly in major production regions like Spain, the U.S., India, Iran, and China. However, pomegranate research continues to encounter several current and future challenges that need to be addressed.

Table 11.4 Major diseases of pomegranate fruit

Diseases	Causal agent(s)	Mode of infection	Symptoms	References
Anthracnose	*Colletotrichum* spp.	Latent	Light to dark brown sunken and necrotic lesions. The lesions can be circular, subcircular or angular, concentric, and may coalesce. The disease can expand into arils.	Cara et al., 2020; Huang and Tsai, 2023; Xavier et al., 2019
Gray mold rot	*Botrytis* spp. (*B. cinerea)*	Latent	Small, brown, and soft spots start in the crown area and later cover the entire fruit, and the spots become darker and softer.	Testempasis et al., 2020
Brown rot	*Cytospora punicae, Monilinia* spp.,	Wounds	Small creamy-brownish or brownish-yellow and circular to oval lesions, and centrally darker that start in the crown area to cover the entire fruit, leading to the softening of the arils.	Mincuzzi et al., 2022; Venter et al., 2017
Black spot	*Alternaria alternata*	Latent	Black spots on fruit surface (no effect on edible parts). Symptoms appear on both leaves and fruit.	Ezra et al., 2010
Black Heart or Heart rot	*Aspergillus* spp., *Alternaria* spp. (*A. alternata)*	Latent	Infected fruit exhibits any noticeable damage (or only a small proportion that is hard to detect visibly) but presents brown to black and brown to dry rot of some/all arils.	Aloi et al., 2021; Ezra et al., 2015; Luo et al., 2017
Wilt	*Ceratocystis* spp.	Preharvest disease	Yellowing and wilting of leaves on one to several branches and brown to black lesions on roots can lead to the death of plants.	Huang et al., 2003; Xu et al., 2011
Inner fruit decay or Internal Breakdown	*Gluconobacter oxydans*	Latent	Creamy bright growth of bacteria in the mesocarp layer and decayed both arils and seeds. There is no external symptom	Armanious, 2020

Continued

Table 11.4 Major diseases of pomegranate fruit—*cont'd*

Diseases	Causal agent(s)	Mode of infection	Symptoms	References
Fruit rot, Green/blue mold	*Penicillium* spp., *Emericella variecolor*	Wounds	Reddish brown in the center of necrosis on fruit surface that becomes blue-gray fungal growth and covered by mold over time. Some authors reported small circular and dark brown necrotic spots. The lesions of soft, brown tissues can be covered with globose yellowish or dark greenish-brown pycnidia.	Labuda et al., 2004; Palou et al., 2010; Quaglia et al., 2016
Bacterial blight	*Xanthomonas axanopodis pv punicae; Pseudomonas* spp.	Wounds	The first symptoms are isolated water-soaked lesions and later apparition of necrosis with small cracks, leading to blemishing or bursting the fruit.	Jagdale et al., 2018; Petersen et al., 2010
Rhizopus rot or soft rot	*Rhizopus stolonifer*	Wounds	Water-soaked lesions on the flowers and flower buds. Fruit infection starts with the internal tissues through the wounds then when the cuticle is breached, infected fruits become covered by a mass of coarse gray mycelium with black macroscopic sporangia. *R. stolonifer* is one the most destructive pathogens because of its quick spread, rotting an entire store within a few days.	Bautista-Baños et al., 2014; Jamadar et al., 2011
Fruit rot	*Pilidiella granati (synonym Coniella granati)* *Fusarium oxysporum*	Wounds	Circular and brownish-yellow spots, starting from the calyx, can expand to entire fruit, including arils. Light brown lesions on the fruit surface and brown lesions with white mycelium inside the fruit.	Mincuzzi et al., 2016

i. There is a pressing need to identify standard reference genes for gene expression studies to enhance the accuracy and reliability of genomic research.
ii. Advancements in gene mapping are required to identify suitable gene candidates linked to economically important traits, including high yield, superior fruit quality, and resistance to biotic and abiotic stresses.
iii. Further research is necessary to integrate advanced tools such as machine learning and deep learning into postharvest technology frameworks. Additionally, many of these technologies have not been tested on a pilot or large scale, and there is a paucity of data on their economic feasibility.
iv. There is a notable gap in research concerning the determination of optimal harvesting times for different pomegranate cultivars to maximize quality and shelf-life.
v. Comprehensive information on global production, cultivated areas, and sales (export and import) is lacking due to the absence of systematic data collection by international agencies.
vi. Establishing robust traceability standards (e.g., fruit quality labeling) and safety standards (e.g., permissible residue levels of treatment products) is imperative for enhancing the export quality of pomegranates.

Given the identified challenges, the following recommendations for future research are proposed:

i. Future studies should focus on the comprehensive mapping of pomegranate genomes, emphasizing identifying genes associated with key agronomic traits. The development of molecular markers for these traits can facilitate marker-assisted selection in breeding programs.
ii. Research should explore the application of innovative postharvest technologies, including intelligent packaging systems that dynamically monitor and control the internal atmosphere. Integrating sensors and IoT (Internet of Things) devices can optimize storage conditions in real-time.
iii. Incorporating machine learning and deep learning algorithms to predict harvesting time and improve postharvest quality and shelf-life can provide actionable insights for farmers and distributors.
iv. Conducting detailed economic analyses of various postharvest technologies to evaluate their cost-effectiveness and potential return on investment at commercial scales is essential.
v. Developing standardized protocols for harvesting times based on cultivar-specific studies can help optimize fruit quality and marketability.
vi. Establishing international collaborations to systematically collect and share data on pomegranate production, trade, and consumption can provide valuable insights into global market trends.
vii. Implementing comprehensive safety and traceability frameworks supported by international standards can enhance the global competitiveness of pomegranate exports.

References

Abad, A. R. K. K., Barzinpour, F., & Pishvaee, M. S. (2023). Toward circular economy for pomegranate fruit supply chain under dynamic uncertainty: A case study. *Computers & Chemical Engineering, 178*, 108362. https://doi.org/10.1016/j.compchemeng.2023.108362

Adiba, A., Razouk, R., Charafi, J., Haddioui, A., & Hamdani, A. (2021). Assessment of water stress tolerance in eleven pomegranate cultivars based on agronomic traits. *Agricultural Water Management, 243*, 106419. https://doi.org/10.1016/j.agwat.2020.106419

Alcaraz-Mármol, F., Calín-Sánchez, Á., Nuncio-Jáuregui, N., Carbonell-Barrachina, Á. A., Hernández, F., & Martínez, J. J. (2015). Classification of pomegranate cultivars according to their seed hardness and wood perception. *Journal of Texture Studies, 46*, 467–474. https://doi.org/10.1111/jtxs.12145

Aloi, F., Riolo, M., Sanzani, S. M., Mincuzzi, A., Ippolito, A., Siciliano, I., et al. (2021). Characterization of alternaria species associated with Heart rot of pomegranate fruit. *Journal of Fungi, 7*, 172. https://doi.org/10.3390/jof7030172

Amiri, A., Ramezanian, A., Mortazavi, S. M. H., Hosseini, S. M. H., & Yahia, E. (2021). Shelf-life extension of pomegranate arils using chitosan nanoparticles loaded with Satureja hortensis essential oil. *Journal of the Science of Food and Agriculture, 101*, 3778–3786. https://doi.org/10.1002/jsfa.11010

Ammar, M. I., & El-Naggar, M. A. (2014). Screening and characterization of fungi and their associated mycotoxins in some fruit crops. *International Journal of Advanced Research, 2*, 1216–1227.

Arendse, E., Fawole, O. A., & Opara, U. L. (2014). Influence of storage temperature and duration on postharvest physico-chemical and mechanical properties of pomegranate fruit and arils. *CyTA - Journal of Food, 12*, 389–398. https://doi.org/10.1080/19476337.2014.900114

Armanious, H. A. H. (2020). Pomegranate (Punica granata L.) inner decay caused by gluconobacter oxydans bacterium. *Asian Journal of Agricultural and Horticultural Research*, 14–29. https://doi.org/10.9734/ajahr/2020/v7i430101

Artés, F., Villaescusa, R., & Tudela, J. A. (2000). Modified atmosphere packaging of pomegranate. *Journal of Food Science, 65*, 1112–1116. https://doi.org/10.1111/j.1365-2621.2000.tb10248.x

Asadi, E., Ghehsareh, A. M., Moghadam, E. G., Hoodaji, M., & Zabihi, H. R. (2019). Improvement of pomegranate colorless arils using iron and zinc fertilization. *Journal of Cleaner Production, 234*, 392–399. https://doi.org/10.1016/j.jclepro.2019.06.129

Ayhan, Z., & Eştürk, O. (2009). Overall quality and shelf life of minimally processed and modified atmosphere packaged "ready-to-eat" pomegranate arils. *J Food Sci, 74*, C399–C405. https://doi.org/10.1111/j.1750-3841.2009.01184.x

Banda, K., Caleb, O. J., Jacobs, K., & Opara, U. L. (2015). Effect of active-modified atmosphere Packaging on the respiration rate and quality of pomegranate arils (cv. Wonderful). *Postharvest Biology and Technology, 109*, 97–105. https://doi.org/10.1016/j.postharvbio.2015.06.002

Bautista-Baños, S., Bosquez-Molina, E., & Barrera-Necha, L. L. (2014). Rhizopus stolonifer (soft rot). In S. Bautista-Baños (Ed.), *Postharvest decay* (pp. 1–44). San Diego: Academic Press. https://doi.org/10.1016/B978-0-12-411552-1.00001-6

Belay, Z. A., Caleb, O. J., Mahajan, P. V., & Opara, U. L. (2018). Design of active modified atmosphere and humidity packaging (MAHP) for 'wonderful' pomegranate arils. *Food Bioprocess Technol, 11*, 1478–1494. https://doi.org/10.1007/s11947-018-2119-0

Ben-Arie, R., & Or, E. (1986). The development and control of husk scald on 'wonderful' pomegranate fruit during storage. *Journal of the American Society for Horticultural Science, 111*, 395–399. https://doi.org/10.21273/JASHS.111.3.395

Berger, B. A., Kriebel, R., Spalink, D., & Sytsma, K. J. (2016). Divergence times, historical biogeography, and shifts in speciation rates of Myrtales. *Molecular Phylogenetics and Evolution, 95*, 116–136. https://doi.org/10.1016/j.ympev.2015.10.001

Bhagat, A., Meshram, D., & Gholkar, M. (2022). Water management for pomegranate orchards. *JART, 47*, 34–38. https://doi.org/10.56228/JART.2022.47107

Buluc, O., & Koyuncu, M. A. (2022). Ozone: A promising alternative to prochloraz for cold storage of pomegranate. *J Food Sci Technol, 59*, 2731–2740. https://doi.org/10.1007/s13197-021-05295-y

Çandir, E., ÖzdemiR, A. E., & Aksoy, M. C. (2019). Effects of modified atmosphere packaging on the storage and shelf life of Hicaznar pomegranate fruits. *Turk J Agric For, 43*, 241–253. https://doi.org/10.3906/tar-1801-119

Caleb, O. J., Mahajan, P. V., Manley, M., & Opara, U. L. (2013). Evaluation of parameters affecting modified atmosphere packaging engineering design for pomegranate arils. *International Journal of Food Science & Technology, 48*, 2315–2323. https://doi.org/10.1111/ijfs.12220

Caleb, O. J., Opara, U. L., & Witthuhn, C. R. (2012). Modified atmosphere packaging of pomegranate fruit and arils: A review. *Food Bioprocess Technol, 5*, 15–30. https://doi.org/10.1007/s11947-011-0525-7

Cara, M., Mincuzzi, A., Merkuri, J., Vrapi, H., Cara, O., Ippolito, A., et al. (2020). Colletotrichum gloeosporioides sensu stricto as causal agent of anthracnose on pomegranate fruit in Albania. *Crop Protection, 137*, 105291. https://doi.org/10.1016/j.cropro.2020.105291

Chang, L., Wu, S., & Tian, L. (2019). Effective genome editing and identification of a regiospecific gallic acid 4-O-glycosyltransferase in pomegranate (Punica granatum L.). *Horticulture Research, 6*, 123. https://doi.org/10.1038/s41438-019-0206-7

Chavan, N. P., Pandey, R., Nawani, N., Nanda, R. K., Tandon, G. D., & Khetmalas, M. B. (2016). Biocontrol potential of actinomycetes against Xanthomonas axonopodis pv. punicae, a causative agent for oily spot disease of pomegranate. *Biocontrol Science and Technology, 26*, 351–372. https://doi.org/10.1080/09583157.2015.1116057

Chen, D., Shao, Q., Yin, L., Younis, A., & Zheng, B. (2019). Polyamine function in plants: Metabolism, regulation on development, and roles in abiotic stress responses. *Frontiers in Plant Science, 9*. Article 1945.

Colantuono, A., Vitaglione, P., Manzo, N., Blaiotta, G., Montefusco, I., Marrazzo, A., et al. (2018). Evaluation of microfiltration and heat treatment on the microbiological characteristics, phenolic composition and volatile compound profile of pomegranate (Punica granatum L.) juice. *Journal of the Science of Food and Agriculture, 98*, 3324–3332. https://doi.org/10.1002/jsfa.8836

Davarpanah, S., Tehranifar, A., Abadía, J., Val, J., Davarynejad, G., Aran, M., et al. (2018). Foliar calcium fertilization reduces fruit cracking in pomegranate (*Punica granatum* cv. Ardestani). *Scientia Horticulturae, 230*, 86–91. https://doi.org/10.1016/j.scienta.2017.11.023

Deshmukh, S., Bharad, S. G., Deshmukh, P. L., & Gholap, S. V. (2019). Effect of different media and time of cutting on rooting and growth of pomegranate cuttings. *PKV Research Journal, 43*.

Dokhanieh, A. Y., Aghdam, M. S., & Sarcheshmeh, M. A. A. (2016). Impact of postharvest hot salicylic acid treatment on aril browning and nutritional quality in fresh-cut pomegranate. *Hortic. Environ. Biotechnol, 57*, 378–384. https://doi.org/10.1007/s13580-016-0087-8

Drogoudi, P. D., Tsipouridis, C., & Michailidis, Z. (2005). Physical and chemical characteristics of pomegranates. *HortScience, 40*, 1200–1203.

Drogoudi, P., Pantelidis, G. E., & Vekiari, S. A. (2021). Physiological disorders and fruit quality attributes in pomegranate: Effects of meteorological parameters, canopy position and acetylsalicylic acid foliar sprays. *Front. Plant Sci., 12*, 645547. https://doi.org/10.3389/fpls.2021.645547

Drogoudi, P., Pantelidis, G., & Manganaris, A. (2012). *Morphological and physiological characteristics in pomegranate cultivars with different yields, in: Options Mediterraneennes, A* (pp. 67–69).

El-Beltagi, H. S., Al-Otaibi, H. H., & Ali, M. R. (2023). A new approach for extending shelf-life of pomegranate arils with combined application of salicylic acid and methyl jasmonate. *Horticulturae, 9*, 225. https://doi.org/10.3390/horticulturae9020225

El-Bolok, T. K., El-Soda, A. S., & El Barbary, M. G. (2023). Impact of pruning and training systems on the growth and productivity of pomegranate trees. *Horticulture Research Journal, 1*, 197–207. https://doi.org/10.21608/hrj.2023.309578

Elango, E., Sridharan, S., Saravanan, P. A., & Balakrishnan, S. (2019). Efficacy of biopesticides against pomegranate sucking pests under laboratory condition. *Journal of Biopesticides, 12*, 30–35.

Ezra, D., Gat, T., Skovorodnikova, Y., Vardi, Y., & Kosto, I. (2010). First report of Alternaria black spot of pomegranate caused by *Alternaria alternata* in Israel. *Plant Disease Notes, 5*(1). https://doi.org/10.1071/DN10001

Ezra, D., Kirshner, B., Hershcovich, M., Shtienberg, D., & Kosto, I. (2015). Heart rot of pomegranate: Disease etiology and the events leading to development of symptoms. *Plant Disease, 99*, 496–501. https://doi.org/10.1094/PDIS-07-14-0707-RE

Fawole, O. A., Atukuri, J., Arendse, E., & Opara, U. O. (2020). Postharvest physiological responses of pomegranate fruit (cv. Wonderful) to exogenous putrescine treatment and effects on physico-chemical and phytochemical properties. *Food Science and Human Wellness, 9*, 146–161. https://doi.org/10.1016/j.fshw.2020.02.007

Fawole, O. A., Kaseke, T., & Opara, U. L. (2022). Pomegranate fruit quality and seed drying method: Effect on the chemical composition and bioactivities of the extracted oil. *Processes, 10*, 3. https://doi.org/10.3390/pr10010003

Fawole, O. A., & Opara, U. L. (2013a). Harvest discrimination of pomegranate fruit: Postharvest quality changes and relationships between instrumental and sensory attributes during shelf life. *Journal of Food Science, 78*, S1264–S1272. https://doi.org/10.1111/1750-3841.12176

Fawole, O. A., & Opara, U. L. (2013b). Effects of storage temperature and duration on physiological responses of pomegranate fruit. *Industrial Crops and Products, 47*, 300–309. https://doi.org/10.1016/j.indcrop.2013.03.028

Fawole, O. A, & Opara, U. L. (2013c). Effects of maturity status on biochemical content, polyphenol composition and antioxidant capacity of pomegranate fruit arils (cv. 'Bhagwa'). *South African Journal of Botany, 85*, 23–31.

Fawole, O. A., Opara, U. L., & Theron, K. I. (2013). Influence of fruit developmental and maturity stages on chemical, phytochemical and antioxidant properties of pomegranate juice. *Acta Hortic*, 461–469. https://doi.org/10.17660/ActaHortic.2013.1007.53

Ferrara, G., Cavoski, I., Pacifico, A., Tedone, L., & Mondelli, D. (2011). Morpho-pomological and chemical characterization of pomegranate (*Punica granatum* L.) genotypes in Apulia region, Southeastern Italy. *Scientia Horticulturae, 130*, 599–606. https://doi.org/10.1016/j.scienta.2011.08.016

Ferrara, G., Farrag, K., & Brunetti, G. (2012). The effects of rock fragmentation and/or deep tillage practices on soil skeletal material and chemical properties in a Mediterranean climate. Soil Use and Management. *Soil Use and Management, 28*. https://doi.org/10.1111/j.1475-2743.2012.00423.x

Ferrara, G., Giancaspro, A., Mazzeo, A., Giove, S. L., Matarrese, A. M. S., Pacucci, C., et al. (2014). Characterization of pomegranate (*Punica granatum* L.) genotypes collected in Puglia region, Southeastern Italy. *Scientia Horticulturae, 178*, 70–78. https://doi.org/10.1016/j.scienta.2014.08.007

Ferrara, G., Palasciano, M., Sarkhosh, A., Cossio, F., Babu, K. D., & Mazzeo, A. (2020). Orchard establishment and tree management., in: The pomegranate: Botany, production and uses, botany, production and uses. *CABY Digital Library*, 247–284.

Ferrara, G., Porfido, C., Terzano, R., Sarkhosh, A., & Mazzeo, A. (2023). A study on the characteristics of buds and flowers in pomegranate: Differences among cultivars. *Horticulturae, 9*, 117. https://doi.org/10.3390/horticulturae9010117

Gözlekçi, S., Saraçoğlu, O., Onursal, E., & Ozgen, M. (2011). Total phenolic distribution of juice, peel, and seed extracts of four pomegranate cultivars. *Pharmacogn Mag, 7*, 161–164. https://doi.org/10.4103/0973-1296.80681

Galindo, A., Rodríguez, P., Collado-González, J., Cruz, Z. N., Torrecillas, E., Ondoño, S., et al. (2014). Rainfall intensifies fruit peel cracking in water stressed pomegranate trees. *Agricultural and Forest Meteorology, 194*, 29–35. https://doi.org/10.1016/j.agrformet.2014.03.015

Ghasemi-Soloklui, A. A., Kordrostami, M., & Gharaghani, A. (2023). Environmental and geographical conditions influence color, physical properties, and physiochemical composition of pomegranate fruits. *Sci Rep, 13*, 15447. https://doi.org/10.1038/s41598-023-42749-z

Ghasemnezhad, M., Zareh, S., Rassa, M., & Sajedi, R. H. (2013). Effect of chitosan coating on maintenance of aril quality, microbial population and PPO activity of pomegranate (Punica granatum L. cv. Tarom) at cold storage temperature. *Journal of the Science of Food and Agriculture, 93*, 368–374. https://doi.org/10.1002/jsfa.5770

Giancaspro, A., Giove, S. L., Marcotuli, I., Ferrara, G., & Gadaleta, A. (2023). Datasets for genetic diversity assessment in a collection of wild and cultivated pomegranates (*Punica granatum* L.) by microsatellite markers. *Data in Brief, 49*, 109346. https://doi.org/10.1016/j.dib.2023.109346

Giancaspro, A., Mazzeo, A., Giove, L. S., Zito, D., Marcotuli, I., Gallotta, A., et al. (2017). Exploiting DNA-based molecular tools to assess genetic diversity in pomegranate (Punica granatum L.) selections and cultivars. *Fruits, 72*, 292–305. https://doi.org/10.17660/th2017/72.5.5

Graham, S. A., & Graham, A. (2014). Ovary, fruit, and seed morphology of the Lythraceae. *International Journal of Plant Sciences, 175*, 202–240. https://doi.org/10.1086/674316

Hasheminejad, N., & Khodaiyan, F. (2020). The effect of clove essential oil loaded chitosan nanoparticles on the shelf life and quality of pomegranate arils. *Food Chemistry, 309*, 125520. https://doi.org/10.1016/j.foodchem.2019.125520

Hiremath, A., Patil, S., Hipparagi, K., Gandolkar, K., & Gollagi, S. (2018). Influence of pruning intensity on growth and yield of pomegranate (Punica granatum L.) CV. super bhagwa under organic conditions. *Journal of Pharmacognosy and Phytochemistry, 7*, 1027−1031.

Holland, D., & Bar-Ya'akov, I. (2018). Pomegranate (Punica granatum L.) breeding. In J. M. Al-Khayri, S. M. Jain, & D. V. Johnson (Eds.), *Advances in plant breeding strategies: Fruits* (Vol. 3, pp. 601−647). Springer International Publishing, Cham. https://doi.org/10.1007/978-3-319-91944-7_15

Holland, D., Hatib, K., & Bar-Ya'akov, I. (2009). Pomegranate: Botany, horticulture, breeding. In *Horticultural reviews* (pp. 127−191). John Wiley & Sons, Ltd. https://doi.org/10.1002/9780470593776.ch2

Huang, C.-J., & Tsai, S.-H. (2023). First report of Colletotrichum grevilleae causing fruit anthracnose of pomegranate in taiwan. *Plant Disease*. https://doi.org/10.1094/PDIS-02-23-0395-PDN. PDIS-02-23-0395-PDN.

Huang, Q., Zhu, Y. Y., Chen, H. R., Wang, Y. Y., Liu, Y. L., Lu, W. J., et al. (2003). First report of pomegranate Wilt caused by ceratocystis fimbriata in yunnan, China. *Plant Disease, 87*. https://doi.org/10.1094/PDIS.2003.87.9.1150B, 1150−1150.

Huang, Y., & Shi, S. (2002). Phylogenetics of Lythraceae sensu lato: A preliminary analysis based on chloroplast rbcL gene, psaA-ycf3 spacer, and nuclear rDNA internal transcribed spacer (ITS) sequences. *International Journal of Plant Sciences, 163*, 215−225. https://doi.org/10.1086/338392

Hussain, I., Khattak, A. M., Amin, N. U., Aman, F., & Sajid, M. (2012). Response of different pomegranate cuttings types to different environmental conditions. *Sarhad Journal of Agriculture, 28*, 15−18.

Hussein, A. S., Abeed, A. H. A., Usman, A. R. A., & Abou-Zaid, E. A. A. (2023). Conventional vs. nano-micronutrients as foliar fertilization for enhancing the quality and nutritional status of pomegranate fruits. *Journal of the Saudi Society of Agricultural Sciences*. https://doi.org/10.1016/j.jssas.2023.09.008

Hussein, Z., Caleb, O. J., & Opara, U. L. (2015). Perforation-mediated modified atmosphere packaging of fresh and minimally processed produce—a review. *Food Packaging and Shelf Life, 6*, 7−20. https://doi.org/10.1016/j.fpsl.2015.08.003

Intrigliolo, D. S., Wang, D., Pérez-Gago, M. B., Palou, L., Ayars, J., Puerto, H., et al. (2020). Water requirements and responses to irrigation restrictions., in: The pomegranate: Botany, production and uses, botany, production and uses. *CABY Digital Library*, 320−343.

Izadi, A., Dehestani-Ardakani, M., Meftahizadeh, H., Gholamnezhad, J., & Hatami, M. (2023). Edible coatings based on guar gum and peppermint essential oil alter the quality enhancement of Zagh pomegranate arils during the postharvest supply chain. *Biomass Conv. Bioref.* https://doi.org/10.1007/s13399-023-04966-4

Jacobo, C. M. (2012). Breeding Mexican pomegranates to improve productivity and quality and increase versatility of uses. In P. Melgarejo, & D. Valero (Eds.), *II international symposium on the pomegranate* (pp. 61−66). Zaragoza.

Jagdale, S. B., Sonawane, M. S., & Kapadnis, B. P. (2018). A new bacterial blight of pomegranate caused by Pseudomonas sp. in Maharashtra, India. *Australasian Plant Dis. Notes, 13*, 27. https://doi.org/10.1007/s13314-018-0311-8

Jain, K., & Desai, N. (2018). Pomegranate the cash crop of India: A comprehensive review on agricultural practices and diseases. *International Journal of Health Sciences, 8*(5), 315−336.

Jamadar, M. M., Jawadagi, R. S., Sataraddi, A. R., Patil, D. R., & Patil, R. V. (2011). Status of pomegranate diseases of Northern Karnataka in India. *Acta Horticulturae, 890*, 501−507.

Jithender, B., Vyas, D. M., Abhisha, S., & Rathod, P. J. (2017). Determination of physical and mechanical properties of pomegranate (var. Bhagwa) fruit and aril. *Int J Curr Microbiol App Sci, 6*, 879–885. https://doi.org/10.20546/ijcmas.2017.610.105

Kader, A. A. (2013). Postharvest technology of horticultural crops - an overview from farm to fork. *Ethiopian Journal of Applied Science and Technology*, 1–8.

Kader, A. A., Chordas, A., & Elyatem, S. (1984). Responses of pomegranates to ethylene treatment and storage temperature. *California Agriculture, 38*, 14–15.

Kahramanoğlu, İ., Aktaş, M., & Gündüz, Ş. (2018). Effects of fludioxonil, propolis and black seed oil application on the postharvest quality of "Wonderful" pomegranate. *PLoS One, 13*, Article e0198411. https://doi.org/10.1371/journal.pone.0198411

Kahramanoğlu, I. (2019). Trends in pomegranate sector: Production, postharvest handling and marketing. *International Journal of Agriculture Forestry and Life Sciences, 3*, 239–246.

Kalantari Khalil Abad, A. R., Barzinpour, F., & Pishvaee, M. S. (2023). Toward circular economy for pomegranate fruit supply chain under dynamic uncertainty: A case study. *Computers & Chemical Engineering, 178*, 108362. https://doi.org/10.1016/j.compchemeng.2023.108362.

Karimi, H. R. (2011). Stenting (cutting and grafting) – a technique for propagating pomegranate. *Journal of Fruit and Ornamental Plant Research, 19*, 73–79.

Karimi, H. R., & Nowrozy, M. (2017). Effects of rootstock and scion on graft success and vegetative parameters of pomegranate. *Scientia Horticulturae, 214*, 280–287. https://doi.org/10.1016/j.scienta.2016.11.047

Kaur, M., & Kaur, A. (2021). Influence of rooting media and time of air layering on the rooting in air layers of pomegranate (Punica granatum). *Research on Crops, 22*, 849–857. https://doi.org/10.31830/2348-7542.2021.140

Kawhena, T. G., Opara, U. L., & Fawole, O. A. (2022). Effects of gum Arabic coatings enriched with lemongrass essential oil and pomegranate peel extract on quality maintenance of pomegranate whole fruit and arils. *Foods, 11*, 593. https://doi.org/10.3390/foods11040593

Khodabakhshian, R., Emadi, B., Khojastehpour, M., & Golzarian, M. R. (2021). Developmental changes in ripeness indexes and physico-chemical properties of pomegranate fruit during maturity on tree. *Erwerbs-Obstbau, 63*, 215–225. https://doi.org/10.1007/s10341-021-00562-w

Kumar, S., Singh, R., Asrey, R., & Nangare, D. D. (2012). Techno-economic evaluation of integrating canal water harvesting and drip irrigation for pomegranate production in a dry eco-region. *Irrigation and Drainage, 61*, 366–374. https://doi.org/10.1002/ird.658

López-Rubira, V., Conesa, A., Allende, A., & Artés, F. (2005). Shelf life and overall quality of minimally processed pomegranate arils modified atmosphere packaged and treated with UV-C. *Postharvest Biology and Technology, 37*, 174–185. https://doi.org/10.1016/j.postharvbio.2005.04.003

Labuda, R., Hudec, K., Piecková, E., Mezey, J., Bohovič, R., Mátéová, S., et al. (2004). Penicillium implicatum causes a destructive rot of pomegranate fruits. *Mycopathologia, 157*, 217–223. https://doi.org/10.1023/B: MYCO.0000020599.95040.c6

Lazare, S., Lyu, Y., Yermiyahu, U., Heler, Y., Ben-Gal, A., Holland, D., et al. (2020). Optimizing nitrogen application for growth and productivity of pomegranates. *Agronomy, 10*, 366. https://doi.org/10.3390/agronomy10030366

Leila, A., Khalil, A., Manouchehr, J., & Ziyaoddine, T. (2011). The effect of different mediums and cuttings on growth and rooting of pomegranate cuttings. *Iranian Journal of Plant Physiology, 1*, 199–203.

Levin, G. M. (2006). *Pomegranate*. Third Millennium Pubns.

Lorente-Mento, J. M., Guillén, F., Valverde, J. M., Valero, D., Badiche, F., Robles, A., et al. (2024). Chilling injury control in pomegranate fruit with compostable stretchable skin film. *Scientia Horticulturae, 323*, 112480. https://doi.org/10.1016/j.scienta.2023.112480

Lufu, R., Ambaw, A., & Opara, U. L. (2023). Determination of moisture loss of pomegranate cultivars under cold and shelf storage conditions and control strategies. *Sustainable Food Technol, 1*, 79–91. https://doi.org/10.1039/D2FB00017B

Lufu, R., Ambaw, A., & Opara, U. L. (2021). Functional characterisation of lenticels, micro-cracks, wax patterns, peel tissue fractions and water loss of pomegranate fruit (cv. Wonderful) during storage. *Postharvest Biology and Technology, 178*, 111539. https://doi.org/10.1016/j.postharvbio.2021.111539

Luo, Y., Hou, L., Forster, H., Pryor, B., & Adaskaveg, J. E. (2017). Identification of alternaria species causing Heart rot of pomegranates in California. *PLANT DIS, 101*, 421–427. https://doi.org/10.1094/PDIS-08-16-1176-RE

Maghoumi, M., Amodio, M. L., Cisneros-Zevallos, L., & Colelli, G. (2023a). Prevention of chilling injury in pomegranates revisited: Pre- and post-harvest factors, mode of actions, and technologies involved. *Foods, 12*, 1462. https://doi.org/10.3390/foods12071462

Maghoumi, M., Amodio, M. L., Fatchurrahman, D., Cisneros-Zevallos, L., & Colelli, G. (2022). Pomegranate husk scald browning during storage: A review on factors involved, their modes of action, and its association to postharvest treatments. *Foods, 11*, 3365. https://doi.org/10.3390/foods11213365

Maghoumi, M., Fatchurrahman, D., Amodio, M. L., Quinto, M., Cisneros-Zevallos, L., & Colelli, G. (2023b). Is pomegranate husk scald during storage induced by water loss and mediated by ABA signaling? *Journal of the Science of Food and Agriculture, 103*, 2914–2925. https://doi.org/10.1002/jsfa.12385

Mahajan, P. V., Oliveira, F. A. R., Montanez, J. C., & Frias, J. (2007). Development of user-friendly software for design of modified atmosphere packaging for fresh and fresh-cut produce. *Innovative Food Science & Emerging Technologies, 8*, 84–92. https://doi.org/10.1016/j.ifset.2006.07.005

Maity, A., Khayyat, M., Azarmi-Atajan, F., Agehara, S., & Sarkhosh, A. (2020). Soil and nutrition, in: The pomegranate: Botany, production and uses, botany, production and uses. *CABY Digital Library*, 285–319.

Martínez, J. J., Melgarejo, P., Hernández, F., Salazar, D. M., & Martínez, R. (2006). Seed characterisation of five new pomegranate (*Punica granatum* L.) varieties. *Scientia Horticulturae, 110*, 241–246. https://doi.org/10.1016/j.scienta.2006.07.018

Martínez-Romero, D., Castillo, S., Guillén, F., Díaz-Mula, H. M., Zapata, P. J., Valero, D., et al. (2013). *Aloe vera* gel coating maintains quality and safety of ready-to-eat pomegranate arils. *Postharvest Biology and Technology, 86*, 107–112. https://doi.org/10.1016/j.postharvbio.2013.06.022

Meighani, H., Ghasemnezhad, M., & Bakhshi, D. (2015). Effect of different coatings on post-harvest quality and bioactive compounds of pomegranate (Punica granatum L.) fruits. *J Food Sci Technol, 52*, 4507–4514. https://doi.org/10.1007/s13197-014-1484-6

Melgarejo, P., Martınez, J. J., Hernández, F., Martınez-Font, R., Barrows, P., & Erez, A. (2004). Kaolin treatment to reduce pomegranate sunburn. *Scientia Horticulturae, 100*, 349–353. https://doi.org/10.1016/j.scienta.2003.09.006

Mincuzzi, A., Garganese, F., Ippolito, A., & Sanzani, S. (2016). *First report of Pilidiella granati causing postharvest fruit rot on pomegranate in southern Italy, 98*. https://doi.org/10.4454/JPP.V98I2.045

Mincuzzi, A., Sanzani, S. M., Palou, L., Ragni, M., & Ippolito, A. (2022). Postharvest rot of pomegranate fruit in southern Italy: Characterization of the main pathogens. *JoF, 8*, 475. https://doi.org/10.3390/jof8050475

Molaei, S., Soleimani, A., Rabiei, V., & Razavi, F. (2021). Impact of chitosan in combination with potassium sorbate treatment on chilling injury and quality attributes of pomegranate fruit during cold storage. *Journal of Food Biochemistry, 45*, Article e13633. https://doi.org/10.1111/jfbc.13633

Moradinezhad, F., & Ranjbar, A. (2023). Foliar application of fertilizers and plant growth regulators on pomegranate fruit yield and quality: A review. *Journal of Plant Nutrition, 47*, 797–821. https://doi.org/10.1080/01904167.2023.2280152

Morton, J. F. (1987). *Fruits of warm climates.* Miami, F. L.

Mphahlele, R. R., Stander, M. A., Fawole, O. A., & Opara, U. L. (2014). Effect of fruit maturity and growing location on the postharvest contents of flavonoids, phenolic acids, vitamin C and antioxidant activity of pomegranate juice (cv. Wonderful). *Scientia Horticulturae, 179*, 36–45. https://doi.org/10.1016/j.scienta.2014.09.007

Munhuweyi, K., Lennox, C. L., Meitz-Hopkins, J. C., Caleb, O. J., & Opara, U. L. (2016). Major diseases of pomegranate (Punica granatum L.), their causes and management—a review. *Scientia Horticulturae, 211*, 126–139. https://doi.org/10.1016/j.scienta.2016.08.016

Mustafa, G., Rajwana, I. A., Faried, H. N., Haq, T. U., & Bashir, M. A. (2023). The influence of pruning intensity on physical and biochemical fruit attributes of pomegranate. *Erwerbs-Obstbau, 65*, 1849–1859. https://doi.org/10.1007/s10341-022-00824-1

Nagar, D., Meena, K., Mukherjee, S., & Garg, S. (2018). Effect of pruning intensity, crop load and fertigation on growth attributes in Pomegranate var. Bhagwa under Jaipur condition. *Journal of Pharmacognosy and Phytochemistry, 7*, 83–85.

Nanda, S., Sudhakar Rao, D. V., & Krishnamurthy, S. (2001). Effects of shrink film wrapping and storage temperature on the shelf life and quality of pomegranate fruits cv. Ganesh. *Postharvest Biology and Technology, 22*, 61–69. https://doi.org/10.1016/S0925-5214(00)00181-2

Niazi, Z., Kumar, D. T. S., Kumar, D. A. K., Joshi, D. V., & Sunil, D. N. (2020). Effect of pruning and growth regulators on quality of pomegranate Cv. Bhagwa. *J Pharmacogn Phytochem, 9*, 352–357.

Nigro, L., & Spagnoli, F. (2018). Pomegranate (Punica granatum L.) from Motya and its deepest oriental roots. Vicino Oriente XXII (2018). pp. 49-90. *Vicino Oriente, 22*, 49–90. https://doi.org/10.53131/VO2724-587X2018_4

Niu, J., Cao, D., Li, H., Xue, H., Chen, L., Liu, B., et al. (2018). Quantitative proteomics of pomegranate varieties with contrasting seed hardness during seed development stages. *Tree Genetics & Genomes, 14*, 14. https://doi.org/10.1007/s11295-018-1229-1

O'Grady, L., Sigge, G., Caleb, O. J., & Opara, U. L. (2014). Effects of storage temperature and duration on chemical properties, proximate composition and selected bioactive components of pomegranate (*Punica granatum* L.) arils. *LWT - Food Science and Technology, 57*, 508–515. https://doi.org/10.1016/j.lwt.2014.02.030

Okatan, V., Akca, Y., Ercisli, S., & Gozlekci, S. (2015). Genotype selection for physico-chemical fruit traits in pomegranate (Punica granatum l.) in Turkey. *Acta Scientiarum Polonorum Hortorum Cultus, 14*, 123–132.

Opara, U. L., Atukuri, J., & Fawole, O. A. (2015). Application of physical and chemical postharvest treatments to enhance storage and shelf life of pomegranate fruit—a review. *Scientia Horticulturae, 197*, 41–49. https://doi.org/10.1016/j.scienta.2015.10.046

Ozdemir, A. E., & Candir, E. (2021). Impact of hot water and modified atmosphere packaging treatments on the postharvest quality of pomegranate fruit (*Punica granatum* cv. 'hicaznar'). *J. Agric. Sci.-Tarim Bilim. Derg, 27*, 304–311. https://doi.org/10.15832/ankutbd.661496

Pala, H., Tatli, A., Yilmaz, C., & Özgüven, A. I. (2009). Important diseases of pomegranate fruit and control possibilities in Turkey. *Acta Hortic*, 285–290. https://doi.org/10.17660/ActaHortic.2009.818.42

Palou, L., Guardado, A., & Montesinos-Herrero, C. (2010). First report of *Penicillium* spp. and *Pilidiella granati* causing postharvest fruit rot of pomegranate in Spain. *New Disease Reports, 22*. https://doi.org/10.5197/j.2044-0588.2010.022.021, 21–21.

Pathare, P. B., Al-Dairi, M., Al-Yahyai, R., & Al-Mahdouri, A. (2023). Physiological response of stored pomegranate fruit affected by simulated impact. *Foods, 12*, 1122. https://doi.org/10.3390/foods12061122

Petersen, Y., Mansvelt, E. L., Venter, E., & Langenhoven, W. E. (2010). Detection of Xanthomonas axonopodis pv. punicae causing bacterial blight on pomegranate in South Africa. *Australasian Plant Pathology, 39*, 544–546. https://doi.org/10.1071/AP10034

Pienaar, L. (2021). *The economic contribution of South Africa's pomegranate industry.*

Pierce, B., & Kader, A. (2003). Responses of "wonderful" pomegranates to controlled atmospheres. *Acta Hortic*, 751–757. https://doi.org/10.17660/ActaHortic.2003.600.115

Pontonio, E., Montemurro, M., Pinto, D., Marzani, B., Trani, A., Ferrara, G., et al. (2019). Lactic acid fermentation of pomegranate juice as a tool to improve antioxidant activity. *Front. Microbiol, 10*. https://doi.org/10.3389/fmicb.2019.01550

Porat, R., Kosto, I., & Daus, A. (2016). Bulk storage of "Wonderful" pomegranate fruit using modified atmosphere bags. *Israel Journal of Plant Sciences, 63*, 45–50. https://doi.org/10.1080/07929978.2016.1152839

Quaglia, M., Moretti, C., Cerri, M., Linoci, G., Cappelletti, G., Urbani, S., et al. (2016). Effect of extracts of wastewater from olive milling in postharvest treatments of pomegranate fruit decay caused by Penicillium adametzioides. *Postharvest Biology and Technology, 118*, 26–34. https://doi.org/10.1016/j.postharvbio.2016.03.012

Rajkumar, R., Gora, J. S., Kumar, R., Singh, A., Kumar, A., & Gajender, G. (2017). Effect of different growing media on the rooting of pomegranate (Punica granatum L.) cv. "Phulear-akta" cuttings. *Journal of Applied and Natural Science, 9*, 715–719. https://doi.org/10.31018/jans.v9i2.1263

Raut, V. M., Kanitkar, S., Acharij, R., & Borawake, A. (2019). Efficacy of milastin-k (bacillus subtilis ktsb 1015 1.5% a.s.) a bio-fungicide for management of bacterial blight (xanthomonas axonopodis pv. punicae), alternaria and cercospora fruit spot diseases of pomegranate. *Pestology, 43*, 48–56.

SaeiAhagh, H., Sharifani, M., Seifi, E., Mohseni, A., & Akbarpour, V. (2015). How fruit traits influence cracking of pomegranate (Punica granatum L.). *Acta Horticulturae, 1099*, 815–818. https://doi.org/10.17660/ActaHortic.2015.1099.103

Sahel, N. A., Krishna, H., Suresh, G., Bhuvaneswari, S., Anjaneya Reddy, B., & Foshanji, A. (2018). Effect of modified atmosphere package on biochemical properties of pomegranate (Punica granatum L.) fruits. *Journal of Pharmacognosy and Phytotherapy, 7*, 3174–3176.

Saurabh, C. K., Gupta, S., & Variyar, P. S. (2022). Development of grape pomace extract based edible coating for shelf life extension of pomegranate arils. *Food Measure, 16*, 590–597. https://doi.org/10.1007/s11694-021-01186-0

Sayyari, M., Babalar, M., Kalantari, S., Martínez-Romero, D., Guillén, F., Serrano, M., et al. (2011). Vapour treatments with methyl salicylate or methyl jasmonate alleviated chilling injury and enhanced antioxidant potential during postharvest storage of pomegranates. *Food Chemistry, 124*, 964–970. https://doi.org/10.1016/j.foodchem.2010.07.036

Sedaghat, S., Rahemi, M., & Aslmoshtaghi, E. (2022). Effect of hot air treatment on physiochemical properties of pomegranate arils. *International Journal of Horticultural Science and Technology, 9*, 85–95. https://doi.org/10.22059/ijhst.2020.309085.383

Sharma, A., Dogra, R. K., Singh, G., & Bodh, S. (2020). Changing paradigms in pomegranate breeding: A review. *Int.J.Curr.Microbiol.App.Sci., 9*, 1878–1887. https://doi.org/10.20546/ijcmas.2020.909.237

Sharma, D. P., & Singh, N. (2018). Effect of rejuvenation pruning on the growth, productivity and disease incidence in declining trees of pomegranate (Punica granatum L.) cv. Kandhari Kabuli. *Journal of Applied and Natural Science, 10*, 358–362. https://doi.org/10.31018/jans.v10i1.1630

Sidhu, H. S., Díaz-Pérez, J. C., & MacLean, D. (2019). Controlled atmosphere storage for pomegranates (Punica granatum L.): Benefits over regular air storage. *HortScience, 54*, 1061–1066. https://doi.org/10.21273/HORTSCI13796-19

Silva, F. S. B., & Silva, F. A. (2020). A low cost alternative, using mycorrhiza and organic fertilizer, to optimize the production of foliar bioactive compounds in pomegranates. *Journal of Applied Microbiology, 128*, 513–517. https://doi.org/10.1111/jam.14477

Singh, A., Shukla, A. K., & Meghwal, P. R. (2020). Fruit cracking in pomegranate: Extent, cause, and management – a review. *International Journal of Fruit Science, 20*, S1234–S1253. https://doi.org/10.1080/15538362.2020.1784074

Still, D. W. (2006). *Pomegranates: A botanical perspective, in: Pomegranates: Ancient roots to modern medicine* (pp. 199–209).

Tano, K. (1997). QUEBEC. *Emballages de fruits et légumes sous atmosphère modifiée: Impact des fluctuations de température pendant l'entreposage*. Université de Lavel.

Teksür, P. K. (2015). Alternative technologies to control postharvest diseases of pomegranate. *Stewart Postharvest Review, 11*. https://doi.org/10.2212/spr.2015.4.3

Testempasis, S., Puckett, R. D., Michailides, T. J., & Karaoglanidis, G. S. (2020). Genetic structure and fungicide resistance profile of Botrytis spp. populations causing postharvest gray mold of pomegranate fruit in Greece and California. *Postharvest Biology and Technology, 170*, 111319. https://doi.org/10.1016/j.postharvbio.2020.111319

Torshiz, A. O., Goldansaz, S. H., Motesharezadeh, B., Askari, M. A., & Zarei, A. (2020). The influence of fertilization on pomegranate susceptibility to infestation by ectomyelois ceratoniae. *International Journal of Fruit Science, 20*, S1156–S1173. https://doi.org/10.1080/15538362.2020.1778602

Valero, D., Mirdehghan, S. H., Sayyari, M., & Serrano, M. (2015). Chapter 23-vapor treatments, chilling, storage, and antioxidants in pomegranates. In V. Preedy (Ed.), *Processing and impact on active components in Food* (pp. 189–196). San Diego: Academic Press. https://doi.org/10.1016/B978-0-12-404699-3.00023-8

Vandana, Waghmare, G. M., & Solanki, S. P. S. (2017). Influence of chelated Ca, micronutrient and method of application on vegetative growth of pomegranate (Punica granatum L.) cv. Bhagwa. *Int.J.Curr.Microbiol.App.Sci, 6*, 135–147. https://doi.org/10.20546/ijcmas.2017.611.018

Varasteh, F., Arzani, K., Barzegar, M., & Zamani, Z. (2012). Changes in anthocyanins in arils of chitosan-coated pomegranate (*Punica granatum* L. cv. Rabbab-e-Neyriz) fruit during cold storage. *Food Chemistry, 130*, 267–272. https://doi.org/10.1016/j.foodchem.2011.07.031

Venter, E., Lennox, C. L., & Meitz-Hopkins, J. C. (2017). First report of Cytospora punicae causing post-harvest fruit rot on pomegranate in South Africa. *Plant Disease, 101*, 631. https://doi.org/10.1094/PDIS-08-16-1157-PDN

Verma, N., Mohanty, A., & Lal, A. (2010). Pomegranate genetic resources and germplasm conservation: A review. *Fruit, Vegetable and Cereal Science and Biotechnology, 4*, 120–125.

Wan, R., Song, J., Lv, Z., Qi, X., Feng, Z., Yang, Z., et al. (2023). Effects of 1–MCP treatment on postharvest fruit of five pomegranate varieties during low-temperature storage. *Horticulturae, 9*, 1031. https://doi.org/10.3390/horticulturae9091031

Waskar, D. P., Khandare, V. S., Kalalbandi, B. M., & Shelke, P. S. (2015). Effect of polyamines on storability and quality of pomegranate fruit (Punica granatum L.) cv. Bhagwa. *Journal of Horticultural Sciences, 10*, 48–53. https://doi.org/10.24154/jhs.v10i1.153

Weerakkody, P., Jobling, J., Infante, M. M. V., & Rogers, G. (2010). The effect of maturity, sunburn and the application of sunscreens on the internal and external qualities of pomegranate fruit grown in Australia. *Scientia Horticulturae, 124*, 57–61. https://doi.org/10.1016/j.scienta.2009.12.003

Wetzstein, H. Y., Ravid, N., Wilkins, E., & Martinelli, A. P. (2011). A morphological and histological characterization of bisexual and male flower types in pomegranate. *Journal of the American Society for Horticultural Science, 136*, 83–92. https://doi.org/10.21273/JASHS.136.2.83

Xavier, K. V., Kc, A. N., Peres, N. A., Deng, Z., Castle, W., Lovett, W., et al. (2019). Characterization of Colletotrichum species causing anthracnose of pomegranate in the southeastern United States. *Plant Disease, 103*, 2771–2780. https://doi.org/10.1094/PDIS-03-19-0598-RE

Xavier, K. V., Kc, A. N., & Vallad, G. E. (2020). Fungicide application timing essential for the management of leaf spot and fruit rot on pomegranate (Punica granatum L.) in Florida. *Plant Disease, 104*, 1629–1637. https://doi.org/10.1094/PDIS-10-19-2224-RE

Xu, B., Zheng, X. H., Guo, W. X., Zhou, X. P., & He, P. (2011). First report of pomegranate Wilt caused by ceratocystis fimbriata in sichuan province. *Plant Disease, 95*. https://doi.org/10.1094/PDIS-02-11-0146, 776–776.

Yazici, K., & ÇeviK, M.Ş. (2022). Development of DNA markers associated with sunburn resistance in pomegranate (Punica granatum L.) using bulk segregant analysis. *Turkish Journal of Agriculture and Forestry, 46*, 955–965. https://doi.org/10.55730/1300-011X.3056

Yousuf, B., & Srivastava, A. K. (2017). Flaxseed gum in combination with lemongrass essential oil as an effective edible coating for ready-to-eat pomegranate arils. *International Journal of Biological Macromolecules, 104*, 1030–1038. https://doi.org/10.1016/j.ijbiomac.2017.07.025

CHAPTER

Almond[†]

12

María José Rubio-Cabetas[1,2], María Teresa Espiau Ramírez[1], Kenneth A. Shackel[3], Themis J. Michailides[4], Raquel Sánchez-Pérez[5], and Thomas Gradziel[3]

[1]*Department of Plant Science, Agrifood Research and Technology Centre of Aragon (CITA). Avda. Montañana 930, 50059 Zaragoza, Spain;* [2]*Instituto Agroalimentario de Aragón—IA2 (CITA-Universidad de Zaragoza), Zaragoza, España, Spain;* [3]*Department of Plant Sciences, UC Davis, Davis, CA, United States;* [4]*Department of Plant pathology, UC Davis, Kearney Agricultural Research & Extension Center, Parlier, CA, United States;* [5]*Plant Breeding Department, CEBAS-CSIC Campus Universitario de Espinardo, Murcia, Spain e-mail address: mjrubioc@cita-aragon.es, mespiau@cita-aragon.es, kashackel@ucdavis.edu, tjmichailides@ucanr.edu, rsanchez@cebas.csic.es, tmgradziel@ucdavis.edu*

Chapter outline

[†] For Chapter Almond (Prunus duclis), Kenneth A. Shackel part of contribution is in the public domain.

Temperate Tree Fruits and Nuts. https://doi.org/10.1016/B978-0-443-29137-1.00012-8

1. Origin and crop evolution

The almond tree has its origins in Central Asia, as with other stone fruit trees, such as peaches, apricots, and various plum trees. Even today, many endemic populations are found in all the mountains stretching from Tian Shan to Kurdistan, via Turkestan, Afghanistan, and Iran. There are theories that suggest that the cultivated almond tree has its origin in the crossing between species such as *P. fenzliana, P. bucharica* and possibly others. Knowing these species, it seems that among the species that could contribute would be *P. kuramica*, as morphologically it is the closest to the cultivated species. It is likely that frequent hybridizations between these species occurred in the places where they cohabited.

According to Kester et al. (1990), another possible origin of cultivated almond could be the domestication of populations of *P. communis* L.; the similarities between this species and the cultivated almond in terms of tree size, seed and fruit characteristics, suggests direct selection of this species could have led to cultivated almond. The proximity of natural wild populations to centers of civilization in the mountains of Central Asia made their cultivation possible since ancient times. Dissemination to different Asian countries was aided by the fact that the seed was both the unit of propagation and the edible part. Thus, it was distributed throughout Persia, Mesopotamia, and, through trade routes, throughout early Mediterranean and Central Asian civilizations. Toward the East, the almond tree did not spread, probably due to the less favorable climate of the vast intercontinental regions of Central Asia (Figure 12.1).

Until the 19th century, the cultivation of the almond tree evolved little, being propagated almost exclusively by seed and constituting orchards in the areas most

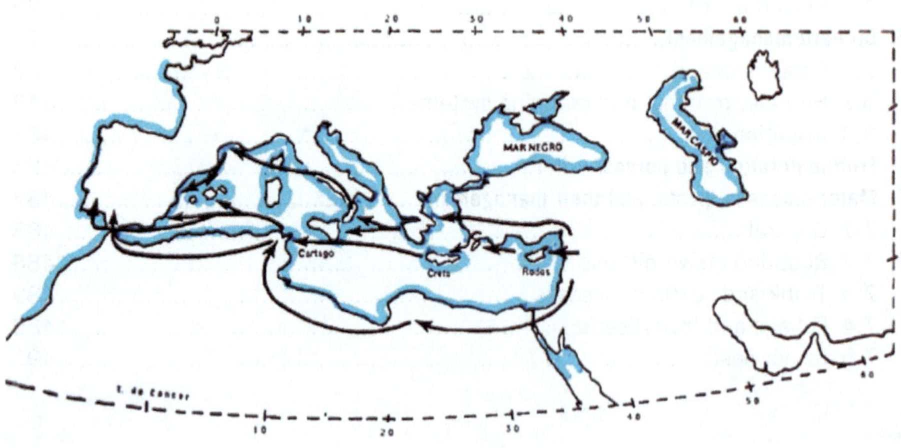

FIGURE 12.1

Dissemination routes followed by almond cultivation in the countries surrounding the Mediterranean Sea.

favorable to its needs: Sicily, the Balearic Islands, the Aegean Islands, etc. Due to repeated multiplication by seed, subjected to natural selection within regional environments, it became domesticated much like that of a forest species. In this way, man also contributed to its evolution by eliminating plants that had unfavorable characteristics and produced few or bad fruits. The large size of the fruit was a trait that has always been considered favorable for the conservation and multiplication of the plants that produce them. It seems likely that it was in Western Asia that the confluence of domestication and early almond cultivation occurred. Normally, almond trees grew mixed with other crops and with minimal cultivation practices, conditions that remained unchanged for centuries. These plantations have been subjected throughout history to the effect of natural and human selection, giving rise to different ecotypes and local varieties that can still be recognized today.

Since the beginning of the domestication of the almond tree, a key goal was the selection of sweet kernel genotypes. Early cultivation was usually relegated to marginal areas, relying on the hardiness that characterizes the species. Due to this resilience, almond has until recently primarily been considered for more marginal areas that did not receive special care because the maximization of its production was not sought. Any harvest was well received as a complement to the family economy. In many cases, these marginal soils were also planted with low-input/low-return crops of other species. The creation of well-planned farms and the application of modern cultivation techniques revealed the almond's true productive potential, so that it has come to be understood that the almond tree can be an important cash crop when planted in favorable environmental conditions and modern cultivation techniques are applied.

The almond tree has been cultivated in Mediterranean countries since ancient times, having been propagated almost exclusively by seeds. These seeds were the result of cross-pollination due to almond's self-incompatibility, resulting in heterogeneous offspring. Although grafting was known and used since ancient times, it was not widely used in Spain, France, Tunisia, and Italy until the 19th century, where plantations created by sowing seeds are still common. As the use of grafting became more widespread in these countries, local selections adapted and associated with production areas with specific environmental conditions were established as regionally improved varieties.

In California, the cultivation of almond began in a different way than in Mediterranean countries. It seems to have been mainly originated from seeds from varieties of French origin, already in some way selected for adaptability. As a result of cross-pollination between a relatively small number of these first California selections, the Californian varieties were selected, being similar to each other and adapted to the climatic conditions of cultivation and industry of the region.

In California, plant material and cultivation techniques evolved in parallel and, because plantings were on rich, irrigated, and intensively cultivated soils, the production of these farms increased, significantly exceeding that obtained in traditional producing countries. As almond became an important commercial crop, studies were initiated on its physiology, behavior with different types of rootstocks, genetic improvement programs, etc., which, in turn, has led in a relatively short time to a much better understanding of the tree and its cultivation. Consequently, the traditional view of almond productivity has been renewed, no longer being considered as just another low-input crop.

As with other crops, the almond tree is subject to the problems associated with the environmental conditions of the regions in which it grows. In the Mediterranean countries, the main problem has been the low yield of traditional plantations. The underlying causes include the lack of knowledge of almond physiology and development of almond dependence on rainfall for water, which often resulted in pronounced water stress. In addition to the low-input nature of Mediterranean production, other contributors for productivity included poor cross-pollination because of insufficient pollinizer trees, poor overlap in flowering time between the producing and pollinating varieties, insufficient pollinators damage from various diseases due to the lack of phytosanitary protection, and vulnerability to late frosts that reduce final crops. In California and Australia, the intensive management of almond orchards raised its production per hectare to levels that did not seem imaginable using traditional techniques. However, there are also different problems in these areas, including excessive irrigation and poor drainage of some fields leading to suffocation, toxicity due to salts, and increased frequency of fungal diseases. New varieties were often based on a very reduced genetic base, which causes inbreeding problems to appear.

2. Economic importance

Almond is the most important fruit nut crop worldwide, with a cultivated area of about 2.3 million ha and an annual in-shell production exceeding 3.6 metric tons (FAOSTAT). The economic importance of almonds is the result of multiple factors related to the regions where they are cultivated, their nutritional value, the diversity of products created by the food industry, the export market, and last but not least, the resilience of production in a climate-change scenario. Almonds are healthy due to their wide spectra of compounds: healthy fats, proteins, fiber, vitamins, and minerals. They are consumed as a snack, in bakery, dairy, confectionery, and even used in cosmetics.

Almond is cultivated in a wide range of countries, with a 75%-85% of production located in California, followed by Spain and Australia, that together produce up to 10% (Figure 12.2). Spain has the most area harvested with more than 700,000 ha. However, California with more than 500,000 ha has the highest yield of approximately 25,000 kg/ha.

Consequently, almonds are widely exported. Again, the U.S. is the dominant country exporting almonds primarily to the EU and also to Asian countries. India is the largest importer (more than 60% total), followed by Spain (15%) and Germany (14%). Regarding per capita consumption, United Arab Emirates ranks first with 4.5 kg, followed by Spain with 2 kg, and other countries like Tunisia, the Netherlands, Australia, and the U.S. consuming about 1.3 kg/capita. The almond

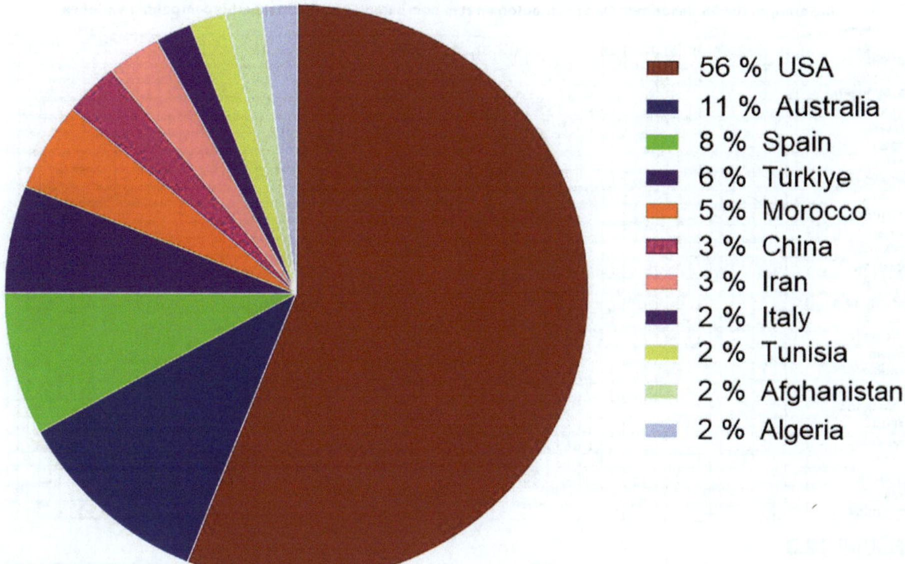

FIGURE 12.2

Top 15 worldwide countries producing almonds. The global distribution of almond cultivation is concentrated in Spain and U.S. (California), where a majority of almond production occurs.

market size is expected to reach $10.12 billion in 2024 and grow at a compound annual growth rate of 5.55% to reach an expected $13.27 billion in 2029.

3. Botany and related species

The cultivated almond has been botanically cited in scientific literature with different denominations: *Prunus dulcis, Prunus amygdalus, Prunus communis*, and even *Amygdalus communis.* Its botanical classification is: Family: Rosaceae, Subfamily: Prunoideae, Genus: *Prunus,* Subgenus: *Amygdalus.*

The *Amygdalus communis* denomination is attributed to Linnaeus and has been adopted by many botanists. *Prunus dulcis* was first used in 1768 to refer to the cultivated almond having a sweet kernel. The use of this denomination implies that *Amygdalus* (Mill.) is considered a subgenus of *Prunus* instead of a genus itself. In 1801, Batsch used the denomination *Prunus amygdalus* (Mill) (derived from the Greek term "amygdalo") to refer to almond cultivars. Some years later, in 1882, Archangeli used the term *Prunus communis.*

After this, some botanists grouped the almond cultivars within the subgenus *Amygdalus* (Mill.) as *Prunus dulcis*, while others preferred to use the denomination *P. amygdalus*. In 1964, due to the existing discrepancy, the International Association

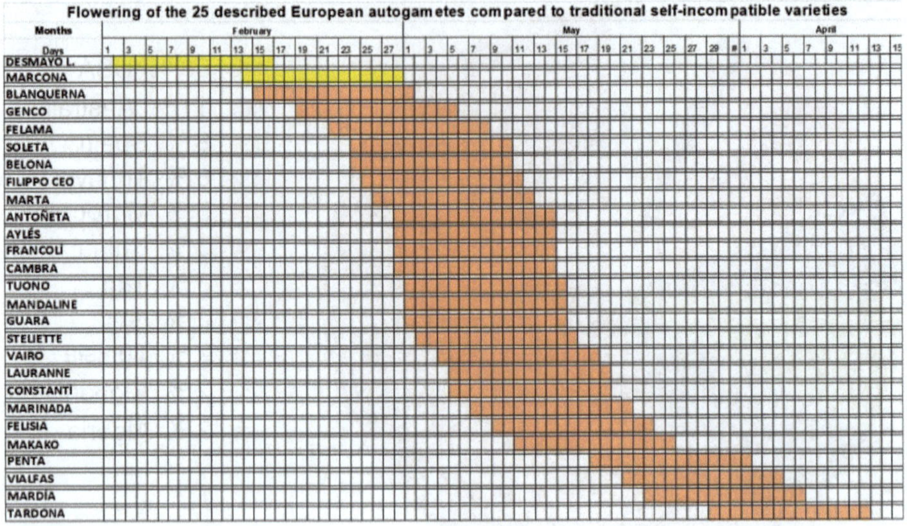

FIGURE 12.3

Flowering times of self-fruitful varieties compared to traditional varieties (estimated data of averages for cold and early zones).

for Plant Taxonomy decided to propose the denomination *Prunus dulcis* (Mill.) for the sweet-kernel cultivated almond, accepting the denomination *P. amygdalus* as a synonym, and later considering *P. communis* a wild ancestor of the present cultivated almond. Practically all the names mentioned before can be found in the scientific literature.

Within the *Amygdalus* subgenus, more than 30 different species of wild almond have been described, including the cultivated almond. They show different morphological traits and can be found in different natural areas of distribution, mainly distributed in Southeastern and Central Asia. These species are classified in five sections.

The section *Euamygdalus* Spach. groups the modern cultivated almond with the species that have been considered as its wild ancestors. They are very close genetically, having the ability to cross among themselves spontaneously, but their geographical distribution is not the same. The majority of these species originated in Central and Western Asia, and they can still be found in the wild in those areas. Their botanical identification is complex due to the fact that most of them are self-incompatible, which on one hand generates a high variability inside the species and on the other hand favors highly frequent hybridizations among them. The species included in this section are: (i) *P. dulcis* (Miller, D. A. Webb), (ii) *P. bucharica* Korschinsky, (iii) *P. communis* (L.) Archangeli, (iv) *P. fenzliana* Fritsch, (v) *P. kuramica* Korschinsky, (vi) *P. orientalis* (Mill.) = *P. argentea* (Lam), (vii) *P. kotschii* (Boissier & Hohenm. [Nab.] & [Rehd.]), (viii) *P. korschinskii* Hand.-Mazz, (ix) *P. webbii* (Spach) Vierh., and (x) *P. zabulica* Serafimov. The main interest in these species is not only because of their contribution in the genealogy of the cultivated almond but because of their potential use as parentals for genetic improvement, including their use as rootstocks. The resilience of many of these species could be a very interesting trait for nonfavorable environments, as well as their potential resistances to biotic stresses. *P. webbii* (Spach) is the only almond wild relative that is naturally present in the north Mediterranean countries, from the west of Turkey to the center of the Iberian Peninsula (Toledo Province). In all these areas, numerous hybrids between this species and the cultivated almond can be found since they often cross naturally. It is not uncommon to find species with self-fertile individuals. Is very probable that the self-fertile almond cultivars have inherited this trait from *Prunus webbii*, since the origination of these cultivars happened in areas where *Prunus webbii* is naturally present (for example, Apulia in Italy and Toledo in Spain).

3.1 The germplasm collections

Over the last decades, the release of new commercial cultivars with desirable productive traits has caused a generalized decrease of the use of local cultivars and hence a high risk of genetic defects. This has increased awareness of the importance of the conservation of the plant genetic resources either in situ or ex situ, for their potential use in the future. The first almond collections were established in the 19th and the 20th centuries. The almond collection at Yalta (Crimea) originated at the end of the 19th century. This collection included more than 200 varieties and wild species. At the same time in the U.S., Wood described 150 almond cultivars in California, most of them coming from seedlings brought in by gold miners and other immigrants, thus establishing the unique Californian ecotype. In Europe, the National Institute of Agricultural Research (INRA) collection in France started in 1951 and gathered over 600 introductions by 1974 that represented 400 cultivars from all over the world. In Spain, different centers (IRTA in Catalonia, CITA in Aragon, and at CEBAS-CSIC in Murcia) started their own collections in the 1960s and 1970s. Other important collections in Europe were established in Italy, Greece, and Portugal by the end of the 20th century. Tunisia and Morocco in Africa and Argentina and Australia in the South Hemisphere keep their own local almond germplasm collections as well. These collections have provided remarkably diverse plant material for the study of the behavior and adaptability of the cultivars in each region and, overall, represent the starting point for genetic improvement programs pursuing different interests and objectives depending on the challenges of each specific region.

4. Breeding and propagation

4.1 Rootstock breeding

The cultivar/rootstock constitutes the complete binary plant in an almond tree, and both should be chosen carefully considering a plantation life that lasts decades. For the grower, choosing the right cultivar(s) is not easy, but the choice is even more complicated because you must also choose the most appropriate rootstock for your specific conditions. The diversity of soil types and pathogens combined with the need to withstand the adverse environmental conditions have led to the search for rootstocks adapted to a wide range of situations. As it is not easy to achieve this goal in the first attempts, thus new selections have been incorporated that include some advantages over preexisting ones. Work on genetic improvement of stone fruit rootstocks continues to produce innovations that expand the choices of those available. Active rootstock breeding programs are underway in the Mediterranean basin as well as in the United States. Rootstocks are generally propagated by two methods: sexually, by sowing seeds with the resulting plants called seedling rootstock, or vegetatively or asexually with rootstocks obtained in this way called clonal rootstocks because they are clones, being derived from a single initial plant (Rubio-Cabetas et al. 2017).

Seedling rootstocks: Self-seed from a seedling rootstock produce plants with characteristics that are more similar to each other than those derived from different rootstocks. Experience has shown that rootstock seed give rise to offspring with an acceptable regularity of behavior, though seedling rootstock never achieved the degree of homogeneity obtained by vegetative propagation.

Clonal rootstocks: Propagation that ensures the maintenance of a single original genotype is called clonal propagation. The resulting rootstock have identical genetic background and so express a very similar behavior. Each clonal rootstock has its own adaptability and requirements in terms of the soil and environmental conditions. A wide range of clonal rootstocks are now commercially available that allow the successful cultivation of a given scion variety in a diverse range of soil types and environments.

The demand for almond rootstocks has increased considerably in Spain as well as North Africa, where the Spanish nurseries are also selling trees. The actual number of hectares planted is difficult to calculate since, unlike the U.S., some new orchards are being planted under intensive systems using dwarfing rootstocks at row and tree spacings of 3.5 m and 1.0-1.5 m, respectively (2222 to 2857 trees/ha). The demand for almond rootstocks has exceeded the supply in Spain, in such a way that most Garnem and GF-677 liners ready for peach grafting had been used for almond grafting. Research on high vigor almond by peach hybrids conducted from the Spanish breeding program of CITA since 1998 led to the GN series from the cross of almond genotypes Garfi as the female parent, chosen for its good morphological characteristics and easy vegetative propagation, and peach (*P. persica* [L.] Batsch) Nemared as the male parent, selected for its resistance to root-knot nematodes (RKN) and its distinctive red color that allows it to be easily distinguished. The selected clones were released as the well-known Garnem, Felinem, and Monegro rootstocks (Esmenjaud et al., 2009, Felipe, 2009). A private breeding program from Agromillora released in 2010 selected rootstocks of low vigor, good soil adaptation, and high productivity. Mainly, three rootstocks with different vigor were released: Rootpac 20, Rootpac 40, and Rootpac R (Pinochet, 2010). Recently, a new low-vigor rootstock also from the GN series was released (Felipe et al., 2022). Other rootstocks with different levels of vigor were also developed, and tolerance to calcareous soils and resistance to RKN continue to be studied. Among the most prominent characteristics of these new rootstocks are low vigor, which results in greater productivity of the grafted variety, and its apical growth habit, its easy vegetative propagation and good graft compatibility, its requirement for a shorter winter rest period, its rapid entry into production, and its better water use efficiency (WUE) (Bielsa et al., 2018). Montesinos et al. (2021, 2022) reported that cultivars grafted onto Pilowred have shorter internodes compared to those that would be grafted onto more vigorous rootstocks such as Garnem and Rootpac 40. In France, Myrotop, a myrobalan rootstock, has recently been released by INRA,with tolerance to Armillaria root rot, resistance to nematodes, and waterlogging tolerance. Its graft compatibility with Ferragnès, Lauranne, and Nonpareil is under study. Similarly, in the United States a range of new rootstocks have become available that are primarily interspecific hybrids. Species parents include almond and peach but also plum and even apricots (Vahdati et al., 2021).

Considering that the current demand for almond rootstocks often exceeds supply, many new rootstock releases have been widely planted without extensive field evaluation. In California, almond growers are using clonal hybrid *Prunus* spp. rootstocks including almond by peach (Nickels, Brights Hybrid 5, Cornerstone, NanoPac), peach by *P. davidiana* (Empyrean 1 a.k.a. Barrier 1), plum by peach (Krymsk 86), plum by almond (Rootpac R), and complex hybrids (Viking, Atlas). Many of the recently released rootstocks require additional testing with almond cultivars to determine their optimal adaptation as well as potential problems such as rootstock incompatibility. In addition, the peach-almond hybrid Hansen 536 and the peach seedling Guardian rootstock are currently being planted on saline/high pH soils and bacterial canker sites, respectively.

An ideal rootstock is not yet available for every growing situation. The hybrids between almond and peach trees generally give good results for almond cultivation under either rainfed or irrigated conditions. The most important requirements for a rootstock could be summarized in five main criteria. The first criterion is the ease of propagation either through *in vitro* propagation or utilizing seed with high germination rates and producing good trueness to type in the resulting plants. Graft compatibility is a very important criterion for compatibility with regionally important cultivars. A third requirement is resistance to biotic stresses such as nematodes (root-knot and Gall), bacteria (i.e., Agrobacterium, Phytophthora fungi *[Armillaria, Verticillium]*), or insects (Capnodis). The resistance to abiotic stresses including adaptation to problematic soils, i.e., including high soil calcium content, drought, and waterlogging, is an additional criterion. A final requirement would be good field survival and orchard growth after transplanting (Rubio-Cabetas et al., 2017).

4.2 Cultivar breeding

Almond cultivars are very diverse in morphology, physiology, and agronomic behavior as well as fruit and nut composition. This diversity represents a genetic richness that allows continued improvement leading to the emergence of new varieties, though often with common ancestors (Perez de los Cobos et al., 2021). The choice of cultivar is largely determined by its compatibility with local growth conditions and markets. The production and commercial structures of the most major producing countries currently focus on a relatively small number of varieties of well-defined commercial types, many of which were originally selected for regional preference.

Since consistent productivity is required over many decades for a production orchard to be profitable, the choice of cultivar, as with rootstock, is crucial. A mistake in that choice will carry over for the entire life of the plantation. Any intervention to correct it, when feasible, represents an additional cost to the grower, reducing its profitability. At the time of planting, the choice of varieties largely determines the production capacity and the marketing possibilities of the crop, while the choice of rootstock provides the ability to adapt to the soil conditions and cultivation techniques adopted. These considerations also affect the planting system: row distance and within-row density, and the distribution of varieties in the field. It is very important to look for reliable information about the varieties (Figure 12.3) and rootstocks (Figure 12.4) from regions or environmental conditions representative of the orchard. In major production regions of Spain and Portugal, the occurrence of late spring frosts and scarcity of bees make self-fruitful and late-flowering cultivars desirable. The current availability of these new varieties, which are more productive and more consistent in production, allows modern farms to succeed under more competitive conditions. In California, self-fruitful varieties reduce the dependency on honeybee pollinizers while also reducing farm management and orchard work since key processes, such as chemical sprays and harvest, only have to be performed once rather than the many times required for the multicultivar, self-unfruitful orchards. Almost half of current California acreage, however, remains planted to traditional, self-sterile varieties such as Nonpareil because of their high and consistent levels of productivity and market value.

4.3 Selection criteria

Depending on the region, almond breeding programs will have different objectives. For example, in Spain, due to the market necessities, almond breeding programs are focused on obtaining late-flowering cultivars to escape the latespring frosts, while retaining a sweet kernel, hard shell, self-compatibility, and of course high productivity. Whereas, in U.S. breeding programs, the objectives are more focused on high productivity and market value given the high input costs of the more intensive orchard management practice. Environmental and food safety concerns are also becoming more important as almond moves beyond being primarily a snack food to the development of a range of new almond food products such as almond milk, which have made it an increasingly important component of consumer nutrition.

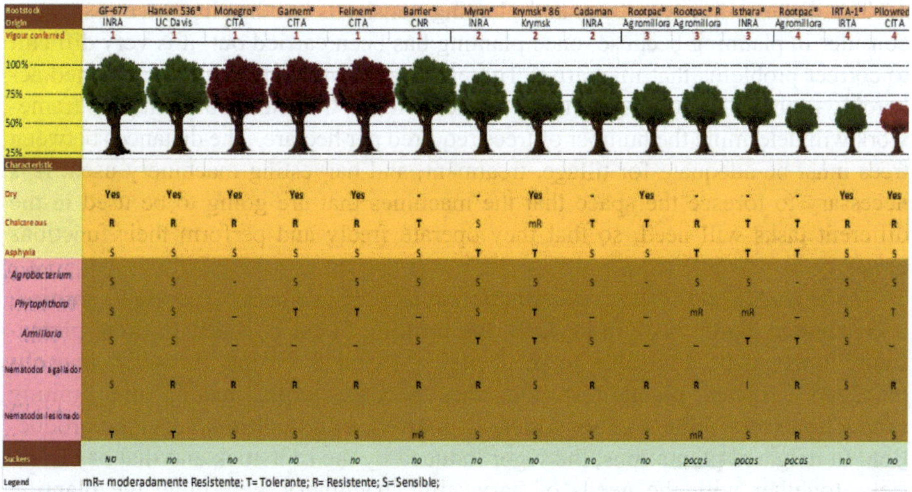

FIGURE 12.4

Hybrid rootstock traits in Mediterranean countries, U.S., and Australia.

Almond is also emerging as a major orchard crop in Europe and North Africa, becoming increasingly competitive with traditional high-value orchard crops such as peach and nectarine. In addition, the increasing cost of field labor combined with the decreasing availability of skilled orchard workers has encouraged greater mechanization. Furthermore, the positive impact of almond production on local farming communities can be further enhanced by promoting local industries that increase the added value of their products and provide additional jobs. Improved job opportunities are also developed within the tourism and entertainment industries as demonstrated by the numerous regional tours and community celebrations developed around the almond bloom season.

5. Orchard management

5.1 Establishment

Self-incompatible cultivars need pollinators that flower at the same time to achieve cross-pollination that is mainly done by bees. This aspect affects the choice of varieties and their distribution in the plantation. Two or more intercompatible varieties that fully coincide with the flowering season are necessary to achieve consistent cross-pollination. Pollen exchange is increased if the number and distribution of bees and pollinizer trees is high. Fifty percent of trees should be pollinizers, and bee density should be between 2.5 and 5.0 hives/ha (Socias i Company et al., 1994; Felipe 2000).

The distance between trees and planting density is another important aspect to consider in planning because, once planting has been carried out, it is very difficult to correct problems that may arise. The planting framework should be planned according to the cultivation system to be practiced and the plant material. This framework will determine the number of trees required per hectare. The distances between trees must be adequate for tillage, treatment, and harvesting machinery used. It is necessary to foresee the space that the machines that are going to be used in the different tasks will need, so that they operate freely and perform their functions correctly.

In the traditional system, a distinction must be made between dryland, rainfed, and irrigated plantations. In dryland, the limiting factor is water. In these plantations, it would be advisable to plant at 7 x 7 and 8 x 8 m. However, in well-managed irrigated plantations, the trees become large, making the limiting production factor light and not water, as is the case in traditional dryland production. In irrigated plantations, the vigor induced by the rootstock and that of the variety, together with the needs of harvesting machinery, determine the planting framework. With the current varieties of moderate vigor, the tree spacing can be from 7 x 7 m, 7 x 6 m, 6 x 6 m, i.e., between 200 and 277 trees per ha. This framework will always depend on the agroclimatic conditions of the area, the availability of irrigation water, and the type of harvesting machinery. With varieties in moderate vigor and faster entry into production, slightly smaller planting spacings of 6 x 6 m, 6 x 5 m, or 5 x 5 m can be used, i.e., between 277 and 333 trees/ha.

The semi-intensive system is based on the same structure as the traditional model, with goblet-shaped trees, although there are other considerations such as planting in hedgerows with much narrower planting frames. The objectives of this model are to bring forward the entry into production of the plantation and reduce the pruning workforce. The planting densities to be used in this model are in the order of 6 x 4 m, 5 x 5 m, 5 x 4.5 m, or 5 x 4 or 3.6 m. The number of trees per ha ranges from 200 to 666 trees/ha, with production greatly increased in the first years of planting, reaching the optimum production much earlier.

The high-density model is based on narrowing the planting densities and changing the tree formation system from a classic basin or hedge to an axis or fruit wall. The limiting factor in this type of plantation is the harvest machine. The volume of the tree is limited to the capacity of the harvest machine. Trees must be reduced in size depending on the machinery available. The planting patterns used consider a tree height of no more than 2.7 to 2.9 m, established at 3.5 to 3.0 m between lanes, and 1.0 to 1.20 m between trees. In this way, we went from 1,502,800 trees/ha to 2300 trees/ha. At latitudes closer to the 30° parallel, both in the Northern and Southern Hemispheres, the distance between lanes is reduced by 3.0 m, since as we approach the equator, the solar arc is sharper and the light rays fall more perpendicularly, so that the projection of shadows between rows is lower. With this higher density of plants, there has been a gain in ease of management and early production, aspects that are difficult to achieve in wider frames that have also been tried, such as the 4 x 2 m, which forces the trees to form according to a false palmette similar to that used in the cultivation of fruit trees. Planting frames of 4 m between lanes have been observed to decrease productive potential. This proximity between plants allows for a continuous vegetative wall by the third year, facilitating and increasing the efficiency of the harvesting machines, which, because they do not encounter empty spaces, wear out and damage to a lesser extent (Torrens, 2015; Felipe et al., 2022).

5.2 Pruning, training, and cropping systems

Short pruning: The main objective of this type of pruning is to achieve a good structure of the tree, although delaying entry into production. It is mainly based on the reduction of the main and secondary branches inside and close to the light, and those located in the canopy plane after the primary branches. Thinning of the secondary branches is done to invigorate the branches chosen to form the major production scaffolds. This type of pruning is mainly carried out in under-rainfed conditions with planting densities of traditional plantations.

Long pruning: This type of pruning has the same criteria as short pruning but without the reduction of the major branches. This type of pruning prioritizes the early entry into production and the reduction of costs over the structure of the tree. Suckers and secondary branches close to the interior or at steep angles are usually removed. The main branches are only pruned back if they have little secondary branching or if they grow excessively. The location of the cut is high, removing less than 25% of a branch.

Long or minimal pruning: The level of pruning is low. The structural branches (primary and secondary) are not typically pruned back, removing only the inner and intersecting branches, but also incorporating a slight thinning if there is an excessive density.

Short or heavy pruning: The level and severity of pruning is high. All structural branches (primary and secondary) are pruned back in the first 2 to 3 years and almost all dying branches are removed. Short pruning, starting from the first 2 years, requires many more hours of work than long pruning or minimal pruning. These differences will become more pronounced over the years.

Method 4.0, minimal pruning: Considering the dissemination of a pruning system developed in Caspe that was first applied in Cerezo under irrigation (Poblador) and is currently very widespread by Abel's nursery as "Method 4.0" (Poblador and Laya, 2023), this pruning method differs appreciably from some of the traditional pruning methods and is being applied in different almond-growing areas with successful results. Method 4.0 refers to interventions that are carried out only up to the fourth branch. In principle, it is an amazingly simple pruning technique applicable to any variety, resulting in a hedge tree more similar to an ornamental than to a traditional goblet tree. It consists of the following phases:

Phase 1: Trimming to support a single terminal bud. In windy areas, a lateral pruning is done to favor the face that suffers from the prevailing wind.

Phase 2: From the formation to the first harvest, pruned back with a hedge trimmer.
Phase 3: From the first harvest to the maximum branching, pruned to leave the trunk free at a height of 90 cm for the harvesting machinery.
Phase 4: Prune back only when the tree shows signs of needing more light.

Types of training system: Three types of almond cultivation can be considered: traditional, intensive, and superintensive. In the traditional type, plantations are in areas where some of their environmental requirements are limiting factors (water, soil type and depth, nutrients, etc.); for example, in rain-fed conditions, on mountain slopes, or on limestone soils that are often shallow. This type of cultivation is characterized by producing low and alternating harvests. This is the system that has traditionally existed in most Asian and Mediterranean areas.

In the intensive system, cultivation practices aimed at maximizing production are used with the goal of eliminating any limiting factors. The soils in which the plantations are located are rich and the optimal irrigation and fertilization are applied, as well as the appropriate phytopathological treatments. Production per hectare is much higher, but so are the costs. This system is mainly practiced in California and Australia. It is also practiced in modern plantations in Mediterranean countries. At present, it is also necessary to consider superintensive or high-density plantations, which have recently been put into practice in Spain using new plant material and are being extended to other producing countries.

Topping: With defined planting densities (3.0−3.5 x 1.0−1.2 m) and current harvesting machines, the vegetation height should be limited to 2.6 to 2.8 m. Improved machine designs will make it possible in the near future to raise this height to 3.5 m. This height, if the training guidelines have been followed and there have been no climatological adversities or pest damage, is usually reached between the second and third year of the almond tree's development. In the summer (July-August), a cut at the desired height should be made to encourage lateral growth and accelerate the formation of a vegetation hedge. This cutting can be done mechanically with the help of a rotating cutting disc. The working speed of these topping machines is 4−6 km/hour, so one hectare takes between 45 and 60 minutes, with a cost of around 45−60 €/ha. The companies that offer this service include the tractor driver, the cutting disc, and the tractor in the price.

Cutting at the desired height in summer should be considered one more operation to be carried out every year. This pruning, unlike winter pruning, stops vegetative growth at the top of the tree, which contributes to the growth of the lateral shoots and promotes the overall balance of the hedge. The hedging machines are fully hydraulically, driven by individual hydraulic motors on each of the discs and therefore with no undesirable pulleys and belts. They have a high performance due to their ease of handling. Cutting positions with more than 250 degrees of travel also improves worker safety. They usually have individual protection for the engines and the rest of the hydraulic components. The configuration of these machines allows them to cut on both sides. They can be positioned indistinctly on the right or left side of the tractor, with adjustable vertical cuts in a angle favorable to the formation of the branches (up or down), and in horizontal cuts throwing the cut branches away from the tree. This type of hedger can be installed on tractors with a minimum power of 60 hp.

There are companies that optionally incorporate some interesting accessories in their hedgers such as cut disinfection systems that apply products that help the healing of the cut at the same time as pruning is being carried out. Another option is special blades for green cutting. These are propeller discs that generate an outward current of air, preventing the air generated by the standard cutting discs from separating the small interior branches. Such systems are ideal for light vertical cuts.

Mechanical side pruning: Side mechanical pruning should begin once the productive tree structure has formed (typically by year 4) both in height and width of the vegetative wall. It is advisable to apply some copper-based fungicide treatment immediately after pruning to heal the multiple cuts made. The tractor with the side hedger travels along the plantation row at a constant speed. The cuts are usually parallel to the axis or with a slight incline.

Side pruning is essential to maintain light penetration and maintain fruiting buds year after year, as well as allow greater aeration of the hedge, which favors better fruiting of the hedge. This type of pruning is increasingly used in superintensive farms. The width of the hedge should be between 65 to 75 cm. The pruning program of each orchard will depend to a large extent on the degree of renewal obtained and other factors such as rootstock, variety, availability of water and fertilizers, soil fertility, and hedge management. It is important to recognize the importance of irrigation programming and fertilization as an additional measure to control vegetative development. The more vegetative development obtained, the more frequent pruning will need to be done.

Side pruning can be carried out on an individual hedge wall or simultaneously hedging both walls of the orchard alley, depending on the topography of the plot and the type of machinery available. When the plots are not very flat, it is difficult to cut both adjacent hedge walls simultaneously. The machine has one or two hydraulic arms, each of which has 5–6 rotating cutting discs. If the ground is completely flat, the machine with two arms allows pruning of both sides of an alley with a single pass for a total time of 1 hour per hectare. In irregular topographies, each individual hedge wall is pruned in a separate pass. In summer, light pruning is also carried out before harvest if there is an excess of vegetation. Through this pruning, we achieve the appropriate width of vegetation, avoiding the formation of thick branches and the associated lack of internal sunlight that causes defoliation and reduced fruit set.

5.3 Irrigation

Currently the most accepted method for determining irrigation requirements in almonds is to estimate crop evapotranspiration (Allen et al., 1998) using weather data at or nearby the orchard site. To use this method for scheduling irrigation (i.e., when to irrigate and how much to apply), a soil-water balance must be calculated. There are a number of soil characteristics that are required for this calculation. These include soil available water, allowable water depletion, and effective rooting depth. Even if these characteristics can be determined, a key limitation in this approach is that soil water is assumed to be "equally available" to the plant over some range of soil moisture. Denmead & Shaw (1962) showed that this assumption was incorrect, even when soil and plant properties were known. For instance, at the same level of soil moisture (e.g., 30% VWC, corresponding to 57% available water in this soil), they found that there was no plant stress if the evaporative demand was moderate (4.3 mm day^{-1}) but substantial plant stress if the demand was high (e.g., 6 mm d^{-1}). In this study, plant (corn) stress was measured by the degree of water-use restriction when compared to the water use of plants maintained at high (near field capacity) soil moisture availability. Plant water potential is recognized as a direct thermodynamic measure of water availability within the plant (Swaef et al., 2022), but its value for irrigation management was not recognized until McCutchan & Shackel (1992) demonstrated that the water potential of nontranspiring (covered) leaves (i.e., stem water potential [SWP]; Fulton et al., 2001) could be used to indicate conditions of plant water stress reliably and sensitively. Since this time, there has been relatively widespread adoption of SWP in almonds and other crops. A sensor for continuous and automated measurement of SWP has been used successfully in a number of woody perennial crops, including almond (Lakso et al., 2022). SWP can be measured at any time of day, but typically the highest level of water stress (most negative water potential) occurs around midday, and this time is commonly used in irrigation studies (Shackel, 2011). A key advantage of SWP in almond and other crops is that a reference (nonsoil water limited) SWP (termed baseline SWP; Shackel et al., 2021) has been found to have a nearly linear dependence on atmospheric vapor pressure deficit (VPD). Hence, from the standpoint of irrigation management, SWP values at or near the baseline value indicate that soil moisture is not limiting plant water availability and hence irrigation is not needed. Reductions in SWP have been consistently associated with reductions in important short-term plant physiological processes such as stomatal conductance and photosynthesis in almond (Spinelli et al., 2016) as well as longer-term processes such as fruit size in pear (Shackel, 2007). For young almond orchards, trees that were maintained near baseline SWP for 3 years (average of about -0.8 MPa) had the highest growth (average trunk cross-sectional area (TCA) of 51 cm^2) and a strong linear reduction in growth of about 73% per MPa reduction in SWP (Shackel, 2007).

Regulated deficit irrigation (RDI) during certain portions of the crop cycle is a common practice in almonds for a number of reasons. For instance, irrigation is typically reduced or suspended prior to harvest to allow for mechanical harvest, as well as to reduce the potential for mechanical injury to the bark during shaking. In addition, mild to moderate water stress (SWP of -1.4 to -1.8 MPa) during the hull split period can be used to advance the timing and rate of hull split without reducing yield (Stewart et al., 2011). Advancing hull split also reduces the period of susceptibility of the nut to insect damage and the incidence of hull rot disease (Teviotdale et al., 2001). The irrigation frequency and amount required to accomplish mild to moderate stress in almonds, however, varies widely with soil type and soil moisture content. For instance, within a single irrigation block of the same orchard in Corning, CA, in 2000, trees on gravel soil had reached 91% hull split by September 14, but trees on silt soil were substantially delayed and had only reached 40% on this date (Shackel et al., 2003). Full synchronization (i.e., a significant advancement of hull split for trees on the silt soil) was accomplished using RDI on this soil from 2001 to 2003, but this required substantial differences in both irrigation timing and amount compared to the gravel soil. For trees on the silt soil, irrigation was substantially reduced beginning 2 months prior to harvest, and only 34% of calculated seasonal ET_c was required to reach the target SWP during hull split. In contrast, for trees on the gravel soil, irrigation was only discontinued a few days prior to harvest, and 100% of calculated seasonal ET_c was required to reach the same SWP target (Shackel et al, 2003).

Almonds are widely regarded as drought resistant, and most studies on the response of almond yield to drought have focused on the relation of long-term yield to the seasonal total of water applied. For instance, in a study by Goldhamer et al. (2006), a strong linear relation between applied water and yield was the case for the control treatment (considered 100% ET_c) and for treatments that applied a fixed fraction (deficit) of ET_c throughout the season (Figure 12.5). In this study, treatments that applied variable fractions depending on the crop growth stage (e.g., preharvest, postharvest) had lower yields for the same amount of water as the fixed fraction treatment. This indicates that the timing as well as the amount of deficit are

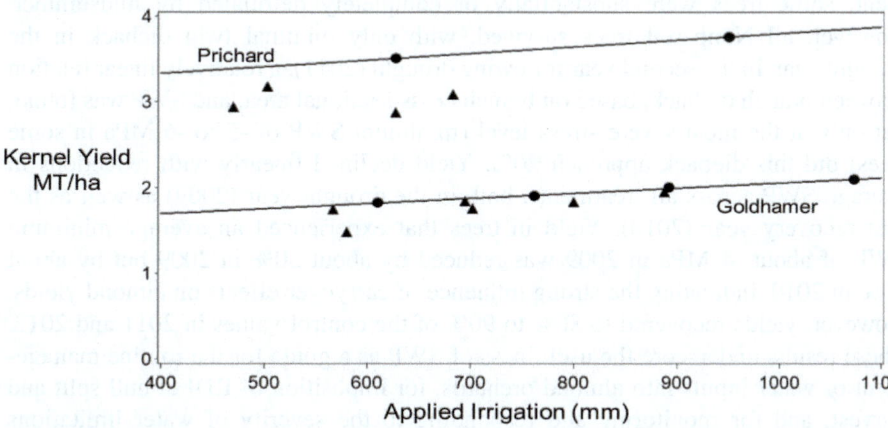

FIGURE 12.5

Relation of average kernel yield to average applied water over 4 years for different levels and timings of deficit irrigation in two separate studies. *Dots* at the highest applied irrigation in each study were the control treatment, and additional dots represent either a fixed fraction of ET (Goldhamer), or a plant-based irrigation regime.

(From Prichard.)

important determinants of yield in almond. Consistent with this were the results of Prichard (1996), who compared a plant-based threshold for irrigation (predawn leaf water potential of -1.2 MPa) to different fractions and timings of an ET-based deficit (Figure 12.5). In this case, plant-based RDI had a yield that was 93% of the control with 66% of the applied water and substantially higher yields compared to other treatments at the same level of irrigation deficit. Plant-based approaches to irrigation management, particularly RDI, represent a paradigm shift from traditional approaches and should allow growers to adjust irrigation more accurately than ET- or soil-based approaches on a site- and year-specific basis. In addition to manual measurement of SWP with a pressure chamber (Fulton et al., 2001), new technologies have also been recently developed for automated and continuous measurement of SWP (Lakso et al., 2022).

SWP will also provide valuable guidance for the efficient use of limited water during periods of drought. A 4-year experiment was performed in a mature Nonpareil almond orchard to document the short- and long-term effects of a single drought year (2009) on tree survival and productivity (Shackel et al., 2012). A severe drought treatment (rainfed) only received 53 mm of in-season rain, compared to a total of 836 mm in the control, with two partially irrigated treatments (145 or 236 mm). Additional drought adaptation strategies of a 50% canopy reduction with and without application of reflective sprays (kaolin) were applied to half of the trees in the nonirrigated treatment, and reflective sprays were also applied to half of the trees in the partially irrigated treatments. This gave a total of seven drought strategy treatments imposed to evaluate treatment effects on tree survival and productivity. The orchard was on a shallow (60 to 120 cm) gravel/loam soil with low water-holding capacity, overlaying a restricting clay (Andreu et al., 1997). In the rainfed treatment, some trees were substantially or completely defoliated by midsummer. However, all Nonpareil trees survived, with only minimal twig dieback in the drought year. In the second year following drought (2011), a relatively linear relation between branch dieback, based on branch cross-sectional area, and SWP was found, but only at the most severe stress level (minimum SWP of -5 to -6 MPa in some trees) did this dieback approach 20%. Yield declined linearly with reductions in average SWP across all treatments, both in the drought year (2009) as well as the first recovery year (2010). Yield in trees that experienced an average minimum SWP of about -4 MPa in 2009 was reduced by about 50% in 2009 but by about 90% in 2010, indicating the strong influence of carryover effects on almond yields. However, yields recovered to 80% to 90% of the control values in 2011 and 2012. These results underscore the usefulness of SWP as a guide for the routine management of water inputs into almond orchards, for imposition of RDI at hull split and harvest, and for monitoring and responding to the severity of water limitations imposed by a drought.

6. Fruit maturation and harvest criteria

Figure 12.6 shows the phases of the annual cycle of the almond tree. In the initial phase, which spans from January to February, the appearance of flowering and the

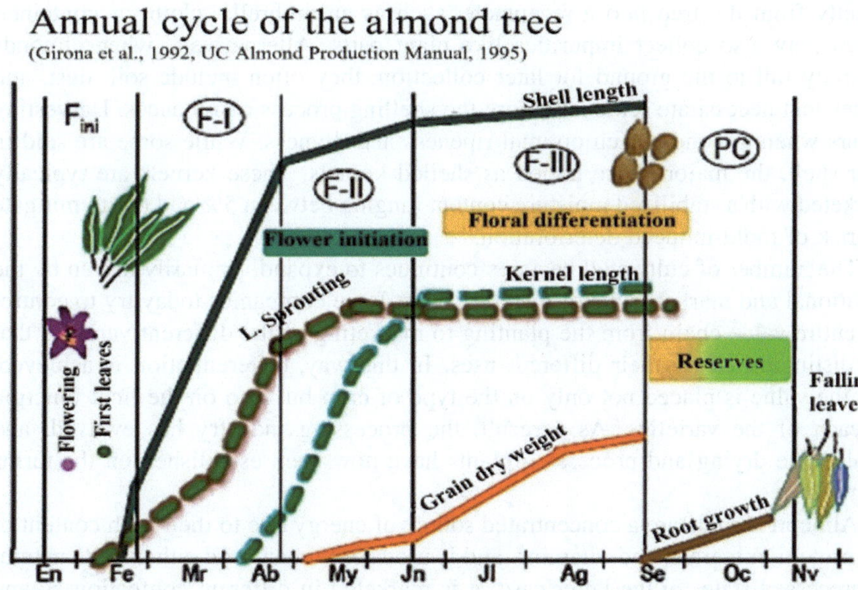

FIGURE 12.6

Annual cycle of almond tree growth and production.

(From J. Girona.)

first leaves begins. In phase I, which spans from the end of February to the end of April, there is a significant increase in the size of the buds and the advancement of flowering. This leads to the beginning of phase II, which extends from the end of April to mid-June. During this phase, the initial stages of flowering and the growth of the length of the seed occur. In phase III, which spans from mid-June to mid-September, there is floral differentiation, stabilization of kernel length, a slight increase in the husk, significant growth of the kernel dry weight, and a decrease in the need for water consumption. In the PC phase, the need for water consumption increases because it is the time of growth of the reserves and roots of the almond tree (Girona et al., 1993; UC Almond Manual,1996).

Once the fruit ripens, the mesocarp dries and splits to reveal the nut inside. The nut contains the shell or endocarp, the embryo (bordered by the two cotyledons), surrounded by the testa, which was formed by the seminal envelopes or integuments and, inside these, the remains of the nucellus and endosperm (Sánchez-Pérez et al. 2008). In contrast to other *Prunus* species, the edible and commercial part of the tree is not the fruit mesocarp but the seed (the kernel of the almond).

Field: The way in which almonds are harvested directly influences the quality of subsequent shelling, selection, and processing. During harvest, depending on the method employed, dust, dirt, and stones may accumulate, impeding efficient shelling, cleaning, and kernel selection. In Europe, almonds are typically gathered directly from the tree into a receptacle, such as an umbrella, cloth, or container, which may also collect impurities like plant parts. Alternatively, when almonds naturally fall to the ground for later collection, they often include soil, dust, and stones that necessitate removal before the shelling process commences. Harvesting occurs when almonds reach optimal ripeness and dryness. While some are sold in their shell, the majority are traded as shelled kernels. These kernels are typically marketed with a stabilized moisture content ranging between 5% and 6% to mitigate the risk of mold-induced deterioration.

The number of cultivated hectares continues to expand, primarily driven by the nutritional and market value of the crop. Most large companies today try to control the entire value chain from the planting to marketing of the different varieties that are distinguished by their different uses. In this way, differentiation is achieved, and the value is placed not only on the type of crop but also on the final function of each of the varieties. As a result, the processing industry has evolved, and small-scale drying and processing plants have now been established on the farms themselves.

Almond kernels are a concentrated source of energy due to their high content of fat, protein, minerals, and vitamins, and is used for human food either in its natural or processed state. In the latter case, it is marketed in different confectionery and pastry presentations or to manufacture nougat, as well as to process almonds into other foods such as almond milk, which has diversified their nutritional value and markets.

The green or fleshy part that makes up the mesocarp contains carbohydrates and fibers that can be used for livestock feed, either in their natural state at the time of harvesting or, once dried and ground, as a constituent part of prepared compound feed. They are also rich in sugars but do not contain protein.

Once stripped of the shell and processed, almonds come in many different forms: natural or peeled, salted or sugared, fried, toasted, seasoned in various ways, and mixed with other nuts. They are also marketed as a nondairy milk, ground into flours, chopped in various presentations of various thicknesses, slices, cubes, or sticks, etc.

Processing: The objectives of the almond processing industry are to ensure continuity of supply and offer a quality product of good preservation. In addition, these specific objectives are contingent on the (1) plant material used (varieties and rootstocks), (2) production techniques (rainfed or irrigated, fertilizers, health, etc.), and (3) the industrial processes applied.

Reception: The reception of the almonds at the processing site is usually done in large hoppers from which samples are usually extracted to determine the product quality. The fruit normally go through a drying process to attain an optimal moisture content, both for the process of removing the hull and shell, and for their storage and preservation.

Hulling: The goal is to remove the fruit hull, which can be done both on small movable machines and on processing lines already associated with crop drying. There are also sweeper hoods and industrial peelers. These processes are associated with conveyor systems (elevators, belts, trays), silos, hoppers, containers, and weighing and traceability systems.

Shelling: The shelling is carried out with shell-splitters that have to adjust to a wide range of sizes, textures, or hardnesses. Almond varieties have different types of endocarps, which require adaptations in the shelling system. The most common types are hard shell (compact or double-shelled) or soft shell. Within both types, there are different degrees of hardness. Depending on the hardness, different procedures are used to shell the seeds, including separating broken pieces. Another varietal characteristic is the presence of single or double seeds. It is advisable to separate the misshapen almonds as well as those of different sizes to market a homogeneous final product. Almonds with noncommercial shapes will be chopped and used in different products such as pastries, nougat, ice creams, and chocolates, etc. these processes are also associated with the sorting lines with separators, cleaning, and precleaning of possible waste from the previous step (cleansers, roughers, and vacuum cleaners).

Peeling: It is the separation of the seedcoat from the seed. The steps of the process are:

- **Hot water:** By immersing the almonds in hot water at a temperature close to the boiling point (blanching) for a short period of time, the seedcoat or integument is softened, allowing it to be easily separated from the grain.
- **Rubbing:** Once the almonds have been blanched, they are subjected to a rubbing process by which the seedcoat is separated from the seed. Rubbing is done by opposing forces acting at certain speeds that rapidly separate the seedcoat from the cotyledons.

The machines can also incorporate calibrating, peeling, and cutting (sheets, sticks, cubes, granules, 3D cutting). There are also mills and refiners for almonds and nuts (fine and ultrafine flour), as well as various types of dryers, coolers, and roasters for nuts. Other specialized equipment includes pasteurization systems, almond blenders (almond milk), and post-processed (paste, pulp, almond butter).

Kernel taste: Wild almonds produce bitter seeds. This bitter content is due to the accumulation of the cyanogenic diglucoside amygdalin. When amygdalin is degraded by enzymes, β-glucosidases, glucose, benzaldehyde (bitter compound), and the toxic hydrogen cyanide is liberated. Currently, most of the varieties that are cultivated are sweet or slightly bitter. A relatively small market remains for the bitter almond varieties.

Although bitter almonds were known since antiquity, only in 2018 was the amygdalin biosynthetic pathway characterized (Thodberg et al. 2018). In this pathway, four genes were identified. A year later, the gene responsible for bitterness, the *Sweet kernel* (*Sk*) gene, encoding a basic helix loop helix (bHLH) transcription factor, was identified (Sánchez-Pérez et al. 2019). The *Sk* gene regulates the synthesis of amygdalin, activating the transcription of the first two genes, two cytochromes P450, the PdCYP79D16 and the PdCYP71AN24. In sweet varieties, the bHLH transcription factor carries a dominant mutation which prevents the expression of the two P450s in the seed. Consequently, prunasin, the precursor of amygdalin, is not synthesized. In bitter (recessive) varieties, the bHLH is able to activate the first two genes. Consequently, prunasin is accumulated in the seedcoat/tegument, and amygdalin is synthesized in the endosperm and, finally, in the cotyledons. Recently, two alleles (*Sk-1*—the most frequent one—and *Sk-2*) for the sweet kernel phenotype have been characterized, and KASP and dual-label functional markers have been developed (Lotti et al., 2023). These markers can be used in any almond breeding program by a high-throughput system to bring only the sweet almond trees to the field. With this big stepping stone, breeders can find the kernel genotype of an almond seedling in a few days, without waiting through the long juvenile phase (3-4 years) that almond has.

7. Major diseases, pests, and their management

7.1 General

Superintensive plantations under high humidity levels may be more affected by certain diseases caused by fungi and/or bacteria than traditional plantations due to less orchard aeration if the vegetation is too vigorous and dense. Certain pests may be of greater or lesser incidence in plantations, especially in specific regions or in certain seasons, due to the endemic nature of various pests. The climatology of each region is the most important factor that facilitates the incidence of most fungal and bacterial diseases. In general, in hot and dry climates, the level of disease is lower than in humid climates. Currently, most of the pests and diseases associated with almond cultivation are efficiently controlled with the application of phytosanitary programs and fungicidal treatments. Almond is vulnerable to fungal, bacterial, and virus diseases. The fungal diseases can be divided into those below ground (root rots and crown diseases), at the trunk and scaffold (canker diseases), and foliar and fruit diseases.

7.2 Root and crown disease

Two major diseases are the cause of severe losses in many locations, Phytophthora root and crown rot and Verticillium wilt. Several species of *Phytophthora* can infect almond grown under excess soil moisture. Thus, the disease becomes severe in wet years and in field sites where waterlogging is common. Surface water can bring propagules of infection into almond orchards. Infection initiates at the wet crown and/or wounded roots and moves into the roots to the trunk to produce distinct cankers. Trunk cankers on the upper margins can show zonation of light alternating with dark brown zonation, and amber-color sap often oozes on the bark surface of the cankered tissues. Infections can begin either in the fall or spring, and as the season's temperatures increase, young trees will wilt and die. In the U.S., there are several fungicides that can control Phytophthora root and crown rot and are listed at http://ipm.ucanr.edu. Good drainage of the soil helps prevent Phytophthora diseases.

In addition to the root and crown Phytophthora disease, almond pruning wounds, trunk cracks, and the tree crotch can be infected by aerial Phytophthora. The pathogen is *Phytophthora syringae.*

Another root disease that can kill young trees is Verticillium wilt caused by *Verticillium dahliae.* Microsclerotia of the fungus surviving in the soil are triggered to germinate by root exudates and infect the roots. The systemic growth of the pathogen plugs the xylem vessels with occlusions, prevents water flow to the upper parts of the trees, and kills the affected tree portions. One to three scaffolds of the tree can be infected, wilt, and die. Disease is severe when almond plantings follow crops susceptible to Verticillium, such as tomatoes, cotton, olive, potatoes, etc. Management of the disease is achieved by planting in fields previously containing crops not susceptible to Verticillium and/or fields without *Verticillium* microsclerotia. Crown gall is a bacterial disease caused by *Agrobacterium tumefaciens* that infects the crown and roots, resulting in the creation of galls. The bacterium needs a wound to infect. Preventative treatment is achieved by dipping the roots or spraying before planting with Galltrol, which is a preparation of the biological control agent *A. tumefaciens* K-84 that produces an antibiotic that can reduce/eliminate infection. Another disease attacking older almond trees is the oak root rot fungus caused by *Armillaria mellea,* a wood decay basidiomycete pathogen. The fungus moves from tree to tree and survives in root parts for several years. Management is achieved with resistant rootstocks and/or particularly good cleaning of the site where killed trees are removed.

7.3 Trunk and scaffold diseases

The trunk of almond can be infected by various canker-causing fungi. Canker diseases of almond include band canker caused by at least eight Botryosphaeriaceae fungi, Cytospora canker caused by *Cytospora plurivora* (syn. *C. leucostoma),* Ceratocystis canker caused by *Ceratocystis fimbriata*, and other less frequent diseases such as Neoscytalidium canker caused by *Neoscytalidium dimidiatum* and Eutypa canker caused by *Eutypa lata.* Avoiding wetting the trunk or using very vigorous rootstocks that result in growth cracks or other practices to reduce wounding should reduce canker disease incidence. The trunk of young almond trees is also susceptible to bacterial canker caused by *Pseudomonas syringae.* Freezing temperatures with rain favor bacterial canker development in all *Prunus* species.

7.4 Foliage and fruit diseases

The canopy parts of almond trees can be infected by a multitude of diseases. Among these, brown rot caused by *Monilinia fructicola* and *M. laxa* can be very devastating in late winter/early spring when wet conditions prevail, and trees are at bloom. The pathogens overwinter in infected tissues from the previous season that produce new sporulation in late winter/early spring. The anthers and stigma of flowers are infected, resulting in blossom blight with the infection moving into the twig, causing twig blight. New sporulation can develop on the flower calyx and surrounding tissues resulting in secondary infection of blossoms of late-blooming varieties. Protective fungicides, starting at the pink tip stage, are highly effective in controlling brown rot. Other blossom diseases include grey mold caused by *Botrytis cinerea,* and bacterial blast caused by *Pseudomonas syringae.* Bacterial blast can become very severe when wet and freezing temperatures occur in late winter/early spring. In essence, the blossoms melt onto the spurs. Other diseases of fruit include jacket rot, which involves the decay of young, green almond fruit and is caused by *B. cinerea*, and scab, caused by *Fusicladium carpophilum* that forms black round lesions on immature fruit. Early-season protective fungicide sprays can easily control these fruit diseases. A very devastating disease of fruit is hull rot. *Rhizopus stolonifer, Aspergillus niger,* and *Monilinia fructicola* can cause hull rot, with *Rhizopus* and *Aspergillus* being most common. In the early 2020s, a new hull rot fungus was found to be caused by *N. dimidiatum,* especially in areas where the inoculum of this pathogen became airborne at the time when fruit were at the early hull split stage. Infections by these fungi occur when the hulls start splitting. The fungi produce abundant sporulation between the shell and the inner surface of the hulls. Infections by Rhizopus result in spur and shoot blight due to secretion of fumaric acid that kills spurs and shoots. Less spur blight and withered leaves occur from infections by *Aspergillus* and none from infections by *Monilinia*. The aboveground foliar and fruit diseases are managed mainly with fungicide sprays in the spring. In addition, a critical spray for hull rot management is done just as hull split begins (http://ipm.ucanr.edu) and by managing irrigation and fertilization. Deficit irrigation and reduced N fertilization reduce disease risk. Practices that keep the orchard under low relative humidity (RH) also reduce disease. Orchards with dense canopy result in increased RH, which favors hull rot. Although deficit irrigation may reduce hull rot caused by Rhizopus, it may create conditions conducive to infection by *Aspergillus*. Another disease of almond fruit is anthracnose caused by *Colletotrichum gloeosporioides* and *C. karsii.* This disease occurs when early spring conditions are very wet. The *Colletotrichum* fungi rot the immature fruit, the fruit shrivels, and the pathogen produces profuse salmon pink sporulation on the fruit hulls.

The leaves of almond can also be infected by diseases. These include shot hole caused by *Wilsonomyces carpophilus,* Alternaria leaf spot caused by *A. alternata* and *A. arborescens,* rust caused by *Tranzschelia discolor*, and leaf blight caused by *Seimatosporium lichenicola*. All these diseases cause necrotic lesions on almond leaves and result in defoliation. There is also a bacterial disease, bacterial spot, that affects both fruit and leaves caused by *Xanthomonas arboricola* pv. *pruni*. Protective fungicides and bactericides provide good control of these diseases (see earlier link). Golden death is another bacterial disease that can kill trees caused by *Xylella fastidiosa*. The bacterium is transmitted to the almond by a small leafhopper when it feeds in the tender shoots of the tree. Apparently, this bacterium cannot survive winters when temperatures are below 0°C. Management of the disease is done only by removing infected trees and controlling the vector.

7.5 Insect pests

Major insect pests of almond include navel orangeworm *(Amyelois transitella)*, carob moth *(Ectomyelois ceratoniae)*, Anarsia, peach twig borer *(Anarsia lineatella)*, flatheaded borer, stink bugs (*Hemiptera* spp.), aphids, and mites. The navel orangeworm (NOW) is a lepidopteron that can cause direct damage to the crop and also increases the risk for aflatoxin contamination of nuts. The insect overwinters as larvae and/or pupae in the nuts that remain on trees after harvest (mummies). After egg hatching, the larvae chew their way in and feed on the kernel. Insect wounds can be infected by *Aspergillus* fungi, *Aspergillus flavus,* or *A. parasiticus* and thus increase the risk for aflatoxin contamination. To break the cycle of NOW, the best approach is sanitation, i.e., removal of mummies from the trees and grinding them to orchard floor cultivation to destroy the NOW in them. Good orchard sanitation is critical for reducing NOW damage. Carob moth is also a lepidopteran insect and occurs in almond grown in Australia and the countries of the Mediterranean basin. Carob moth has one generation per growing season while NOW has three or four generations. Among the stink bugs, the most damaging is the leaf-footed bug (*Leptoglossus* spp.) in California and false tiger and yellow and red frog in Europe. Almonds are susceptible to nematodes with *Meloidogyne incognita*, the root knot nematode, being the most destructive. In the U.S., preplant fumigation is a common practice to control nematodes in new planting sites.

8. Current and future challenges in main production regions

In California, the main production challenge is the loss of good-quality irrigation water as well as favorable soils. Consequently, breeding priorities have shifted toward greater water use efficiency (WUE) and associated tolerances to less favorable soils including those with higher levels of accumulated salts. In addition, more environmentally friendly disease and pest control is being pursued through more integrated pest/disease management that includes improved genetic resistance in varieties and rootstocks. Reduced environmental damage is also being pursued such as reducing mechanized traffic in orchards with its associated generation of atmospheric dust. The reduction of dust generated during harvest (sweeping and gathering nuts from the ground) in dry climates, such as in California, is required to reduce air pollution.

In Europe, the development of self-fruitful varieties and late-flowering cultivars from different breeding programs has allowed for the expansion of almond crops to different valleys and inner areas with no previous history of almond crop cultivation but with greater water availability. This expansion has increased knowledge concerning the performance of the different cultivars to areas where they were not originally elected. In Spain, the almond crop is moving from Mediterranean climates to more Atlantic climates such as Portugal. California has also started planting single-variety self-fruitful orchards that can reduce both orchard traffic and associated dust since orchard management practices such as spraying and harvesting can now be performed just once rather than the multiple spray and harvest practices previously required for traditional multicultivar orchards.

A major challenge in all the areas, especially in Europa due to the Green Deal, is the use of softer materials to control pests and diseases. These materials, in general, have lower efficacy, and one would expect higher future losses from pests and diseases than when using conventional materials. Finally, plantations of almonds in new landscapes and environments make them prone to invasive pests and diseases. Ongoing climate change and associated water scarcity will become more severe requiring a better understanding concerning which almond cultivars and rootstock are more efficient in water usage and climate resilience.

References

Allen, R. G., Pereira, L. S., Raes, D., & Smith, M. (1998). *FAO irrigation and drainage paper 56*. Rome, Italy: Food and Agriculture Organization of the United Nations, 92-5-104219-5.

Andreu, L., Hopmans, J. W., & Schwankl, L. J. (1997). Spatial and temporal distribution of soil water balance for a drip-irrigated almond tree. *Agricultural Water Management, 35*, 123–146.

Bielsa, B., Bassett, C., Glenn, D. M., & Rubio-Cabetas, M. J. (2018). Assessing field Prunus genotypes for drought responsive potential by carbon isotope discrimination and promoter analysis. *Agronomy, 8*, 42.

Denmead, O. T., & Shaw, R. H. (1962). Availability of soil water to plants as affected by soil moisture content and meteorological conditions. *Agronomy Journal, 54*, 385–390.

Esmenjaud, D., Voisin, R., van Ghelder, C., Bosselut, N., Lafargue, B., di Vito, M., et al. (2009). Genetic dissection of resistance to root-knot nematodes Meloidogyne spp. in plum, peach, almond, and apricot from various segregating interspecific Prunus progenies. *Tree Genet Genom, 5*, 279–289.

Felipe, A. J. (2009). 'Felinem', 'Garnem', and 'Monegro' almond × peach hybrid rootstocks. *HortScience, 44*, 196–197.

Fulton, A., Buchner, R., Olson, B., Schwankl, L., Gilles, C., Bertagna, N., et al. (2001). Rapid equilibration of leaf and stem water potential under field conditions in almonds, walnuts, and prunes. *HortTechnology, 11*, 609–615.

Girona, J., Marsal, J., Cohen, M., Mata, M., & Miravete, C. (1993). Physiological, growth and yield responses of almond (*Prunus dulcis* L.) to different irrigation regimes. *Acta Hortic, 335*, 389–398.

Goldhamer, D. A., Viveros, M., & Salinas, M. (2006). Regulated deficit irrigation in almonds: Effects of variations in applied water and stress timing on yield and yield components. *Irrigation Science, 24*, 101–114.

Kester, D. E., Gradziel, T. M., & Graselly, C. (1990). Almonds (Prunus). *Acta Hortic, 209*, 699–758.

Lakso, A. N., Zhu, S., Santiago, M., Shackel, K., Volkov, V., & Stroock, A. D. (2022). A microtensiometer sensor to continuously monitor stem water potentials in woody plants - design and field testing. *Acta Hortic, 1335*, 317–324.

Lordan, L., Zazurca, L., Maldonado, M., Torguet, L., Alegre, S., & Miarnau, X. (2019). Horticultural performance of 'Marinada' and 'Vairo' almond cultivars grown on a genetically diverse set of rootstocks. *Sci Hortic, 256*, Article 108558.

Lotti, C., Minervini, A. P., Delvento, C., Losciale, P., Gaeta, L., Sánchez-Pérez, R., et al. (2023). Detection and distribution of two dominant alleles associated with the sweet kernel phenotype in almond cultivated germplasm. *Frontiers in Plant Science, 14*, Article 1171195.

McCutchan, H., & Shackel, K. A. (1992). Stem-water potential as a sensitive indicator of water stress in prune trees (Prunus domestica L. cv. French). *Journal of the American Society for Horticultural Science, 117*(4), 607–611.

Montesinos, Á., Grimplet, J., & Rubio–Cabetas, M. J. (2022). Proleptic and sylleptic shoot formation is affected by rootstock genotype in two-year-old branches of almond trees. *Agronomy, 12*, 2006.

Montesinos, Á., Thorp, G., Grimplet, J., & Rubio-Cabetas, M. (2021). Phenotyping almond orchards for architectural traits influenced by rootstock choice. *Horticulturae, 7*, 159.

Pérez de los Cobos, F., Martínez-García, P. J., Romero, A., Miarnau, X., Eduardo, I., Howad, W., et al. (2021). Pedigree analysis of 220 almond genotypes reveals two world mainstream breeding lines based on only three different cultivars. *Horticulture Research, 8*, 11.

Pinochet, J., Angles, M., Dalmau, E., Fernandez, C., & Felipe, A. (1996). Prunus rootstock evaluation to root-knot and lesion nematodes in Spain. *Journal of Nematology, 28*, 616–623.

Pinochet, J. (2010). "Replantpac" (Rootpac R), a plum-almond hybrid rootstock for replant situations. *HortScience, 45*, 299–301.

Prichard, T. (1996). Residual effects of water deficits and irrigation strategies on almonds. *Recuperado de*. https://rd.almondboard.com/data/A.1995.95-M7.Prichard.Residual%20Effects%20of%20Water%20Deficits%20and%20Irrigation%20Strategies%20on%20Almonds.pdf.

Rubio-Cabetas, M. J., Felipe, A. J., & Reighard, G. L. (2017). Rootstock development. In En R. Socias i Company, & T. M. Gradziel (Eds.), *Almonds: Botany, production and uses* (pp. 209–227). CABI.

Sánchez-Pérez, R., Jørgensen, K., Olsen, C. E., Dicenta, F., & Møller, B. L. (2008). Bitterness in almonds. *Plant Physiol, 146*, 1040–1052.

Sánchez-Pérez, R., Pavan, S., Mazzeo, R., Moldovan, C., Aiese Cigliano, R., Del Cueto, J., et al. (2019). Mutation of a bHLH transcription factor allowed almond domestication. *Science, 364*, 1095–1098.

Shackel, K. A. (2011). A plant-based approach to deficit irrigation in trees and vines. *Horticultural Science, 46*, 173–177.

Shackel, K. A. (2007). Water relations of woody perennial plant species. *Journal International des Sciences de la Vigne et du Vin, 41*, 121–129.

Shackel, K., Buckner, R., et al. (2003). Deficit irrigation management during hull-split. *Recuperado de*. https://rd.almondboard.com/data/A.2003.03-KS-01.Shackel.Deficit%20Irrigation%20Management%20During%20Hull-Split.pdf.

Shackel, K., Edstrom, J., et al. (2012). Drought survival strategies for established almond orchards on shallow soil. *Recuperado de*. https://rd.almondboard.com/data/P.2012.12-HORT8-Shackel.Shackel.Drought%20Strategies%20for%20Established%20Almond%20Orchards%20on%20Shallow%20Soil.pdf.

Shackel, K., Moriana, A., Marino, G., Corell, M., Pérez-López, D., Martin-Palomo, M. J., et al. (2021). Establishing a reference baseline for midday stem water potential in olive and its use for plant-based irrigation management. *Frontiers in Plant Science, 12*. https://doi.org/10.3389/fpls.2021.79171

Socias i Company, R, Espada, J. L., & Felipe, A. J. (1994). The effect of orchard design on almond fruit set. *HortTechnology, 4*(4).

Spinelli, G. M., Snyder, R. L., Sanden, B. L., & Shackel, K. A. (2016). Water stress causes stomatal closure but does not reduce canopy evapotranspiration in almond. *Agricultural Water Management, 168*, 11–22.

Stewart, W. L., Fulton, A. E., Krueger, W. H., Lampinen, B. D., & Shackel, K. A. (2011). Regulated deficit irrigation reduces water use of almonds without affecting yield. *California Agriculture, 65*, 90–99.

Swaef, T., Pieters, O., Appeltans, S., Borra-Serrano, I., Coudron, W., Couvreur, V., et al. (2022). On the pivotal role of water potential to model plant physiological processes. *Silico Plants, 4*, Article diab038.

Teviotdale, B. L., Goldhamer, D. A., & Viveros, M. (2001). Effects of deficit irrigation on hull rot disease of almond trees caused by *Monilinia fructicola* and *Rhizopus stolonifer*. *Plant Disease, 85*, 399.

Thodberg, S., Del Cueto, J., Mazzeo, R., Pavan, S., Lotti, C., Dicenta, F., et al. (2018). Elucidation of the amygdalin pathway reveals the metabolic basis of bitter and sweet almonds (Prunus dulcis). *Plant Physiology, 178*, 1096–1111.

Vahdati, K., Sarikhani, S., Arab, M. M., Leslie, C. A., Dandekar, A. M., Aletà, N., et al. (2021). Advances in rootstock breeding of nut trees: Objectives and strategies. *Plants, 10*, 2234.

CHAPTER

Walnut

13

George D. Nanos[1], Pat J. Brown[2], Davide Neri[3], Kourosh Vahdati[4], Andreas Westphal[5]

[1]*Department of Agriculture, Crop Production & Rural Environment, University of Thessaly, Volos, Greece;* [2]*Department of Plant Sciences, University of California, Davis, CA, United States;* [3]*Department of Agricultural, Food and Environmental Sciences, Polytechnic University of Marche, Ancona, Italy;* [4]*Department of Horticulture, College of Aburaihan, University of Tehran, Tehran, Iran;* [5]*Department of Nematology, University of California, Riverside, CA, United States*

Chapter outline

1. Origin and spread

Wild walnuts are native to a wide range of areas worldwide, including Central Asia, and extending throughout North, Central, and South America along the Andes Mountains to Argentina. The genus *Juglans* is divided into four sections: Trachycaryon (butternut), Rhysocaryon (black walnut), Cardiocaryon (Asiatic heartnut), and Dioscaryon (Persian walnut). Fossil evidence and molecular phylogenetic studies indicate that the center of origin for walnuts in the Trachycaryon and Rhysocaryon sections is North America. These plants spread from North America via the Bering land bridge during the Middle Tertiary period and the North Atlantic land bridge in the late Eocene period. Furthermore, during the late Miocene period, adaptations to the cool climate of the Neogene led to expansions into Southeastern Europe and Central and South America. The Cardiocaryon section is native to East Asia (Aradhya et al., 2007; Bernard et al., 2018; Vahdati et al., 2019).

Temperate Tree Fruits and Nuts. https://doi.org/10.1016/B978-0-443-29137-1.00013-X

Persian (English or common) walnuts (*Juglans regia* L.) belong to the Dioscaryon section and have a long history as a food source, dating back to ancient Persia around 7000 BCE, according to historical references. Central Asia and the Near East are recognized as primary centers of origin for Persian walnuts. In these regions, the history of walnuts intersects with biogeography and culture (Akça et al., 2020). As such, walnut trees are not only cultivated for economic benefits, but they also hold special significance as sacred trees that provide financial support for newborns. Walnuts have been grown across Asia, from Xinjiang province in Western China to the Caucasus region, and have survived the Last Glacial Maximum. Contrary to previous beliefs, it appears that the distribution of walnuts in these areas was partially influenced by ancient human activities. By trading walnuts along ancient routes, such as the Silk Road and the Persian Royal Road, humans overcame geographic obstacles, facilitating the exchange of walnuts across Asia (Vahdati, 2014; Pollegioni et al., 2015).

Molecular marker RFLP studies have demonstrated that European germplasm primarily originated from Iran, while Asian germplasm is linked to regions near the Himalayan Mountains (Fjellstrom et al., 1994). Given the presence of natural walnut populations in Europe and the Balkans following the last glacial period, it appears that Persian walnuts were introduced to Europe from Asia during the Early Bronze Age (Pollegioni et al., 2017). Archaeological evidence suggests that walnuts were transported from ancient Persia by traders to Turkey and then to Greece via the Silk Road (Vahdati et al., 2014). Notably, the distribution of Persian walnuts was significantly aided by the efforts of Alexander the Great during the era of Ancient Greece. A further surge in human-mediated dispersal of walnuts to Portugal, Spain, and North Africa occurred during the Roman Empire, thus facilitating the spread of walnuts between Western Europe and the Balkans in Eastern Europe (Pollegioni et al., 2017; Bernard et al., 2018).

In the 16th century, English traders transported Persian walnut trees to South America, specifically to Chile, which is why they are also referred to as English walnuts. In the 19th century, Persian walnuts were introduced to California, where a significant portion of genetic improvement and commercial cultivars, including Chandler, have been developed (Vahdati et al., 2023). Currently, walnuts are cultivated across a broad geographic range in both the Northern and Southern Hemispheres, with a primary focus on regions between 30 and 55 degrees latitude in the Northern Hemisphere (Fallah et al., 2022). Countries such as Chile, Argentina, South Africa, Australia, and New Zealand have emerged as key players in walnut cultivation in the Southern Hemisphere (Vahdati et al., 2019).

2. Economic importance

Different species of walnuts hold significant economic value as trees appreciated for both their timber and edible nuts. Traditionally foraged from the wild, walnuts are now primarily cultivated in plantations. Most walnut species, apart from *Juglans regia* L., are grown for their high-quality timber. Certain species and interspecific hybrids, such as Black Walnut and Paradox (a cross of Black and English Walnut), serve as suitable rootstocks for Persian walnuts. Walnut wood is highly regarded for crafting classical, high-quality furniture due to its durability, workability, color variation, attractive appearance, unique grain patterns, and smooth texture. The trunk, roots, and branches are commonly used as sawn timber or veneer for this purpose. Currently, the global demand for walnut wood exceeds its supply. However, precise statistics regarding the market size of walnut wood are lacking. As of now, the global walnut furniture market is valued at $9.3 billion and is projected to reach $13.7 billion by 2030, with a compound annual growth rate (CAGR) of 5.6% from 2024 to 2030. This market encompasses tables, chairs, and various other furniture items.

The Persian walnut is one of the most important nuts worldwide, known for its multifunctional uses. It is cultivated in horticulture for its edible nuts, in forestry for its valuable timber, in pharmaceuticals for its medicinal properties, and as an ornamental tree in parks (Vahdati et al., 2018; Habibi et al., 2022). Walnuts are utilized in numerous ways, including the preparation of local sweets like walnut halva, the production of medicinal oil, dyeing with the walnut green pericarp, and in various pharmaceutical products. Overall, the cultivation, utilization, and processing of walnuts generate employment opportunities because this crop offers a wide range of applications (Vahdati et al., 2014).

Nut trees, in general, hold a substantial share of global trade due to their high nutritional value and relative ease of storage and handling (Hassani et al., 2020; Vahdati, 2024). The estimated supply value of nut trees is projected to reach $36 billion by 2023 at a consistent increasing rate of $1.14 billion USD per year over the past decade. Walnuts, as the second most important nut tree globally (accounting for 22% of total nut production), are a significant contributor to the overall production and market of nut trees worldwide (INC, 2023). Notably, walnuts have experienced the highest annual growth rates among nut trees over the past decade, at 9%.

According to FAO statistics, the total area harvested for walnuts worldwide spans 1.25 million hectares, with China, Turkey, and the United States accounting for 28.6%, 13.3%, and 13.0%, respectively (FAO, 2023). Over the past decade, the harvested area for walnuts has increased by 48%, with an average annual growth rate of 0.04 million hectares. The world's walnut production and supply are approximately 3.87 million tons and $3.58 billion USD, respectively, with average annual growth rates of 5.86% and 2.94%. China, the United States, Iran, Turkey, and Mexico are the leading countries in walnut production, collectively accounting for more than 76% of the world's total output (Table 13.1, FAO, 2023).

A significant limitation to the expansion of walnut cultivation is the lack of irrigation water for walnut plantations in many regions worldwide, compounded by inadequate winter chill required for proper and timely blooming due to climate change. Additionally, early leafing cultivars may be susceptible to spring frosts, presenting another challenge to expanding walnut plantations in various temperate areas.

Table 13.1 World walnut production, area harvested, supply, and respective compound annual growth rates (CAGR) in 2022 (FAO, 2023)

		Value	CAGR (%)	Leading countries and their share
Production (M T)		3.87	5.86	China (36%), U.S. (18%), Iran (9%)
Area harvest (M ha)		1.25	4.82	China (29%), Turkey (13%), U.S. (13%)
Supply	(Shelled)	2332.05	4.40	U.S. (38%), Mexico (20%), Chile (9%)
	(In shell)	1251.11	0.89	U.S. (32%), Chile (21%), China (15%)

3. Botany and relevant species

The walnut tree is monoecious with male catkins growing in last year's buds and female flowers at the tip of the new spring growth while the leaves unfold (Figure 13.1). Most cultivars release their pollen before the female flowers are receptive (Figure 13.2). Thus for isolated young orchards, at least two cultivars are required in a walnut plantation for cross-pollination. When mostly only one mixed bud on the top of the previous season's growth produces a shoot and, at its top, female flowers are in most of the growing points of the tree, the cultivar is terminal bearing. If most of the last year's shoots have two mixed buds at the top and other mixed buds along their growth, then the cultivar is lateral bearing.

In the Northern Hemisphere, the nut grows substantially after bloom (April) to late June—early July, when the nut reaches its final size, and the endocarp has hardened (see Figure 13.2). During the summer and September, the kernel (seed) is filling up with lipids, protein, and minerals up to ripening, usually late September to late October, depending on cultivar and local climate.

The genus *Juglans* includes more than 20 species of dioecious or monoecious deciduous trees with pinnately-compound leaves and is divided into three sections. Section Dioscaryon (also called section Juglans) includes two species: *J. regia*,

FIGURE 13.1

Walnut last year's shoot in March (end of the winter) with mixed buds and male buds.

(Photo from G.D. Nanos.)

FIGURE 13.2

Female flowers at almost full bloom *(left)* and nuts grown substantially in June *(right)*.

(Photos from G.D. Nanos.)

the Persian walnut cultivated worldwide, and *J. sigillata*, cultivated locally in Western China. Section Rhysocaryon includes 16 species of "black walnuts" distributed throughout the Americas; several of these species are used as rootstocks. Section Cardiocaryon includes three species distributed across East Asia: *J. mandshurica*, *J. cathayensis*, and *J. ailantifolia*. A fourth section, Trachycaryon, is sometimes included containing a single species, the butternut *J. cinerea*. Chloroplast DNA evidence clearly groups *J. cinerea* within Rhysocaryon (Dong et al., 2017), whereas nuclear DNA groups *J. cinerea* within Cardiocaryon (Stanford et al., 2000), suggesting that the butternut may have arisen from introgression of Cardiocaryon nuclear genes into Rhysocaryon cytoplasm. Nuts from section Dioscaryon are clearly distinguished from other species in the genus by two characters: a hull that becomes dehiscent at maturity, which leads to "hull split" and facilitates harvest; and the presence of a papery septum or "packing tissue" between the kernel halves, which facilitates kernel removal. In sections Rhysocaryon and Dioscaryon, hulls are indehiscent and the septum is hard and shell-like.

Iran and China are both centers of genetic diversity. Very old trees are found along the Silk Road suggesting that walnuts have been dispersed by humans for millennia (Vahdati, 2014). Western European populations of walnut show evidence of a genetic bottleneck. The one other species in section Dioscaryon, *J. sigillata*, overlaps with *J. regia* in the Himalayas and has been viewed as either a sister species or merely an ecotype of *J. regia*. A comprehensive recent study of *Juglans* diversity in the Himalayan region (Yan et al., 2024) classified 65 populations as either *J. regia* or *J. sigillata* based on morphological and genetic analysis, revealing both extensive hybridization between taxa and the association of *J. sigillata* populations with areas of higher precipitation. *J. sigillata* is being investigated as a source of tolerance to walnut blight *(Xanthomonas arboricola* pv. *juglandis)*, a foliar pathogen with significant impacts on global walnut production.

4. Breeding

Scion breeding targets: Fruitfulness on lateral buds, or "lateral bearing," is an important trait for all modern walnut cultivars and is defined as the capacity to produce female flowers from axillary buds. Older cultivars of *J. regia*, as well as all species in sections Rhysocaryon and Cardiocaryon, are "terminal bearing," developing female flowers only within terminal buds. Lateral bearing is thought to have arisen in the Xinjiang province of Western China and is associated with both increased early yield and precocity. Several studies have reported a major locus for lateral bearing (LB1) on chromosome 11 of the Chandler v2 assembly (Marrano et al. 2019; Bernard et al. 2020), though modifiers may exist.

Walnut shows remarkable variation in phenology that reflects its broad adaptation to a range of latitudes and altitudes; different cultivars may have leafing dates as much as 8 weeks apart. Late leafing is desirable both for avoidance of spring frost and, in many growing regions, for avoidance of walnut blight, because the blight bacterium is spread by early spring rains. Much of the variation of walnut phenology in Californian and European germplasm appears to be controlled by a major locus on chromosome 1 of the Chandler v2 assembly (Marrano et al. 2019; Bernard et al. 2020).

Kernel yield is defined as the proportion of the nut weight that is made up of kernel. High kernel yield is economically important, and thicker shells reduce kernel yield; however, very thin shells are associated with increased susceptibility to mold, breakage, and quality deterioration during storage. The ideal walnut shell is one that is thin but strong with a solid seal between the two hemispheres. Woody shell protrusions into the packing tissue or septum of the walnut may interfere with kernel removal and are selected against.

The color of the pellicle, the thin layer of maternal tissue rich in phenolic compounds that surrounds the embryo or "kernel," is economically important for walnuts traded in the international market. Extra-light pellicle color is an indicator of quality and freshness and commands a higher price. Pellicle (packing tissue) darkening may be caused by prolonged storage or by oxidative stress before, during, or after harvest (Ortiz et al. 2019). However, different varieties display a range of pellicle colors when fresh, ranging from very light to very dark, and a dominant mutation confers red pellicle color. The perceived "sweetness" of some walnut kernels results from reduced levels of astringent catechins and epicatechins in the pellicle.

Scion cultivars: Franquette is an old French cultivar that is terminal-bearing, late leafing, displays superior taste and shelf-life, and is still grown extensively in Europe (Bernard et al., 2018). The California breeding program was founded to combine the late leafing and nut quality of Franquette with the lateral bearing of Chinese cultivars. The most famous release from this program is Chandler, the world's most popular walnut cultivar (Figure 13.3). Chandler is lateral-bearing, late leafing, and is particularly prized for its reliable production of extra-light kernels across a range of climatic and management conditions. Persia is one of four recently introduced cultivars in Iran and has garnered significant interest from walnut growers due to its late-leafing, midharvesting, lateral bearing and high kernel quality characteristics (Hassani et al., 2020). Wen 185 is a Chinese cultivar with superior kernel yield and thin shells that may not be suitable for mechanized harvest. The most recent California release, UC Wolfskill, combines the superior pellicle color of Chandler with earlier harvest and a kernel yield intermediate to that of Chandler and Wen 185.

Rootstock breeding targets, clones, and seedling populations: A good rootstock displays resistance or tolerance to soilborne pests and pathogens, adequate vigor, and resilience to abiotic stresses including salt, cold, flooding, and water deficit. Much of the walnut industry worldwide relies on genetically variable seedlings of *J. regia* as rootstocks. Early cultivation in Northern California used a native black walnut species, *J. hindsii*, as a rootstock. American horticulturalist Luther Burbank discovered that *J. hindsii* × *J. regia* hybrids (Paradox) showed superior vigor, and these were soon widely adopted. Development of *in vitro* propagation and rooting methods for walnut rootstocks (Leslie et al., 2005) allowed the commercialization of several hybrid clones, including Vlach, VX211, and Grizzly (*J. hindsii x J. regia* hybrids) as well as RX1 (*J. microcarpa x J. regia* hybrid). VX211 and Grizzly are both tolerant of walnut root lesion nematode, and RX1 shows strong resistance to Phytophthora. Hybrids of *J. regia* with species from Section

FIGURE 13.3

Mature trees of cv. Chandler *(left)* and cv. Franquette *(right)* in April.

(Photo from G.D. Nanos.)

Cardiocaryon have been reported to possess nematode resistance but have been little exploited in rootstock breeding. Likewise, *J. hindsii* × *J. nigra* hybrids are reported to tolerate flooding and root anoxia but have not been widely used. In general, most of the diversity in Rhysocaryon—the 16 species of black walnuts that populate the Americas from Canada to Patagonia—remains unexplored for rootstock breeding. *J. australis* × *J. regia* hybrids are reportedly used in Argentina.

Cherry leaf roll virus (CLRV) is the only virus known to infect the genus *Juglans*. *J. regia* is asymptomatic and tolerates the virus, whereas black walnuts mount a hypersensitive response; the response of other *Juglans* species is not known. CLRV is spread by pollen between *J. regia* individuals and prefers cool conditions. When a CLRV-infected *J. regia* scion is grafted to a black walnut or hybrid rootstock, the graft disorder known as "blackline" disease develops, triggered by a hypersensitive response in the rootstock tissue at the graft union (Mircetich et al., 1980). Possible genetic solutions to blackline disease include the development of hypersensitive scions and/or the development of tolerant rootstocks. Adoption of new hybrid rootstocks (e.g., VX211 or RX1) in growing regions where *J. regia* seedling rootstocks are currently used should proceed with caution until the incidence of asymptomatic CLRV is fully understood.

5. Propagation

Commercial walnut trees are produced by budding or grafting the scion onto a properly grown rootstock (Figure 13.4). Traditionally, rootstocks have been seedlings and rarely cuttings or micropropagated. Walnuts for seedling production should be harvested carefully to avoid contamination from soilborne diseases, stratified

FIGURE 13.4

Mature walnut trees budded on *J. nigra* seedling rootstock.

(Photo from G.D. Nanos.)

in sand or refrigerated for 6 to 8 weeks to grow properly in the spring (Vahdati et al., 2012). Traditional walnut grafting is a reliable method for walnut propagation and is influenced by environmental and internal factors (Dehghan et al., 2009; Thapa et al., 2021), hypocotyl, epicotyl, and micro-/minigrafting techniques have been introduced as highly efficient methods (Vahdati and Zareie, 2005), but the optimization of these methods needs more research work. The cuttings and stool layering are cheap methods for walnut propagation; however, there is no significant success in these propagation methods, and they are not commercially used (Vahdati et al. 2022).

Cuttings for rootstock or self-rooted cultivars are rarely used due to low propagation success rates. Micropropagation of selected rootstocks is used, while micropropagation of self-rooted cultivars seems to be successful but not yet widely used. The lack of success of commercial micropropagation programs is dependent on the recalcitrance of most of the genotypes to be *in vitro* cultured. Although multiplication rates (MRs) of the established clones do not represent a great limitation for commercial micropropagation, the increasing of MRs undoubtedly contributes to the improvement of rooting and survival during acclimatization, which would reduce the production costs of *in vitro* plants (Vahdati et al., 2009; Vahdati et al. 2022). Starting from 2006 in Italy, commercial micropropagation of Chandler has become widespread (Navacchi et al. 2021). The cumulative production per plant until now has been equivalent or higher in micropropagated Chandler orchards than in grafted ones. This result was obtained with less pruning and greater vegetative growth of micropropagated plants, which made the training of a structured central axis easier (Figure 13.5). The most planted type of micropropagated plants was the dormant bare root type. These plants can be obtained in one growing season in the nursery, after early spring acclimation ex vitro. This type of plant was between 0.2 and 0.6 m in height and easier to manage during the first year of growth in the orchard compared to potted plants. Micropropagated plants have a higher growth rate and more homogeneity with minimal pruning to train the structured central axis. While

FIGURE 13.5

New orchard with micropropagated Chandler trees in Emilia Romagna region (Italy).

(Photo from D. Neri.)

FIGURE 13.6

Rootstock preparation for budding *(left)* and cultivar patch *(right)* to be inserted in the rootstock in May.

(Photos from G.D. Nanos.)

a delay in production of female flowers was found in micropropagated Chandler trees during the first 3 or 4 years, the cumulative production per plant was higher than in grafted plants 8 years after planting. A draw-back of these self-rooted plants may be susceptibility to soilborne diseases.

In nurseries, walnuts are commonly propagated with budding or grafting the cultivar onto a walnut rootstock. Depending on the growing area, once the rootstock is grown to a diameter similar to the budwood, it can be budded or whip-and-tongue grafted in June with budwood refrigerated from February or with newly developed shoots (Figure 13.6). In the successfully budded plant, the scion bud will grow up to 40 to 50 cm during the rest of the season and can be sold by December of the same year. Otherwise, properly grown rootstock is patch or T-budded late August to mid-September. The successfully budded trees will grow next spring and summer to give a tree with 1-to-1.4-m high scion. Grafting is done in spring. Care needs to be taken to avoid graft losses due to sap flowing from the rootstock. In more traditional approaches, the rootstock is grown for one season, grafted at the end of the first dormant season, and cultivated for another growing season to give rise to a nursery-completed bare-root tree.

Another walnut propagation method is the hot callusing technique. The rootstock is bench grafted, and the grafting point is kept at warm temperature to hasten the development of callus at the graft union. This method usually results in higher percentages of grafting success (Erdogan, 2005).

6. Plantation establishment

Climatic requirements: Walnut trees are not as hardy as most temperate zone perennial species. In winter, the tree can be damaged below -10°C in warm areas

(Ramos, 1998), but in cooler areas, walnut can withstand temperatures even lower than -14°C without damaging tree or bud health. In spring, during early shoot growth (usually April) spring frosts with -3°C to -4°C will damage the young shoots that would carry the pistillate flowers, thus the year's crop will be lost. Regular high relative humidity periods during spring will support bacterial and fungal damage to young shoots, staminate flowers, pistillate flowers, and fruitlets. High summer temperatures (usually >40°C to 42°C) will cause kernel darkening (quality loss) or shriveling (lost commercial value) (Ramos, 1998) (Figure 13.7) and excessive irradiation will damage exposed young trunks (see Figure 13.7). Early fall heavy rains will impair kernel quality due to delayed harvest, darkening, or even disease damage.

On the other hand, walnut buds require a substantial amount of winter chill (the California cultivars >800 h, the French cultivars >1000 h of temperatures between 0°C and 7°C) to grow normally in spring. A study in California observed changes in budbreak timing over the last decades due to rising winter temperatures. During the previous decades, as temperatures were increasing, chill requirements were still fulfilled and walnut tree male flowering and leafing out would take place early. During the last decades, as chill requirements have not been completed during some winters, male flowers still flower early, while leafing out is delayed, causing problems with lack of pollination and reduced fruit set (unpublished data). In another research work, walnut cultivars had similar winter chill requirements but different heat requirements after winter rest completion to initiate budbreak and flowering (Charrier et al., 2011). Therefore, it is crucial for future walnut breeding programs to consider low-chill and high-heat requirement as key traits to address the adverse effects of

FIGURE 13.7

Walnuts damaged from high solar radiation exposure *(left)* and newly planted unprotected walnut tree trunk with sunscald (*right, in blue circle*).

(Photos from G.D. Nanos.)

climate change (Hassankhah et al., 2017; Hassankhah et al., 2020; Hajinia et al., 2021). In short, walnut trees will grow successfully in areas with mild winters (if they received the necessary winter chill) and relatively cool summers of the world. It is recommended to avoid frost-prone areas or planting at a slope.

Soil requirements: Walnut trees have extensive root systems. In some cases under dryland cultivation, root systems can easily reach three meters deep. Thus, deep nonstratified medium-textured soils are preferred for walnut tree successful growth and productivity. On shallow soils, walnut trees will require regular and local irrigation and will remain small stature trees. Walnuts prefer slightly alkaline soils (with pH <7.5) but will grow well in soil pH >5.5. Walnuts will not grow well in soils with high conductivity (especially due to high Na^+, Cl^- ions).

Water requirements: The walnut tree requires large amounts of water to transpire and grow the shoots and nuts. At lower elevations with proper climate for walnut growth, natural precipitation is insufficient to cover the walnut tree water requirements making regular irrigation necessary. Depending on the growing area, at higher elevations there may be sufficient rain during the growing period (from April to October in the Northern Hemisphere) to cover a large part of the tree water needs.

Walnut cultivar and rootstock selection: The choice of the main cultivar, the proper pollinators and the rootstock used in a plantation should be made early in the orchard development process because nurseries require 1 to 2 years to produce the trees. Pollinators must not exceed 3% of the trees planted, as a larger number of pollinators (large pollen volumes) was found to increase pistillate flower abortion (Krueger, 2000). The number of plants ordered will be decided on the basis of cultivar type × training to be used × intensification of the plantation, as different distances will be used for each case. Optimizing plant and row spacing is of constant debate. Cultivars producing mainly on terminal buds should be square-planted at distances of 9 m or more (Ramos, 1998). Cultivars with strong lateral bud fruiting can be planted at 8 × 8 m. Central leader training can reduce the distances between trees in the row. Distances are also decided based on the harvest machines available. Hedgerows can be planted 8 × 5 m.

Soil preparation for planting: Soil preparation will be required to break any hard pans present below the soil surface and develop passive drainage if necessary. If soil analysis has shown that P or K are low or pH is below 5.5, then P, K or lime (calcium carbonate) can be incorporated in the soil to help the growth of the roots and scions during the first years. The soil surface is leveled to support flood irrigation (if used) and to increase uniform surface water drainage (avoid flooding).

Planting: Planting positions today are marked with the use of GPS, which will help to develop a map for the plantation. The planting holes are opened mechanically. New plants are placed in the hole and, perhaps microorganisms (mainly mycorrhiza and microbial antagonists to soilborne diseases) and may be well-composted organic material are mixed with the soil to fill the planting hole and improve soil properties next to the roots.

Young plant care: The trunks of the newly planted trees should be protected from sun-injury with white latex paint (Figure 13.8). Traditionally, the tree is headed

FIGURE 13.8

White-painted trunks of young trees for sun protection.

(Photo from G.D. Nanos.)

depending on the training system of the tree with the main purpose to develop a trunk at least 1.2 m high to enable the use of mechanical shakers during harvest. It is recommended that at planting the scion must be cut leaving 4 to 6 well-developed buds. As the buds grow in spring, the manager manipulates the growth to develop a strong tree for the years to come. Plant protection is also applied vigorously in parallel to irrigation and plant nutrition to have strong healthy shoot growth >1-m-long every year.

Pruning–Training: Young trees may be trained in a cup-shaped tree (or open center system), modified central leader (or delayed open center system), central leader, or hedgerow. The main objective of early years' training is to keep vigorous shoots growing relatively perpendicularly, selecting usually three scaffolds (the basic large branches) coming out from the trunk to grow toward all horizontal directions and from different heights on the trunk (for open center training), keeping the central leader stronger than the 4 to 6 scaffolds (in delayed open center training) (Figure 13.9), keeping the central leader very strong with more horizontal 3 to 4 relatively weak scaffolds for every year's growth (for central leader training with >12 scaffolds on the central leader), or development of central leader and freely grown scaffolds mainly longitudinally to the fruiting wall (in hedgerows). In all cases, trunk and canopy must be high enough to accommodate the shaker equipment to be used.

Thinning cuts retaining the proper scaffolds are used to both terminal- and lateral-fruiting cultivars, but heading cuts during the first years are required for lateral-fruiting cultivars to sustain the selected scaffold growth if it is relatively weak. In general, minimum pruning is practiced by most managers over the first years to support scaffold growth, but many short shoots are left on the tree for early nut production. In a minimal pruning approach, only branches that would encumber orchard access or those that are prone to break are removed.

FIGURE 13.9

Three-year-old tree trained to central leader.

(Photo from G.D. Nanos.)

Pruning mature productive walnut trees: (a) controls tree size; (b) maintains vigor, light, and fruitfulness; (c) increases nut size; (d) stimulates new fruiting wood to replace old less productive branches; (e) removes weak or dead wood and interfering branches; and (f) helps harvesting operations (Figure 13.10). Pruning also reduces tree height and spread, so that equipment can move through the plantation and plant protection sprays can reach the whole canopy. Pruning mature trees involves removing whole branches (thinning cuts) and reducing the length of fruiting branches (heading cuts). Thinning cuts are useful for both terminal-bearing cultivars and lateral-bearing ones. The removal of whole branches opens the canopy to air and light and keeps the lower part of the canopy healthy and productive. Heading cuts are important for lateral-bearing cultivars, as, with cropping, the branches are becoming more horizontal or hanging. This may interfere with equipment movement and the tree may lose vigor (yearly growth is very weak, often less than 5 cm long). Removing part of the branch invigorates the tree and new growth and gradually new fruiting wood develops, while the remaining part of the branch has sufficient light to produce more and bigger nuts. Heading cuts pruning (as an example heading 30-40 branches with diameter usually 3-4 cm per tree) has to be applied yearly or, even better, biennially (with more headed branches per tree) to reduce pruning costs. Biennial pruning is done after a low production season to reduce alternate bearing.

FIGURE 13.10

Mature tree properly trained for shaking the nuts and good light penetration.

(Photo from G.D. Nanos.)

FIGURE 13.11

Hedgerow at maturity.

(Photo from G.D. Nanos.)

Mechanical hedging is a cheaper alternative to reduce tree height and head cut the canopy on one side of the tree every year. The purpose of mechanical or manual pruning to improve light penetration of the canopy but mainly reaching the edges of the canopy from top to bottom (leaving the light reach the area between the canopies of adjacent trees), so productivity is maintained throughout the canopy.

Walnut hedgerows are mechanically topped and side worked (one alternating side every year) to keep the fruiting wall in a total depth of <3 m and tree height at around 8 m (Ramos, 1998) (Figure 13.11). This operation is fast and economic, and keeps the fruiting wall under control. Hedgerow advantages are the early entry to production and fast maximum productivity of the plantation, fewer productivity

FIGURE 13.12

Walnut trees 2 years after rejuvenation pruning.

(Photo: from G.D. Nanos.)

losses due to tree death from Phytophthora, and easier to cover with plant protection sprays. The main disadvantage is the high initial cost of purchasing higher numbers of trees, staking them, and taking the initial care to develop properly. The higher tree number also increases harvesting costs of tree shaking. Finally, abandoned or very crowded trees can be cut down to the scaffolds to rejuvenate the trees (Figure 13.12). It will take a few years before productivity reaches economic levels, but it is the easiest way to make the trees strongly growing and productive again.

Plant nutrition: A decision on which nutrients and how much of each nutrient should be applied in a walnut plantation is based on: (a) age and yield of the trees, (b) appearance of the trees (new shoot growth, green leaves, nut size), (c) soil analysis (this mainly helps to decide if a nutrient should be applied and in what format the fertilizers should be), (d) leaf analysis (extremely important to decide which and how much of each nutrient should be applied), and (e) a balance of inputs and outputs of nutrients in the plantation (Olsen, 2006).

Young plants during the first years need minor quantities of balanced N-P-K fertilizers with three to four split fertigations yearly depending on the shoot growth. If shoot growth exceeds 1 m, further fertilization may not be required. Foliar analysis (using 100 healthy leaflets per block and cultivar collected late July) is strongly recommended for planning yearly fertilizer application and should be done every 1 to 3 years. Adequacy levels: N 2.31%-2.80%, P 0.14%-0.50%, K 1.21%-2.50%, Ca 1.11-2.50, Mg 0.25-0.60, Fe 51-400 ppm, Zn 16-60 ppm, B 86-150 ppm (Ramos, 1998).

Outputs include nutrient losses to air and deep percolation, nuts harvested, and prunings. Prunings are usually trimmed and remain in the plantation and are not outputs. One thousand kg of kernel + shell (seed + endocarp, respectively) remove 15 kg N, 2 kg P_2O_5, 5 kg K_2O, and 20 g B. A mature walnut plantation produces around 5000 kg of nuts (kernel + shell) per hectare. The pericarp also contains some nutrients, i.e., per dry matter 0.84% N, 0.78% P, and 0.56% K, basically minor quantities compared to kernels (3.16% N, 0.43% P, and 0.52% K). Thus, with 5000 kg of nuts per ha around 100 kg of N, <20 kg of P_2O_5 and around 30 kg of K_2O are removed per ha from the plantation.

Inputs may include irrigation water, cover crop, fertilizers, and organic matter added. Irrigation water may contain substantial amounts of N, Ca, Mg, and should be subtracted from the required inputs as irrigation water volume may be high, and these nutrients are readily available to roots.

Based on all of this, N fertilization is important with yearly applications of large quantities. A total of 140 to 160 kg N applied per ha are reported in the literature in mature plantations, always depending on the yield and leaf analysis results. N should be applied mainly in March with three-fifths or two-thirds of the total required, followed by one-fifth to one-sixth in May, and the rest in July. Fertigation is more efficient than broadcast application. But N fertilization is largely area-specific around the globe.

It seems that P fertilization in walnuts is not useful as the tree has never shown deficiency symptoms or reacted to applied P. On the contrary, K fertilization should include around 70-80 kg K per ha per year to maintain yields with application either in March and June or only in June-July. When foliar K is at adequate levels, K fertilization could be reduced. N and K are better applied with fertigation or on the irrigated soil as almost all functioning roots throughout the season are there. Mild (foliar Zn below 15 ppm and B below 80 ppm) or severe Zn and B deficiencies are common in alkaline soils around the world in walnut. B and Zn deficiencies are corrected with foliar applications at the end of pistillate flower bloom (stigma has turned brown) and 2 weeks later with around 200 g B/ton and 1000 g Zn/ton of expected yield. These treatments must be applied yearly. B deficiency can be corrected with soil applications of borax not exceeding 150-200 g/tree every 2 to 3 years. Be aware that excess B can easily be toxic to roots and scion. Organic matter application on the irrigated soil can improve soil properties, reducing the needs for fertilizer inputs, especially for P and micronutrients, but mainly improves microorganism populations and biodiversity improving altogether the soil-root ecosystem and the plant resilience to stresses. Biostimulants are lately used extensively in fruit production and should be tried in walnut plantations to reduce stress-related yield losses (water scarcity, high temperatures) (Lotfi et al., 2022; Behrooz et al., 2019). Foliar applications of homobrassinolide at pistillate full bloom almost doubled yield due to increased fruit set (Sotomayor et al., 2022).

Irrigation: For 1 kg of nuts (and 5000 kg/ha), the walnut tree requires 2440 L of water (Ramos, 1998). Thus, it is clear that the walnut tree needs a lot of water. Actually, reduced availability of water results in smaller nut size, more sunburn, and higher mite pressure. Some walnut plantations are situated in high elevations with at least 800 mm of rain with a good part of it during the growing period. In these cases, irrigation may not be necessary, although yield losses may be encountered due to dry periods. Most of the plantations are grown in drier areas and irrigation with ample water quantities is required. The irrigation water quantity needed depends on the area shaded by the plantation tree canopy, the climate of the area, the soil and various other less important factors. A very rough estimation of yearly water needs for a mature walnut plantation in the Mediterranean climate is at least 8000 m^3/ha/year depending on the irrigation method used.

But there are ways to better estimate the required irrigation water for each plantation every week and during the season. One commonly used method is to calculate the amount of water a plantation uses every week, called the water budget method. With the use of meteorological station data (sensors for total light, temperature, wind, relative humidity and rain volume), soil characteristics and crop growth and period of the year, crop evapotranspiration values are calculated weekly and these water quantities should be applied. This method alone is risky, although easy to implement. To improve precision of this method soil moisture and plant stress should be monitored.

Soil moisture is measured with various sensors, most commonly tensiometers, but also electrical resistance sensors or neutron probes. The sensors are set in a couple of sites in the required depth throughout the plantation and data from such sensors are collected continuously with proper electronics together with climatic data. Plant stress evaluation requires specific instrumentation to observe (field cameras or Unmanned Aerial Vehicles [UAVs]) or sensors to follow the movement of water in the trunk or the microshrinkage of the trunk in a tree. These are expensive and give localized information. A combination of the water budget method with soil water availability is nowadays the safest economic method to precisely calculate the walnut plantation water needs over the season. A pressure bomb can also assist when measuring the stem water potential of the trees as an indication of the water status.

Soil water should be at its capacity from April to end of June as during this period the nut is growing to its maximum size and all vegetation is taking place. Slight deficit in water availability during the months of May and June may reduce nut size but will definitely reduce shoot growth and increase mixed buds (thus, pistillate flowers) for next spring. During the rest of summer and up to fruit ripening, walnut trees require sufficient irrigation to keep leaves at their full photosynthetic productivity to fill the kernel successfully and reduce the risk of sunburn.

Irrigation methods used for walnuts include: (a) flood and sprinkler irrigation, (b) microsprinkler irrigation below the canopy, (c) drip irrigation on the soil surface, and (d) subsurface drip irrigation. Flood irrigation is the cheapest method, but water use efficiency is very low (Figure 13.13). It is used in areas with copious and inexpensive supplies of irrigation water. Sprinkler irrigation also has low irrigation water use efficiency due to large amounts of water evaporated from all wet areas during irrigation. Microsprinkler irrigation (usually two microsprinklers per tree) seems to be the most efficient way to supply the large amounts of water needed in the

FIGURE 13.13

Walnut plantation flood irrigated *(left)* and drip irrigated walnut plantation with double dripper line *(right)*.

(Photos from G.D. Nanos.)

FIGURE 13.14

Mature walnut plantation microsprinkler irrigated *(in red circle)*.

(Photo from G.D. Nanos.)

high-demanding cultivation areas with hot and dry summers (Figure 13.14), as water use efficiency of this method is relatively high (>80%, depending on the time of the day irrigation takes place and the wind speed) and a relatively large area of soil surface and volume is irrigated. In cooler areas drip irrigation is the most efficient method and, depending on how much soil volume is irrigated, can be a very useful method (see Figure 13.13). Subirrigation (water delivery below the soil surface) has the highest water use efficiency and does not interfere with the other management practices especially weed management and harvest operations. The number of drippers and their volume output per hour per tree depends on the water availability and the soil characteristics. With any irrigation method it is important that the soil will not remain flooded for long periods of time and the trunk is not getting wet regularly, as the lower part of the trunk and the root system may suffer from soilborne diseases and stress from anoxia. One very important factor is that walnut is sensitive to saline water (water containing Na^+ and Cl^- ions), and only high-quality water should be used for walnut plantation irrigation.

In newly established orchards young trees must be irrigated close to the trunk to supply water regularly to the existing root system. Care should be taken to avoid wetting the tree crown for long periods of time. In this way, the root system of the young tree is working continuously with high irrigation water efficiency. Depending on orchard vigor, after the third to fourth winter in the field, the permanent irrigation system is installed. It should be kept in mind that wherever irrigation water is applied on the soil surface, weeds will grow continuously and will consume irrigation water (but will also cool the plantation). Their regular management increases management costs.

When deficit irrigation is applied (due to low water availability), mycorrhiza together with bacteria *Azotobacter chroococcum* helped the tree to sustain water stress by helping the root system to use soil water and nutrients (Zou et al., 2023).

Kaolin is sprayed commonly onto tree canopies to reduce heat stress during the summer. This will reduce sunburn and water needs with the reflection of solar radiation (Gharaghani et al., 2018; Mahmoudian et al., 2021). This reflected radiation partially reduces photosynthetic productivity of the kaolin-covered leaves, but this light is spread in the shaded areas of the canopy increasing their photosynthetic productivity (Rosati et al., 2007). Thus, sprayed kaolin often increases whole tree productivity with less irrigation water. Delaying the initiation of irrigation in late spring can help the walnut roots grow deeper, extract more soil water, and make the tree more resilient to drought. But this technique needs measurements of stem water potential, a time-consuming measurement, to find out the stress experienced by the tree and to properly initiate irrigation.

Weed management: Weed management in a walnut plantation is important because weeds compete for nutrients, water, and, in young trees, light; growing weeds may lead to increased rodent populations; weeds could interfere with harvest operations among other less significant reasons. A key activity is the recording of weed species present in the plantation. In late winter and late spring, weed species and abundance should be noted over the years to determine if any changes in weed populations occurred due to weed management. This will help to support or disrupt those changes manipulating the management techniques used or the herbicides selected.

In young walnut plantations, weeds should not grow next to young trees, as competition for water and nutrient will be significant, while rodents and soilborne diseases may reduce tree growth or kill trees. Various herbicides can be used with young trees to control vegetation next to the trunk without damaging it (see Figure 13.13). During the first years of the plantation a lot of free space between the rows is available to plant cover crops. Care being taken that cover crop species chosen do not support the build-up of plant-parasitic nematodes that infect walnut. These crops can produce substantial amounts of N and biomass and improve soil conditions but will suppress weed species abundance (will reduce biodiversity). In mature walnut plantations, weed management is also important in the irrigated area of the soil to reduce water consumption by the weeds. At the remaining soil surface, weeds are mulched (mowed) periodically. It is important to notice that in mature walnut plantations the light reaching the soil is minor, thus few weeds may grow under those low light conditions. For walnut shaking and harvesting from the ground, the soil must be free of weeds and various debris before harvest. This may require soil cultivation followed by firming the soil ahead of harvest. Surface soil cultivation is also applied with soils developing a surface crust which inhibits water infiltration. Herbicides are used at the irrigated area usually in less than one-third of the soil surface of the plantation. Herbicides pre- and postemergent are applied in fall or at the end of winter, and contact herbicides will be repeatedly applied during May and September if weeds are growing in the irrigated areas.

Environmentally friendly weed management moves toward increasing weed biomass produced and increasing biodiversity of species present in the plantation. Weeds on the alleys can help increase water infiltration, reduce soil compaction, reduce soil erosion in sloping orchards, increase soil organic matter content, cool the orchard, and provide habitat for beneficial insects. In spring, weeds could be allowed to grow to maturity and set seeds for next fall (under these conditions rodent populations may increase), as long as frost danger is not high. If spring frosts appear regularly in the area, weeds should be herbicide-killed under the canopy and cut close to the ground at the alleys during the frost-prone period to keep the plantation as warm as possible. With ample quantities of irrigation water weeds should be allowed to grow even during the summer to help minimize heat stress and produce biomass. With September rains weeds should be allowed to grow as long as they do not interfere with harvest operations. In Europe, soil free of weeds during fall and winter is not allowed on agricultural land. During this period, weeds will extract the remaining N from the soil and will keep erosion low or absent during heavy rain incidents.

Fruit maturation and harvest criteria

The walnut fruit grows substantially during the first 2 to 2.5 months after pistillate flower set reaching its final size at the end of June-early July in mild climate regions. During the summer months and up to the beginning of ripening, sometime in September-October (in Northern Hemisphere) dry matter accumulates in all nut parts but basically imported carbohydrates and amino acids in the kernel are stored as fat and protein. Ripening starts in all cultivars with the darkening of the packing tissue in the area between the two halves of the kernel. At that time in some cultivars, the pericarp is still immature and tightly closed, and in others it starts rupturing. Weather conditions also influence pericarp dehiscence (Figure 13.15). In cooler areas the pericarp may dehisce earlier close to the seed pellicle color change, but in warmer areas pericarp dehiscence is delayed. So, the best quality walnut kernel would be if harvested when pellicle changes color in the area around the two halves, but the walnuts may not be able to shake and harvest.

FIGURE 13.15

Nut pericarp dehisced, ready to be harvested.

(Photo from G.D. Nanos.)

Harvest finally starts around 15 days later when most pericarps on the tree have cracked. At that time, the nut is easily separated from the pericarp. In cultivars like Chandler, the packing tissue color does not darken substantially with delayed harvest. Nevertheless, ethephon is commonly applied to walnut trees for earlier harvest. Ethephon will hasten pericarp dehiscence and will help harvest more walnuts with the first shaking of the tree or scaffolds. Ethephon is applied when all nuts are at the stage of brown packing tissue and can bring harvest up to 1 week earlier.

Harvesting is conducted in different ways. Nuts can be harvested by knocking the individual small branches with poles so the nuts will fall onto spread tarps. Mechanical shakers for branches are also used to drop the nuts on tarps. Tree mechanical shakers are used for large plantations and nuts can fall on tarps or on the ground. The trees to be shaken must be properly trained to have tall trunks (>1 m high) and a canopy without hanging branches. In large operations, the nuts are dislodged from the tree by mechanically shaking at the trunk. Fallen on the orchard floor, they are wind-rowed and picked up with specialized harvest equipment.

7. Postharvest physiology and technology

Immediately after harvest, the walnuts must be removed from the plantation and not be exposed to the sun. Exposing the nuts for a few hours to the sun will detrimentally affect kernel color, taste, and possibly nutritional quality. Thus, walnuts are taken for pericarp removal on the same day (Figure 13.16) and for drying in hot air-drying facilities. Walnuts contain 10% to 30% water at the time of harvest. This water must be removed immediately after harvest and the kernel must have 8% water content to

FIGURE 13.16

Pericarp removal in the presence of water to keep endocarp light colored *(left)* and drying the walnuts with hot air without sun exposure *(right)*.

(Photos from G.D. Nanos.)

safely store. Drying is taking place with hot air dryers at temperatures of up to 43°C (see Figure 13.16). This drying procedure may take from 24 to 36 hours depending on the nut water content, the volume of nuts to be dried (the lower the depth of the walnut volume to be dried, the faster and more uniform the drying procedure), and the capacity of the drying equipment (increasing the air flow rate decreases drying time).

The walnut kernel contains substantial amounts of lipids and protein. But walnuts contain many polyunsaturated fatty acids, which are excellent for human health but are easily oxidized giving the undesirable taste of rancidity. Thus, the dried in-shell nuts can be stored for a few months at ambient temperature, but the lower the storage temperature the better the quality is maintained. Usually after 2 to 4 months at room temperature (and before the end of winter), the nuts should be stored in refrigerated rooms as kernels (Ramos, 1998). And even better in rooms or packages with very low O_2 (<1%) after flushing with nitrogen. Kernels can be stored at -20°C for up to 24 months with <1% O_2.

8. Walnut main diseases and pests and their management

Scion breeding needs to consider the overwhelming importance of fruitfulness, fruit quality, quantity and other parameters limiting the selection pool to *Juglans regia* germplasm to safeguard quality traits but will limit the diversity of breeding material used for selecting superior scion cultivars. For soilborne pathogens, genetic material for rootstocks with beneficial resiliency traits can be selected from different *Juglans* species providing a large selection base. Aboveground pests and diseases include fungal, bacterial, and viral diseases, and pest insects

Several species of the fungi **Botryosphaeria and Phomopsis canker disease** infect bearing branches and cause necrotic dieback of the shoots. Stress conditions, e.g., severe heat, can increase symptom development and severity. Depending on the infestation level of the orchard, spray schedules are necessary to reduce the disease spread. Pruning dead branches in the dormant season and possibly removing them as inoculum source from the orchard is one additional mitigation strategy (Adaskaveg et al., 2017).

Walnut blight caused by *Xanthomonas arboricola* pv. *juglandis* is a widespread foliar and fruit disease of walnut (Figure 13.17). Under disease-conducive conditions, rigorous spray programs are necessary to suppress the disease. The pathogen overwinters on the tree and thorough monitoring is necessary to schedule a treatment program. In California, disease is more severe on early-leafing cultivars and in general in more northern areas (Adaskaveg et al., 2017).

Deep bark canker caused by *Brenneria rubrifaciens* occurs on stressed trees. Trees respond with bleeding from infected areas on the trunk that can expand 30 to 60 cm up- and downward of the trunk per year. The interference of the infection with the conductive tissue reduces the vigor of the infected tree. A cure has not been identified. Mitigating stress conditions, optimizing irrigation by avoiding over- or underwatering can slow the impact of the disease.

Thousand cankers disease is a complex disease of the fungus *Geosmithia morbida* that is introduced into the bark layers of the tree by the walnut twig beetle (*Pityophthorus juglandis*). The insect pierces the bark, transmitting the fungus. As a result, a small pin hole is at the center of each sub-bark canker. From the outside, multiple holes visible on the bark surface provide indication for a possible Thousand cankers infection. The disease seems to be limited in severity but has been identified in almost all walnut growing areas in California (Adaskaveg et al., 2017).

Walnut husk fly, *Rhagoletis completa*, is widespread. This one generation-per-year insect feeds damaging in the mid to late season. The adult lays eggs into the husk of the walnut fruit where maggots develop that deteriorate the husk tissue by

FIGURE 13.17

Nuts damaged from walnut blight.

(Photo from G.D. Nanos.)

their feeding. The resulting blackening of the nutshell reduces the quality of the fruit. Insecticide spray applications are made, timing being determined by trapping the fly (Grant et al., 2017).

The Codling moth, *Cydia pomonella*, overwinters under loose bark of walnut trees in infested orchards. It has multiple flights during the season monitored with pheromone traps (Figure 13.18). The most damaging is the flight that allows gravid females to lay eggs into developing walnut fruit. Only one larva per fruit enters and develops while feeding on the developing kernel. For insecticide sprays at the most critical time of egg laying, the biofix, i.e., the time that winged males are caught, needs to be monitored. Mating disruption is another strategy to reduce the damage potential of these insects (Grant et al., 2017).

Navel orangeworm *(Amyelois transitella)* causes somewhat similar damage. This insect overwinters in mummy nuts on the tree or the ground. It also enters the walnut to feed on the walnut kernel. It has become a widespread pest in California nut crops. Management has relied on insecticide applications, but recently mating disruption strategies have become much more feasible (Grant et al., 2017).

Different species of scales, e.g., San Jose Scale *(Diaspidiotus perniciosus)* and Frosted Scale *(Parthenolecanium corni)*, colonize the trees during the dormant season and at the beginning of the new vegetation period. Low population densities appear to have limited damage potential, but at high population densities, the sucking of plant saps may diminish critical plant nutrients, and the plant may be weakened and fruit size reduced (Grant et al., 2017) (Figure 13.19). Spray oils in the dormant season can frequently reduce infestations (Grant et al., 2017).

FIGURE 13.18

Pheromone trap to monitor codling moth presence.

(Photo from G.D. Nanos.)

FIGURE 13.19

Scales grown on walnut leaf.

(Photo from G.D. Nanos.)

The Pacific flatheaded borer, *Chrysobothris mali*, occurs during May and June in California, or depending on temperature may be even earlier in the year. Its adults are attracted to weakened (diseased or injured by sunburn) trees. The adults lay the eggs that then develop into larvae and create a tunnel system below the bark with their feeding activities. Young trees are especially vulnerable because portions of the bark may die, and the feeding activities may girdle the tree (Grant et al., 2017). Affected branches need to be pruned and, in severe cases, insecticide applications may be considered.

Belowground diseases are caused by nematodes, fungi, and bacteria. The severity of soilborne diseases is predetermined by the susceptibility of the rootstock used in the orchard.

Plant-parasitic nematodes: *Pratylenchus vulnus* and *Meloidogyne* spp. cause nematode damage. Most current rootstocks are susceptible to these nematodes. All nematodes can easily spread with planting material. In California, the need for cleanliness from nematode infection is recognized and has been addressed in the Nursery Inspection Procedures Manual—7. This document provides preplant soil treatment protocols for nursery soil that qualify them for certification for nematode cleanliness. Where permitted in commercial orchards, preplant soil fumigation with 1,3-dichloropropene (1,3-D), often in combination with chloropicrin, has become standard practice to reduce populations of soilborne nematodes and pathogens. Regulatory stringency makes such applications more difficult, and in some countries, applications are not allowed. Because of the severe damage potential of the root lesion nematode, *Pratylenchus vulnus,* monitoring for its presence is especially important. At planting, 1 nematode per 250 cm^3 of soil can cause damage to currently used walnut rootstocks (Westerdahl et al., 2017; Buzo et al., 2009). When population densities have resurged, nonfumigant chemical or biological nematicides are used that prevent population density increases and protect yield. Infections with *Meloidogyne* spp. may be of regional importance but are more likely to occur on susceptible Paradox and English walnut rootstocks. In California, the root lesion nematode-tolerant rootstocks VX211 and Grizzly are available.

The replant disease occurs when walnut follows walnut and can become damaging. Hypothetically in this disorder of unknown etiology, detrimental microorganisms left behind by the prior crop of walnut colonize the roots of new plants that grow much slower than plants in non-walnut soil or after soil fumigation. Chloropicrin (in combination with 1,3-D) is given particular importance in this process because of its microbial reducing efficacy. Plant-parasitic nematodes are probably important in these replanting situations, but there appears a microbial component contributing to the disorder (McKenry, 1999).

Phytophthora root and crown rot caused by *Phytophthora pini, P. citricola* and possibly other species is a severe disease that can subliminally infect trees for multiple years before it has deteriorated root function and tissues to such extent that the tree collapses (Figure 13.20). Incidence of the disease can be reduced by careful watering, avoiding excessive soil water saturation periods of >24 h (Adaskaveg et al., 2017). Currently the clonal rootstock RX1 carries high levels of resistance that can protect walnut plantings. Phosphite-containing materials can suppress disease incidence but must be applied with great care to avoid residues in the harvested nuts that could occur because of the systemic movement potential of the chemical.

Paradox canker is a disease of unknown etiology that seems to increase in frequency in California growing areas. So far, it has only been observed on Paradox hybrid rootstocks. Necrotic regions at the lower trunk of affected trees are visible after removal of the bark. Identifying causal agent(s) of this disease has not been possible and its management is remains difficult.

FIGURE 13.20

Walnut tree infected by soilborne disease.

(Photo from G.D. Nanos.)

FIGURE 13.21

Crown gall growing on the bottom of the trunk.

(Photo from G.D. Nanos.)

Crown gall infections by *Agrobacterium tumefaciens* introduce genetic information into the plant cell inducing uncontrolled growth that results in rough galls (Figure 13.21). If the tumor growth goes unchecked, these structures will eventually infect the circumference of the tree and girdle the tree. After initial stunting, this disruption of conductive tissue will kill the tree. Various mechanical and chemical treatment options for Crown gall suppression are employed. Frequently surgical removal of the galls is combined with antibacterial wound treatments. This labor-intensive protocol is challenged for cost efficacy. In California, orchards established using clonal rootstocks show reduced incidence of crown gall compared to orchards on seedling rootstocks.

Armillaria root rot (oak root fungus), *Armillaria mellea,* is a severe disease of tree roots. Roots exhibit white mycelial mats underneath the bark of infected trees. The pathogen survives on dead roots. Often diseased areas are circular within an infested field. Disease spread is difficult to control. Management via cleanliness and maintaining soil at moderate moisture levels can mitigate spread and disease severity. No remedy is known.

Blackline disease is caused by incompatibility of walnut rootstocks with the pollen-transmitted CLRV. This virus infects the scion part of the tree via the flowers and then colonizes branches of the tree. When the viral infection grows down in the tree trunk and passes the graft union into the rootstock, the name-giving symptoms occur when the rootstock resists the viral infection. Masses of necrotic tissue create the visible black line when peeling back the bark. Growers in heavily infested areas have tried to avoid the disease by planting own-rooted English walnut. For example, Chandler has been propagated for this purpose. Because of the high susceptibility to other soilborne pathogens, such own-rooted trees suffer severe damage from those pathogens and succumb to their infections quickly. In the virtual absence of other management tools, current efforts in rootstock development look for virus-tolerant genotypes or breeding of virus-resistant scion cultivars.

9. Current and future challenges in main production regions

Walnuts growing in valleys experience serious problems with climate change/crisis due to lack of winter chill required for proper bloom and fruit set and due to hot summer spells, which will reduce yield and nut quality. Treatments to alleviate the negative influence of climate change are needed. Rootstocks resistant or tolerant to soil diseases and nematodes are lacking around the world. Genetic material with resistance to walnut diseases should be identified and used in new cultivar development. Walnut trees require large quantities of low conductivity irrigation water to grow successfully and produce large-size high-quality nuts; rootstocks that maintain yield under decreased water quality or quantity have not yet been identified. Dry cultivation gradually becomes not viable due to reduced rain events and volume especially during the walnut growing period. Plant protection is getting more problematic in Europe and other areas. Most insecticides are not allowed, only Cu products are available to manage diseases, and these are under certain restrictions. Alternatives to chemical insecticides are needed. The widespread plant-parasitic nematode *Pratylenchus vulnus* is a major challenge in California where fumigation options are becoming more restricted. Biological methods to manage plant-parasitic nematodes are developed. A holistic and agroecological approach should be pursued to help facing climate and social changes in a more sustainable way.

Acknowledgments

We wish to express our gratitude to Saadat Sarikhani for his collaboration in preparing the initial draft of the sections on origin, spread, and economy.

References

Adaskaveg, J. E., Buchner, R. P., Browne, G. T., Gubler, W. D., Michailides, T. J., Hasey, J. K., et al. (2020). *UC IPM management guidelines: Walnut, 3471*. UCANR Publication.

Adaskaveg, J. E., Gubler, W. D., & Michailides, T. J. (2020). *Fungicides, Bactericides, and Biologicals for Deciduous Tree Fruit, Nut, Strawberry, and Vine Crops*. University of California Online Statewide IPM Program. http://ipm.ucanr.edu/PDF/PMG/fungicideefficacytiming.pdf.

Akça, Y., Yuldaşulu, Y. B., Murad, E., & Vahdati, K. (2020). Exploring of walnut genetic resources in Kazakhstan and evaluation of promising selections. *International Journal of Horticultural Science and Technology, 7*, 93–102.

Aradhya, M. K., Potter, D., Gao, F., & Simon, C. J. (2007). Molecular phylogeny of *Juglans (juglandaceae)*: A biogeographic perspective. *Tree Genetics & Genomes, 3*, 363–378.

Behrooz, A., Vahdati, K., Rejali, F., Lotfi, M., Sarikhani, S., & Leslie, C. (2019). Arbuscular mycorrhiza and plant growth-promoting bacteria alleviate drought stress in walnut. *HortScience, 54*, 1087–1092.

Bernard, A., Lheureux, F., & Dirlewanger, E. (2018). Walnut: Past and future of genetic improvement. *Tree Genetics & Genomes, 14*(1). https://doi.org/10.1007/s11295-017-1214-0

Bernard, A., Marrano, A., Donkpegan, A., Brown, P. J., Leslie, C. A., Neale, D. B., et al. (2020). Association and linkage mapping to unravel genetic architecture of phenological traits and lateral bearing in Persian walnut (*Juglans regia* L.). *BMC Genomics, 21*(1), 203.

Buzo, T., McKenna, J., Kaku, S., Anwar, S. A., & McKenry, M. V. (2009). VX211, a vigorous new walnut hybrid clone with nematode tolerance and a useful resistance mechanism. *Journal of Nematology, 41*, 211–216.

Charrier, G., Bonhomme, M., Lacointe, A., & Ameglio, T. (2011). Are budburst dates, dormancy and cold acclimation in walnut trees (*Juglans regia* L.) under mainly genotypic or environmental control? *Int. J. Biometeorol, 55*, 763–774.

Dehghan, B., Vahdati, K., Rezaee, R., & Hassani, D. (2009). Persian walnut (*Juglans regia* L.) grafting as influenced by different bench grafting methods and scion cultivars. *J. Appl. Hortic., 11*, 56–58.

Dong, W., Xu, C., Li, W., Xie, X., Lu, Y., Liu, Y., et al. (2017). Phylogenetic resolution in Juglans based on complete chloroplast genomes and nuclear DNA sequences. *Front. Plant Sci., 8*. https://doi.org/10.3389/fpls.2017.01148

Erdogan, V. (2005). Use of hot callusing cable in walnut propagation. *Acta Hortic, 705*, 313–317. https://doi.org/10.17660/ActaHortic.2005.705.41

Fallah, M., Vahdati, K., Hasani, D., Rasouli, M., & Sarikhani, S. (2022). Breeding of Persian walnut: Aiming to introduce late-leafing and early-harvesting varieties by targeted hybridization. *Scientia Horticulturae, 295*, Article 110885.

FAO. (2023). FAO statistical yearbook. Agricultural production. *Food and Agriculture Organization of the United Nations.*

Fjellstrom, R. G., Parfitt, D. E., & McGranahan, G. H. (1994). Genetic relationships and characterization of Persian walnut (Juglans regia L.) cultivars using restriction fragment length polymorphisms (RFLPs). *Journal of American Society for Horticultural Science, 119*, 833–839.

Gharaghani, A., Javarzari, A. M., & Vahdati, K. (2018). Kaolin particle film alleviates adverse effects of light and heat stresses and improves nut and kernel quality in Persian walnut. *Scientia Horticulturae, 239*, 35–40. https://doi.org/10.1016/j.scienta.2018.05.024

Grant, J. A., Hasey, J. K., Coates, W. W., Van Steenwyk, R. A., Symmes, E. J., Seybold, S. J., et al. (2017). *UC IPM pest management guidelines, 3471*. Walnut UCANR Publication.

Habibi, A., Hamedpour-Darabi, M., & Vahdati, K. (2022). Local cultural values of Persian walnut in Iran. In W. Jeffrey (Ed.), *The cultural value of trees*. Routledge.

Hajinia, Z., Sarikhani, S., & Vahdati, K. (2021). Exploring low-chill genotypes of Persian walnut (*Juglans regia* L.) in west of Iran. *Genetic Resources and Crop Evolution, 68*, 2325–2336.

Hassani, D., Sarikhani, S., Dastjerdi, R., Mahmoudi, R., Soleimani, A., & Vahdati, K. (2020). Situation and recent trends on cultivation and breeding of Persian walnut in Iran. *Scientia Horticulturae, 270*, 109369.

Hassankhah, A., Rahemi, M., Ramshini, H., Sarikhani, S., & Vahdati, K. (2020). Flowering in Persian walnut: Patterns of gene expression during flower development. *BMC Plant Biology, 20*, 136.

Hassankhah, A., Vahdati, K., Rahemi, M., Hassani, D., & Sarikhani Khorami, S. (2017). Persian walnut phenology: Effect of chilling and heat requirements on budbreak and flowering date. *International Journal of Horticultural Science and Technology, 4*(2), 259–271.

INC. (2023). *Nuts and dried fruits statistical yearbook 2022/2023* (p. 79). Spain: International Nut and Dried Fruit Council.

Krueger, W. H. (2000). Pollination of English walnuts: Practices and problems. *HortTechnology, 10*, 127–130.

Leslie, C. A., Hackett, W. P., Bujazha, D., Hirbod, S., & McGranahan, G. H. (2005). Adventitious rooting and clonal plant production of hybrid walnut (Juglans) rootstock selections. *Acta Hortic, 705*, 325–328. https://doi.org/10.17660/ActaHortic.2005.705.43

Lotfi, N., Soleimani, A., Çakmakçı, R., Vahdati, K., & Mohammadi, P. (2022). Characterization of plant growth-promoting rhizobacteria (PGPR) in Persian walnut associated with drought stress tolerance. *Scientific Reports, 12*(1), Article 12725.

Mahmoudian, M., Rahemi, M., Karimi, S., Navid, Y. N., Sarikhani, S., & Vahdati, K. (2021). Role of kaolin on drought tolerance and nut quality of Persian walnut. *Journal of the Saudi Society of Agricultural Sciences, 20*, 409–416.

Marrano, A., Sideli, G. M., Leslie, C. A., Cheng, H., & Neale, D. B. (2019). Deciphering of the genetic control of phenology, yield, and pellicle color in Persian walnut (Juglans regia L.). *Front Plant Sci, 10*, 1140. https://doi.org/10.3389/fpls.2019.01140

McKenry, M. V. (1999). *The Replant Problem and its Management*. Fresno, CA: Catalina Publishing. https://ucanr.edu/sites/default/files/2011-04/86495.pdf.

Mircetich, S. M., Sanborn, R. R., & Ramos, D. E. (1980). Natural spread, graft transmission and possible etiology of walnut blackline disease. *Phytopathology, 70*, 962–968.

Navacchi, O., Bastianelli, M., Neri, D., & Cozzolino, E. (2021). Micropropagation of walnut: A real opportunity. *Acta Hortic, 1318*, 63–70.

Neri, D., Cozzolino, E., Sirri, S., Massetani, F., & Giovannini, D. (2021). Shoot growth and bud topology in 'Chandler' walnut. *Acta Hortic, 1318*, 97–107.

Ortiz, C. M., Vicente, A. R., Fields, R. P., Grilo, F., Labavitch, J. M., Donis-Gonzalez, I., et al. (2019). Walnut (*Juglans regia* L.) kernel postharvest deterioration as affected by pellicle integrity, cultivar and oxygen concentration. *Postharvest Biology and Technology, 156*, Article 110948.

Olsen, J. (2006). *Growing walnuts in Oregon*. Oregon State University Extension Service. EM 8907.

Pollegioni, P., Woeste, K., Chiocchini, F., Del Lungo, S., Ciolfi, M., Olimpieri, I., et al. (2017). Rethinking the history of common walnut (*Juglans regia* L.) in Europe: Its origins and human interactions. *PloS One, 12*(3), Article e0172541.

Pollegioni, P., Woeste, K. E., Chiocchini, F., Del Lungo, S., Olimpieri, I., Tortolano, V., et al. (2015). Ancient humans influenced the current spatial genetic structure of common walnut populations in Asia. *PloS One, 10*(9), Article e0135980.

Ramos, D. E. (1998). *Walnut production manual, 3373* (p. 320). U.C. D.A.N.R. Publ..

Rosati, A., Metcalf, S. G., Buchner, R. P., & Lampinen, B. D. (2007). Effects of kaolin application on light absorption and distribution, radiation use efficiency and photosynthesis of almond and walnut canopies. *Annals of Botany, 99*, 255–263.

Sotomayor, C., Mandava, N. B., & Mandava, S. (2022). Effects of brassinosteroids on walnut trees fruit set, nut weight and pollen germination. *Acta Hortic, 1344*, 115–122.

Stanford, A. M., Harden, R., & Parks, C. R. (2000). Phylogeny and biogeography of Juglans (Juglandaceae) based on matK and ITS sequence data. *American Journal of Botany, 87*(6), 872–882.

Thapa, R., Thapa, P., Ahamad, K., & Vahdati, K. (2021). Effect of grafting methods and dates on the graft take rate of Persian walnut in open field condition. *International Journal of Horticultural Science and Technology, 8*(2), 133–147.

Vahdati, K., Aslamarz, A. A., Rahemi, M., Hassani, D., & Leslie, C. (2012). Mechanism of seed dormancy and its relationship to bud dormancy in Persian walnut. *Environmental and Experimental Botany, 75*, 74–82.

Vahdati, K., Razaee, R., & Mirmasoomi, M. (2009). Micropropagation of some dwarf and early mature walnut genotypes. *Biotechnology, 8*(1), 171–175.

Vahdati, K., Arab, M. M., Sarikhani, S., Sadat-Hosseini, M., Leslie, C. A., & Brown, P. J. (2019). Advances in Persian walnut (*Juglans regia* L.) breeding strategies. In J. Al-Khayri, S. Jain, & D. Johnson (Eds.), *Advances in plant breeding strategies: Nut and beverage crops*. Springer.

Vahdati, K., Hassani, D., Rezaee, R., Jafari Sayadi, M. H., & Sarikhani Khorami, S. (2014). Walnut footprint in Iran. In D. Avanzato, G. H. McGranahan, K. Vahdati, M. Botu, L. Iannamico, & J. V. Assche (Eds.), *Following walnut footprints (Juglans regia L.) cultivation and culture, folklore and history, traditions and uses* (Vol. 17, pp. 187–201). Scripta Horticulturae.

Vahdati, K., Sadeghi-Majd, R., Sestras, A. F., Licea-Moreno, R. J., Peixe, A., & Sestras, R. E. (2022). Clonal propagation of walnuts (Juglans spp.): A review on evolution from traditional techniques to application of biotechnology. *Plants, 11*, 3040.

Vahdati, K., Sarikhani Khorami, S., & Arab, M. M. (2018). Walnut: A potential multipurpose nut crop for reclaiming deteriorated lands and environment. *Acta Horticulturae, 1190*, 95–100.

Vahdati, K., Sheikhi, A., Arab, M. M., Sarikhani, S., Habibi, A., & Ataee, H. (2023). Cultivars and genetic improvement. In M. M. Mir, M. U. Rehman, U. Iqbal, & S. A. Mir (Eds.), *Temperate nuts*. Singapore: Springer.

Vahdati, K. (2024). *Targeted breeding innovations in nut tree cultivars*. Heal news. Spain: International Nut and Dried Fruit Council.

Vahdati, K. (2014). Traditions and folks for walnut growing around the silk road. *Acta Hortic, 1032*, 19–24.

Vahdati, K., & Zareie, N. (2005). Evaluation of side-stub and hypocotyl grafting efficiency for walnut propagation in Iran. *Acta Hortic, 705*, 347–350.

Walnut world production by country https://www.statista.com/statistics/675974/walnut-production-worldwide-by-country/ (accessed May 17, 2024)

Westerdahl, B. B., Westphal, A., & Hasey, J. K. (2017). *Walnut nematodes. UC Pest Management Guidelines (updated 6/17)*. UCIPM University of California Agriculture and Natural Resources, Statewide Integrated Pest Management Program. http://ipm.ucanr.edu/PMG/r881200111.html.

Westerdahl, B. B., Westphal, A., & Hasey, J. K. (2020). *UC IPM pest management guidelines: Walnut, 3471*. UCANR Publication.

Yan, L. J., Fan, P. Z., Wambulwa, M. C., Qi, H. L., Chen, Y., Wu, Z. Y., et al. (2024). Human-associated genetic landscape of walnuts in the himalaya: Implications for conservation and utilization. *Diversity and Distributions, 30*, Article e13809. https://doi.org/10.1111/ddi.13809

Zou, Y. N., Qin, Q. Y., Ma, W. Y., et al. (2023). Metabolomics reveals arbuscular mycorrhizal fungi-mediated tolerance of walnut to soil drought. *BMC Plant Biology, 23*, 118. https://doi.org/10.1186/s12870-023-04111-3

CHAPTER

Pistachio

14

Giulia Marino[1], Petros A. Roussos[2], Giorgio Gusella[3], Themis J. Michailides[4], Louise Ferguson[1], Tiziano Caruso[5]

[1]*Department of Plant Sciences, University of California Davis, Davis, CA, United States;* [2]*Laboratory of Pomology, Agricultural University of Athens, Athens, Greece;* [3]*Dipartimento di Agricoltura, Alimentazione e Ambiente, Università di Catania, Catania, Italy;* [4]*Department of Plant Pathology, University of California Davis, Davis, CA, United States;* [5]*Department of Agricultural, Food and Forest Sciences, University of Palermo, Palermo, Italy*

Chapter outline

1. Origin and spread

Pistachio is considered one of the oldest plant species. Archaeological excavations and fossil studies suggest its presence around 80 million years ago (Sheikhi et al., 2019). Additionally, some studies indicate that pistachios were first consumed by Neanderthals approximately 300,000 years ago (Nezami and Gallego, 2023). The most significant species within the genus is *Pistacia vera* L., which is highly valued for its edible nuts (Parfitt et al., 2012) and is believed to be the most ancestral one, from which all others have evolved (Sheikhi et al., 2019). The origin of pistachio was and to some extent is still not clear. There are two possible centers of origin: (1) Syria, eastern Anatolia, and Central Asia, and (2) southern Central Asia (Hormaza and Wünsch, 2007, Mir-Makhamad et al., 2022). According to Zeng et al. (2019),

Temperate Tree Fruits and Nuts. https://doi.org/10.1016/B978-0-443-29137-1.00014-1

during the Pleistocene period, pistachios were grown in ice-free areas across southern Central Asia where they formed a wild population. According to botanists, wild *P. vera* existed in three ecological sites distributed in (1) the Northern Tian Shan region of Northeastern Kyrgyzstan, as well as in Uzbekistan and Kazakhstan; (2) in Southwestern Tajikistan, Southern Uzbekistan, and Eastern Turkmenistan; and (3) in Southwestern Turkmenistan (Mir-Makhamad et al., 2022).

Many attempts to pinpoint the origin of pistachios based on shells or archaeological remains should be treated with caution. These artifacts may indicate areas of cultivation, potential origins, or simply commercial commodities. The center for *P. vera* diversity by the majority of scholars is considered the area from Northern to Northeastern Iran, Southern Turkmenistan, and northern parts of Afghanistan, through Uzbekistan, Tajikistan, Kazakhstan, and Kyrgyzstan, i.e., South-Central Asia, where natural wild populations can be found (Mir-Makhamad et al., 2022, Parfitt et al., 2012).

Based on current evidence, the domestication and cultivation of pistachio began *ca.* 3000 years ago in Iran (Nezami and Gallego, 2023). The etymology of the word pistachio originates from the word *pstk* (pistag or pista) from the ancient Persian language Avestan (Hormaza and Wünsch, 2007). This word constitutes the root of the modern name of pistachio in several languages such as *pistakia, pistacia, fistuk, fustak, fistik, pisutachio, pistazie,* etc.

From this area, it seems pistachios expanded to Syria, possibly facilitated by the expansion of the Persian Empire, by the end of the first millennium BCE and from there slowly spread westward. It seems that pistachio was cultivated in Syria for a long time before being dispersed across the Mediterranean basin, as both Dioscorides and Pliny the Elder describe Syria as a center of pistachio cultivation by the first century AD. Archaeobotanical remains suggest that Syria was the first area of pistachio cultivation outside of its area of origin, already by the first millennium AD.

According to Pliny the Elder, the pistachio was likely brought to Italy during the Roman Empire (1st century AD) by Vitellius (Hormaza and Wünsch, 2007). At the same time, it was introduced to Spain by Pompeius Flavius, a knight of Rome under the orders of Vitellius. During the following centuries, pistachio cultivation has expanded to other Mediterranean countries as well as to the U.S., China, and lately to Australia (Hormaza and Wünsch, 2007, Mir-Makhamad et al., 2022).

2. Economic importance

The U.S. (California), Iran, and Turkey are the main pistachio-producing countries in the world, producing 67%, 17%, and 11% of the world's production, respectively (Mir-Makhamad et al., 2022). Collectively, these countries produce 95% of the world's pistachios. Greece, Syria, Afghanistan, and other countries produce the remaining 5%.[3]

California's approximately 50-year-old pistachio industry consisted of 255,578 total hectares; 191,320 bearing and 64,184 nonbearing hectares.[1] In 2023, California produced 647,865,982 metric tons.[2] The impact on California's economy is significant. California's pistachio industry, processors, and growers employed 55,100 workers full time, spent U.S. $3.5 billion producing and processing the crop, and paid U.S. $230.9 million in business taxes, a total economic impact of $6.4 billion, or U.S. $17.6 million per day on California's economy. The United States exported almost 60% of their pistachio production (390,000 metric tons), followed by Iran, Turkey, and Syria. Main importers are China (160,000 metric tons) followed by the European Union, Turkey, and India.[5]

In 2020, it required an estimated U.S. $47,675 to establish one hectare of pistachio, independent of land costs, and an average of 6 years to reach beginning bearing, and 8 to 10 years to achieve full bearing. Thereafter, it requires an estimated U.S. $7736 per hectare of operating costs to produce pistachios.[4] In the past 5 years, 2019-2023 when yields averaged 2929 kg per ha and growers were paid an average U.S. $5.79 per kg, it was profitable to grow pistachio. An average gross return was U.S. $16,511 per hectare and a net return was U.S. $8775 per hectare. However, in 2023, the price dropped by 40%, and water costs, among the major production costs in pistachio production, have risen precipitously. These costs, combined with increased competition and 25% of the pistachio orchards in California yet to enter bearing, will probably lower the profitability of California pistachio production. The California industry will need to lower production costs, primarily water, diversify its products beyond dried and salted split in-shell nuts, and develop new markets beyond Europe and Southeast Asia.

In the rainfed or supplementary irrigated pistachio orchards in the Mediterranean countries, yearly production is more variable and lower with respect to the fully irrigated Californian orchards. Average dried in the shell per tree ranges between 3 to 10 kg, corresponding to an average of ~1600 kg/ha. Despite the significantly lower yield efficiency compared to the United States, the pistachio is still an economically important crop in these areas and often referred to as the "golden tree" or the "green gold" (Mir-Makhamad et al., 2022; Marino and Marra, 2019). Indicatively, Sicilian pistachios average price per kilogram were priced for 13.20 € in 2019, higher with respect to an average price of 8-9 € paid in the European market for pistachios of non-Sicilian origin.

3. Botany and relevant species

The pistachio belongs to the order Terebinthales, family Anacardiaceae, genus *Pistacia*. Within this genus, there are at least 11 species; four of them, *P. atlantica* Desf., *P. integerrima* Stewart, and *P. terebinthus* L., are used as rootstocks or for creating rootstocks, and only one, *P. vera*, is used to produce edible fruits. Other species like *P. lentiscus* and *P. terebinthus* L. are also important for resin production.

The tree has slow growth when grown under natural arid conditions and/or on marginal lands, but under suitable soil and appropriate cultivation management, it can exceed 10 m in height. The bark is reddish on young 1-year-old branches, ashy gray on older branches, and dark gray on the trunk. In young wood, the lenticels are evident and abundant. The leaves are comprised of three to five green leaflets, the terminal one generally larger than the lateral ones. At the base of each leaf, there is an axillary bud for flowers or wood. The genus is characterized by marked amphistomy with 60% of the stomata found on the lower leaf surface and the remaining 40% on the upper surface.

Pistachio is dioecious with male and female reproductive organs in separate individuals, being considered as anemophilous in terms of pollination (Figure 14.1). Female flowers are apetalous and clustered in panicles, borne on 1-year-old branches. Female inflorescences typically contain an average of 100 to 400 flowers. Male flowers, clustered in panicles borne on 1-year-old branches, have 5 to 7 bracts, with 5 stamens carrying large yellow anthers that open longitudinally, emitting abundant pollen. The fruit, grouped in infructescences, is a monospermic ovoid drupe, with the epicarp and mesocarp (pericarp) forming the pulp, which at maturity is not very consistent and varies in color from whitish-pink to yellow-cream or reddish-purple, depending on the cultivar and fruit ripeness level. The endocarp is woody and ivory-colored; at maturity, it may exhibit various degrees of dehiscence depending on the cultivar, fruit load, and water regime (Goldhamer et al., 1987). The vegetative habit of the plant is influenced by pronounced apical dominance.

New vegetation is almost exclusively produced by the budding of the apical bud of each branch. This leads the plant to have long, generally prostrate branches, bare in the basal and middle sections, with the productive zone limited to the outer area of the canopy. The plant is very hardy, xerophilous, tolerating long periods of drought and high temperatures, from flowering to fruit maturity (April-September); at the same time, it can survive in environments with winter rainfall between 250 and 400 mm. During winter dormancy, it can withstand temperatures of -20°C to -30°C, while in summer, it can tolerate up to 50°C, provided there is low atmospheric humidity. It is extremely vulnerable to spring frosts when the bud scales

FIGURE 14.1

Female *(left)* and male *(right)* inflorescence of pistachio at full bloom stage.

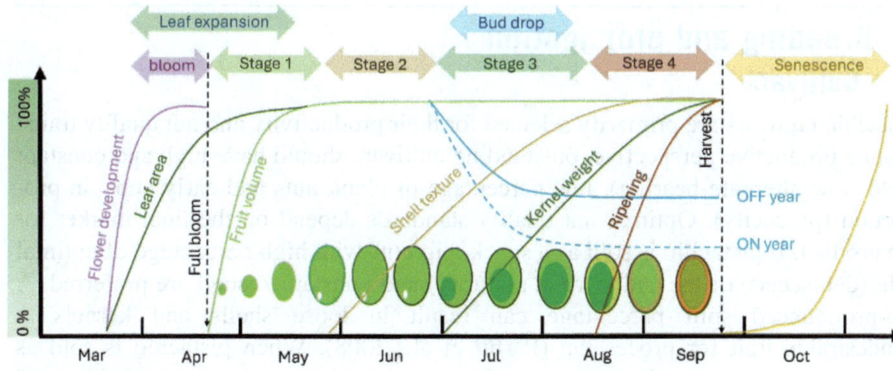

FIGURE 14.2

Phenological stages of pistachio during the growing season.

(From G. Marino)

begin to spread. Pedologically, it adapts to soils of various types, exhibiting moderate resistance to moderately high pH values, but it does not tolerate waterlogging (Gusella et al., 2021). All species of the *Pistacia* genus are very sensitive to high soil moisture, even for short periods. Depending on the cultivar and the area of origin, the species has a chill requirement ranging from 600 to 1000 chilling units.

A scheme of the phenological progression of pistachio is provided in Figure 14.2. Vegetative growth begins at the end of March or early April with the opening of flower buds on branches formed the previous year, followed by the opening of apical wood buds. The flowering patterns in males and females highlight the importance of choosing the correct pollinator. Within 15 to 20 days after full flowering, rapid shoot growth of the apical bud begins, completing at 30 to 35 days after full flowering (Barone et al., 1995).

During this period, approximately 60% of the flowers shed, and after pollination, about 55% of the small fruits drop. Normally, only 15% to 20% of potential fruits reach maturity. The different parts of the fruit—pericarp (pulp), endocarp (shell), embryo, and cotyledons (seed)—have different growth rhythms, and dry matter accumulation in the fruit follows a typical double sigmoid pattern consisting of three stages. Stage 1: Nut expansion: pulp and shell cells elongate and thicken (between April and May, until 500 growing degree days [GDDs]). Stage 2: Shell hardening: the shell lignifies, and drupe development is completed but the seed is still not growing (during June, between 500 and 1000 GDDs). Stage 3: Kernel fill, progressive embryo growth (in July and August, between 1000 and 2000 GDDs) (Crane and Iwakiri, 1981). Six weeks before harvest, the separation of the suture line begins to appear, which may become more pronounced as maturation progresses, leading to seed dehiscence (Polito and Pinney, 1999). Two to three weeks before harvest (in September, after 2000 GDDs), the ripening process starts (Stage 4).

4. Breeding and propagation

4.1 Cultivars

Pistachio cultivars are primarily selected for their productivity and nut quality traits. From a productive perspective, outstanding cultivars should have high and constant yield (low alternate-bearing), low percentage of blank nuts and early entry in production (precocity). Optimal nut quality standards depend on the final market for the product. If pistachio is sold as a snack, big nuts with high percentage of optimal split (dehiscence of the endocarp at maturity) and nonstained shell are preferred. A too-pronounced split percentage can result in loose shells and kernels, a nondesirable trait for processing (Parfitt et al., 2008). When pistachio is sold as kernel to the processing industries (confectionery for ice cream, pastry, and bagged meat), then nuts with appealing organoleptic characteristics are the most desirable, such as taste, kernel greenness, and aromatic profile. All these characteristics, in addition to being impacted by genetics, are affected by the environmental conditions and management practices.

The California pistachio industry has been dominated for a long time by the female cultivar Kerman and the pollinizer Peters. Kerman was selected by the USDA Plant Introduction Station in Chico, California, for its high yields and large nuts, and its main drawbacks are the high tendency to alternate bearing, the high percentage of blank and nonsplit nuts, and the mild organoleptic properties.

In the last 30 years, the University of California (UC) Pistachio Breeding Program has focused on selecting new cultivars with earlier flowering and harvest time than Kerman. This trait would allow the growing industry to expand the harvest season and reduce susceptibility to disease associated with late harvests. The new cultivars Golden Hills and Lost Hills that bloom and ripen 7 to 14 days before Kerman were released in 2005. Their pollinizer Randy showed higher pollen viability compared to Peters (75% versus 45%) (Parfitt et al., 2010).

Currently, Golden Hill is the most planted female cultivar in California: it yields similarly to Kerman and produces a processed nut similar in size (about 1.5 g/nut) but with higher split edible in-shell nuts (87%-89%) than Kerman (74%-79%) (Parfit et al., 2007). On the negative side, Golden Hills branches show a higher tendency of breaking with respect to Kerman and leaves are more susceptible to boron toxicities (Kallsen and Parfitt, 2008). Lost Hills showed less alternated-bearing pattern than Kerman and Golden Hills and a slightly larger nut (1.43 g/nut) but also a higher percentage of loose shell and kernels.

The female cultivar Gumdrop was released in 2016, and it can be harvested 10 to 12 days before Golden Hills and 24 days before Kerman (Kallsen and Parfitt, 2017), and its main pollinator is Tejon (Kallsen and Parfitt, 2019). The male cultivar Famoso, together with a new male that is under the process of commercial release, have been developed as additional pollinizer for Kerman to cover potential mismatch in bloom with Peters, which is quite common in low chill years (Kallsen and Parfitt, 2019).

Kerman has been historically the most planted cultivar also in the Iberian Peninsula, where it is now being replaced by Sirora, Larnaka, Lost Hills, Avdat, Kastel, and Golden Hills. Aeginis and Mateur are other cultivars used but in a much smaller proportion. The male cultivars Peters, C-especial, Askar, Nazar, Egino, Randy, and Chaparrillo are used as pollenizers.

In Greece, the two local cultivars are Aeginis and Pontikis. Both have a good split percentage (80%) and are being used for the snack market. Pontikis kernel has a deep green color and is considered tastier than Aeginis. Currently there are four types of male cultivars, the A-, B-, C-, and D-type, which are classified based on pollen release date (A-type the most precocious one) along with Peters, to fully cover female pollination needs in all growing areas of Greece.

In Italy, the pistachio is mainly grown in Sicily, where the main cultivar Bianca (also called Bronte or Napoletana) has been selected in the past by growers for its intense green color of the cotyledons. It produces small nuts with a very low percentage of split (41%-45%), but it is not an issue since it is sold for the kernel market (Marino et al., 2018). The Department of Arboriculture of the University of Palermo selected three local male genotypes (M1, M4, and M5) as the best option for pollination (Marino and Marra, 2019).

In Australia, the most used cultivar Sirora was bred at the Commonwealth Scientific and Industrial Research Organization (CSIRO) to have a low chilling requirement, higher yield and earlier ripening, better organoleptic properties, and smaller nuts than Kerman (Sheikhi et al., 2019). Breeding programs in Iran and California are currently focusing on the development of cultivars with lower chilling requirements.

4.2 Rootstocks

In traditional rainfed plantations in the Mediterranean countries such as Italy, Spain, and Greece, the species *P. terebinthus* L. has been mostly used as rootstock due to its rusticity and drought resistance. This rootstock is highly susceptible to *Verticillium dahliae* and shows very slow growth and long unproductive period. In Iran, the *P. vera* is used as rootstock because it is vigorous and tolerant to salinity and drought. However, *P. vera* is relatively sensitive to nematodes and very sensitive to *Phytophthora*, which has limited its use when tested under fully irrigated conditions.

When pistachio started to expand in California, *P. atlantica* was initially used as rootstock, but due to its high susceptibility to *Verticillium,* it was soon substituted by *P. integerrima* (Pioneer Gold I [PGI]), which is tolerant. The interspecific hybrid obtained from open pollination between *P. integerrima* x *P. atlantica*, called Pioneer Gold II (PGII), was largely planted in the 1990s. The *P. atlantica* parentage conferred the rootstock higher freeze and salinity resistance (Ferguson et al., 2005).

Slightly after, a controlled cross between two selected *P. atlantica* (female) and *P. integerrima* (male) individuals, called University of California Berkeley 1 (UCB1), was developed. UCB1 has high vigor and productivity, salinity and freezing tolerance (Ferguson et al., 2002), as well as resistance to *Verticillium*, along with better resistance to *Phytophthora* and *Armillaria* (Vahdati et al., 2021). Recent studies demonstrated that UCB1 rootstocks are more tolerant to saline conditions since they can retrieve sodium from the sap, reducing the accumulation of salt in the leaves (Godfrey et al., 2021). Since then, UCB1's superior characteristics overshadowed other rootstocks (Ferguson et al., 2005; Sheikhi et al., 2019).

In recent years, with the development and expansion of micropropagation techniques at a commercial scale, several superior clones of UCB1 have been developed by nurseries and are now available in the market.

Current rootstocks breeding efforts in California are focusing on selecting superior UCB1 among the existing seedling population, with a focus on higher salinity tolerance, vigor, and productivity. In addition, the UC Davis breeding program has developed some new *P. vera* x *P. integerrima* hybrids rootstocks that are still under evaluation but showed very promising results in terms of reduced boron toxicity, reduced suckering and vigor, timing of leaf out and entry into dormancy closer to the *P. vera* scions, and phytophthora resistance (Kallsen and Parfitt, 2022).

In Iran, the Nazari Business Group and Iran's Pistachio Research Institute breeding focuses on rootstocks tolerance to abiotic stresses. They recently developed a new interspecific pistachio hybrid rootstock Arota *(P. integerrima* × *P. atlantica)* that demonstrated improved salinity stress tolerance (Akbari et al., 2020). They are also exploring the salt tolerance of several native *P. vera* individuals (Akbari et al., 2018; Aliakbarkhani et al., 2015; Hajiboland et al., 2014; Momenpour and Imani, 2018).

4.3 Propagation

Pistachio trees can be propagated through both sexual and asexual methods, each offering distinct advantages and disadvantages (Nezami and Gallego, 2023). Being a dioecious species, sexual propagation results in a mix of male and female trees, making it an impractical method for the commercial multiplication of elite cultivars. Moreover, seedlings have different vigor, resulting in heterogeneous orchards, and a lengthy juvenile period, posing a significant challenge for modern pistachio cultivation. Despite these drawbacks, seedlings are often used as rootstocks in various countries because *Pistacia* spp. are recalcitrant to clonal propagation and *in vitro* techniques are needed to produce clonal rootstocks.

Sexual propagation (seed propagation) is a straightforward and cost-effective method. To achieve high germination rates, nuts must undergo a scarification process using concentrated sulfuric acid. After harvest, the pericarp is removed by rubbing the nuts in sieves and rinsing them under running tap water. The nuts, now reduced to the endocarp and seed, are soaked in water; any floating nuts are discarded. The remaining seeds are immersed in concentrated sulfuric acid (96%), using roughly three volumes of acid per seed volume, and stirred occasionally (Kostas et al., 2021). The duration of acid exposure depends on the endocarp's density, which is periodically tested by cutting the seed in half to ensure the acid has not reached the actual seed. Typically, this process lasts 10 to 20 min. Following scarification, the seeds are thoroughly rinsed under running water with continuous stirring to eliminate excess acid. Seeds are then planted in a suitable substrate at elevated temperatures (23°C-25°C) or sown in spring as temperatures rise. Seedlings are grown in individual pots until they are large enough for budding.

To circumvent the challenges associated with the use of seedlings, various **asexual propagation** techniques have been explored, though with contradictory results (Nezami and Gallego, 2023). Propagation by cuttings, either softwood or hardwood, has been attempted in pistachio but generally shows inconsistent outcomes. Indole-3-butyric acid is commonly used as a rooting hormone, with concentrations ranging from 5 mg/L to 80 g/L (Almehdi et al., 2002). However, UCB1 rootstock has demonstrated low rooting percentages (just above 40% under optimum conditions), rendering them unsuitable for commercial production (Nezami and Gallego, 2023; Almehdi et al., 2002).

In vitro propagation techniques have advanced significantly over the past two decades, facilitating the commercialization of initially experimental methods (Marín et al., 2017). Numerous studies have successfully demonstrated the efficient *in vitro* propagation of pistachio using basal media such as modified Driver-Kuniyuki medium for walnut (Driver and Kuniyuki, 1984) and Murashige-Skoog, particularly during the establishment and multiplication stages (Benmahioul et al., 2012; Onay, 2003). Rooting can be accomplished *ex vitro*, enhancing the survival rates of *in vitro* propagated plantlets (Benmahioul et al., 2012). This method has become the preferred approach for producing UCB1 plants on a large scale, ensuring uniformity and avoiding the variability inherent in seed-derived plants.

As propagated plants reach adequate trunk girth, **grafting** with T-budding is performed using buds from the desired cultivar (Nezami and Gallego, 2023). The budding height varies by region, influenced by the harvesting method employed—higher positions are used, where trunk shakers are used, to prevent damage to the budding zone. Budding can be performed in summer (June-July) in rootstocks already established in the field. Otherwise, growers can purchase already budded trees. Recently some nurseries have been using dormant buds collected the previous year, stored at low temperature during winter. Despite the fact that no experimental data are available, this method looks promising since it allows for budding as early as May-June leading to longer first season growth. In the nursery, budding is typically performed from early August to mid-September. *In vitro* micrografting has been explored experimentally (Onay et al., 2007), but it has not become a standard practice, mainly due to its relevance to local cultivars in specific pistachio-producing regions. Budded pistachio plants are grown in pots or nursery bags, receiving appropriate fertilization and phytosanitary care to reach the desired height of 1.0 to 1.5 m before being sold to growers. In some countries, rootstocks in nurseries are grown in the ground and sold as bare root trees.

5. Orchard management

Orchard establishment: The soil profile is assessed with backhoe pits prior to planting; one pit per 8 ha with soil samples at 0.5 m to 2.0 m. The land is tilled with a slip-plow down the tree row to 2 m deep. The field is then ripped with a one shank ripper in between the slip-plow (in between the tree rows) at the same depth. Deep tillage is only done on the area that will be drip irrigated. Ripping is followed by one pass with a stubble disc, twice, and the ground is floated to smooth the surface. This is followed by the installation of a single- or double-line drip system with fertigation capacity. Irrigation system capacity is designed to deliver 1270 mm per ha per year.

Trees and planting density: Usually, the rootstocks are planted first, and the scion cultivar is field budded to the rootstock. However, prebudded nursery trees are becoming more popular, particularly for replants. Because pistachios are dioecious, male trees are planted at 1 male tree to every 24 female trees in a regular every fifth row every fifth tree. California pistachios are planted on 5.2 x 6.0 m spacing with 308 trees per ha (Figure 14.3).

Year	Applied Nitrogen (kg/ha)	Applied water (mm)
0	11	127
1	17	254
2	22	368
3	34	495
4	45	648
5	62	813
6	84	965
7	151	1118
8+	196	1257

FIGURE 14.3

Recently budded pistachio orchard, with protective sleeves placed over the plants and tied to the stakes. The scion *(the lighter color shoot on the right)* is growing and the rootstock *(darker apical shoot)* has still not been removed. On the *right* is a table describing the nitrogen and water requirement of young pistachio orchards (0 to 8 years).

Seedling rootstocks, grown 1 year in the nursery, topped at 0.6 m before delivery, are planted in February through April. Before planting the unbudded rootstocks, the field is surveyed, tree sites marked, and holes are dug. Immediately after planting, new 5 x 5 cm x 1.8 m wooden or metal stakes are spread in the field and installed on the upwind side after planting. Protective sleeves are placed over the plants and tied to the stakes. Stakes are normally removed prior to the first mechanical harvest. In late July of the first year, the trees are commercially budded 0.7 to 0.8 m above ground level, preferably into new shoot growth. Failed buds on surviving rootstock are rebudded with two buds per tree in September. No attempt is made to push the rebudded trees because they are susceptible to killing frost in November or early December. In the second year, failed buds should be rebudded as soon as possible. The budding failure rate should be no more than 5% in the first year and 2% in the second. A rootstock failing to take a bud after two attempts should be replaced in March of the second year in the ground. In years 1 and 2, replanted tree loss in each year is 1% or less. Nursery budded trees or unbudded trees are replanted as soon as possible. Replants placed in older orchards normally have a higher failure rate due to improper irrigation and inadequate vertebrate pest control.

During spring of the first year (called "first leaf"), a work crew selects and ties the new rootstock growth destined to be budded to the stake and removes any growth that is within 20 cm of the ground to facilitate herbicide applications. A second pass is made in June to tip or eliminate shoots competing with the one destined for scion. The trees are typically budded from late June through July. Five to seven days after budding, the crew partially girdles (notches) the rootstock directly above the scion bud. Simultaneously, 50% to 70% of the rootstock growth is also removed using heading cuts. Both operations encourage uniform and vigorous growth of the scion bud. Three weeks later (August), or when 15 to 20 cm of the scion growth is achieved, a fourth pass is made to begin training the shoot to grow up by attaching it to the stake and to continue suppression of new rootstock growth by pinching off its terminals. The crew makes three more tree-training passes at 14-day intervals during August and September. This results in a total of six passes through the orchard by the end of September (two suckering and tying, notching, and heading after budding, and four training). Dormant scion shoots are headed to 1.1 m, and the rootstock lateral branches are removed in January (first dormant pruning). Both the rootstock and the dormant scion shoots are then tied to the stake. If trees are planted and grafted early in the first season and growth warrants it, this first heading cut may be accomplished during the first growing season to begin formation of primary branches. However, there must be sufficient time in the remaining season to develop strong primaries at least 0.5 m long.

Young tree irrigation and fertilization regimes are described in Figure 14.3. In the early years, nitrogen is needed for tree growth. Potassium fertilization begins in the fourth year in the ground to develop an available potassium soil pool for nut growth. The micronutrient boron is applied through the lines in the establishment years if soil levels are below 2 ppm. Boron (3.7 kg per ha) is applied through the drip system during the first year. In the following years, boron is foliar applied,

0.75 kg per ha after bloom, in early May in combination with copper EDTA, 0.6 kg per ha and zinc, 2.25 kg zinc 36% per ha at 50% leaf expansion.

Training and pruning: Training a young tree is a laborious and time-consuming practice, yet highly significant for the rest of the tree's lifespan. The goal of training is to establish a desired tree form and should consider factors like rootstock and cultivar traits, soil fertility, irrigation capabilities, and cultural management practices, with the primary focus on facilitating nut harvest. The training of pistachio trees may last at least 4 to 5 years, but in a way, it continues throughout the tree's life. Pistachios are usually trained as open vase trees, although different training systems have been tested, with a trunk height depending mostly on the harvesting method. For mechanical tree trunk shakers, a minimum trunk height of 80 to 100 cm is required. For scaffold shakers, knapsack shakers, or traditional sticks, the trunk height can be lower.

Pistachios are usually pruned in winter (dormant pruning) and in summer. Winter pruning is more invigorating than the summer one and usually heavier. During the first season after tree planting (full-grown budded tree), the scion (cultivar) is headed back to approximately 100 to 110 cm to induce lateral branching the following spring near the cutting edge, which will be the future primary scaffolds (Beede and Ferguson, 2005). If the desired tree height is lower (based mainly on the harvest method to be used), then the heading cut can be done lower at approximately 80 cm height. The new shoots are then selected to grow down the trunk at intervals of 15 to 20 cm, properly oriented to form the primary scaffolds. If they are vigorous enough, they can be pinched (shoot tip removed) by early to mid-June (this depends on the cultivar and the climatic conditions of the site) to force the formation of secondary scaffolds. The removal of the blade of the second leaf below the cut forces the bud to open, thus improving the branching of the secondaries. During that time, tying of the developing shoots is necessary to allow us to form an up straight open vase and avoid excessive opening. Formation of tertiary scaffolds within the same year is better to be avoided, since there will be a large depletion of assimilates to support the new growth and this will negatively affect overall tree performance as shoots have not attained the desired girth to support the growth of further laterals (Beede and Ferguson, 2005). At the same time, any suckers grown from the rootstock can be removed, while laterals grown on the trunk can be headed back to one or two nodes leaving the leaves intact, as these may provide assimilates for further trunk growth.

During the second dormant pruning, two or three secondary branches are kept on each primary scaffold, in desired orientation, avoiding branches on the same side or opposite to each other (Beede and Ferguson, 2005). It is better to choose branches growing along the primary scaffold at spaces of about 8 to 10 cm and head them back to approximately 50 cm. This will force the sprouting of buds the following spring, which will form the tertiary branches. These should be distributed evenly on the secondary branch, allowing them to grow in all directions. Again, two to three tertiary branches will be selected, based on their position and orientation on the secondary branch. If necessary, they will be tied up to retain tree shape. If these branches are vigorous enough and their growth exceeds 100 cm, they can be headed back during early to mid-June to enhance lateral formation.

In subsequent years, dormant pruning focuses on maintaining tree shape and removing any suckers from the rootstock. Some tying may be necessary if the tree's upright habit is not retained, while vigorous shoots with good girth can be cut to approximately 50 to 60 cm. As the pistachio canopy is not a dense one, the excessive opening of the center of the tree does not significantly improve light distribution (shading is not an issue in pistachio). When trees start to bear fruits, vegetative growth slows down, so it is necessary to have completed the training of the tree by the fifth to seventh year after planting, to support a high and qualitative yield.

Pruning mature pistachio trees: The pruning of mature pistachio trees, while less demanding than other fruit species, requires careful consideration due to several unique characteristics of this species. Firstly, and most important, pistachio is characterized by alternate bearing, i.e., the production of a heavy crop one year followed by a light crop the next year. Secondly, its shoots are characterized by strong apical dominance with few lateral branches formed, unless forced to (Crane and Iwakiri, 1985; Nikpeyma, 2020; Beede and Ferguson, 2005). Any vegetative bud located further down from the branch tip may sprout but eventually develop into a short spur, with limited fruiting capability (although it has been found that in Kerman these shoots collectively produced 55%-60% of the total yield) (Spann et al., 2009). A third characteristic is that pistachio does not have a very dense canopy that can produce shading sites, thus limiting photosynthesis. Fourthly, as the tree enters the productive phase, the vegetative buds are mainly located close to the branch tip, while occasionally some may be also found on its base (Crane and Iwakiri, 1985). Lastly, pistachios are now commonly harvested mechanically with shakers, which means that during mature tree pruning, the upright habit of the tree should be retained to facilitate mechanical harvesting.

Based on these aspects, the following simple rules should be applied to efficiently prune pistachio trees. The first is to maintain the growing habit of the tree, removing (either through heading or thinning cuts) any branch growing downward, which will eventually be a problem for the harvesting machines. The second is to thin out crowded shoots to enhance light penetration and improve spray efficiency. The third is the ongoing removal of suckers from the rootstock, as well as of damaged or diseased branches. The last one is to gradually rejuvenate the tree and renew the fruiting sites. Pistachio bears fruits on 1-year-old wood, laterally from single flower buds. To stimulate the formation of new shoots (whip shoots) on long branches, a heading cut should be performed, thus resulting in at least doubling the number of fruiting branches of the tree. Such heading cuts should be performed around the canopy of the tree to uniformly stimulate lateral branching. Care should be taken though to perform the heading cut above at least one vegetative bud, to avoid the dying back of the branch after the current season's fruiting. The heading back cut can also be done above an already grown lateral, to reduce tree height and facilitate harvesting. Thinning cuts are also performed to reduce branch density and increase canopy exposure to sunlight and possibly fruitfulness.

Furthermore, thinning cuts are also performed to the outer parts of the canopy when the branches are getting horizontal and expand beyond the allotted space of the tree (Beede and Ferguson, 2005). Heavy pruning, however, should be approached with caution as it can drastically reduce shell dehiscence, negatively affecting the percentage of split nuts. In California, pistachio is mechanically pruned by hedging and topping the trees that reduces labor costs (Ferguson et al., 1995; Beede and Ferguson, 2005). Male pistachio trees are pruned even lighter compared to female ones, aiming at maintaining the upright habit of the tree, removing any dead wood and facilitating tractor's movement in the orchard.

Fertilization: Nitrogen (N), for early tree growth, and later potassium (K) for nut production are the major macronutrients required for pistachio production. Fertilization of mature pistachio is directed primarily by the previous year's midseason leaf analysis. Leaves are sampled in late July to early August, after the nuts have fully filled (see Table 14.1). As part of a nitrogen and general nutrient fertilization and amendment program, particularly in saline-sodic soils, soil nutrients are sampled in the winter after any salt leaching is performed. An additional early current season crop load estimate is used to more precisely direct fertilization. Brown and Siddiqui (2012) determined 1120 kg of dry in-shell split nuts per ha removes 12.6 kg active ingredient (AI) N per ha, 11.3 kg per ha K, and 1.4 kg per ha AI phosphorus (P). Therefore, after determining if the previous year's leaf sample determines the orchard's nutritional status, the fertilization program is adjusted to support the current year's visually estimated crop load. An additional 11.3 kg N per ha is also added for tree growth. A typical mature orchard would receive 224 kg per ha AI N and 140 kg per ha AI K. The macronutrient fertilization program is accompanied by micronutrient foliar applications; boron (B), zinc (Zn), and copper (Cu). Nitrogen fertilization begins in May and is applied into August. The frequency of micronutrient

Table 14.1 Leaf nutrient levels for California pistachio production

Element	Critical value	Adequate range
Nitrogen (N)	1.8%	2.2%–2.5%
Phosphorus (P)	0.14%	0.14%–0.17%
Potassium (K)	1.6%	1.8%–2.2%
Calcium (Ca)	2.0%	2.1%–4.0%
Magnesium (Mg)	0.45%	0.5%–1.2%
Sodium (Na)	–	–
Chlorine (Cl)	–	0.1%–0.3%
Manganese (Mn)	30 ppm	30–80 ppm
Boron (B)	90 ppm	150–250 ppm
Zinc (Zn)	7 ppm	10–15 ppm
Copper (Cu)	4 ppm	6–10 ppm

Source: Weinbaum et al. (1988, 1995) for nitrogen; Brown et al. (1999) for potassium; Uriu (1984) and Brown et al. (1993a,b) for boron; Uriu and Pearson (1981, 1983, 1984, 1986) for zinc; Uriu et al. (1989) for copper; Brown and Siddiqui (2012) for magnesium and calcium.

application strongly depends upon soil type and cropping history. For example, boron uptake varies greatly by soil texture and pH. Boron is foliar applied (3.4 kg Solubor per ha) before bloom at bud-swell (mid-March) and also after bloom at 50% leaf expansion (early May) in combination with copper EDTA (0.56 kg per ha) and zinc (1.8 kg zinc 36% per ha). In many areas of the southern San Joaquin Valley, soils are naturally high in boron and boron fertilization is not necessary.

Irrigation: Pistachio is a drought- and salt-tolerant species traditionally cultivated in arid regions on marginal soils under rainfed conditions (Goldhamer, 1995), where it can survive but also produce profitable yield. Irrigation increases pistachio yield substantially (Goldhamer et al., 1985; Goldhamer, 2005; Iniesta et al., 2008) and also reduces alternate bearing. For this reason, modern pistachios are irrigated when water is available. When irrigated, pistachio transpiration is surprisingly high, with stomatal conductance values reaching values as high as 600 mmol m^{2} s^{-1} and a mature orchard being able to evapotranspire 8.1 mm of water per day (200 l day^{-1} $tree^{-1}$) in peak season (Iniesta et al., 2008; Goldhamer et al., 1985). The suggested crop coefficient for a mature pistachio orchard increases from 0.4 in spring to 1.19 in July (Goldhamer 2005), resulting in total seasonal crop evapotranspiration (ETc) of about 1300 mm (Figure 14.4).

Irrigation does not need to match this volume to ensure maximum productivity. In fact, regulated deficit irrigation techniques have been tested successfully in this species. During Stage 2 of nut development (June-July, during endocarp lignification), pistachio shows very high drought tolerance. In this period, changes in leaf osmotic potential and elastic modulus confer the leaves the capability to reduce the water potential while still maintaining high level of gas exchanges (Gijón et al., 2011). Irrigating with only 50% of ETc in this stage does not impact productivity (Goldhamer and Beede, 2004; Gijón et al., 2009). Water stress during Stage 1 of fruit

FIGURE 14.4

Double-line drip irrigation system delivering water in a pistachio orchard and seasonal crop coefficient of pistachio.

(From G. Marino; Data from Goldhamer, 2005)

development (April-May, during endocarp lignification) can affect nut quality at harvest. Goldhamer and Beede (2004) reported increased shell splitting (endocarp dehiscence) at harvest. Sedaghatib and Alipour (2005) report that water stress in April-early June increased early split, an undesired phenomenon where hull and shell split around July exposing the kernels to disease. The impact of water stress on early split has not been observed in other trials (Gijon et al., 2009). Stage 3 of fruit development (July-August, during kernel filling) is very sensitive to water stress: limited water will impact kernel growth and extremize competition with reproductive buds, thus increasing bud drop and alternate bearing. During this period, irrigation should match ETc, and water stress reduces nut size, shell splitting, nut removal and yield, while increasing the percentage of blank nuts (Goldhamer and Beede, 2004).

Research on the use of plant-based irrigation management in pistachio is limited. Midday stem water potential (SWP) of nonstressed trees would range between -0.7 and -1.5 MPa, depending on environmental conditions (Memmi et al., 2016). The relation between tree water status and gas exchanges is strongly influenced by phenology (Marino et al., 2018; Marra et al., 2017), but roughly a midday SWP below -1.5 MPa has been associated to reduced gas exchanges and suggested as a threshold to irrigate pistachio in Stage 2 (Memmi et al 2016).

Pistachio is also very salt tolerant and can maintain productivity in soil with electrical conductivity of saturated soil paste extract (ECe) as high as 8 dS m^{-1} (Sanden et al., 2013). Under high salinity, vegetative growth will be affected, resulting in 30% to 60% smaller trees in the long term (Figure 14.5A), depending on the soil salinity concentration and type (Marino et al., 2019). Salinity will reduce tree water uptake dynamically through the season (Marino et al., 2021); osmotic stress will be

FIGURE 14.5

Impact of salinity in pistachio: (A) reduced growth, (B) visual ion toxicity impacting pistachio leaves at the end of the season, and (C) tree dieback in high sodic soil with impacted structure and limited infiltration.

the main limiting factor to water uptake in the first part of the season, while specific ion accumulation in the leaves will impair stomatal conductance and further reduce water use in the second part of the season, as they accumulate in leaves (Figure 14.5B). Finally, pistachio is extremely sensitive to water logging, resulting in hypoxia and poor oxygen diffusion from the bulk soil to the root surface (Figure 14.5C). For this reason, in saline sodic soils with impacted soil structure and reduced water infiltration, pistachio productivity is strongly impacted (Marino et al., 2019).

Crop load and thinning: Pistachio is considered an alternate-bearing species, since high yields induce a crop load control mechanism that reduces current and following year production. The mechanism of alternate bearing in pistachio is very peculiar: in spring, the trees develop new shoots with the reproductive buds, and then part of these buds drops during the summer (Caruso et al., 1993). The drop percentage depends on the crop load, reaching 80% during a heavy crop year, compared to only 20% during off years (Takeda et al., 1980).

The physiological mechanism associated with bud drop is still not fully understood. Hormonal signals are involved in the final dehiscence phenomenon (Crane and Iwakiri, 1987); however, external applications of plant growth regulators do not clearly increase bud retention (Takeda and Crane, 1980). Research evidence correlated bud drop with a branch-level sink-source balance. The growing fruits compete with other organs for energy, metabolites, and nitrogen (Crane and Nelson, 1971; Crane et al., 1973; Caruso et al., 1995; Marra et al., 2009; Rosecrance et al., 1998; Spann et al., 2008) and deplete reserves of the shoot, triggering alternate bearing. The temporal coincidence of bud drop and kernel growth supports this theory.

High branch fruit set may also induce fruit drop resulting in a lower number of fruits per cluster and smaller nuts. Finally, imbalanced branch sink-source levels can anticipate leaf senescence by several months, with up to 70% reduction of photosynthesis in the second part of the season (Marino et al., 2023). The phenomenon is very localized at the branch level, with several studies reporting higher carbohydrates accumulation in nonbearing branches (Barone et al., 1995; Nzima et al., 1997). However, when extrapolated at the entire tree level, the off-year does not provide any advantage in terms of extra storage of resources. This suggests that in pistachio the alternate-bearing phenomenon could be more an adaptive mechanism of this crop to ecologically favor long-term survival over short-term reproduction (Stevenson and Shackel, 1997). This is why management practices supporting maximum carbon assimilation and storage would reduce but not eliminate alternance in pistachio. In rainfed orchards alternate bearing is more pronounced than in fully irrigated ones, and supplemental irrigation would reduce the bud drop under rainfed conditions (Marino et al., 2018). However, fully irrigated and properly managed pistachio will still have some level of alternance.

The use of chemical thinning agents during bloom has shown promising results; however, only a few research trials have been performed (Rahemi and Ramezanian, 2007; Kamiab et al., 2020), and thinning is not a commercially adopted management practice. Overall, studies showed that ethephon and NAA applied 2 weeks after full bloom may improve floral bud retention, nut quality traits (size, split, and blank percentages), vegetative growth, branch nonstructural carbohydrate levels, and eventually following year production.

6. Fruit maturation and harvest criteria

Ripening: Pistachio ripening progression will depend on the genotype, management, crop load, and environmental conditions. For the cultivar Kerman, ripening starts at about 2000 growing degree days (GDD), corresponding to mid-August in California, and full maturity for harvest is at about 2500 GDD, or mid-September. The following quality parameters will change during ripening:

(1) **Nut weight and moisture:** Kernel maturation is characterized by dehydration, with moisture decreasing constantly. Panahi and Khezri (2011) reported a decrease from about 35%-40% in August to 30%-35% in September with cultivar-specific difference being dominant versus the time progression. Labavitch et al (1982) reported in Kerman a reduction from 55%-70% to 45%-50% from mid-August to the end of September with differences associated to the year. While moisture decreases, fresh and dry weight of the kernel increases. In Kerman, kernel dry weight increased from mid-August to October from 0.4 to 0.8 g and the fresh weight from 1.1 to 1.3 g (Labavitch et al., 1982). Esmaeilpour and Shakerardekani (2018) reported fresh and dry weight of entire nuts cv. Ohadi increased from August 16 to September 6 by 8% and 16%, respectively, and dry weight of the kernel by 21%.

(2) **Shell split** increases linearly during ripening until reaching a plateau with maximum percentage of split being determined by the interaction genotype-environment-management. Kerman starts to split in mid-August. Panahi and Khezri (2011) reported the percentage of split nuts significantly increased in the cultivar Ahmad-Aghaii from 50% on August 23 to 87% on October 11 for four different Iranian cultivars, while no significant differences after September 20 in the percentage of split were monitored.

(3) **Hull color and texture:** Hull turns from green to red or yellow color, with up to 60% discoloration in September with respect to late July (Esmaeilpour and Shakerardekani, 2018). The hull becomes softer as ripening progresses and eventually cracks.

(4) **Fat content** increases with ripening peaking to a maximum of about 50%-70% of total nut weight between August and September, with effect of season and cultivar (Labavitch et al., 1982; Panahi and Khezri 2011).

(5) **Chlorophyll content** decreases during ripening, turning the kernels from green to yellow-green (Kunter et al., 1995). Esmaeilpour and Shakerardekani (2018) showed a 60% reduction in kernel total chlorophyll from end-July (0.4 mg/g) to end-August (0.15 mg/g).

(6) Nutritional properties of the kernel change with ripening. Protein content declines from about 30% in late July to mid-August to about 20%-22% at harvest (Labavitch et al., 1982). Phosphorus, potassium, and magnesium are not impacted by harvest time, while nitrogen and calcium decrease over time (Esmaeilpour and Shakerardekani, 2018). Sugars increase during maturation until a peak in August-September; individual fatty acid composition is constant through ripening, with palmitic oleic and linoleic being predominant fatty acids (Labavitch et al., 1982).

Harvest timing: Both early and late harvesting time decrease the quality of pistachio nuts. Early harvested nuts are mostly nonsplit and immature, often including an undeveloped kernel. Harvestability will also be lower, with a higher number of filled nuts left on the trees. Pistachios harvested too late are vulnerable to hull cracking, shell staining, hull and kernel deterioration, fruit shedding, mechanical injuries, and insects' attacks (Ferguson et al., 2005). Of special concern with late harvesting are: (1) *Aspergillus* spp. contamination of the early-split and hull cracked nuts and (2) *Amyelois transitella*, navel orangeworm (NOW), third generation emerges in early to mid-August and lays eggs in the sutures of the early-split nuts. Early split are nuts where the shells split before the hulls mature causing the hull to split along the suture and expose the kernel to insect infestation and fungal infection (Figure 14.6). The most used indicator for harvest readiness is the hull color and lucidity/opaqueness, and hull tattering (when the shell nut ejects easily from the hull when the nut is held between the thumb and forefinger and pressure is applied to the long axis of the nut) (Thompson and Kader, 2016).

FIGURE 14.6

Pistachio cluster with early split nuts *(left)* susceptible to suture staining *(center)*, and stained shell *(right)*, a common defect associated to late harvest, Alternaria late blight infection of the hulls, and temperature and holding time between harvesting and processing.

Harvest criteria: Pistachios are generally mechanically harvested; however, manual harvest was commonly used in the past and is still adopted in some areas where mechanization is not possible for multiple reasons, including orography, ancient and nonrational plantations, and in young trees. For harvesting manually, tarps are spread under both sides of the tree and nuts are removed by knocking the tree or striking the branch behind the clusters with a pole (Thompson and Kader, 2016).

In California, side-by-side harvesters are used with two separate machines, one containing the shaker head and the other moving in parallel on the other side of the row to form a continuous collection surface under the tree canopy (Figure 14.7). The tree trunk is clamped about 60 cm above the ground and shaken for 6 to 8 s. After shaking, there is an air separator to remove light debris, then the nuts can be collected in vented bins in the bin harvesting machines or on a larger cart receiver in the case of the bulk harvesting system (Figure 14.8). This system is extremely

FIGURE 14.7

Trunk shaker with reverse umbrella *(left)* and side by side harvester *(right)*.

FIGURE 14.8

After shaking, the nuts are collected on a larger cart receiver in the case of the bulk harvesting system *(left)* or in vented bins in the bin harvesting machines *(right)*.

fast; it is considered a continuous harvest, and it takes 1.8 min per tree. When side-by-side harvesters are not available, other types of mechanical harvesting methods can be used, such as limb and trunk shakers with the nuts gathered on tarps or on a reversed umbrella. The efficiency of these methods is about 5 min per tree (Polat et al., 2007).

7. Postharvest physiology and technology

Transportation: Shell staining is a common defect in pistachios that becomes evident after harvest and can strongly decrease the nut commercial value. Although several factors appear to be involved in the appearance of shell stains, the combination of temperature and holding time between harvesting and processing is of primary importance. Thompson et al. (1997) reported that pistachio nuts (cv. Kerman) kept at 40°C before hulling can show signs of shell stain as fast as 8 to 16 h after harvest. They also found that unhulled pistachio nuts varied in their susceptibility to staining, possibly due to differences in nut maturity. Based on this study and other evidence, handling guidelines were developed to indicate that processing the nuts within a 24-h time frame after harvest could prevent shell stains in Kerman.

Grading: Once delivered to the processing plant, nuts are generally graded to calculate grower payout (Figure 14.9). A sample is collected, weighted, dehulled, and dried to calculate the fresh-to-dried-weight ratio. The percentage of nuts with defect is also calculated as the sum of non-split nuts, nuts with hull adhering to the shell, stained shells, damage by other means (deformity caused by various factor), insects damage, blank percentage, and small sizes.

FIGURE 14.9

Once nuts are delivered to the transformation facilities, gross weight is measured *(left)*, then a subsample of nuts (library samples) are graded for assessing growers' payout *(right)*.

Processing: Once nuts reach the transformation facility, they are separated from the hulls and dried to 4% to 6% moisture content (Figure 14.9). Nuts separated from shell, small nuts, severely stained, insect damaged, and floating nuts should be removed prior to drying to reduce aflatoxin contamination (Doster and Michailides, 1994; Thompson and Kader, 2016). Nuts can be sun-dried by spreading them on the ground for 4 or 5 days under protective nets. The major disadvantage to this method is the potential for fungal growth and increased oxidative process during the conservation. Use of drying machines helps to concentrate operations in a short time window, improved management of harvesting operations, and reduced cost of harvesting and the risk of pathogens. In California, a two-stage drying process is currently used (Figure 14.10): nuts are first dried to 12% to 13% water content in about 3 h at temperatures near 70°C and then further dried to 4% to 6% moisture with unheated, forced air, or air heated to less than 48°C for about 24 to 48 h (Thompson and Kader, 2016). Temperatures above 55°C can lead to chlorophyll degradations, decreasing the values of pistachio sold as kernel to the confectionery industry (Marino and Marra, 2019). Temperatures above 80°C cause shells to split open so widely that the kernel drops out (Thompson and Kader, 2016).

Storage: Dried nuts can be held at 20°C and 65% to 70% relative humidity for up to 1 year (see Figure 14.10). Storage at low temperatures and controlled atmosphere storage or packaging can be used to limit the development of oxidative rancidity. Fumigation, insecticidal-controlled atmospheres, irradiation, and insect-proof packaging are all methods to control insects in stored pistachios (Thompson and Kader, 2016).

FIGURE 14.10

Nuts conveyers over an air leg to remove debris *(left)*, driers *(center)* and storage units *(right)*.

8. Main diseases and pests and their management

Fungi: Two major root diseases can infect pistachio and even kill trees. Verticillium wilt is caused by the soilborne fungus *Verticillium dahliae,* resulting in killing individual branches initially. The disease severity depends on the rootstock susceptibility, the soil density of microsclerotia (surviving resistant structures of the pathogen), and the history of the field. Pistachio orchards planted in fields where susceptible crops were previously grown will suffer by Verticillium wilt. Symptoms of the disease include yellowing of leaves and wilt of individual branches. The disease progresses slowly and eventually trees lose productivity and die. The use of resistant and tolerant rootstocks such as hybrids of *P. atlantica,* × *P. integerrima* [UCB1], *P. integerrima,* and clonal rootstocks originated from selected resistant individuals) has minimized this problem (Morgan et al., 1992). A second disease is a crown and root rot disease caused by *Phytophthora* spp. This disease develops in lower areas of orchards where water can accumulate, in orchards with soil infiltration problems, or when irrigation lines (drip or microsprinklers) make the base of young trees excessively wet. Good drainage of the soil and avoiding wetting the trunk crown (especially in young trees) is essential to manage this disease. Early infections by *Phytophthora* can be diminished with registered fungicides specific to these diseases. Other disease associated with specific regions and being more sporadic include root rots, such as Armillaria root rot (caused by *Armillaria mellea*), Phymatotrichopsis root rot caused by *Phymatotrichopsis omnivora*, and Rosellinia root rot caused by *Rosellinia necatrix.*

Among the aboveground diseases attacking pistachio, the most important in terms of yield loss are the Botryosphaeria panicle and shoot blight, the Alternaria late blight, the Septoria leaf spot, the anthracnose blight, and Botrytis blossom and shoot blight. Botryosphaeria panicle and shoot blight (Figure 14.11) is caused by more than 10 members of the *Botryosphaeriaceae* fungal family (Michailides, 1991a,b; Michailides and Morgan, 1992).

All these fungi produce pycnidia (flask-like structures that contain the fungal spores). The pycnidiospores become water-splashed and can infect all the current growth tissues. During the fall and winter, the spores land on dormant vegetative and flower buds and are trapped among the basal bud scales. Under conducing conditions, the first blighted structures are the blighted leaves from infections at the base of the petiole, or blighted leaflets that can defoliate. Infections at the base of young developing shoots from contaminated buds turn black resulting in leaf wilting, and the blight of the entire shoot. These shoots retain their killed leaves that will turn a bright brown color, very distinct among the green, healthy foliage. In years with severe disease, multiple blighted shoots can develop per tree. But the most devastating blight phase of the disease is the killing of the fruit clusters (panicle blight). The blighting of the pistachio cluster can occur either from infection of bud or latent infection of the fruit. Infection of the fruiting bud can lead to girdling of the cluster since the infection will develop at the rachis base, and it is usually very quick (1 week). In contrast, fruit latent infections will develop during summer and blight

FIGURE 14.11

Symptoms of Botryosphaeria panicle and shoot blight; *left,* blighted panicle (cluster); *upper right,* a canker initiated from infected male flower; *lower right,* cross-section of blighted shoot showing pycnidia (the spore-producing structures of the pathogen).

the fruit. The infection will move from the fruit into the rachis to kill either part of the cluster or the entire cluster, depending on when the latent infection is triggered to express disease symptoms. Blight of a pistachio cluster can also occur when large Hemiptera insects (*Leptoglossus* spp., leaffooted bug [LFB], *Chlorochroa* spp., etc.) puncture the fruit to feed and in this process inoculate the fruit with conidia. Infection will start from one or two fruits, grow through the peduncle, and move into the cluster rachis (main stem) to cause cluster blight. After reaching the base of the rachis, the infection will move into the last year's shoot to cause a distinct canker. The blighted rachis and the cankers play a significant role in the life cycle of the pathogen and the disease because portions of these tissues are covered with pycnidia of the pathogen, surviving onto the trees. If a late season rain occurs and temperatures are still above 20°C, flower buds can get infected and become entirely or partially blighted, directly impacting next year's crop. Once the blighted shoots, clusters, and buds develop, it is too late to control the disease. However, protecting the green tissues from latent infections in the spring can control the disease. Sprays are done starting at bloom and finish by the end of July; spray(s) in August and/or later are not effective. If a grower has low pressure of the disease and desires to apply only one spray, the best timing would be the first part of June.

A second disease is the Alternaria late blight (Figure 14.12), caused by three species of *Alternaria*, i.e., *A. alternata*, *A. tenuissima*, and *A. arborescens.* Among these species, *A. alternata* is the most common. This disease has become severe in

FIGURE 14.12

Alternaria late blight of pistachio. *Upper left,* leaf lesions with black sporulation; *lower left,* infection of fruit hulls; *right,* severe premature defoliation.

California in response to the increase of the surface of irrigated orchards, and the planting density. Furthermore, the new cultivar Golden Hills is more susceptible to Alternaria late blight than Kerman. Infections by *A. alternata* develop in the spring as latent infections. However, rain in the spring or sprinkler irrigation wetting the fruit results in pin-sized black spots associated with the lenticels. Some of these small black spots will enlarge in size as the fruit grows and develop black rot lesions with a red halo. This red halo makes these lesions distinguishable from Botryosphaeria lesions that lack the red halo. As the leaves start senescing they become susceptible to Alternaria lesion development. Such lesions start as angular brown necrotic lesions which enlarge to roundish shape and can reach a few cm in diameter. If the relative humidity is high in the orchard, then the pathogen will cover the central portion of the lesion with black sporulation. Later in the fall, the lesions are populated, and small- and large-size lesions appear in the same leaflet. Factors that are conducing to the disease include cover crop between the tree rows, flood irrigation and soil of poor infiltration, dense canopies slowing down the ventilation of the orchard, and cultivar susceptibility.

Alternaria late blight of pistachio can promote excess defoliation, creating major problems during mechanical harvest. The disease will infect the nuts when the harvest is delayed or a rain occurs in the summer. Infection of fruit mainly results in shell staining, and infection of fruit can result in yield losses (up to 25% of the crop). Premature defoliation presumably reduces the photosynthetic area of pistachio with probable detrimental effects in kernel growth. Kernel mold is a direct yield loss. Factors affecting Alternaria late blight are the relative humidity and the physiological stage of the crop. Inoculum is always present since this pathogen can also grow saprophytically on plant debris and can easily be recovered from the surface of the trees. When the duration of relative humidity above 95% increases from 3 to 5 hours, lesions per leaf increase from 20 to 100 (Figure 14.13). Delaying harvest also results in deterioration of the hulls that become very susceptible to Alternaria colonization and rotting and staining of the shells. Management of Alternaria involves an integrated approach. Cultural practices that can limit the infection are increasing infiltration of the soil to avoid water puddles; hedging trees to increase orchard ventilation; when planting a new orchard in areas conducive to Alternaria late blight, orient rows parallel with the prevailing winds; and having bare soil in the orchard (no cover crops). Fungicidal control is very effective when sprays include compounds for which resistance of the pathogen has not been detected. However, the majority of the fungicides are single mode of action, and resistant strains among the population are easily and quickly selected (two to three seasons). A 2022-2023 study showed high levels of resistance among the populations of Alternaria isolates obtained directly from leaf lesions to fungicide groups FRAC 7 (carboximides) and FRAC 11 (strobilurins). Fungicides of FRAC 3 (triazoles) should be included in a fungicide program in a rotation with other groups and/or premixtures that include a FRAC 3 fungicide for effective control of Alternaria late blight (http://ipm.ucanr.edu/PDF/PMG/fungicideefficacytiming.pdf). For one spray program, the best timing is during the first week of July.

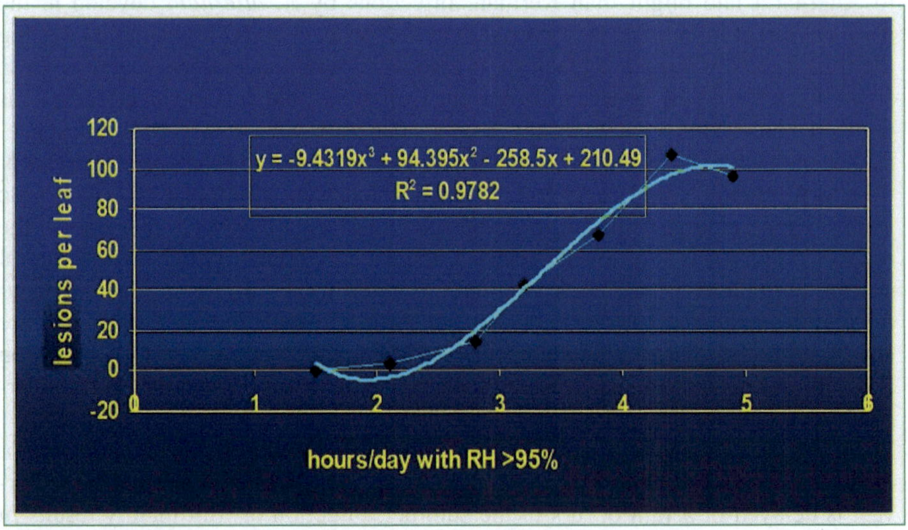

FIGURE 14.13

Effect of duration of high (>95%) relative humidity on severity of Alternaria late blight leaf lesion density.

FIGURE 14.14

Septoria leaf spot. *Left,* leaf lesions: *upper Septoria pistaciarum; lower, Septoria pistacia; right,* severe Septoria leaf spot leading to preharvest excessive defoliation.

Another foliar and fruit disease is the Septoria leaf spot (Figure 14.14). Causal agents of this disease have been known for a long time as *S. pistaciarum, S. pistaciae,* and *S. pistacina*. This disease has been reported in the Mediterranean countries, Iran, and the U.S. Progress in phylogeny and taxonomy helped the reclassification of these foliar pathogens, belonging to three distinct genera: *Cylindroseptoria*, *Pseudocercospora,* and *Septoria*. Depending on the *Septoria* sp. and the country, the disease can cause severe defoliation. Symptoms of *S. pistaciarum* on the leaves are characterized by small, angular reddish to dark brown lesions on both sides of the leaves. The first symptoms appear at the end of May progressing with the summer. At the end of the summer, infected leaves show the presence of large, coalesced necrotic blotches. Under high humidity oozing white cirri (mass of spores) appear from the pycnidia of leaf lesions. The spores of *S. pistaciarum* are distinguishable in size and shape from all the other known Septoria-like pathogens infecting pistachio and can be used as a microscopic morphological feature for disease diagnosis. A careful examination of leaf lesions reveals the presence of black pycnidia in and around the lesions. The life cycle of the Septoria pathogens attacking pistachio is poorly understood; however, it is known that the asexual pycnidiospores cause secondary infection cycles during the summer. As in other Septoria diseases, it is likely that the Septoria attacking pistachio overwinters in leaf debris under the canopy. In Mediterranean countries, sprays with FRAC 7 and FRAC 11 fungicides effectively control the main pistachio canopy diseases, including Septoria spot disease. In

Arizona where Septoria defoliates the trees prematurely, control is achieved with the same groups of fungicides.

Botrytis blossom and shoot blight, caused by *Botrytis cinerea*, has been reported in California pistachios, but it can occur in any place where pistachios are growing under conditions that favor the development of the disease. Botrytis needs wet and cold (10°C to 20°C) weather in the spring during leaf out and bloom. Botrytis can live among the scales of the flowers and vegetative buds and can be easily detected plating the buds on agar media. As the buds expand, Botrytis will grow initially in the dead scales (necrotrophic fungus) and from there will invade the base of the shoots or the flower. Buff color sporulation of Botrytis can develop within a few days after infection. The infection of the shoot causes wilting and the shoot forms a "shepherd's hook" symptom. Infections of male flowers lead to small cankers that develop around the base of the flower. Male cultivars that have bulkier (and in general larger) flowers are more susceptible to infection by *Botrytis*. In fact, it was shown that residual pollen among the anthers boosts the development of Botrytis. If a cool rain occurs in late May/early June, then partial blight of clusters can occur. These infections stop developing once the temperatures start increasing, resulting in only partially blighted clusters. Control of Botrytis blight can be done with fungicides (http://ipm.ucanr.edu/PDF/PMG/fungicideefficacytiming.pdf). Usually, one spray when the terminal shoots are about 3 to 5 cm can control the disease. If conditions keep being wet and cool (10°C-20°C) in the spring for 2 to 3 weeks, a second spray may be needed. Empirically, when there are 10 shoot blights per tree as an average, a treatment is justified.

Anthracnose of pistachio is caused by species of *Colletotrichum*. The two species *C. fioriniae* and *C. karstii* were reported in California in the cultivars Red Aleppo and Kerman, and one species *C. gloeosporioides* in the cultivar Sirora in Australia. The disease develops when conditions are wet and warm during fruit maturation, although symptoms can occur earlier if conditions were favorable. The pathogens overwinter on the tree canopy in rachises, shoots, and hanging fruit covered with sporulation of the fungus. In the spring lesions on fruit and leaves are initially black but turn pink after sporulation matures on them. The conidia are water-splashed spread around during spring and summer rains. Among the main cultivars, Kerman seem to be very tolerant, while Red Aleppo and Sirora are very susceptible. Isolates of *Colletotrichum* were found to be resistant to quinol oxidation inhibitor (QoIs), succinate dehydrogenase inhibitor (SDHIs), and benzimidazole fungicide groups. Because of the lack of anthracnose management, growers should avoid planting susceptible cultivars.

A disease that can cause significant damage to the cv. Kerman is Phoma blight that was found in climates where monsoon rains can occur during the period close to harvest. Phoma blight is caused by *Phoma fungicola*. It is a pycnidia-producing fungus with minute spores which can develop mainly on mature hulls of pistachio. The pathogen also causes light brown leaf lesions with very chlorotic margins that result in large necrotic blotch areas. However, the major damage by the disease is the slimy rotting of the hulls at harvest time. Fungicides controlling the disease include the premixture fluxapyroxad + pyraclostrobin (Merivon) and metconazole (Quash).

Aspergillus blight has major implications on aflatoxin and ochratoxin contamination. *A. flavus* and *A. parasiticus* are the causes of aflatoxin contamination, and *A. westerdijkiae*, *A. ochraceus*, *A. melleus*, and *A. bridgeri* are the causes of ochratoxin contamination (Singh et al., 2024). To reduce aflatoxin contamination, growers can use either AF36 Prevail or Afla-Guard GR, two atoxigenic strains that when applied in the orchard floor after an irrigation sporulate displacing the toxigenic strains and thus usually reducing aflatoxin contamination. Bacterial and viral diseases are not common nor of a major importance in pistachio.

Pests: Major insect pests of pistachio include Lepidoptera, Hemiptera, Coleoptera, and Homoptera insects. For instance, in California, the major and most devastating insect is the navel orangeworm *(Amyelois transitella)*. The navel orangeworm (NOW) is a lepidopteran that can cause direct damage to the crop by damaging the kernel of the nuts and also increase the risk for aflatoxin contamination of nuts. This insect could overwinter as larvae and/or pupae in the nuts that remain on trees after harvest or fall on the ground (mummies), if they remain intact and devoid of decay or predation by orchard insects, rodents, and birds. After egg hatching, the NOW larvae chew their way in and feed on the kernel. Insect wounds can be infected by *Aspergillus* fungi, *A. flavus,* or *A. parasiticus* and thus contaminate the nuts with aflatoxins. To break the cycle of NOW, the best approach is sanitation, i.e., removal of mummies from the trees (winter shaking), blowing the mummies off the berms, and grinding them. Multiple insecticide sprays are also necessary just before each of the four flights of this insect. Other insects that attack pistachios include: carob moth *(Ectomyelois ceratoniae)*, twig borer *(Palumbina guerinii)*, flatheaded borer *(Capnodis tenebrionis)*, a bug feeding on pistachio buds *(Acrantus vestitus)*, stink bugs (*Hemiptera* spp.), scales, aphids, and mites. Carob moth is also a lepidopteran insect and occurs in pistachios grown in Australia and the countries of the Mediterranean basin. Carob moth has one generation per growing season, surviving in mummy nuts while NOW has three or four generations. Among the stink bugs, small stink bugs (like *Lygus, Leptocoris, Neurocopus*) cause damage to the nuts before lignification of the shell, while large stink bugs (like *Leptoglossus, Thiantha, Chlorochroa* spp.) can damage the nut even after lignification of the shell. They cause symptoms like epicarp lesion, kernel necrosis, and stigmatomycosis (when they transmit a yeast into the kernel) (Figure 14.15). In the Mediterranean areas, species belonging to Hemiptera and Heteroptera and species of Hymenoptera, Eurytomidae, and Torymidae represent pests for this crop. Another devastating insect that is a problem in pistachios in the Mediterranean basin is the pistachio seed chalcid *(Megastigmus pistaciae)* which develops into an adult inside the shell and the larva destroys the kernel. As far as we know, pistachios are not susceptible to nor are damaged by nematodes.

FIGURE 14.15

Stigmatomycosis, a soft decay of kernel caused by *Eremothecium coryli* or *Aureobasidium pullulans* transmitted by large hemiptera insects (i.e., leaffooted bug, *Leptoglossus zonatus* and others).

9. Current and future challenges in main production regions

United States of America: The pistachio industry will face two main challenges: one is economic, and the other is environmental.

Economic challenges: Because of the large expansion of pistachio acreage, in 2023 the price per pound dropped by 40% and major production costs in pistachio production, including that of water, have risen precipitously. These costs, combined with the increase in competition from the 25% of the newly planted pistachio orchards yet to enter bearing, will probably lower the profitability of California pistachio production. The California industry will need to lower production costs, primarily water, diversify its products beyond dried and salted split in-shell nuts, and develop new markets beyond Europe and Southeast Asia.

Environmental challenges: Recurring droughts and environmental regulations are drastically reducing the amount of water available for agriculture uses, particularly in the areas where pistachio is mainly grown in California. Improving irrigation practices to increase water use efficiency and water productivity through precision agriculture and the adoption of new technologies is an important first step but not the solution. To maintain profitability in the future, the industry will need to reconsider the entire management practices format. The current focus—more canopy, more vigor, more yield—will not be sustainable without abundant irrigation water. A more comprehensive change will imply pruning differently to reduce water loss, adapt irrigation to actual crop load, change tree spacing, move to other growing regions, adopt new rootstocks and cultivars more resilient to drought stress, focus on maximize nut quality, develop profitable bioproduct from hulls and shells, etc.

Another challenge will be the increase in temperature extremes. High winter temperatures will be more frequent, leading to bloom anomalies and reduced yield.

New genotypes with lower chill requirements are needed to maintain yield under a warming climate. Finally, erratic temperatures at bloom and during fruit set and early growth can strongly impact production and nut quality, further impacting profitability. Current understanding of these impacts is limited, and more research will be needed to study how to mitigate the impact of these events on productivity.

Europe and other Mediterranean countries: The greatest challenge in European pistachio cultivation is the limited number of registered phytosanitary products. Compared to other fruit tree species, which have a wide variety of available pesticides, pistachio and other nut crops have far fewer options. This is quite concerning, as the pressure from pests and diseases, both old and new, is increasing and climate change may exacerbate these pressures. At the same time, climate change impacts on rainfall quantity and distribution could pose significant problems. Although pistachios are a xerophytic species and can tolerate water shortages, the quantity and quality of the nuts is diminished significantly with reduced water quality and quantity. Growers in many pistachio areas are now, and increasingly in the future, forced to use brackish water. At the same time, good-quality water is used to cover the increasing needs of popular tourist sites during the summer period, leaving low-quality water available for pistachio irrigation (mainly water with high electrical conductivity, an example is Aegina Island in Greece). Climate change might also disrupt efficient flower pollination in several ways: partial fulfillment of chilling requirements could lead to earlier or later pollen release, mismatched with the receptivity of female flowers; drought can increase dust during flowering reducing pollen quality; sudden and extreme rainfalls during the flowering period could wash away the pollen, reducing nut set and therefore subsequent yields.

Inappropriate postharvest handling and storage conditions also pose significant challenges for growers and retailers. Inadequate drying or prolonged drying times can exacerbate the problem of aflatoxins, a critical issue, particularly for nuts imported into Europe from third-party countries. The reliance on a small number of cultivars is another potential issue, as new enemies or unusual climatic conditions could threaten national pistachio production. Growers are often reluctant to test new cultivars, preferring to rely on traditional ones. In Greece, for example, the majority of pistachio production comes from a single cultivar. Nonetheless, this reluctance is understandable, given the long juvenile phase of pistachio trees, which increases farmers' apprehension about the unknown. However, this reliance on a single cultivar also increases the risk of future yield losses.

References

Akbari, M., Hokmabadi, H., Heydari, M., & Ghorbani, A. (2020). 'Arota': a new interspecific hybrid pistachio rootstock. *HortScience, 55*, 965–966.

Akbari, M., Mahna, N., Ramesh, K., Bandehagh, A., & Mazzuca, S. (2018). Ion homeostasis, osmoregulation, and physiological changes in the roots and leaves of pistachio rootstocks in response to salinity. *Protoplasma, 255*, 1349–1362.

Aliakbarkhani, S. T., Akbari, M., Hassankhah, A., Talaie, A., & Moghadam, M. F. (2015). Phenotypic and genotypic variation in Iranian pistachios. *Journal of Genetic Engineering and Biotechnology, 13*, 235–241.

Almehdi, A. A., Parfitt, D. E., & Chan, H. (2002). Propagation of pistachio rootstock by rooted stem cuttings. *Scientia Horticulturae, 96*, 359–363.

Barone, E., Caruso, T., Marra, F. P., & Motisi, A. (1995). Vegetative growth and inflorescence bud abscission in bearing and non-bearing pistachio trees. *Acta Hort, 419*, 29–36.

Beede, R., & Ferguson, L. (2005). In L. Ferguson (Ed.), *Pruning mature bearing trees. Pistachio Production Manual* (4ª Ed, pp. 97–102). Davis: University of California.

Benmahioul, B., Dorion, N., Kaid-Harche, M., & Daguin, F. (2012). Micropropagation and ex vitro rooting of pistachio (*Pistacia vera* L.). *Plant Cell, Tissue and Organ Culture, 108*, 353–358.

Brown, P., Ferguson, L., & Picchioni, G. (1993a). Boron nutrition of pistachio: Final report. In *California pistachio industry annual report 1993* (pp. 57–59).

Brown, P. L., & Siddiqui, M. I. (2012). Development of leaf sampling and interpretation methods for pistachio and development of a nutrient budget approach to fertilizer management in pistachio. In *California pistachio industry annual report 2012* (pp. 3–29).

Brown, P., Zhang, Q., & Beede, R. (1993b). Effect of foliar fertilization on zinc nutritional status of pistachio trees. In *California pistachio industry annual report 1993* (pp. 77–80).

Brown, P., Zhang, Q., Huang, Z., Holtz, B., & Hornung, C. (1999). Agronomic and economic responses of mature 'Kerman' pistachio trees to potassium applications in California. In *California pistachio industry annual report 1999* (pp. 84–85).

Caruso, T., Di Marco, L., Marra, F. P., & Motisi, A. (1993). Effects of crop load and adjustment time on alternate bearing of pistachio (Pistacia vera L.) cv. Bianca. In *Proceedings IX GREMPA meeting pistachio*Bronte (CT) - Sciacca.

Caruso, T., Fabbri, A., & Giovannini, D. (1995). Inflorescence bud growth, development and abscission in shoots of bearing and disbudded Bianca pistachio trees. *Journal of Horticultural Science, 70*, 857–866.

Crane, J. C., Al-Shalan, I. M., & Carlson, R. M. (1973). Abscission of pistachio inflorescence buds as affected by leaf area and number of nuts. *Journal of the American Society for Horticultural Science, 98*, 591–592.

Crane, J. C., & Iwakiri, B. T. (1981). Morphology and reproduction of pistachio cultivars, California. Hortic. *Rev., 3*, 376–393.

Crane, J. C., & Iwakiri, B. T. (1987). Reconsideration of the cause of inflorescence bud abscission in pistachio. *HortScience, 22*, 1315–1316.

Crane, J. C., & Iwakiri, B. T. (1985). Vegetative and reproductive dominance in pistachio. *HortScience, 20*, 1092–1093.

Crane, J. C., & Nelson, M. M. (1971). The unusual mechanism of alternate bearing in the Pistachio. *HortScience, 6*, 489–490.

Crous, P. W., Quaedvlieg, W., Sarpkaya, K., Can, C., & Erkılıç, A. (2013). Septoria-like pathogens causing leaf and fruit spot of pistachio. *IMA Fungus, 4*, 187–199.

Doster, M. A., & Michailides, T. J. (1994). *Aspergillus* molds and aflatoxins in pistachio nuts in California. *Phytopathology, 84*, 583–590.

Driver, J. A., & Kuniyuki, A. H. (1984). *In vitro* propagation of Paradox walnut rootstock. *HortScience, 19*, 507–509.

Esmaeilpour, A., & Shakerardekani, A. (2018). Effects of early harvest times on nut quality and physiological characteristics of pistachio (*Pistacia vera*) trees. *Fruits, 73*, 110–117.

Ferguson, L., Beede, R. H., Reyes, H., Sanden, B. L., Grattan, S. R., & Epstein, L. (2005). Pistachio rootstocks. *Pistachio Production Manual, 3545*, 65–74.

Ferguson, L., Maranto, J., & Beede, R. (1995). Mechanical topping mitigates alternate bearing of' Kerman pistachios (*Pistacia vera* L.). *HortScience, 30*, 1369–1372.

Ferguson, L., Poss, J. A., Grattan, S. R., Grieve, C. M., Wang, D., Wilson, C., Donovan, T. J., et al. (2002). Pistachio rootstocks influence scion growth and ion relations under salinity and boron stress. *Journal of the American Society for Horticultural Science, 127*, 194–199.

Gijón, M. C., Gimenez, C., Perez-López, D., Guerrero, J., Couceiro, J. F., & Moriana, A. (2011). Water relations of pistachio (*Pistacia vera* L.) as affected by phenological stages and water regimes. *Sci. Hortic., 128*, 415–422.

Gijón, M. C., Guerrero, J., Couceiro, J. F., & Moriana, A. (2009). Deficit irrigation without reducing yield or nut splitting in pistachio (*Pistacia vera* cv Kerman on Pistacia terebinthus L.). *Agricultural Water Management, 96*, 12–22.

Godfrey, J. M., Ferguson, L., & Zwieniecki, M. A. (2021). Sodium retrieval from sap may permit maintenance of carbohydrate reserves in mature xylem tissues of a salt-tolerant hybrid pistachio rootstock exposed to 100 mM NaCl. *Journal of the American Society for Horticultural Science, 146*, 224–232.

Goldhamer, D. A. (1995). In L. Ferguson (Ed.), *Irrigation management. Pistachio production* (pp. 71–81).

Goldhamer, D. A., Phene, B. C., Beede, R., Sherlin, L., Mahan, S., & Rose, D. (1987). Effects of sustained deficit irrigation on pistachio tree performance. In *Annual report, crop year 1986–1987* (pp. 61–66). Fresno, CA: California Pistachio Industry.

Goldhamer, D. A., Kjelgren, R., Beede, R., Williams, L., Moore, M. J., Lan, J., et al. (1985). Water Use requirements of pistachio trees and response to water stress. *Annual Report of the California Pistachio Commission.*

Goldhamer, D. A. (2005). Tree water requirements and regulated deficit irrigation. In L. Ferguson (Ed.), *Pistachio production manual* (ed 4, pp. 103–116).

Goldhamer, D., & Beede, R. (2004). Regulated deficit irrigation effects on yield, nut quality and water-use efficiency of mature pistachio trees. *The Journal of Horticultural Science and Biotechnology, 79*, 538–545.

Gusella, G., Aiello, D., Michailides, T. J., & Polizzi, G. (2021). Update of pistachio leaf spot caused by *Septoria pistaciarum* in light of new taxonomic advances in Italy. *Fungal Biology, 125*, 962–970.

Hajiboland, R., Norouzi, F., & Poschenrieder, C. (2014). Growth, physiological, biochemical and ionic responses of pistachio seedlings to mild and high salinity. *Trees, 28*, 1065–1078.

Hormaza, J. I., & Wünsch, A. (2007). Pistachio. In C. Kole (Ed.), *Fruits and nuts*. Berlin, Heidelberg: Springer Berlin Heidelberg.

Iniesta, F., Testi, L., Goldhamer, D. A., & Fereres, E. (2008). Quantifying reductions in consumptive water use under regulated deficit irrigation in pistachio (*Pistacia vera* L.). *Agricultural Water Management, 95*, 877–886.

Kallsen, C. E., & Parfitt, D. E. (2008). Differences in leaf boron concentration among four pistachio genotypes. *HortScience, 43*, 1283.

Kallsen, C. E., & Parfitt, D. E. (2018). Famoso', a new male pistachio cultivar to replace 'Peters. *HortScience, 52*, 1829–1833.

Kallsen, C. E., & Parfitt, D. E. (2019). Tejon' male pistachio: An early blooming pollenizer for 'Gumdrop. *HortScience, 53*, 1719–1721.

Kallsen, C. E., & Parfitt, D. E. (2017). 'Gumdrop', a new early harvest pistachio cultivar. *HortScience, 52*, 310–312.

Kallsen, C. E., & Parfitt, D. E. (2022). Leaflet boron concentration reduced with hybrid Pistacia vera rootstocks. *HortScience, 57*, 65–71.

Kamiab, F., Rasouli, P., & ZamaniBahramabadi, E. (2020). Application of some bloom thinning techniques to overcome alternate bearing in pistachio (*Pistacia vera* L.). *Horticulture, Environment, and Biotechnology, 61*, 31–39.

Kostas, S., Hatzilazarou, S., Pipinis, E., Vasileiadis, A., Magklaras, P., Smyrnioudis, I., et al. (2021). Propagation of *Pistacia lentiscus* var. Chia genotypes and determination of their ornamental traits combined with a genetic analysis using ISSR markers. *Agronomy, 11*, 205.

Kunter, B., Gülsen, Y., & Ayfer, M. (1995). Determination of the most suitable harvest time for green color and high kernel quality of pistachio nut (Pistacia vera L.). *Acta Hortic, 419*, 393–398.

Labavitch, J. M., Heintz, C. M., Rae, H. L., & Kader, A. A. (1982). Physiological and compositional changes. *J. Amer. Soc. Hort. Sci, 107*, 688–692.

Marin, J., Garcia, E., Lorente, P., Arbeloa, A., & Andreu, P. (2017). Propagation of pistachio applying *in vitro* culture techniques. *Acta Horticulturae, 1155*, 321–326.

Marino, G., Zaccaria, D., Snyder, R. L., Lagos, O., Lampinen, B. D., Ferguson, L., et al. (2019). Actual evapotranspiration and tree performance of mature micro-irrigated pistachio orchards grown on saline-sodic soils in the San Joaquin Valley of California. *Agriculture, 9*, 76.

Marino, G., Di Martino, S., Roxas, A. A., Caruso, T., Ferguson, L., Barone, E., et al. (2018). Sustainability of pistachio production (*Pistacia vera* L.) under supplemental irrigation in a Mediterranean climate. *Scientia Horticulturae, 241*, 260–266.

Marino, G., Guzmán-Delgado, P., Caruso, T., & Marra, F. P. (2022). Modeling seasonal branch carbon dynamics in pistachio as a function of crop load. *Scientia Horticulturae, 296*, Article 110875.

Marino, G., Guzmán-Delgado, P., Santos, E., Adaskaveg, J. A., Blanco-Ulate, B., Ferguson, L., et al. (2023). Interactive effect of branch source-sink ratio and leaf aging on photosynthesis in pistachio. *Frontiers in Plant Science, 14*, Article 1194177.

Marino, G., La Mantia, M., Caruso, T., & Marra, F. P. (2018a). Seasonal dynamics of photosynthesis and total carbon gain in bearing and nonbearing pistachio (*Pistacia vera* l.) shoots. *Photosynthetica, 56*(3), 932–941.

Marino, G., & Marra, F. P. (2019). Horticultural management of Italian Pistachio orchard systems: Current limitations and future prospective. *Italus Hortus, 26*, 14–26.

Marino, G., Zaccaria, D., Lagos, L. O., Souto, C., Kent, E. R., Grattan, S. R., et al. (2021). Effects of salinity and sodicity on the seasonal dynamics of actual evapotranspiration and surface energy balance components in mature micro-irrigated pistachio orchards. *Irrigation Science, 39*, 23–43.

Marra, F. P., Amico Roxas, A., Lo, B. R., Marino, G., & Caruso, T. (2017). Water status and gas exchange of pistachio trees under different irrigation levels. *Acta Horticulturae, 1177*, 281–288.

Marra, F. P., Barone, E., LA Mantia, M., & Caruso, T. (2009). Toward the definition of a carbon budget model: Seasonal variation and temperature effect on respiration rate of vegeta-tive and reproductive organs of pistachio trees (*Pistacia vera*). *Tree Physiology, 29*(9), 1095–1103.

Memmi, H., Gijón, M. C., Couceiro, J. F., & Pérez-López, D. (2016). Water stress thresholds for regulated deficit irrigation in pistachio trees: Rootstock influence and effects on yield quality. *Agric Water Manag, 164*, 58–72.

Michailides, T. J. (1991b). First report of Septoria spot in California pistachios. In *California pistachio industry annual report crop year 1990-91* (p. 93). Fresno, CA, USA.

Michailides, T. J., & Morgan, D. P. (1992). Effects of temperature and wetness duration on infection of pistachio by *Botryosphaeria dothidea* and management of disease by reducing duration of irrigation. *Phytopathology, 82*, 1399–1406.

Michailides, T. J. (1991a). Pathogenicity, distribution, sources of inoculum, and infection courts of *Botryosphaeria dothidea* on pistachio. *Phytopathology, 81*, 1114–1123.

Mir-Makhamad, B., Bjorn, R., Stark, S., & Spengler, I. I. I.,R. N. (2022). Pistachio (P*istacia vera* L.) domestication and dispersal out of Central Asia. *Agronomy, 12*, 1758.

Momenpour, A., & Imani, A. (2018). Evaluation of salinity tolerance in fourteen selected pistachio (*Pistacia vera* L.) cultivars. *Advances in Horticultural Science, 32*, 249–264.

Morgan, D. P., Epstein, L., & Ferguson, L. (1992). Verticillium wilt resistance in pistachio rootstock cultivars: Assays and an assessment of two interspecific hybrids. *Plant Disease, 76*, 310–313.

Nezami, E., & Gallego, P. P. (2023). History, phylogeny, biodiversity, and new computer-based tools for efficient micropropagation and conservation of pistachio (*Pistacia* spp.) germplasm. *Plants, 12*, 323.

Nikpeyma, Y. (2020). The effect of decapitation treatment on lateral shoot and flower bud induction in mature pistachio trees (cv Siirt). *Sylwan, 164*.

Nzima, M. D. S., Martin, G. C., & Nishijima, C. (1997). Seasonal changes in total nonstructural carbohydrates within branches and roots of naturally "off" and "on" 'Kerman' pistachio trees. *Journal of the American Society for Horticultural Science, 122*, 856–862.

Onay, A. (2003). *Micropropagation of pistachio. Micropropagation of Woody trees and fruits*. Springer.

Onay, A., Tilkat, E., Isikalan, C., & Namli, S. (2007). *Micrografting of pistachio (Pistacia vera L. cv. Siirt). Protocols for micropropagation of woody trees and fruits* (pp. 289–298).

Panahi, B., & Khezri, M. (2011). Effect of harvesting time on nut quality of pistachio (*Pistacia vera* L.) cultivars. *Scientia Horticulturae, 129*, 730–734.

Parfitt, D. E., Kallsen, C. E., Holtz, B. A., & Maranto, J. (2008). Lost Hills' a new pistachio cultivar for California. *HortScience, 43*, 247–249.

Parfitt, D. E., Kallsen, C. E., Holtz, B. A., & Maranto, J. (2010). Randy' male pistachio. *HortScience, 45*, 1113–1115.

Parfitt, D. E., Kallsen, C., Maranto, J., & Holtz, B. (2007). Golden Hills' pistachio. *HortScience, 42*, 694–696.

Parfitt, D., Kafkas, S., Battle, I., Vargas, F. J., & Kallsen, C. E. (2012). *Pistachio. Fruit breeding* (pp. 803–826).

Polat, R., Gezer, I., Guner, M., Dursun, E., Erdogan, D., & Bilim, H. C. (2007). Mechanical harvesting of pistachio nuts. *Journal of Food Engineering, 79*, 1131–1135.

Polito, V. S., & Pinney, K. (1999). Endocarp dehiscence in pistachio (*Pistacia vera* L. *International Journal of Plant Sciences, 160*, 827–835.

Pryor, B. M., & Michailides, T. J. (2002). Morphological, pathogenic, and molecular characterization of Alternaria isolates associated with Alternaria late blight of pistachio. *Phytopathology, 92*, 406–416.

Rahemi, M., & Ramezanian, A. (2007). Potential of ethephon, NAA, NAD and urea for thinning pistachio fruitlets. *Scientia Horticulturae, 111*(2), 160–163.

Rosecrane, R. C., Weinbaum, S. A., & Brown, P. H. (1998). Alternate bearing affects nitrogen, phosphorus, potassium and Starch storage pools in mature pistachio trees. *Annals of Botany, 82*, 463–470.

Sanden, B. L., Ferguson, L., & Corwin, D. L. (2013). Development and long-term salt tolerance of pistachios from planting to maturity using saline groundwater. *Acta Hortic, 1028*, 327–332.

Sedaghati, N., & Alipour, H. (2005). The effect of different time of irrigation on occurrence of early split (ES) of pistachio nuts. *IV International Symposium on Pistachios and Almonds, 726*, 582–586.

Sheikhi, A., Arab, M. M., Brown, P. J., Ferguson, L., & Akabri, M. (2019). Pistachio (*pistacia* spp.) breeding. In J. M. Al-khayri, S. M. Jain, & D. V. Johnson (Eds.), *Advances in plant breeding strategies: Nut and beverage crops* (Vol. 4). Cham: Springer International Publishing.

Singh, P., Jaime-Garcia, R., Puckett, D. R., Lake, J., Papangelis, A., & Michailides, T. J. (2024). Ochratoxin A contamination of California pistachios and identification of causal agents. *Plant Disease, 108*. PDIS-06.

Spann, T. M., Beede, R. H., & DeJong, T. M. (2009). Contributions of short-and long-shoots to yield of 'Kerman' pistachio (*Pistacia vera* L.). *Scientia Horticulturae, 121*, 495–500.

Spann, T. M., Beede, R. H., & DeJong, T. M. (2008). Seasonal carbohydrate storage and mobilization in bearing and non-bearing pistachio (*Pistacia vera*) trees. *Tree Physiology, 28*, 207–213.

Stevenson, M. T., & Shackel, K. A. (1997). Alternate bearing in pistachio as a masting phenomenon: construction cost of reproduction versus vegetative growth and storage. *International Symposium on Pistachios and Almonds, 470*, 340–348.

Takeda, F., & Crane, J. C. (1980). Abscisic-acid in pistachio as related to inflorescence bud abscission. *J. Am. Soc. Hortic. Sci., 105*, 573–576.

Thompson, J. F., & Kader, A. A. (2016). Preparing for harvest and harvest timing. *Pistachio Production Manual, 3545*, 189.

Thompson, J. F., Rumsey, T. R., & Spinoglio, M. (1997). Maintaining quality of bulk-handled, unhulled pistachio nuts. *Applied Engineering in Agriculture, 13*, 65–70.

Uriu, K., & Pearson, J. (1981). Diagnosis and correction of nutritional problems. In , *Vol. 30. California pistachio industry annual report 1981*.

Uriu, K., & Pearson, J. (1983). Diagnosis and correction of nutritional problems, including the crinkle leaf disorder. In , *Vol. 79. California pistachio industry annual report 1983*.

Uriu, K., & Pearson, J. (1984). Diagnosis and correction of nutritional problems, including the crinkle leaf disorder. In *California pistachio industry annual report 1984* (pp. 49–50).

Uriu, K., & Pearson, J. (1986). Zinc deficiency in pistachio-diagnosis and correction. In *California pistachio industry annual report 1986* (pp. 71–72).

Uriu, K., Teranishi, R., Beede, R., & Pearson, J. (1989). Copper deficiency in pistachio. In , *Vol. 77. California pistachio industry annual report 1989*.

Vahdati, K., Sarikhani, S., Arab, M. M., Leslie, C. A., Dandekar, A. M., Aletà, N., Bielsa, B., Gradziel, T. M., Montesinos, Á., Rubio-Cabetas, M. J., & Sideli, G. M. (2021). Advances in rootstock breeding of nut trees: objectives and strategies. *Plants, 10*, 2234.

Zeng, L., Tu, X.-L., Dai, H., Han, F.-M., Lu, B.-S., Wang, M.-S., et al. (2019). Whole genomes and transcriptomes reveal adaptation and domestication of pistachio. *Genome Biology, 20*, 1–13.

Links

(1) https://acpistachios.wpenginepowered.com/wp-content/uploads/2024/01/2023-Pistachio-Statistics.pdf

(2) https://americanpistachios.org/about-us/pistachio-power-unshelled/press-releases/california-pistachio-growers-and-processors

(3) https://worldpopulationreview.com/country-rankings/pistachio-production-by-country

(4) https://coststudyfiles.ucdavis.edu/uploads/cs_public/e4/bd/e4bd9353-25a1-4258-b5b1-42a203bc09ad/2020pistachiosjvsouth.pdf

(5) Tree Nuts: World Markets and Trade February 15, 2024 World Production, Markets, and Trade Report https://apps.fas.usda.gov/psdonline/circulars/TreeNuts.pdf

Index

Note: Page numbers followed by *f* indicate figures and *t* indicate tables.